Clifford Algebra and Spinor-Valued Functions

Mathematics and Its Applications

Volume 53

Clifford Algebra and Spinor-Valued Functions

A Function Theory for the Dirac Operator

by

R. Delanghe

Department of Mathematics,
University of Ghent,
Ghent, Belgium

F. Sommen

National Fund for Scientific Research,
Belgium

and

V. Souček

Charles University,
Prague, Czechoslovakia

Related REDUCE software

by

F. Brackx

Department of Mathematics,
University of Ghent,
Ghent, Belgium

and

D. Constales

National Fund for Scientific Research,
Belgium

SPRINGER-SCIENCE+BUSINESS MEDIA, B.V.

Library of Congress Cataloging-in-Publication Data

Delanghe, Richard.
Clifford algebra and spinor-valued functions : a function theory for the Dirac operator / by R. Delanghe, F. Sommen, and V. Souček ; related REDUCE software by F. Brackx and D. Constales.
p. cm. -- (Mathematics and its applications ; v. 53)
Includes bibliographical references and index.
ISBN 978-94-010-5297-9 ISBN 978-94-011-2922-0 (eBook)
DOI 10.1007/978-94-011-2922-0
1. Clifford algebras. 2. Spinor analysis. 3. Dirac equation. I. Sommen, F. II. Souček, V. III. Title. IV. Series: Mathematics and its applications (Kluwer Academic Publishers) ; v. 53.
QA199.D46 1992
512'.57--dc20 92-6900

ISBN 978-94-010-5297-9

Printed on acid-free paper

To

Renée

Brigitte

and

Věra

SERIES EDITOR'S PREFACE

'Et moi, ..., si j'avait su comment en revenir, je n'y serais point allé.'
Jules Verne

The series is divergent; therefore we may be able to do something with it.
O. Heaviside

One service mathematics has rendered the human race. It has put common sense back where it belongs, on the topmost shelf next to the dusty canister labelled 'discarded nonsense'.
Eric T. Bell

Mathematics is a tool for thought. A highly necessary tool in a world where both feedback and nonlinearities abound. Similarly, all kinds of parts of mathematics serve as tools for other parts and for other sciences.

Applying a simple rewriting rule to the quote on the right above one finds such statements as: 'One service topology has rendered mathematical physics ...'; 'One service logic has rendered computer science ...'; 'One service category theory has rendered mathematics ...'. All arguably true. And all statements obtainable this way form part of the raison d'être of this series.

This series, *Mathematics and Its Applications*, started in 1977. Now that over one hundred volumes have appeared it seems opportune to reexamine its scope. At the time I wrote

> ''Growing specialization and diversification have brought a host of monographs and textbooks on increasingly specialized topics. However, the 'tree' of knowledge of mathematics and related fields does not grow only by putting forth new branches. It also happens, quite often in fact, that branches which were thought to be completely disparate are suddenly seen to be related. Further, the kind and level of sophistication of mathematics applied in various sciences has changed drastically in recent years: measure theory is used (non-trivially) in regional and theoretical economics; algebraic geometry interacts with physics; the Minkowsky lemma, coding theory and the structure of water meet one another in packing and covering theory; quantum fields, crystal defects and mathematical programming profit from homotopy theory; Lie algebras are relevant to filtering; and prediction and electrical engineering can use Stein spaces. And in addition to this there are such new emerging subdisciplines as 'experimental mathematics', 'CFD', 'completely integrable systems', 'chaos, synergetics and large-scale order', which are almost impossible to fit into the existing classification schemes. They draw upon widely different sections of mathematics.''

By and large, all this still applies today. It is still true that at first sight mathematics seems rather fragmented and that to find, see, and exploit the deeper underlying interrelations more effort is needed and so are books that can help mathematicians and scientists do so. Accordingly MIA will continue to try to make such books available.

If anything, the description I gave in 1977 is now an understatement. To the examples of interaction areas one should add string theory where Riemann surfaces, algebraic geometry, modular functions, knots, quantum field theory, Kac-Moody algebras, monstrous moonshine (and more) all come together. And to the examples of things which can be usefully applied let me add the topic 'finite geometry'; a combination of words which sounds like it might not even exist, let alone be applicable. And yet it is being applied: to statistics via designs, to radar/sonar detection arrays (via finite projective planes), and to bus connections of VLSI chips (via difference sets). There seems to be no part of (so-called pure) mathematics that is not in immediate danger of being applied. And, accordingly, the applied mathematician needs to be aware of much more. Besides analysis and numerics, the traditional workhorses, he may need all kinds of combinatorics, algebra, probability, and so on.

In addition, the applied scientist needs to cope increasingly with the nonlinear world and the extra mathematical sophistication that this requires. For that is where the rewards are. Linear models are honest and a bit sad and depressing: proportional efforts and results. It is in the nonlinear world that infinitesimal inputs may result in macroscopic outputs (or vice versa). To appreciate what I am hinting at: if electronics were linear we would have no fun with transistors and computers; we would have no TV; in fact you would not be reading these lines.

There is also no safety in ignoring such outlandish things as nonstandard analysis, superspace and anticommuting integration, *p*-adic and ultrametric space. All three have applications in both electrical engineering and physics. Once, complex numbers were equally outlandish, but they frequently proved the shortest path between 'real' results. Similarly, the first two topics named have already provided a number of 'wormhole' paths. There is no telling where all this is leading - fortunately.

Thus the original scope of the series, which for various (sound) reasons now comprises five subseries: white (Japan), yellow (China), red (USSR), blue (Eastern Europe), and green (everything else), still applies. It has been enlarged a bit to include books treating of the tools from one subdiscipline which are used in others. Thus the series still aims at books dealing with:

- a central concept which plays an important role in several different mathematical and/or scientific specialization areas;
- new applications of the results and ideas from one area of scientific endeavour into another;
- influences which the results, problems and concepts of one field of enquiry have, and have had, on the development of another.

As I have had occasion to write before, spinors (and twistors) and Clifford algebras are important in theoretical physics, geometry and topology. Just about every table of contents of e.g. the journals Letters in Mathematical Physics or Journal of Mathematical Physics contains something on this group of entities.

Even so, granted that Clifford algebras are indeed very important, is it not overdoing it to have two books on them in the same series, published nearly within a year of each other. (The other one is A. Crumeyrolle, *Orthogonal and symplectic Clifford algebras: Spinor structures*, KAP, 1990).

The answer is no. Crumeyrolle's book is about Clifford algebras and spinor (and twistor) structures on manifolds. The present volume is about Clifford analysis, more precisely the function theory of the Dirac operator. Thus there is but little overlap limited to the algebraic introductory part of this volume (Chapter 0,1) and the two volumes complement and reinforce each other.

In 1982, two of the three present authors together with F. Brackx published a by now well known research note volume on Clifford analysis. Since then much has happened and here then is the complete up-to-date account of the subject by the experts and originators themselves.

The shortest path between two truths in the real domain passes through the complex domain.

J. Hadamard

La physique ne nous donne pas seulement l'occasion de résoudre des problèmes ... elle nous fait pressentir la solution.

H. Poincaré

Never lend books, for no one ever returns them; the only books I have in my library are books that other folk have lent me.

Anatole France

The function of an expert is not to be more right than other people, but to be wrong for more sophisticated reasons.

David Butler

Bussum, 11 February 1992

Michiel Hazewinkel

Contents

Preface

In the last decades the study of the Dirac operator has been a central theme in many areas of mathematics.

On the one hand, a lot of important global problems have been investigated connected with the study of this operator on manifolds (for an excellent up-to-date account of this subject we refer to [52]). On the other hand, th e study of local properties of nullsolutions of this operator has led to a theory of functions, commonly called now Clifford analysis (for a first account of this approach we refer to [10]).

In this book an attempt is made to describe substantial developments in Clifford analysis which took place in the last ten years, more precisely concerning the major rôle played by the Spin group when studying properties of nullsolutions of the Dirac operator. Although we are not using explicitly the language of representation theory, the reader will feel it present in the way a lot of results are established. This is also one of the reasons why much attention is paid to basic spinor representations in the introductory algebraic part of the book (see Chapter I).

In fact this book deals with three main themes.

The algebraic part is spread over Chapter 0 and Chapter I. It contains classical results on real and complex Clifford algebras – most of all without proofs – (Chapter I) and it shows how lower dimensional real Clifford algebras suit very well for describing basic geometric notions in Euclidean space (Chapter 0).
The second part – which could be easily called a classical analysis part – provides with methods and techniques suitable for practical applications. In Chapter II and Chapter III it is thus illustrated how Clifford analysis extends or refines a lot of computational tools available in, respectively, complex analysis in the plane or harmonic analysis in space.

The third part reveals some connections between Clifford analysis and analysis on manifolds. In Chapter IV the concept of monogenic differential forms is presented in such a way that it may be generalized straightforwardly to the case of Spin-manifolds. In Chapter V analysis on homogeneous spaces is dealt with, thus showing how Clifford analysis may be connected with the Penrose transform.

Finally, in an Appendix some basic results related to algebraic and analytic structures discussed in the Chapters 0 - II are made accessible for computational purposes by means of computer algebra programmes written in REDUCE ; a floppy disk is added.

The formal prerequisites for the reading of this volume amount to a standard background in, respectively, real and complex analysis (Chapters II and III) and abstract algebra (Chapters 0 and I). For Chapters IV and V a good background in differential geometry and a basic knowledge in algebraic topology are indispensable. For the convenience of the reader two appendices have been added : the first one contains a description of the Leray-Norguet residue theory used in Chapter IV and V ; the second one reviews basic properties from analysis on homogeneous spaces.

Each chapter starts with an introduction describing its main purposes and results; it is divided into sections and subsections within which the definitions, results and remarks are numbered decimally. Results referred to within the same chapter are not preceded by the number of this chapter, in contrast with the case where reference is made to results from other chapters. Formulae within a chapter are numbered according to the section they appear in.

The making of this book would not have been completed without help from many sources. The first two authors wish to thank the Belgian National Fund for Scientific Research NFWO for the essential support offered to their research group at the University of Ghent since many years. They are also indebted to all members of the Seminar of Algebra and Functional Analysis for their constant scientific and practical advice. The third author is indebted to a number of people for the support he got while he was working out ideas developed in the book. First of all he wishes to thank Prof. R. Penrose for the possibility to spend a semester at Oxford and to take part in the discussions in his seminar. Ideas brought from Oxford then evolved during his stay at the Department of Mathematics at the University of Bologna, where he was invited by Prof. A.V. Ferreira whom he wants to thank for his warm hospitality. His stays at the Max-Planck Institut für Mathematik at Bonn were very fruitful for the further development of his work and he is very grateful to Prof. F. Hirzebruch for having got the opportunity to stay in such a stimulating place. Furthermore he wishes to thank the University of Ghent and the NFWO which enabled an efficient cooperation with the Clifford analysis group in Gent during the final steps in the preparation of the book. At the same time he would like to express his gratitude to the members of this research group for their hospitality and help during these stays and, especially, to R. Delanghe and his family for the constant support and for the many things making the stays in Gent so pleasant. Last but not least he wants to thank Prof. J. Bureš : the results presented in Chapter V are arising from joint work; his critical reading of parts of the book and the many discussions we had were very helpful.

The authors wish to thank F. Brackx and D. Constales for providing an efficient and helpful computational tool by means of the REDUCE programs they include in this book.[1]

---

[1]See also : *Computer Algebra with LISP and REDUCE. An Introduction to Computer-aided pure Mathematics*, F. Brackx and D. Constales, Kluwer Academic Publishers 1991. (ISBN 0-7923-1441-7)

The authors are also most grateful to Rita Vanhove for her patient and skillful typing of the text and to A. Hamsík for his drawings.

R. Delanghe
F. Sommen
V. Souček

Chapter 0

Clifford algebras over lower dimensional Euclidean spaces

This chapter is meant as an introductory one and we felt there are several reasons for writing it.

First of all, by defining the Clifford algebras $\mathbf{R}_{0,m}$ $(m \leq 4)$ in a rather heuristic way, we wanted the reader to become acquainted as quickly as possible with calculating in these algebras, with describing them as matrix algebras and with characterizing some of their important subalgebras and subgroups. At the same time we wanted to show the geometrical meaning of such algebras, particularly in the cases $m = 2$ and $m = 3$. In these cases it is shown in full detail how the Spingroups Spin(2) and Spin(3) describe rotations in, respectively, $\mathbf{R}^2$ and $\mathbf{R}^3$. Moreover for $m = 3$ we make the connections between vector algebra, the algebra $\mathbf{H}$ of real quaternions and the Clifford algebra $\mathbf{R}_{0,3}$, thus emphasizing how vector algebra fits naturally into these algebras. Finally we also show how rotations in $\mathbf{R}^3$ and $\mathbf{R}^4$ can be described by using the algebra $\mathbf{H}$.

Although the material covered by this chapter is classical, we found it spread over several textbooks or papers and not necessarily presented the same way (see e.g. [6], [23], [36], [35], [63], [83]). An Appendix is added, aimed to recall some basic terminology and results concerning Euclidean m-space $\mathbf{R}^m$ and to make the reader familiar with the notations used throughout Chapter 0. We hope the reader will find this chapter interesting for itself and helpful as an introduction to Chapter 1 and Chapter 2.

1 The algebras C and H

In this section we recall how to make $\mathbf{R}^2$ and $\mathbf{R}^4$ into real linear associative algebras. We also give matrix representations of these algebras.

1.1 The algebra C

Consider the real two-dimensional vector space $\mathbf{R}^2$ and define multiplication on it by putting

$$(a,b)(a',b') = (aa' - bb', ab' + ba').$$

Then $\mathbf{R}^2$ is turned into a real commutative algebra with identity $1 = (1,0)$, called the *algebra* $\mathbf{C}$ *of complex numbers.*

As furthermore the set $\mathbf{C}' = \mathbf{C}\setminus\{0\}$ is a multiplicative abelian group where for each nonzero element (a,b), its inverse is given by

$$(a,b)^{-1} = (\frac{a}{a^2+b^2}, \frac{-b}{a^2+b^2}),$$

we have that $\mathbf{C}, +, .$ is a field. Notice that each element $(a,b) \in \mathbf{C}$ may be written as

$$(a,b) = (a,0)(1,0) + (b,0)(0,1)$$

whence embedding $\mathbf{R}$ into $\mathbf{C}$ by identifying $a \in \mathbf{R}$ with $(a,0)$ in $\mathbf{C}$ and putting $i = (0,1)$, we also have

$$(a,b) = a1 + bi.$$

Observing that $i^2 = -1$, multiplication in $\mathbf{C}$ is then completely governed by the Cayley table

	1	i
1	1	i
i	i	-1

The element $(a,b) = a + ib$ is commonly denoted by

$$z = a + ib.$$

Considering the unique algebra involution of $\mathbf{C}$ different from the identity, namely conjugation $\mathbf{C} \to \mathbf{C}$ with $z = a + ib \to \bar{z} = a - ib$, we obtain that for all $z \in \mathbf{C}$

$$z\bar{z} = |z|^2 = a^2 + b^2$$

where the modulus function $|.|$ satisfies the relation

$$|zz'| = |z||z'|.$$

Furthermore for all $z \neq 0$,

$$z^{-1} = \frac{\bar{z}}{|z|^2}.$$

Calling for $z \neq 0$, $\theta \in [0, 2\pi[$ the (unique) element such that

$$\cos\theta = \frac{a}{|z|} \quad \text{and} \quad \sin\theta = \frac{b}{|z|},$$

we also have that

$$\begin{aligned} z &= a + ib \\ &= |z|(\cos\theta + i\sin\theta) \\ &= |z|e^{i\theta}. \end{aligned}$$

Moreover, since for each $z \in \mathbf{C}$,

$$|z| = 1 \quad \Longleftrightarrow \quad a^2 + b^2 = 1,$$

the subgroup

$$U(1) = \{e^{i\theta} : \theta \in \mathbf{R}\} = \{z \in \mathbf{C} : |z| = 1\}$$

of $\mathbf{C}',$. may be identified with the unit circle S^1 in $\mathbf{R}^2$.

Now take $c = a + ib \in \mathbf{C}$ fixed and associate with it the (real) linear operator Φ_c in $\mathbf{R}^2$ given by

$$\Phi_c(z) = zc.$$

Then with respect to the standard orthonormal basis e of $\mathbf{R}^2$, Φ_c has the matrix representation

$$\left[\Phi_c\right]_e = \begin{pmatrix} a & b \\ -b & a \end{pmatrix}$$

Hence $\mathbf{C}$ is isomorphic to the subalgebra Θ of the (real) algebra $\mathbf{R}(2)$ consisting of (2×2)-matrices of the form $\begin{pmatrix} a & b \\ -b & a \end{pmatrix}$. In such a way

$$|c|^2 = \bar{c}c = \det \begin{pmatrix} a & b \\ -b & a \end{pmatrix}.$$

Now, as for all $z \in \mathbf{C}$, $|zc| = |z||c|$, multiplication by c is a rotation of $\mathbf{R}^2$ if and only if $|c| = 1$ and conversely, it is easily seen that any rotation of $\mathbf{R}^2$ is so induced.

Summarizing we thus obtain :

$$S^1 \cong U(1) \cong \mathrm{SO}(2).$$

As conjugation in $\mathbf{C}$ determines an antirotation in $\mathbf{R}^2$ and any antirotation in $\mathbf{R}^2$ is the composition of any fixed antirotation with a rotation, antirotations in $\mathbf{R}^2$ can thus also be represented by $\mathbf{C}$.

1.2 The algebra **H**

Consider the real four-dimensional vector space $\mathbf{R}^4$. Then multiplication on it can be introduced as follows. First off, put

$$\begin{aligned} \mathbf{1} &= (1,0,0,0) \\ \mathbf{i} &= (0,1,0,0) \\ \mathbf{j} &= (0,0,1,0) \\ \mathbf{k} &= (0,0,0,1) \end{aligned}$$

and define the Cayley table between these elements by

	1	**i**	**j**	**k**
1	**1**	**i**	**j**	**k**
i	**i**	−**1**	**k**	−**j**
j	**j**	−**k**	−**1**	**i**
k	**k**	**j**	−**i**	−**1**

i.e. **1** acts as identity, $\mathbf{ij} = \mathbf{k}$, $\mathbf{ji} = -\mathbf{k}$, etc ...

Then this table tells us how to multiply elements of $\mathbf{R}^4$ if for each $q = (a, b, c, d) \in \mathbf{R}^4$, we write q as

$$q = a\mathbf{1} + b\mathbf{i} + c\mathbf{j} + d\mathbf{k}.$$

$\mathbf{R}^4$ is then turned into a real, non-commutative algebra, called the *algebra* **H** *of real quaternions*. Notice that, as $\mathbf{H}',.$ with $\mathbf{H}' = \mathbf{H}\backslash\{0\}$ is a non-abelian group, $\mathbf{H},+,.$ is a skew-field.

R and $\mathbf{R}^3$ are embedded in **H** by identifying each $a \in \mathbf{R}$ with $(a,0,0,0) = a\mathbf{1}$ and, respectively, each $\underline{b} \in \mathbf{R}^3$ with $(0, b_1, b_2, b_3) = b_1\mathbf{i} + b_2\mathbf{j} + b_3\mathbf{k}$. **C** is embedded in **H** by identifying each $(a, b) \in \mathbf{C}$ with $a\mathbf{1} + b\mathbf{i}$. Putting $|q| = \sqrt{a^2 + b^2 + c^2 + d^2}$ and defining the involution $q = a\mathbf{1} + b\mathbf{i} + c\mathbf{j} + d\mathbf{k} \to \overline{q} = a\mathbf{1} - b\mathbf{i} - c\mathbf{j} - d\mathbf{k}$, we have that

$$q\overline{q} = \overline{q}q = |q|^2, \quad |qq'| = |q||q'|$$

while for $q \neq 0$,

$$q^{-1} = \frac{\overline{q}}{|q|^2}.$$

Now we identify **H** with the (left) complex space $\mathbf{C}^2$ in the following way. First of all observe that any $q \in \mathbf{H}$ can be written as

$$q = a\mathbf{1} + b\mathbf{i} + (c\mathbf{1} + d\mathbf{i})\mathbf{j}$$

thus suggesting that **H**, as a left vector space over **C**, is isomorphic to $\mathbf{C}^2$, the isomorphism being given by

$$(u,v) \to q = u + v\mathbf{j}.$$

Fixing $q \in \mathbf{H}$ and defining the $\mathbf{C}$-linear map $\Phi_q : \mathbf{C}^2 \to \mathbf{C}^2$ by

$$\begin{aligned} \Phi_q(z) &= \Phi_q((z_1, z_2)) \\ &= (z_1 + z_2\mathbf{j})q \end{aligned}$$

we obtain that, with respect to the standard orthonormal basis e of $\mathbf{C}^2$, Φ_q has the matrix representation

$$[\Phi_q]_e = \begin{pmatrix} u & v \\ -\overline{v} & \overline{u} \end{pmatrix}.$$

Hence $\mathbf{H}$ is isomorphic to the subalgebra Φ of the real algebra $\mathbf{C}(2)$ consisting of those complex 2×2 matrices of the form $\begin{pmatrix} u & v \\ -\overline{v} & \overline{u} \end{pmatrix}$. Clearly

$$[\Phi_q]_e[\Phi_q]_e^+ = |q|^2 I_2$$

and

$$|q|^2 = \det \begin{pmatrix} u & v \\ -\overline{v} & \overline{u} \end{pmatrix}$$

whence

$$\begin{aligned} |q| = 1 \quad &\Leftrightarrow \quad \det \begin{pmatrix} u & v \\ -\overline{v} & \overline{u} \end{pmatrix} = 1. \\ &\Leftrightarrow \quad [\Phi_q]_e \in SU(2). \end{aligned}$$

Calling

$$\mathrm{Sp}(1) = \{q \in \mathbf{H} : |q| = 1\}$$

we have that $\mathrm{Sp}(1)$ is a multiplicative subgroup of $\mathbf{H}$ which may be identified with the unit sphere S^3 in $\mathbf{R}^4$.

Summarizing we thus obtain that

$$S^3 \cong \mathrm{Sp}(1) \cong SU(2).$$

Finally let us introduce some more notions related to $\mathbf{H}$. As we already pointed out, $\mathbf{R}$ and $\mathbf{R}^3$ are identified in $\mathbf{H}$ with $\mathbf{R1}$ and $\mathbf{Ri} \oplus \mathbf{Rj} \oplus \mathbf{Rk}$. We therefore write $\mathbf{H}$ as

$$\mathbf{H} = Re\mathbf{H} \oplus Ve\mathbf{H}$$

and call arbitrary elements $q = a\mathbf{1} \in Re\mathbf{H}$ and $q = b\mathbf{i} + c\mathbf{j} + d\mathbf{k} \in Ve\mathbf{H}$, respectively, *real* and *pure quaternions*. In the sequel a real quaternion $q = a\mathbf{1}$ will also be denoted by $q = a$. Clearly a quaternion is real if and only if it commutes with every quaternion whence the center Z of $\mathbf{H}$ is $Re\mathbf{H} = \mathbf{R1}$. Furthermore a quaternion q is pure if and only if its square is a non-positive real number or, equivalently, if and only if $q = -\overline{q}$.

2 The Clifford algebras $\mathbf{R}_{0,0}$ and $\mathbf{R}_{0,1}$

Obviously $\mathbf{R}$ is a real algebra generated by the identity 1. By definition we call it the Clifford algebra $\mathbf{R}_{0,0}$. The multiplicative subgroup S^0 of $\mathbf{R}$ consisting of those elements having absolute value 1 is given by

$$S^0 = \{1, -1\}.$$

Now consider the real vector space $\mathbf{R}^1$ and let $e = \{e_1\}$ be an orthonormal basis for it. Then we wish to construct a real algebra of maximal dimension containing $\mathbf{R}$ and $\mathbf{R}^1$ such that for each element $\underline{x} \in \mathbf{R}^1$, $\underline{x}^2 = -|\underline{x}|^2$.

This implies that for the basic vector $\underline{e}_1$, we are requiring that $\underline{e}_1^2 = -1$. We thus obtain an algebra of dimension 2 having as basis the set of elements 1, $\underline{e}_1$; this algebra is denoted by $\mathbf{R}_{0,1}$ and clearly $\mathbf{R}_{0,1}$ is isomorphic to $\mathbf{C}$. Clearly $\mathbf{R}$ and $\mathbf{R}^1$ are identified, respectively, with the subspaces $\mathbf{R}1$ and $\mathbf{R}\underline{e}_1$ of $\mathbf{R}_{0,1}$.

3 The Clifford algebra $\mathbf{R}_{0,2}$

Let $e = \{\underline{e}_1, \underline{e}_2\}$ be an orthonormal basis of $\mathbf{R}^2$. Then we wish to construct a real algebra of maximal dimension containing $\mathbf{R}$ and $\mathbf{R}^2$ such that for each vector $\underline{x} \in \mathbf{R}^2$, $\underline{x}^2 = -|\underline{x}|^2$. Clearly this requirement implies that for the basic vectors e_j

$$\underline{e}_j^2 = -1 \quad , \quad j = 1, 2$$

and

$$\underline{e}_1 \underline{e}_2 + \underline{e}_2 \underline{e}_1 = 0.$$

We thus obtain an algebra of dimension 4 having as basis the set of elements $1, \underline{e}_1, \underline{e}_2$ and $\underline{e}_1\underline{e}_2$, written henceforth as $1, e_1, e_2, e_{12}$. This algebra is denoted by $\mathbf{R}_{0,2}$ and, as is readily seen, $\mathbf{R}_{0,2}$ is isomorphic to $\mathbf{H}$ since, identifying $\mathbf{1}$ with 1, $\mathbf{i}$ with e_1, $\mathbf{j}$ with e_2 and $\mathbf{k}$ with e_{12}, the multiplication rules in the Cayley table are satisfied (see also 1.2).

Notice that $\mathbf{R}$ and $\mathbf{R}^2$ are identified with, respectively, the subspaces $\mathbf{R}1$ and $\mathbf{R}\underline{e}_1 \oplus \mathbf{R}\underline{e}_2$ of $\mathbf{R}_{0,2}$. Now let $\underline{x} = x_1 e_1 + x_2 e_2$, $\underline{y} = y_1 e_1 + y_2 e_2$ be arbitrary elements of $\mathbf{R}^2$. Then in $\mathbf{R}_{0,2}$

$$\underline{x}\underline{y} = -(x_1 y_1 + x_2 y_2)1 + (x_1 y_2 - x_2 y_1) e_{12}.$$

Writing $\underline{x}\underline{y}$ as

$$\underline{x}\underline{y} = \underline{x}.\underline{y} + \underline{x} \wedge \underline{y}$$

with

$$\underline{x}.\underline{y} = -(x_1y_1 + x_2y_2) = -< \underline{x}, \underline{y} >$$

and

$$\underline{x} \wedge \underline{y} = (x_1y_2 - x_2y_1)e_{12}$$

we have for $\underline{y}\underline{x}$ that

$$\begin{aligned} \underline{y}\underline{x} &= \underline{y}.\underline{x} + \underline{y} \wedge \underline{x} \\ &= \underline{x}.\underline{y} - \underline{x} \wedge \underline{y}. \end{aligned}$$

We thus obtain that

$$\underline{x}.\underline{y} = \frac{1}{2}(\underline{x}\underline{y} + \underline{y}\underline{x})$$

and

$$\underline{x} \wedge \underline{y} = \frac{1}{2}(\underline{x}\underline{y} - \underline{y}\underline{x}).$$

Notice that

$$\underline{x} \perp \underline{y} \Leftrightarrow \underline{x}\underline{y} = -\underline{y}\underline{x} = \underline{x} \wedge \underline{y}$$

and that

$$\begin{aligned} \underline{x} \| \underline{y} &\Leftrightarrow \underline{x} \wedge \underline{y} = 0 \\ &\Leftrightarrow \underline{x}\underline{y} = \underline{y}\underline{x} \end{aligned}$$

The element $\underline{x} \wedge \underline{y} = (x_1y_2 - x_2y_1)e_{12}$ is called a *bivector* and we also say that

$$\mathbf{R}_{0,2} = \mathbf{R}_{0,2}^0 \oplus \mathbf{R}_{0,2}^1 \oplus \mathbf{R}_{0,2}^2$$

where $\mathbf{R}_{0,2}^0 = \mathbf{R}1$, $\mathbf{R}_{0,2}^1 = \text{span}_{\mathbf{R}}\{e_1, e_2\}$ and $\mathbf{R}_{0,2}^2 = \mathbf{R}e_{12}$, their elements being called, respectively, *scalars, vectors* and *bivectors.*

On $\mathbf{R}_{0,2}$ we define the norm

$$|a|^2 = a_0^2 + a_1^2 + a_2^2 + a_{12}^2$$

if $a = a_0 1 + a_1 e_1 + a_2 e_2 + a_{12} e_{12}$. This norm is of course exactly the one which is obtained when identifying $\mathbf{R}_{0,2}$ with $\mathbf{H}$. This identification also leads to introducing a conjugation $a \to \overline{a}$ on $\mathbf{R}_{0,2}$ by putting

$$\begin{aligned} \overline{1} &= 1 \\ \overline{e}_j &= -e_j, \quad j = 1, 2 \\ \overline{e}_{12} &= -e_{12}. \end{aligned}$$

For non-parallel vectors $\underline{x}, \underline{y} \in \mathbf{R}_{0,2}^1$ we then have that

$$|\underline{x} \wedge \underline{y}| = |x_1 y_2 - x_2 y_1|$$

is nothing else but the area of the parallelogram constructed on $\underline{x}$ and $\underline{y}$, whence $\underline{x} \wedge \underline{y}$ may be interpreted as determining the oriented area of this parallelogram. In particular the element

$$e_{12} = \underline{e}_1 \, \underline{e}_2 = \underline{e}_1 \wedge \underline{e}_2$$

determines the oriented area of the unit square, i.e. the square constructed on two unit vectors and hence it determines the direction of the plane spanned by e_1 and e_2.

The same element e_{12} may be given a second interpretation when looking at it as being counterlockwise rotation by a right angle in the plane $\mathbf{R}^2$. To this end consider e_{12} as the linear operator in $\mathbf{R}^2$ defined by

$$e_{12} : \underline{x} \quad \rightarrow \quad e_{12}\underline{x}.$$

As

$$\begin{aligned} e_{12}e_1 &= e_2 \\ e_{12}e_2 &= -e_1 \end{aligned}$$

we indeed have that e_{12} is nothing else but the rotation in $\mathbf{R}^2 \cong \mathbf{R}^1_{0,2}$ through a right angle.

As

$$e_{12}(e_{12}e_1) = e_{12}(e_2) = -e_1$$

and

$$e_{12}(e_{12}e_2) = e_{12}(-e_1) = -e_2$$

two consecutive rotations through right angles reverses the direction of a vector. So the relation

$$e_{12}^2 = -\mathbf{1}$$

indeed yields a second geometric interpretation for the element e_{12}.

This interpretation also inspires us to embed $\mathbf{C}$ into $\mathbf{R}_{0,2}$ by identifying $z = a + i\,b$ with the element $a\mathbf{1} + be_{12}$ in $\mathbf{R}_{0,2}$. Indeed, the element $\mathbf{1}$ acts on $\mathbf{R}^1_{0,2} \cong \mathbf{R}^2$ as the identity operator $\mathbf{1}$ and, as we just saw, e_{12} acts on $\mathbf{R}^1_{0,2}$ as a rotation through a right angle, whence any element of $\mathbf{R}^0_{0,2} \oplus \mathbf{R}^2_{0,2}$ could be interpreted as generating a rotation in $\mathbf{R}^2$.

Let us work out this remark in more detail.

Consider the unit circle in $\mathbf{R}^2$, i.e.

$$S^1 = \{\underline{\omega} \in \mathbf{R}^1_{0,2} : |\underline{\omega}| = 1\}$$

and associate with each $\underline{\omega} \in S^1$ fixed the map $\rho_{\underline{\omega}} : \mathbf{R}^2 \rightarrow \mathbf{R}^2$ given by

$$\rho_{\underline{\omega}}(\underline{x}) = \underline{\omega}\,\underline{x}\,\underline{\omega}.$$

That $\underline{\omega x \omega}$ indeed belongs to $\mathbf{R}^2$ for each $\underline{x} \in \mathbf{R}^2$ is easily seen as follows. Calling

$$\underline{\omega}^{\perp} = \{\underline{y} \in \mathbf{R}^1_{0,2} : \underline{y}.\underline{\omega} = 0\}$$

and taking the decomposition (see Appendix)

$$\begin{aligned} \underline{x} &= \underline{x}_{\parallel} + \underline{x}_{\perp} \\ &= \lambda\underline{\omega} + \underline{x}_{\perp} \end{aligned}$$

with $\lambda = \underline{x}.\underline{\omega}$, we find :

$$\begin{aligned} \underline{\omega}\,\underline{x}\,\underline{\omega} &= -(\underline{x}.\underline{\omega})\underline{\omega}^3 + \underline{\omega}\,\underline{x}_{\perp}\underline{\omega} \\ &= (\underline{x}.\underline{\omega})\underline{\omega} + \underline{x}_{\perp}. \end{aligned}$$

Here we used the properties $\underline{\omega}^2 = -1$ and $\underline{\omega}\,\underline{x}_{\perp} + \underline{x}\,\underline{\omega} = 2(\underline{x}_{\perp}.\underline{\omega}) = 0$. So we indeed obtain that $\underline{\omega}\,\underline{x}\,\underline{\omega} \in \mathbf{R}^1_{0,2}$ for all $\underline{x} \in \mathbf{R}^1_{0,2}$ and, moreover, we have that the linear transformation $\rho_{\underline{\omega}}$ is nothing else but reflection in the line $\underline{\omega}^{\perp}$.

Now taking two elements $\underline{\omega}, \underline{\nu} \in S^1$ and putting $s = \underline{\omega}\,\underline{\nu}$, we get for $\Phi_s : \mathbf{R}^2 \rightarrow \mathbf{R}^2$ with $\Phi_s = \rho_{\underline{\omega}} \circ \rho_{\underline{\nu}}$ that

$$\begin{aligned} \Phi_s(\underline{x}) &= \rho_{\underline{\omega}}(\rho_{\underline{\nu}}(\underline{x})) \\ &= \underline{\omega}(\underline{\nu}\,\underline{x}\,\underline{\nu})\underline{\omega} \\ &= (\underline{\omega}\,\underline{\nu})\underline{x}(\overline{\underline{\omega}\,\underline{\nu}}) \\ &= s\underline{x}\,\overline{s} \end{aligned}$$

is a rotation about the origin. As to the angle of rotation we have, putting

$$\underline{\omega} = \cos\varphi\, e_1 + \sin\varphi\, e_2$$

and

$$\underline{\nu} = -\cos\psi\, e_1 - \sin\psi\, e_2,$$

that

$$s = \underline{\omega}\,\underline{\nu} = \cos(\varphi - \psi) + \sin(\varphi - \psi)e_{12}.$$

Hence, putting $\theta = \varphi - \psi$

$$\begin{aligned}\Phi_s(e_1) &= s e_1 \bar{s} \\ &= \cos 2\theta\, e_1 + \sin 2\theta\, e_2\end{aligned}$$

or Φ_s is a rotation through the angle 2θ.

Conversely, as any rotation in the plane is the composition of two reflections in lines and any line is given by $\underline{\omega}^\perp$ for some appropriate $\underline{\omega} \in S^1$, we have that, following the previous discussions, each rotation in $\mathbf{R}^2$ is determined by some Φ_s with $s = \underline{\omega}\underline{\nu}$, $\underline{\omega}, \underline{\nu} \in S^1$.

Let us now have a closer look at the set

$$\mathrm{Spin}\,(2) = \{\underline{\omega}\,\underline{\nu} \;:\; \underline{\omega}, \underline{\nu} \in S^1\}.$$

Then clearly Spin(2) is a multiplicative subgroup of $\mathbf{R}_{0,2}$ and, as we already pointed out, each of its elements $s = \underline{\omega}\,\underline{\nu}$ can be written as

$$s = \cos\theta + \sin\theta e_{12}.$$

Hence

$$\mathrm{Spin}\,(2) \subset \{a \in \mathbf{R}^0_{0,2} \oplus \mathbf{R}^2_{0,2} \;:\; |a| = 1\}.$$

But, as is easily seen, each element $a \in \mathbf{R}^0_{0,2} \oplus \mathbf{R}^2_{0,2}$ with $|a| = 1$ can be written as

$$a = \underline{\omega}\,\underline{\nu}$$

for some $\underline{\omega}, \underline{\nu} \in S^1$. Consequently, putting

$$\mathbf{R}^+_{0,2} = \mathbf{R}^0_{0,2} \oplus \mathbf{R}^2_{0,2}$$

we have

$$\mathrm{Spin}\,(2) = \{a \in \mathbf{R}^+_{0,2} : |a| = 1\}.$$

Clearly $\mathbf{R}^+_{0,2}$ is a subalgebra of $\mathbf{R}_{0,2}$; it is called the *even subalgebra* of $\mathbf{R}_{0,2}$.

As we already observed, $\mathbf{R}^+_{0,2} \cong \mathbf{C}$ and so we obtain that as a group

$$\begin{aligned}\mathrm{Spin}\,(2) &\cong U(1) \\ &\cong \mathrm{SO}(2).\end{aligned}$$

As a final remark, let us recall that in 1.2 we identified $\mathbf{C}$ with the subalgebra $\mathbf{R}1 \oplus \mathbf{R}i$ in $\mathbf{H}$, this identification being of a pure algebraic nature while in this section we identified $\mathbf{C}$ with $\mathbf{R}^+_{0,2} = \mathbf{R}^0_{0,2} \oplus \mathbf{R}^2_{0,2} \cong \mathbf{R}1 \oplus \mathbf{R}k$, this identification having a more geometrical meaning.

4 The Clifford algebra $\mathbf{R}_{0,3}$

The aim of this section is to give several possible approaches to basic geometric notions related to the 3-dimensional Euclidean space $\mathbf{R}^3$. Special emphasis is put on describing rotations in $\mathbf{R}^3$ by means of, respectively, vector algebra, the algebra $\mathbf{H}$ of real quaternions and the Clifford algebra $\mathbf{R}_{0,3}$.

4.1 *Vector algebra in* $\mathbf{R}^3$

After having recalled some basic notions of vector algebra, we give a classification of the orthogonal transformations in $\mathbf{R}^3$ using matrix techniques. Furthermore rotations in $\mathbf{R}^3$ are described in terms of vector algebra.

4.1.1 The scalar and cross products

In Euclidean 3-space $\mathbf{R}^3$ one also considers, besides the scalar product between vectors (see Appendix), the so-called *cross product* of two vectors, which is defined as follows. Let $\underline{e}_1 = (1,0,0)$, $\underline{e}_2 = (0,1,0)$ and $\underline{e}_3 = (0,0,1)$ be the standard orthonormal basis of $\mathbf{R}^3$. Then, if $\underline{x} = x_1\underline{e}_1 + x_2\underline{e}_2 + x_3\underline{e}_3$ and $\underline{y} = y_1\underline{e}_1 + y_2\underline{e}_2 + y_3\underline{e}_3$, the cross product of $\underline{x}$ and $\underline{y}$ is the vector $\underline{x} \times \underline{y}$ given by

$$\underline{x} \times \underline{y} = \det\begin{pmatrix} x_2 & y_2 \\ x_3 & y_3 \end{pmatrix} \underline{e}_1 + \det\begin{pmatrix} x_3 & y_3 \\ x_1 & y_1 \end{pmatrix} \underline{e}_2 + \det\begin{pmatrix} x_1 & y_1 \\ x_2 & y_2 \end{pmatrix} \underline{e}_3,$$

i.e. $\underline{x} \times \underline{y}$ is obtained by computing symbolically the determinant of the matrix

$$\begin{pmatrix} \underline{e}_1 & x_1 & y_1 \\ \underline{e}_2 & x_2 & y_2 \\ \underline{e}_3 & x_3 & y_3 \end{pmatrix}$$

along the expansion by cofactors down the first column.

Among others, the cross product satisfies the following basic properties

(i) It defines an anti-symmetric bilinear operator from $\mathbf{R}^3 \times \mathbf{R}^3$ into $\mathbf{R}^3$.

(ii) $\underline{x} \times \underline{y}$ is orthogonal to $\underline{x}$ and to $\underline{y}$

(iii) $\underline{x} \times \underline{y} = \underline{0}$ if and only if $\underline{x}$ and $\underline{y}$ are linearly dependent.

(iv) If $\underline{x} \neq \underline{0}$ and $\underline{y} \neq \underline{0}$, then

$$|\underline{x} \times \underline{y}| = |\underline{x}||\underline{y}| \sin\theta$$

where θ is the angle between the vectors $\underline{x}$ and $\underline{y}$.

$$\begin{aligned}
\text{(v)} < \underline{x} \times \underline{y}, \underline{z} > &= < \underline{x}, (\underline{y} \times \underline{z} > \\
&= \det(\underline{x}\,\underline{y}\,\underline{z}) \\
&= \det \begin{pmatrix} x_1 & y_1 & z_1 \\ x_2 & y_2 & z_2 \\ x_3 & y_3 & z_3 \end{pmatrix}
\end{aligned}$$

for all $\underline{x}, \underline{y}, \underline{z} \in \mathbf{R}^3$.
for all $\underline{x}, \underline{y}, \underline{z} \in \mathbf{R}^3$.

Notice that, if $\underline{x}$ and $\underline{y}$ are linearly independent, then the ordered bases $(\underline{e}_1, \underline{e}_2, \underline{e}_3)$ and $(\underline{x}, \underline{y}, \underline{x} \times \underline{y})$ of $\mathbf{R}^3$ are like-oriented. From now on we assume tacitly that the orthonormal basis $(\underline{e}_1, \underline{e}_1, \underline{e}_3)$ of $\mathbf{R}^3$ determines a *righthanded* or *dextral* set of basic vectors. This means that a righthanded screw pointing in the $\underline{e}_3$-direction will advance in that direction when given a counterclockwise rotation in the $(\underline{e}_1, \underline{e}_2)$-plane. $\mathbf{R}^3$ oriented by such a set of vectors is then called *positively oriented.*

4.1.2 Orthogonal transformations in $\mathbf{R}^3$

Let T be an orthogonal transformation in $\mathbf{R}^3$ and let A be its matrix representation relative to the standard orthonormal basis $e = (\underline{e}_1, \underline{e}_2, \underline{e}_3)$ (see Appendix). If $\det A = 1$ (i.e. $T \in \mathrm{SO}(3)$), $\underline{e}'_3$ is a unit eigenvector corresponding to the eigenvalue 1 of A and $(\underline{e}'_1, \underline{e}'_2, \underline{e}'_3)$ is an orthonormal basis of $\mathbf{R}^3$, then relative to the basis e' T has the matrix representation

$$A' = \begin{pmatrix} \cos\theta & -\sin\theta & \\ \sin\theta & \cos\theta & \\ & & 1 \end{pmatrix}$$

where θ is the angle between $\underline{e}'_1$ and $T\underline{e}'_1$.

If $\theta = 0 \bmod (2\pi\mathbf{Z})$ then T is the identity map while for $\theta = \pi \bmod (2\pi\mathbf{Z})$, T is the reflection of $\mathbf{R}^3$ in the line $\mathbf{R}e'_3$. If $\sin\theta \neq 0$ then T is a rotation about e'_3 through the angle θ whence, T restricted to the $(\underline{e}'_1, \underline{e}'_2)$-plane being a rotation, it may be considered as the composition of two reflections in hyperplanes intersecting along $\underline{e}'_3$. If $\det A = -1$ (i.e. $T \in \mathrm{O}\ (3) \setminus \mathrm{SO}(3)$), $\underline{e}'_3$ is a unit eigenvector corresponding to the eigenvalue -1 of A and $e' = (\underline{e}'_1, \underline{e}'_2, \underline{e}'_3)$ is an orthogonal basis of $\mathbf{R}^3$, then relative to this basis T has the matrix representation

$$A' = \begin{pmatrix} \cos\theta & -\sin\theta & \\ \sin\theta & \cos\theta & \\ & & -1 \end{pmatrix}$$

where θ is the angle betweeen $\underline{e}'_1$ and $T\underline{e}'_1$.

If $\theta = 0 \bmod (2\pi\mathbf{Z})$ then T is the reflection of $\mathbf{R}^3$ in the $(\underline{e}'_1, \underline{e}'_2)$-plane while if $\theta \neq 0 \bmod (2\pi\mathbf{Z})$, T is the composition of a rotation about $\underline{e}'_3$ through the angle θ and a reflection in the $(\underline{e}'_1, \underline{e}'_2)$-plane. Hence, as it should be, T is the composition of an odd number (≤ 3) of reflections in hyperplanes.

Let us now take a unit vector $\underline{u}$ in $\mathbf{R}^3$ and determine how to rotate a given vector $\underline{x}$ about $\underline{u}$ through the angle θ to get a vector $\underline{y}$. To do this, we first take the orthogonal projection of $\underline{x}$ on $\underline{u}$ given by $<\underline{x}, \underline{u}>\underline{u}$ and we put

$$\underline{z} = \underline{x} - <\underline{x}, \underline{u}> \underline{u}.$$

Then clearly $<\underline{z}, \underline{u}> = 0$ and of course

$$\underline{x} = \underline{z} + <\underline{x}, \underline{u}> \underline{u}.$$

Setting $\underline{w} = \underline{u} \times \underline{z}$ we have that

$$<\underline{w}, \underline{u}> = 0, \quad <\underline{w}, \underline{z}> = 0 \quad \text{and} \quad |\underline{w}| = |\underline{z}|.$$

Moreover $\underline{z}$ is obviously the orthogonal projection of $\underline{x}$ on the $(\underline{z}, \underline{w})$-plane, the latter coinciding with $\underline{u}^\perp$. Now rotating $\underline{z}$ about $\underline{u}$ through the angle θ we obtain the vector $\underline{z}\cos\theta + \underline{w}\sin\theta$ which is nothing else but the orthogonal projection of $\underline{y}$ on the $(\underline{z}, \underline{w})$-plane. Hence

$$\underline{y} = <\underline{x}, u> \underline{u} + \underline{z}\cos\theta + \underline{w}\sin\theta.$$

Since

$$\begin{aligned} \underline{w} &= \underline{u} \times \underline{z} \\ &= \underline{u} \times \underline{x} - <\underline{x}, \underline{u}> \underline{u} \times \underline{u} \\ &= \underline{u} \times \underline{x} \end{aligned}$$

we finally obtain that

$$\underline{y} = <\underline{x}, \underline{u}> (1 - \cos\theta)\underline{u} + \cos\theta\, \underline{x} + \sin\theta\, \underline{u} \times \underline{x}. \tag{4.1}$$

The formula (4.1), known as Euler's Formula, shows how by means of vector algebra the rotation about an axis through an angle θ may be described.

4.2 The algebra **H** *revisited*

The aim of this subsection is twofold. In 4.2.1. we show how by using properties of the algebra **H** only, the important notions of dot and cross products between vectors in $\mathbf{R}^3$ may be derived. In the same spirit the notions of orientation in $\mathbf{R}^3$ and rotation are introduced. It is then shown how to describe rotations in $\mathbf{R}^3$ by means of quaternions. In 4.2.2 we follow the opposite way and show how vector algebra may be used to define the algebra **H**.

4.2.1 From quaternions to vector algebra

Consider the algebra $\mathbf{H}$ of real quaternions as introduced in 1.2. Using the multiplication rules in $\mathbf{H}$, we may write the product of any two elements $\underline{x}$ and $\underline{y}$ in $\mathbf{R}^3 \cong Ve\,\mathbf{H}$ as

$$\underline{x}\,\underline{y} = -<\underline{x},\underline{y}> + \underline{x} \times \underline{y}$$

where, by definition, we have put

$$\begin{aligned} <\underline{x},\underline{y}> &= -Re\,(\underline{x}\,\underline{y}) \\ \text{and} \qquad & \\ \underline{x} \times \underline{y} &= Ve(\underline{x}\,\underline{y}), \end{aligned} \tag{4.2}$$

i.e.

$$<\underline{x},\underline{y}> = \sum_{t=1}^{3} x_t\, y_t$$

and

$$\underline{x} \times \underline{y} = \det\begin{pmatrix} x_2 & y_2 \\ x_3 & y_3 \end{pmatrix}\mathbf{i} + \det\begin{pmatrix} x_3 & y_3 \\ x_1 & y_1 \end{pmatrix}\mathbf{j} + \det\begin{pmatrix} x_1 & y_1 \\ x_2 & y_2 \end{pmatrix}\mathbf{k}.$$

Using conjugation in $\mathbf{H}$, we get immediately that for all $\underline{x} \in \mathbf{R}^3$,

$$\begin{aligned} <\underline{x},\underline{x}> &= \underline{x}\overline{\underline{x}} \\ &= |\underline{x}|^2. \end{aligned} \tag{4.3}$$

Moreover, as is easily seen, the function $(\underline{x},\underline{y}) \to <\underline{x},\underline{y}>$ satisfies all the properties of the dot product in $\mathbf{R}^3$ and by (4.3), the norm on $\mathbf{H}$ induced on $Ve\,\mathbf{H} \cong \mathbf{R}^3$ coincides with the norm derived from the dot product.

We continue calling $(\underline{x},\underline{y}) \to <\underline{x},\underline{y}> = -Re(\underline{x}\,\underline{y})$ and $(\underline{x},\underline{y}) \to \underline{x} \times \underline{y} = Ve(\underline{x}\,\underline{y})$ the dot and cross products on $\mathbf{R}^3$. As to the function $(\underline{x},\underline{y}) \to \underline{x} \times \underline{y}$, it is also easily seen that it satisfies all the properties of the cross product between vectors. This means that, by embedding $\mathbf{R}^3$ into $\mathbf{H}$ as prescribed, the dot product and cross product between two vectors are obtained simultaneously by taking the product in $\mathbf{H}$ between these vectors. Hence classical notions such as orthogonality, angle, distance etc ... and the basic properties related to them may be derived immediately.

Now let us describe rotations in $\mathbf{R}^3$ using quaternions. To this end let us first introduce a positive orientation in $\mathbf{R}^3$ as follows.

Let $\{\underline{\varepsilon}_1\underline{\varepsilon}_2, \underline{\varepsilon}_3\}$ be an orthonormal basis in $\mathbf{R}^3$. Then, as $\underline{\varepsilon}_2\,\underline{\varepsilon}_3 = \underline{\varepsilon}_2 \times \underline{\varepsilon}_3$ and hence $\underline{\varepsilon}_2\,\underline{\varepsilon}_3 \perp \underline{\varepsilon}_2$ and $\underline{\varepsilon}_2\,\underline{\varepsilon}_3 \perp \underline{\varepsilon}_3$, we have that $\underline{\varepsilon}_2\,\underline{\varepsilon}_3 = \lambda\underline{\varepsilon}_1$ with $\lambda = \pm 1$, whence

$$\underline{\varepsilon}_1\underline{\varepsilon}_2\,\underline{\varepsilon}_3 = \pm\mathbf{1}.$$

We say that the orthonormal basis $\{\underline{\varepsilon}_1, \underline{\varepsilon}_2, \underline{\varepsilon}_3\}$ is orienting $\mathbf{R}^3$ positively if $\underline{\varepsilon}_1 \underline{\varepsilon}_2 \underline{\varepsilon}_3 = -1$. Notice that in such a way, the standard basis $\{\mathbf{i}, \mathbf{j}, \mathbf{k}\}$ defines the positive orientation of $\mathbf{R}^3$ since $\mathbf{i}\,\mathbf{j}\,\mathbf{k} = -1$. More generally, if $\{\underline{x}, \underline{y}, \underline{z}\}$ is a basis for $\mathbf{R}^3$, we say that sgn $Re(\underline{x}\,\underline{y}\,\underline{z})$ (the sgn of the scalar part of $\underline{x}\,\underline{y}\,\underline{z}$) determines the orientation of the basis $\{\underline{x}, \underline{y}, \underline{z}\}$, where the orientation is called positive if $\operatorname{sgn} Re(\underline{x}\,\underline{y}\,\underline{z}) < 0$.

Notice that, putting $e = \{\mathbf{i}, \mathbf{j}, \mathbf{k}\}$ and calling $[\underline{x}\,\underline{y}\,\underline{z}]_e$ the (3×3)-matrix whose columns consist of the coordinate matrices of $\underline{x}, \underline{y}, \underline{z}$ relative to e, we have that for any basis $\{\underline{x}, \underline{y}, \underline{z}\}$ of $\mathbf{R}^3$,

$$Re(\underline{x}\,\underline{y}\,\underline{z}) = -\det[\underline{x}\,\underline{y}\underline{z}]_e.$$

Indeed, on the one hand

$$\begin{aligned} Re(\underline{x}\,\underline{y}\,\underline{z}) &= Re(\underline{x}(- < \underline{y}, \underline{z} > + \underline{y} \times \underline{z})) \\ &= Re(\underline{x}(- < \underline{y}, \underline{z} >)) + Re(\underline{x}(\underline{y} \times \underline{z})) \\ &= Re(\underline{x}(\underline{y} \times \underline{z})) \\ &= - < \underline{x}, \underline{y} \times \underline{z} > \end{aligned}$$

while on the other hand

$$< \underline{x}, \underline{y} \times \underline{z} >= \det[\underline{x}\,\underline{y}\,\underline{z}]_e.$$

Hence

$$\operatorname{sgn} Re(\underline{x}\,\underline{y}\,\underline{z}) = -\operatorname{sgn} \det[\underline{x}\,\underline{y}\,\underline{z}]_e.$$

Now let T be an invertible linear transformation in $\mathbf{R}^3$, i.e. $T \in GL(3; \mathbf{R})$, and call $A = [T]_e$ its matrix relative to the basis e. Then T is orientation preserving if and only if $\det T = \det A > 0$. As may now be expected, a linear transformation T in $\mathbf{R}^3$ is called a *rotation* if and only if it leaves the dot product and the orientation invariant. Of course we then have that T is invertible with $\det T = 1$.

Now we claim that T is a rotation if and only if there exists $q \in \mathbf{H}$ with $|q| = 1$ such that $T = T_q$ where for all $\underline{x} \in \mathbf{R}^3$,

$$T_q(\underline{x}) = q\underline{x}\overline{q}.$$

To this end, take $q \in \mathbf{H}$ fixed with $|q| = 1$ and consider the mapping $T_q : \mathbf{H} \to \mathbf{H}$ with

$$T_q(q') = q\,q'\,\overline{q} \quad , \quad q' \in \mathbf{H}.$$

Then clearly T_q is a linear isomorphism from **H** onto **H** which moreover preserves the product in **H**. In fact, T_q is a so-called inner automorphism on **H**. As for each $\underline{x} \in \mathbf{R}^3$, $y = T_q(\underline{x}) = q\underline{x}\overline{q}$ satisfies $\overline{y} = -y$, we have that y is a pure quaternion and so $y \in \mathbf{R}^3$. Consequently T_q is a linear transformation in $\mathbf{R}^3$.

Moreover, taking an arbitrary orthonormal basis $\{\underline{\varepsilon}_1, \underline{\varepsilon}_2, \underline{\varepsilon}_3\}$ of $\mathbf{R}^3$ we obtain that, as $\underline{\varepsilon}_1\underline{\varepsilon}_2\underline{\varepsilon}_3 \in \mathbf{R}$,

$$\begin{aligned}(T_q\underline{\varepsilon}_1)(T_q\underline{\varepsilon}_2)(T_q\underline{\varepsilon}_3) &= q\underline{\varepsilon}_1\underline{\varepsilon}_2\underline{\varepsilon}_3\overline{q} \\ &= \underline{\varepsilon}_1\underline{\varepsilon}_2\underline{\varepsilon}_3\end{aligned}$$

whence T_q perserves the orientation. Finally, for all $\underline{x} \in \mathbf{R}^3$, we clearly have that

$$|T_q(\underline{x})|^2 = |\underline{x}|^2$$

whence, as for all $\underline{x}, \underline{y} \in \mathbf{R}^3$,

$$< \underline{x}, \underline{y} >= \frac{1}{2}|\underline{x}+\underline{y}|^2 - \frac{1}{2}|\underline{x}-\underline{y}|^2,$$

T_q preserves the dot product. We thus obtain that T_q is a rotation.

Conversely, suppose that T is a rotation in $\mathbf{R}^3$ and let A be the matrix representation of T relative to the orthonormal basis $\{\underline{e}_1, \underline{e}_2, \underline{e}_3\}$. Furthermore let $\underline{\varepsilon}_3$ be a unit eigenvector corresponding to the eigenvalue 1 and choose a positive orthonormal basis $\varepsilon = \{\underline{\varepsilon}_1, \underline{\varepsilon}_2, \underline{\varepsilon}_3\}$ of $\mathbf{R}^3$, i.e. $\underline{\varepsilon}_1\underline{\varepsilon}_2\underline{\varepsilon}_3 = -1$. Then relative to the basis ε, T has the matrix representation

$$A' = \begin{pmatrix} \cos\theta & -\sin\theta & \\ \sin\theta & \cos\theta & \\ & & 1 \end{pmatrix}.$$

Now we claim that $T = T_q$ where

$$q = \cos\lambda + \underline{\varepsilon}_3 \sin\lambda$$

with $\lambda = -\frac{\theta}{2}$. Indeed, first notice that $|q| = 1$. Moreover, taking account of the relations

$$\begin{aligned} \underline{\varepsilon}_t^2 &= -\underline{\varepsilon}_t\overline{\underline{\varepsilon}}_t = -1 \quad , && t = 1,2,3, \\ \underline{\varepsilon}_t\underline{\varepsilon}_s &= \underline{\varepsilon}_t \times \underline{\varepsilon}_s = -\underline{\varepsilon}_s\underline{\varepsilon}_t, && s,t = 1,2,3; s \neq t, \\ \underline{\varepsilon}_1\underline{\varepsilon}_2 &= \underline{\varepsilon}_3 \,;\; \underline{\varepsilon}_2\underline{\varepsilon}_3 = \underline{\varepsilon}_1 \,;\; \underline{\varepsilon}_3\underline{\varepsilon}_1 = \underline{\varepsilon}_2, \end{aligned} \tag{4.4}$$

straightforward computations yield that

$$\begin{aligned} T_q(\underline{\varepsilon}_1) &= \underline{\varepsilon}_1 \cos\theta + \underline{\varepsilon}_2 \sin\theta \\ T_q(\underline{\varepsilon}_2) &= -\underline{\varepsilon}_1 \sin\theta + \underline{\varepsilon}_2 \cos\theta \\ T_q(\underline{\varepsilon}_3) &= \underline{\varepsilon}_3 \end{aligned}$$

whence the matrix representation of T_q relative to ε is exactly A'.

Consequently $T = T_q$.

4.2.2 From vector algebra to quaternions

Consider the 3-dimensional Euclidean space $\mathbf{R}^3$ oriented by some positive orthonormal basis $\{\underline{e}_1, \underline{e}_2, \underline{e}_3\}$. Then by $\mathcal{A}$ we mean the set of pairs $(x_0, \underline{x})$ with $x_0 \in \mathbf{R}$ and $\underline{x} \in \mathbf{R}^3$.

Addition and multiplication on $\mathcal{A}$ are defined by

$$(x_0, \underline{x}) + (y_0, \underline{y}) = (x_0 + y_0, \underline{x} + \underline{y})$$

and

$$(x_0, \underline{x})(y_0, \underline{y}) = (x_0 y_0 - < \underline{x}, \underline{y} >, x_0 \underline{y} + y_0 \underline{x} + \underline{x} \times \underline{y})$$

Furthermore, for $\lambda \in \mathbf{R}$ and $(x_0, \underline{x}) \in \mathcal{A}$ we put

$$\lambda(x_0, \underline{x}) = (\lambda x_0, \lambda \underline{x}).$$

Finally defining for each $q = (x_0, \underline{x}) \in \mathcal{A}$, the anti-involution $q \to \overline{q}$ with $\overline{q} = (x_0, -\underline{x})$, it is readily seen that for all $\lambda \in \mathbf{R}$ and $q, q' \in \mathcal{A}$,

$$\begin{aligned} \overline{(\lambda q + q')} &= \lambda \overline{q} + \overline{q'} \\ \overline{q q'} &= \overline{q'}\,\overline{q} \\ q\overline{q} &= (x_0^2 + x_1^2 + x_2^2 + x_3^2, \underline{0}). \end{aligned}$$

We thus obtain in a straightforward way that $\mathcal{A}, +, .$ is a real linear associative algebra with identity $(1, \underline{0})$, generated by the elements $(0, \underline{e}_1)$ and $(0, \underline{e}_2)$.

Noticing that $(0, \underline{e}_3) = (0, \underline{e}_1)(0, \underline{e}_2)$ and putting henceforth $\mathbf{1} = (1, \underline{0}), \mathbf{i} = (0, \underline{e}_1), \mathbf{j} = (0, \underline{e}_2)$ and $\mathbf{k} = (0, \underline{e}_3)$ we may write any $(x_0, \underline{x}) \in \mathcal{A}$ as

$$\begin{aligned} (x_0, \underline{x}) &= x_0 \mathbf{1} + x_1 \mathbf{i} + x_2 \mathbf{j} + x_3 \mathbf{k} \\ &= x_0 + \underline{x}, \end{aligned}$$

thus identifying each $x_0 \in \mathbf{R}$ with $x_0 \mathbf{1}$ and each $\underline{x} \in \mathbf{R}^3$ with $(0, \underline{x})$ in $\mathcal{A}$. Obviously the algebra $\mathcal{A}$ is isomorphic to the algebra $\mathbf{H}$ of real quaternions through the identification $(x_0, \underline{x}) \to (x_0, x_1, x_2, x_3)$, where the anti-involution on $\mathcal{A}$ is carried over to the conjugation in $\mathbf{H}$. Notice also that for all $\underline{x}, \underline{y} \in \mathbf{R}^3$,

$$< \underline{x}, \underline{y} > = -\frac{1}{2}(\underline{x}\,\underline{y} + \underline{y}\,\underline{x})$$

and

$$\underline{x} \times \underline{y} = \frac{1}{2}(\underline{x}\underline{y} - \underline{y}\underline{x}).$$

Of course the function $q \to |q| = \sqrt{q\overline{q}}$ is a norm on $\mathcal{A}$ satisfying $|qq'| = |q||q'|$ and the set of elements $q \in \mathcal{A}$ such that $|q| = 1$ may be identified with the unit sphere S^3 in $\mathbf{R}^4$. Furthermore observe that each $q \in \mathcal{A}$ may be written as $q = a + b\underline{u}$ with $a, b \in \mathbf{R}$ and $\underline{u} \in \mathbf{R}^3$ such that $|\underline{u}| = 1$. Of course we then have that $|q| = 1$ if and only if $a^2 + b^2 = 1$.

Now we wish to give another proof of the fact that each rotation T in $\mathbf{R}^3$ is of the form $T = T_q$ for some $q \in \mathcal{A}$ with $|q| = 1$, where for each $\underline{x} \in \mathbf{R}^3$, $T_q(\underline{x}) = q\underline{x}\overline{q}$. Indeed, fix $q \in \mathcal{A}$ with $|q| = 1$ and put for $\underline{x} \in \mathbf{R}^3$, $y = T_q(\underline{x}) = q\underline{x}\overline{q}$. Then, as $\overline{y} = -y$, $y = \underline{y} \in \mathbf{R}^3$ and so T_q becomes a linear operator in $\mathbf{R}^3$. Furthermore, as

$$\begin{aligned} \underline{y}\,\overline{\underline{y}} &= (q\underline{x}\overline{q})(q\overline{\underline{x}}\overline{q}) \\ &= \underline{x}\,\overline{\underline{x}}, \end{aligned}$$

we have that $|\underline{y}| = |\underline{x}|$, whence T_q is an orthogonal transformation in $\mathbf{R}^3$.

Moreover, putting $q = a + b\underline{u}$ with $|\underline{u}| = 1$, we have

$$\begin{aligned} T_q(\underline{u}) &= q\underline{u}\overline{q} \\ &= (a + b\underline{u})\underline{u}(a - b\underline{u}) \\ &= (a^2 + b^2)\underline{u} \\ &= \underline{u} \end{aligned}$$

so that T_q leaves $\underline{u}$ invariant or T_q is a rotation about $\underline{u}$. In order to determine the angle of rotation, put $a = \cos\varphi$ and $b = \sin\varphi$. Then by simple calculations we obtain that

$$\begin{aligned} T_q(\underline{x}) &= (a + b\underline{u})\underline{x}(a - b\underline{u}) \\ &= \underline{x}\cos^2\varphi + \underline{u} \times \underline{x}\sin 2\varphi - \underline{u}\,\underline{x}\,\underline{u}\sin^2\varphi \end{aligned}$$

Choosing a positive orthonormal basis $\underline{\varepsilon} = \{\underline{\varepsilon}_1, \underline{\varepsilon}_2, \underline{\varepsilon}_3\}$ in such a way that $\underline{u} = \underline{\varepsilon}_3$, we obtain by using expressions of the form (4.4), that

$$\underline{u}\,\underline{x}\,\underline{u} = \underline{x} - 2 < \underline{x}, \underline{u} > \underline{u}.$$

Hence

$$T_q(\underline{x}) = \underline{x}\cos 2\varphi + < \underline{x}, \underline{u} > (1 - \cos 2\varphi)\underline{u} + \underline{u} \times \underline{x}\sin 2\varphi.$$

Comparing this result with Euler's formula (4.1) we obtain that T_q is a rotation about $\underline{u}$ through the angle $\theta = 2\varphi$.

Conversely, if T is a rotation in $\mathbf{R}^3$ about a unit vector $\underline{u}$ through the angle θ, we get by virtue of the foregoing calculations that $T = T_q$ with $q = \cos\frac{\theta}{2} + \underline{u}\sin\frac{\theta}{2}$.

4.3 The Clifford algebra $\mathbf{R}_{0,3}$

Starting from an orthonormal basis in $\mathbf{R}^3$ we construct the linear associative algebra $\mathbf{R}_{0,3}$, which will be called in Chapter 1 the Clifford algebra for $\mathbf{R}^{0,3}$. Our main purpose is to show how important geometric notions in $\mathbf{R}^3$ may be derived directly from properties valid in the algebra $\mathbf{R}_{0,3}$.

4.3.1 Definitions

Consider the 3-dimensional Euclidean space $\mathbf{R}^3$ and let $e = \{\underline{e}_1, \underline{e}_2, \underline{e}_3\}$ be an orthonormal basis in it. Then we wish to construct a linear associative algebra of maximal dimension containing $\mathbf{R}$ and $\mathbf{R}^3$ such that for each vector $\underline{x} \in \mathbf{R}^3$, $\underline{x}^2 = -|\underline{x}|^2$. This implies that for the basic vectors

$$\underline{e}_j^2 = -1, \quad j = 1,2,3$$

and

$$\underline{e}_i\underline{e}_j + \underline{e}_j\underline{e}_i = 0, \quad i \neq j, \quad i,j = 1,2,3.$$

We thus obtain an algebra of dimension 8 having as basis the set of elements 1, $\underline{e}_1, \underline{e}_2, \underline{e}_3, \underline{e}_1\underline{e}_2, \underline{e}_2\underline{e}_3, \underline{e}_3\underline{e}_1$ and $\underline{e}_1\underline{e}_2\underline{e}_3$.

This algebra is henceforth denoted by $\mathbf{R}_{0,3}$. Moreover we write

$$\begin{aligned} \underline{e}_i &= e_i \quad , \quad i = 1,2,3, \\ \underline{e}_i\underline{e}_j &= e_{ij} \quad , \quad i,j = 1,2,3, \quad i \neq j \end{aligned}$$

and

$$\underline{e}_1\underline{e}_2\underline{e}_3 = e_{123}.$$

An arbitrary element $a \in \mathbf{R}_{0,3}$ may thus be written as

$$a = a_0 + a_1e_1 + a_2e_2 + a_3e_3 + a_{23}e_{23} + a_{31}e_{31} + a_{12}e_{12} + a_{123}e_{123} \tag{4.5}$$

where the coefficients are real numbers.

The subspace spanned by $\{1\}$ is called the space of *scalars* ; it will be identified by $\mathbf{R}$ and its elements will be denoted by $\alpha, \beta, \ldots$ The subspace spanned by $\{e_1, e_2, e_3\}$ is called the space of *vectors* ; it will be identified with $\mathbf{R}^3$ and its elements will be denoted by $\underline{x}, \underline{y}, \ldots$ The subspace spanned by $\{e_{23}, e_{31}, e_{12}\}$ is called the space of *bivectors* ; its elements will be denoted by $\underline{X}, \underline{Y}, \ldots$ Finally, the subspace spanned by $\{e_{123}\}$ is called the space of *trivectors* or *pseudoscalars* ; its elements are of the form $\alpha\, e_{123}$, $\alpha \in \mathbf{R}$; they will be denoted by $A, B, \ldots$

For an element $a \in \mathbf{R}_{0,3}$ written out in the form (4.5), we call, respectively,

$$\begin{aligned}
[a]_0 &= a_0, \\
[a]_1 &= a_1e_1 + a_2e_2 + a_3e_3, \\
[a]_2 &= a_{23}e_{23} + a_{31}e_{31} + a_{12}e_{12} \\
[a]_3 &= a_{123}e_{123}
\end{aligned}$$

the *scalar*, *vector*, *bivector* and *pseudoscalar part* of a.

Now we introduce the following involutions or anti-involutions on $\mathbf{R}_{0,3}$. Putting

$$\begin{aligned}
\hat{1} &= 1, \\
\hat{e}_j &= -e_j, \quad j = 1,2,3, \\
\hat{e}_{ij} &= e_{ij}, \quad i \neq j, \quad i,j = 1,2,3 \\
\hat{e}_{123} &= -e_{123}
\end{aligned}$$

and extending this operator by linearity to the whole of $\mathbf{R}_{0,3}$, we call the function $a \to \hat{a}$, $a \in \mathbf{R}_{0,3}$, the *main involution* on $\mathbf{R}_{0,3}$. Clearly

$$\begin{aligned}
\widehat{\lambda a + b} &= \lambda\hat{a} + \hat{b} \\
\widehat{ab} &= \hat{a}\hat{b} \\
\hat{\hat{a}} &= a
\end{aligned}$$

for all $a, b \in \mathbf{R}_{0,3}$ and $\lambda \in \mathbf{R}$, whence $a \to \hat{a}$ is an involution on $\mathbf{R}_{0,3}$.

Moreover, putting

$$\mathbf{R}_{0,3}^{+} = \{a_0 + a_{23}e_{23} + a_{31}e_{31} + a_{12}e_{12} : a_0, a_{ij} \in \mathbf{R}\}$$

and

$$R_{0,3}^{-} = \{a_1e_1 + a_2e_2 + a_3a_3 + a_{123}e_{123} : a_j, a_{123} \in \mathbf{R}\}$$

we have that

$$\mathbf{R}_{0,3} = \mathbf{R}_{0,3}^{+} \oplus \mathbf{R}_{0,3}^{-}$$

and that

$$\hat{a} = a \text{ for } a \in \mathbf{R}_{0,3}^{+}$$

and

$$\hat{a} = -a \text{ for } a \in \mathbf{R}_{0,3}^{-}.$$

Obviously $\mathbf{R}_{0,3}^{+}$ is a subalgebra of $\mathbf{R}_{0,3}$, called the *even subalgebra*.

Next, putting

$$\begin{aligned}\tilde{1} &= 1 \\ \tilde{e}_j &= e_j, \quad j = 1,2,3, \\ \tilde{e}_{ij} &= -e_{ij}, \quad i \neq j, \quad i,j = 1,2,3 \\ \tilde{e}_{123} &= -e_{123}\end{aligned}$$

and extending this operator by linearity to the whole of $\mathbf{R}_{0,3}$, we call the function $a \to \tilde{a}$, $a \in \mathbf{R}_{0,3}$, the *reversion* on $\mathbf{R}_{0,3}$. Clearly

$$\begin{aligned}(\lambda a + b)^{\sim} &= \lambda\tilde{a} + \tilde{b} \\ (ab)^{\sim} &= \tilde{b}\tilde{a} \\ \tilde{\tilde{a}} &= a\end{aligned}$$

for all $a, b \in \mathbf{R}_{0,3}$ and $\lambda \in \mathbf{R}$, whence $a \to \tilde{a}$ is an anti-involution on $\mathbf{R}_{0,3}$.

Finally, putting for each $a \in \mathbf{R}_{0,3}$

$$\bar{a} = (\hat{a})^{\sim} = (\tilde{a})^{\wedge},$$

we obtain that $a \to \bar{a}$ is an anti-involution on $\mathbf{R}_{0,3}$, called the *conjugation*. By means of the conjugation, a norm $a \to |a|$ may be introduced on $\mathbf{R}_{0,3}$ by setting

$$|a|^2 = [a\bar{a}]_0 = \sum_{A \subset \{1,2,3\}} a_A^2 = [\bar{a}a]_0 .$$

Notice that, if for an element $a \in \mathbf{R}_{0,3}$, $a\bar{a} \in \mathbf{R}^+$ whence in that case $a\bar{a} = |a|^2$, then a is invertible with $a^{-1} = \frac{\bar{a}}{|a|^2}$. This is clearly the case for nonzero scalars, vectors, bivectors and trivectors. This is also the case for each nonzero linear combination of a scalar and a bivector or of a vector and a trivector. Indeed

$$(\alpha + \underline{X})(\alpha - \underline{X}) = \alpha^2 + |\underline{X}|^2$$

while

$$(\underline{x} + \beta e_{123})(-\underline{x} + \beta e_{123}) = |\underline{x}|^2 + \beta^2$$

However condition $a\bar{a} \in \mathbf{R}^+$ is not necessary for the invertibility of $a \in \mathbf{R}_{0,3}$ as is seen from the following counterexample. Take $a = 2 + e_{123}$. Then $a\bar{a} = \bar{a}a = 5 + 4e_{123} \notin \mathbf{R}^+$ although a is invertible with inverse $a^{-1} = \frac{1}{3}(2 - e_{123})$.

Finally notice that for vectors $\underline{x}$ and $\underline{y}$, $|\underline{x}\,\underline{y}| = |\underline{x}||\underline{y}|$ as follows from

$$\begin{aligned}|\underline{x}\,\underline{y}|^2 &= \left[\underline{x}\,\underline{y}\overline{\underline{x}\,\underline{y}}\right]_0 \\ &= \left[\underline{x}\,\underline{y}\bar{\underline{y}}\,\bar{\underline{x}}\right]_0 \\ &= |\underline{x}|^2|\underline{y}|^2.\end{aligned}$$

Remark. At the very beginning we required that the linear associative algebra constructed on $\mathbf{R}^3$ satisfying the basic multiplication rules

$$\underline{e}_i\underline{e}_j + \underline{e}_j\underline{e}_i = -2\delta_{ij}, \quad i,j = 1,2,3,$$

for an orthonormal basis $e = \{\underline{e}_1, \underline{e}_2, \underline{e}_3\}$ of $\mathbf{R}^3$, should have maximal dimension, namely 2^3. This requirement implies that we assume that $\underline{e}_1\underline{e}_2\underline{e}_3 \neq \pm 1$ (see also I.1.3, Theorem). We could indeed also have put e.g.

$$\underline{e}_1\underline{e}_2 = \underline{e}_3, \quad \underline{e}_3\underline{e}_1 = \underline{e}_2, \quad \underline{e}_2\underline{e}_3 = \underline{e}_1.$$

Then $\underline{e}_1\underline{e}_2\underline{e}_3 = -1$, thus producing the algebra of real quaternions.

4.3.2 From $\mathbf{R}_{0,3}$ to vector algebra

The aim of this subsection is to show how vector algebra fits naturally into $\mathbf{R}_{0,3}$. Let $\underline{x}, \underline{y}$ be two vectors. Then their product $\underline{x}\,\underline{y}$ is the sum of a scalar and a bivector, namely

$$\underline{x}\,\underline{y} = \underline{x}.\underline{y} + \underline{x} \wedge \underline{y}$$

with

$$\underline{x}.\underline{y} = -(x_1y_1 + x_2y_2 + x_3y_3) = -<\underline{x}, \underline{y}>$$

and

$$\begin{aligned}\underline{x} \wedge \underline{y} &= (x_2y_3 - x_3y_2)e_{23} + (x_3y_1 - x_1y_3)e_{31} \\ &\quad + (x_1y_2 - x_2y_1)e_{12}.\end{aligned}$$

It is clear that

$$\begin{aligned}\underline{x}.\underline{y} &= \tfrac{1}{2}(\underline{x}\,\underline{y} + \underline{y}\,\underline{x}) \\ \text{and} \qquad & \\ \underline{x} \wedge \underline{y} &= \tfrac{1}{2}(\underline{x}\,\underline{y} - \underline{y}\,\underline{x}).\end{aligned} \tag{4.6}$$

The scalar $\underline{x}.\underline{y}$ and the bivector $\underline{x} \wedge \underline{y}$ are called, respectively, the *inner* and *outer product* between the vectors $\underline{x}$ and $\underline{y}$. From (4.6) it follows immediately that the inner product is commutative while the outer product is anticommutative.

Moreover, as the notation already includes, the inner product is minus the classical dot product whence all properties of the inner product are derived immediately from those of the dot product. In particular, calling $\theta \in [0, \pi]$ the angle between the vectors $\underline{x}$ and $\underline{y}$, we have that

$$\begin{aligned}\left[\underline{x}\,\underline{y}\right]_0 &= \underline{x}.\underline{y} \\ &= -|\underline{x}||\underline{y}|\cos\theta.\end{aligned} \tag{4.7}$$

Furthermore comparing the definitions of the outer product between vectors and their cross product (see also 4.1.1) we find easily that inside $\mathbf{R}_{0,3}$

$$\begin{aligned} \left[\underline{x}\,\underline{y}\right]_2 &= \underline{x} \wedge \underline{y} \\ &= -\underline{x} \times \underline{y}\, e_{123}. \end{aligned} \tag{4.8}$$

Notice that for two vectors $\underline{x}$ and $\underline{y}$, $\underline{x}\,\underline{y} = \underline{x} \wedge \underline{y}$ if and only if $\underline{x}.\underline{y} = 0$, i.e. if and only if $\underline{x}$ and $\underline{y}$ are orthogonal, while $\underline{x}\,\underline{y} = \underline{x}.\underline{y}$ if and only if $\underline{x}$ and $\underline{y}$ are parallel.

So we can say that the product between vectors takes account of their relative position where orthogonality results into anticommutativity and parallelism results into commutativity of their product. In particular we have for the basic elements e_1, e_2 and e_3 that

$$e_i \wedge e_i = 0, \quad i = 1, 2, 3$$

and

$$e_i \wedge e_j = e_{ij}, \quad i \neq j, \quad i, j = 1, 2, 3.$$

The relation (4.8) illustrates wonderfully the role of the pseudoscalar e_{123} in $\mathbf{R}_{0,3}$. Indeed, first of all observe that e_{123} commutes with all elements of $\mathbf{R}_{0,3}$. Moreover by multiplying a vector with e_{123} one gets a bivector and vice-versa, multiplying a bivector with e_{123} yields a vector. In particular, as $e_{123}^2 = 1$, we obtain from (4.8) that for vectors $\underline{x}$ and $\underline{y}$:

$$\underline{x} \times \underline{y} = -\underline{x} \wedge \underline{y}\, e_{123} \tag{4.9}$$

As

$$|\underline{x} \wedge \underline{y}| = |\underline{x}||\underline{y}| \sin\theta$$

we may also interpret $\underline{x} \wedge \underline{y}$ as the oriented parallelogram constructed on the vectors $\underline{x}$ and $\underline{y}$.

Let us now introduce the inner and outer product between a vector $\underline{x}$ and a bivector $\underline{X}$. In order to motivate these definitions, let us first observe that both the products $\underline{x}\,\underline{X}$ and $\underline{X}\,\underline{x}$ contain a vector part and a trivector part and, as simple computations show,

$$\underline{x}\,\underline{X} = [\underline{x}\,\underline{X}]_1 + [\underline{x}\underline{X}]_3$$

and

$$\underline{X}\,\underline{x} = [\underline{X}\,\underline{x}]_1 + [\underline{X}\,\underline{x}]_3$$

where

$$[\underline{x}\,\underline{X}]_1 = -\,[\underline{X}\,\underline{x}]_1$$

and

$$[\underline{x}\,\underline{X}]_3 = [\underline{X}\,\underline{x}]_3 .$$

This prompts us to put

$$\begin{aligned} \underline{x}.\underline{X} &= \tfrac{1}{2}(\underline{x}\,\underline{X} - \underline{X}\,\underline{x}) \\ \text{and} & \\ \underline{x} \wedge \underline{X} &= \tfrac{1}{2}(\underline{x}\,\underline{X} + \underline{X}\,\underline{x}) \end{aligned} \tag{4.10}$$

called respectively the *inner* and *outer product* between $\underline{x}$ and $\underline{X}$.

Similarly we put

$$\begin{aligned} \underline{X}.\underline{x} &= \tfrac{1}{2}(\underline{X}\,\underline{x} - \underline{x}\,\underline{X}) \\ \text{and} & \\ \underline{X} \wedge \underline{x} &= \tfrac{1}{2}(\underline{X}\,\underline{x} + \underline{x}\underline{X}), \end{aligned} \tag{4.11}$$

called respectively the *inner* and *outer* product between $\underline{X}$ and $\underline{x}$. As is readily seen, the outer product between vectors and bivectors is commutative and is a trivector while their inner product is anti-commutative and is a vector.

Moreover we have that

$$\begin{aligned} \underline{x}\,\underline{X} &= \underline{x}.\underline{X} + \underline{x} \wedge \underline{X} \\ \text{and} & \\ \underline{X}\,\underline{x} &= \underline{X}.\underline{x} + \underline{X} \wedge \underline{x} \end{aligned} \tag{4.12}$$

Notice that in the case where the bivector $\underline{X}$ is itself the outer product of two nonparallel vectors $\underline{y}$ and $\underline{z}$ (as will be seen in 4.3.4, Proposition 1 any bivector may be written that way), we obtain, taking account of (4.9) :

$$\begin{aligned} \underline{x} \wedge (\underline{y} \wedge \underline{z}) &= \frac{1}{2}(\underline{x}(\underline{y} \wedge \underline{z}) + (\underline{y} \wedge \underline{z})\underline{x}) \\ &= -\frac{1}{2}(\underline{x}(\underline{y} \times \underline{z}) + (\underline{y} \times \underline{z})\underline{x})e_{123} \\ &= -\underline{x}.(\underline{y} \times \underline{z})e_{123} \\ &= \underline{y} \wedge (\underline{z} \wedge \underline{x}) \\ &= \underline{z} \wedge (\underline{x} \wedge \underline{y}) \end{aligned} \tag{4.13}$$

and

$$\begin{aligned} \underline{x}.(\underline{y} \wedge \underline{z}) &= \frac{1}{2}(\underline{x}(\underline{y} \wedge \underline{z}) - (\underline{y} \wedge \underline{z})\underline{x}) \\ &= -\frac{1}{2}(\underline{x}(\underline{y} \times \underline{z}) - (\underline{y} \times \underline{z})\underline{x})e_{123} \\ &= -x \wedge (\underline{y} \times \underline{z})e_{123} \\ &= \underline{x} \times (\underline{y} \times \underline{z}). \end{aligned} \tag{4.14}$$

We so have that if $\underline{x}, \underline{y}$ and $\underline{z}$ are linearly independent vectors, the outer product between $\underline{x}$ and $\underline{y} \wedge \underline{z}$ determines geometrically the oriented volume of the parallelopiped constructed on $\underline{x}, \underline{y}$ and $\underline{z}$, while the inner product between $\underline{x}$ and $\underline{y} \wedge \underline{z}$ yields the so-called double cross product between $\underline{x}, \underline{y}$ and $\underline{z}$.

Now let again $\underline{x}, \underline{y}$ and $\underline{z}$ be three vectors and consider their product $\underline{x}\,\underline{y}\,\underline{z}$. Then from

$$\underline{x}\,\underline{y}\,\underline{z} = (\underline{x}.\underline{y} + \underline{x} \wedge \underline{y})\underline{z} \tag{4.15}$$

we obtain that

$$\begin{aligned}
\left[\underline{x}\,\underline{y}\,\underline{z}\right]_3 &= \left[(\underline{x} \wedge \underline{y})\underline{z}\right]_3 \\
&= -\left[(\underline{x} \times \underline{y})e_{123}\underline{z}\right]_3 \\
&= -\left[(\underline{x} \times \underline{y})\underline{z}e_{123}\right]_3 \\
&= -\left[(\underline{x} \times \underline{y})\underline{z}\right]_0 e_{123} \\
&= -((\underline{x} \times \underline{y}).\underline{z})e_{123} \\
&= x \wedge y \wedge z.
\end{aligned}$$

Consequently, if $\underline{x}, \underline{y}$ and $\underline{z}$ are linearly independent, the pseudoscalar part of $\underline{x}\,\underline{y}\,\underline{z}$ also determines the oriented volume of the parallelopiped constructed on $\underline{x}, \underline{y}$ and $\underline{z}$. From the relations (4.13) it follows immediately that $\underline{x} \wedge (\underline{y} \wedge \underline{z}) = 0$ if and only if the three vectors $\underline{x}, \underline{y}$ and $\underline{z}$ are linearly dependent in which case

$$\underline{x}(\underline{y} \wedge \underline{z}) = \underline{x}.(\underline{y} \wedge \underline{z}). \tag{4.16}$$

Furthermore, from (4.14) it follows that the inner product between the vector $\underline{x}$ and the bivector $\underline{y} \wedge \underline{z}$ is a vector lying in the plane determined by $\underline{y}$ and $\underline{z}$ and orthogonal to $\underline{x}$. Moreover $\underline{x}.(\underline{y} \wedge \underline{z}) = 0$ if and only if $\underline{x}$ and $\underline{y} \times \underline{z}$ are parallel whence $\underline{x}$ is perpendicular to the plane determined by $\underline{y}$ and $\underline{z}$. In that case

$$\underline{x}(\underline{y} \wedge \underline{z}) = \underline{x} \wedge (\underline{y} \wedge \underline{z}).$$

We end up this subsection with the following

Proposition 1. *Let $\underline{y}$ and $\underline{z}$ be linearly independent vectors, let $\underline{X} = \underline{y} \wedge \underline{z}$ and let W be the plane spanned by $\underline{y}$ and $\underline{z}$. Then for each vector $\underline{x}$, the elements $(\underline{x}.\underline{X})X^{-1}$ and $(\underline{x} \wedge \underline{X})\underline{X}^{-1}$ are the orthogonal projections of $\underline{x}$ on, respectively, W and $W^{\perp}$. Consequently*

$$\underline{x} = (\underline{x}.\underline{X})\underline{X}^{-1} + (\underline{x} \wedge \underline{X})\underline{X}^{-1}.$$

Proof. First observe that by virtue of (4.14), $\underline{x}.\underline{X} \in W$, say $\underline{x}.\underline{X} = \lambda\underline{y} + \mu\underline{z}$. Hence, by means of (4.12), (4.13) and (4.14),

$$\begin{aligned}(\underline{x}.\underline{X})\underline{X}^{-1} &= (\lambda\underline{y} + \mu\underline{z})(-\frac{\underline{X}}{|\underline{X}|^2}) \\ &= -\frac{\lambda}{|\underline{X}|^2}(\underline{y}.\underline{X} + \underline{y}\wedge\underline{X}) - \frac{\mu}{|\underline{X}|^2}(\underline{z}.\underline{X} + \underline{z}\wedge\underline{X}) \\ &= -\frac{\lambda}{|\underline{X}|^2}\underline{y}.\underline{X} - \frac{\mu}{|\underline{X}|^2}\underline{z}.\underline{X}\end{aligned}$$

clearly belongs to W. Obviously

$$\begin{aligned}(\underline{x}.\underline{X})\underline{X}^{-1} + (\underline{x}\wedge\underline{X})\underline{X}^{-1} = \underline{x}\,\underline{X}\,\underline{X}^{-1} \\ = \underline{x}\end{aligned}$$

so that it remains to prove (see Appendix) that $\underline{x} - (\underline{x}.\underline{X})X^{-1} = (\underline{x}\wedge\underline{X})\underline{X}^{-1}$ is orthogonal to $\underline{y}$ and $\underline{z}$. But, as by virtue of (4.9) and (4.13),

$$\begin{aligned}(\underline{x}\wedge\underline{X})\underline{X}^{-1} &= -(\underline{x}\wedge(\underline{y}\wedge\underline{z}))\frac{\underline{y}\wedge\underline{z}}{|\underline{y}\wedge\underline{z}|^2} \\ &= -(\underline{x}.(\underline{y}\times\underline{z}))e_{123}e_{123}\frac{\underline{y}\times\underline{z}}{|\underline{y}\times\underline{z}|^2} \\ &= \frac{-1}{|\underline{y}\times\underline{z}|^2}(x.(\underline{y}\times\underline{z}))\underline{y}\times\underline{z}\end{aligned}$$

we indeed have that $(\underline{x}\wedge\underline{x})X^{-1}$ is orthogonal to $\underline{y}$ and $\underline{z}$. ∎

4.3.3 Algebraic structure of $\mathbf{R}_{0,3}$

In 4.3.1 we introduced the even subalgebra $\mathbf{R}_{0,3}^{+}$ of $\mathbf{R}_{0,3}$ consisting of all linear combinations of scalars and bivectors, i.e.

$$\mathbf{R}_{0,3}^{+} = \{\alpha + a_{23}e_{23} + a_{31}e_{31} + a_{12}e_{12} : \alpha, a_{ij} \in \mathbf{R}\}.$$

It is readily seen that this algebra is isomorphic to the algebra $\mathbf{H}$ of real quaternions by identifying e_{23}, e_{31} and e_{12} respectively with $\mathbf{i}$, $\mathbf{j}$ and $\mathbf{k}$. Moreover, the conjugation in $\mathbf{R}_{0,3}$ restricted to $\mathbf{R}_{0,3}^{+}$ coincides with the conjugation in $\mathbf{H}$. The identification of $\mathbf{H}$ with $\mathbf{R}_{0,3}^{+}$ is the way we consider $\mathbf{H}$ to be a subalgebra of $\mathbf{R}_{0,3}$. Of course, since $\mathbf{H}$ is isomorphic to $R_{0,2}$ (see §3) we have that $\mathbf{R}_{0,3}^{+}$ and $\mathbf{R}_{0,2}$ are isomorphic too. Clearly the splitting

$$\mathbf{R}_{0,3} = \mathbf{R}_{0,3}^{+} \oplus \mathbf{R}_{0,3}^{-}$$

can also be written as

$$\mathbf{R}_{0,3} = \mathbf{R}_{0,3}^{+} \oplus e_{123}\mathbf{R}_{0,3}^{+}.$$

where, as we already remarked several times, the element e_{123} commutes with e_1, e_2 and e_3 and hence belongs to the center Z of $\mathbf{R}_{0,3}$.

Easy calculations show that the only possible elements belonging to Z are linear combinations of scalars and pseudoscalars so that

$$Z = \mathbf{R} \oplus \mathbf{R}e_{123}.$$

Since $e_{123}^2 = 1$ we have that the elements

$$\omega_+ = \frac{1}{2}(1 + e_{123})$$

and

$$\omega_- = \frac{1}{2}(1 - e_{123})$$

are mutually annihilating idempotents in $\mathbf{R}_{0,3}$, i.e.

$$\omega_\pm^2 = \omega_\pm \quad \text{and} \quad \omega_+\omega_- = \omega_-\omega_+ = 0.$$

As moreover

$$\omega_+ + \omega_- = 1$$

we obtain that

$$\mathbf{R}_{0,3} = \omega_+\mathbf{R}_{0,3} \oplus \omega_-\mathbf{R}_{0,3},$$

$\omega_\pm\mathbf{R}_{0,3}$ being two-sided ideals in $\mathbf{R}_{0,3}$. This implies that each $a \in \mathbf{R}_{0,3}$ may be written as

$$a = \omega_+a + \omega_-a.$$

Now we claim that for each $a \in \mathbf{R}_{0,3}$, there exist unique elements $a_+, a_- \in \mathbf{R}_{0,3}^{+} \cong \mathbf{H}$ such that

$$a = \omega_+a_+ + \omega_-a_-.$$

Indeed as,

$$\begin{aligned} e_1\omega_+ &= -e_{23}\omega_+ \\ e_2\omega_+ &= -e_{31}\omega_+ \\ e_3\omega_+ &= -e_{12}\omega_+ \end{aligned}$$

and

$$\begin{aligned} e_1\omega_- &= e_{23}\omega_- \\ e_2\omega_- &= e_{31}\omega_- \\ e_3\omega_- &= e_{12}\omega_- \end{aligned}$$

we have that for each $a \in \mathbf{R}_{0,3}$, there ought to exist $a_+, a_- \in \mathbf{R}_{0.3}^+ \cong \mathbf{H}$ such that $\omega_+ a = \omega_+ a_+$ and $\omega_- a = \omega_- a_-$.

As, conversely, for all $a_+, a_- \in \mathbf{R}_{0,3}^+ \cong \mathbf{H}$, $\omega_+ a_+ \in \omega_+ \mathbf{R}_{0,3}$ and $\omega_- a_- \in \omega_- \mathbf{R}_{0,3}$ we have proved that

$$\mathbf{R}_{0,3} \cong \omega_+ \mathbf{H} \oplus \omega_- \mathbf{H}.$$

In fact, because of $\omega_\pm \in Z$, $\omega_\pm^2 = \omega_\pm$ and $\omega_+\omega_- = \omega_-\omega_+ = 0$, we may conclude that

$$\mathbf{R}_{0,3} \cong \mathbf{H} \oplus \mathbf{H}$$

where, in a classical way, the ring structure on $\mathbf{H} \oplus \mathbf{H}$ is given by

$$(q_1, q_2) + (q_1', q_2') = (q_1 + q_1', q_2 + q_2')$$

and

$$(q_1, q_2)(q_1', q_2') = (q_1 q_1', q_2 q_2').$$

Finally notice that if

$$a = \omega_+ a_+ + \omega_- a_-,$$

then

$$\overline{a} = \omega_+ \overline{a}_+ + \omega_- \overline{a}_-$$

and

$$\begin{aligned} a\overline{a} &= \omega_+ a_+ \overline{a}_+ + \omega_- a_- \overline{a}_- \\ &= \omega_+ |a_+|^2 + \omega_- |a_-|^2 \end{aligned}$$

or $a\overline{a}$ is the sum of a scalar and a pseudoscalar. Consequently

$$|a|^2 = [a\overline{a}]_0 = \frac{1}{2}(|a_+|^2 + |a_-|^2)$$

and $a\overline{a} \in \mathbf{R}$ if and only if $|a_+| = |a_-|$, the latter being the case for instance when $a \in \mathbf{R}_{0,3}^+$ (see also 4.3.1).

4.3.4 The groups Pin(3) and Spin(3)

Recalling that $\mathbf{R}^3$ is embedded in $\mathbf{R}_{0,3}$ by identifying $\underline{x} = (x_1, x_2, x_3) \in \mathbf{R}^3$ with $\underline{x} = \sum_{j=1}^{3} x_j e_j$, we have that the unit sphere S^2 of $\mathbf{R}^3$ becomes the set of elements $\underline{\omega} = \sum_{j=1}^{3} \omega_j e_j$ with $|\underline{\omega}| = 1$.

Now consider the following sets :

$$\mathrm{Pin}\,(3) = \{\Pi_{j=1}^{k} \underline{\omega}_j \quad : \quad \underline{\omega}_j \in S^2, k \in \mathbf{N}\}$$

and

$$\mathrm{Spin}\,(3) = \{\Pi_{j=1}^{2k} \underline{\omega}_j \quad : \quad \underline{\omega}_j \in S^2, k \in \mathbf{N}\}.$$

Then it is easily seen that, provided with the multiplication in $\mathbf{R}_{0,3}$, Pin(3) and Spin(3) are subgroups of $\mathbf{R}_{0,3}$ called, respectively, the *Pin* and *Spin groups.* Clearly Spin(3) is a subgroup of Pin(3).

Now we wish to characterize the elements of Spin(3) and O Pin(3), the latter being by definition the set

$$\mathrm{O\,Pin}\,(3) = \mathrm{Pin}\,(3) \backslash \mathrm{Spin}\,(3)$$

Let us start with Spin(3). First of all notice that if $\underline{\omega}_1, \underline{\omega}_2 \in S^2$, then $\underline{\omega}_1 \underline{\omega}_2 \in \mathbf{R}_{0,3}^+$ with $|\underline{\omega}_1 \underline{\omega}_2| = 1$. Hence, as $\mathbf{R}_{0,3}^+$ is identified with $\mathbf{H}$ and the set of elements in $\mathbf{H}$ having norm equal to 1 is nothing else but the group Sp(1) (see also 1.2), we obtain that Spin(3) may be identified with a subgroup of Sp(1).

Conversely, we prove

Proposition 1. *Each element a of $\mathbf{R}_{0,3}^+$ with $|a| = 1$, i.e. each element of Sp(1), may be written as the product of two unit vectors and hence belongs to Spin(3).*

Proof. Write a as $a = \alpha + \underline{X} = \alpha + a_{23}e_{23} + a_{31}e_{31} + a_{12}e_{12}$ with $\alpha^2 + a_{23}^2 + a_{31}^2 + a_{12}^2 = 1$. Then clearly there exists a vector $\underline{x} = x_1 e_1 + x_2 e_2 + x_3 e_3$ such that

$$\begin{cases} x_1^2 + x_2^2 + x_3^2 & = & 1 \\ x_1 a_{23} + x_2 a_{31} + x_3 a_{12} & = & 0, \end{cases}$$

i.e. $\underline{x}$ is a unit vector lying in the plane through the origin given by the equation

$$x a_{23} + y a_{31} + z a_{12} = 0.$$

We so have that $\underline{x} \wedge \underline{X} = 0$ or $\underline{x}\,\underline{X} = \underline{x}.\underline{X}$. Consequently

$$\underline{x} a = \underline{x}(\alpha + \underline{X}) = \underline{x}\alpha + \underline{x}\,\underline{X}$$

is a vector, say $\underline{y}$, and, as simple computations show, $|\underline{y}| = 1$.

We thus have that

$$a = (\alpha + \underline{X}) = \underline{x}^{-1} \underline{y}$$

where $\underline{x}^{-1}$ and $\underline{y}$ are unit vectors ■

So we obtain

Theorem 1. $\text{Spin}(3) \cong \text{Sp}(1) \cong \{a \in \mathbf{R}_{0,3}^{+} : |a| = 1\}$.

The foregoing considerations also lead to

Theorem 2. $\text{Spin}(3) = \{\underline{\omega}\underline{\nu} : \underline{\omega}, \underline{\nu} \in S^2\}$.

Notice that the factorization of each element of Spin(3) into a product of two unit vectors is not unique. Next we prove

Proposition 2. *Each $u \in$ OPin(3) may be written as the product of at most three unit vectors.*

Proof. First observe that if $u \in \text{OPin}(3)$, then u is the product of an odd number of unit vectors, say

$$u = \prod_{j=1}^{2k+1} \underline{\omega}_j$$

whence

$$u = \left(\prod_{j=1}^{2k} \underline{\omega}_j\right) \underline{\omega}_{2k+1}.$$

Now, $s = \prod_{j=1}^{2k} \underline{\omega}_j \in \text{Spin}(3)$, whence by Theorem 2, there exist $\underline{\omega}, \underline{\nu}$ unit vectors such that $s = \underline{\omega}\,\underline{\nu}$ ■

Another characterization of OPin(3) reads as follows.

Theorem 3. $\text{OPin}(3) = \{a \in \mathbf{R}_{0,3}^{-} : |a| = 1\}$.

Proof. Let $u \in \text{OPin}(3)$. Then $u = \prod_{j=1}^{2k+1} \underline{\omega}_j$ for some $k \in \mathbf{N}$, whence $u = s\underline{\omega}_{2k+1}$ with $s = \prod_{j=1}^{2k} \underline{\omega}_j \in \mathbf{R}_{0,3}^{+}$ and $|s| = 1$. Consequently $u \in \mathbf{R}_{0,3}^{-}$ with $|u| = 1$.

Conversely, let $a \in \mathbf{R}_{0,3}^{-}$ with $|a| = 1$ and take any unit vector $\underline{\omega}$. Then $a\,\underline{\omega} \in \mathbf{R}_{0,3}^{+}$ and $|a\,\underline{\omega}| = 1$ or $a\underline{\omega} \in \text{Spin}(3)$. Hence

$$a = (a\,\underline{\omega})\underline{\omega}^{-1} \in \text{Pin}(3) \backslash \text{Spin}(3) = \text{OPin}(3). ■$$

4.3.5 Pin(3) and Spin(3) as vectorial representations of O(3) and SO(3)

Using the main involution introduced in 4.3.1, we associate with each $c \in$ Pin(3) the following mapping $\chi(c)$ from $\mathbf{R}_{0,3}$ into $\mathbf{R}_{0,3}$:

$$\chi(c)(a) = ca\hat{c}^{-1}, \quad a \in \mathbf{R}_{0,3}.$$

Now we claim that $\chi(c)$ restricted to $\mathbf{R}^3$ is an orthogonal transformation in $\mathbf{R}^3$. In order to prove this, first take $c = \underline{\omega}$, $\underline{\omega}$ being a unit vector in $\mathbf{R}^3$ (or $\underline{\omega} \in S^2$). Then, decomposing each vector $\underline{x} \in \mathbf{R}^3$ as

$$\underline{x} = \underline{x}_{\|} + \underline{x}_{\perp}$$

with $\underline{x}_{\|} = \lambda\underline{\omega} \in \mathbf{R}\underline{\omega}$ and $\underline{x}_{\perp} \in \underline{\omega}^{\perp}$ and noticing that $\hat{\underline{\omega}}^{-1} = \underline{\omega}$, we obtain

$$\begin{aligned} \chi(\underline{\omega})(\underline{x}) &= \underline{\omega}\,\underline{x}\,\underline{\omega} \\ &= \underline{\omega}(\lambda\,\underline{\omega} + \underline{x}_{\perp})\underline{\omega} \\ &= -\lambda\underline{\omega} + \underline{\omega}\underline{x}_{\perp}\underline{\omega} \\ &= -\lambda\,\underline{\omega} + \underline{x}_{\perp}. \end{aligned}$$

Here we used the fact that since $\underline{\omega}\perp\underline{x}_{\perp}$,

$$\begin{aligned} \underline{\omega}\underline{x}_{\perp} + \underline{x}_{\perp}\underline{\omega} &= 2\underline{\omega}.\underline{x}_{\perp} \\ &= 0. \end{aligned}$$

Consequently $\chi(\underline{\omega})(x) \in \mathbf{R}^3$ and moreover $\chi(\underline{\omega})$ is nothing else but the reflection $\rho_{\underline{\omega}}$ in the plane $\underline{\omega}^{\perp}$.

As conversely each plane H (through 0) is of the form $H = \underline{\omega}^{\perp}$ for some unit vector $\underline{\omega}$, we have that the set $\{\chi(\underline{\omega}) : \underline{\omega} \in S^2\}$ yields all possible plane reflections of $\mathbf{R}^3$. Now let $c \in$ Pin(3). Then by the foregoing considerations, $\chi(c)$ restricted to $\mathbf{R}^3$ is a linear transformation in $\mathbf{R}^3$ which moreover is the composition of reflections in planes, whence $\chi(c)$ is an orthogonal transformation. Furthermore, as Pin(3) = Spin(3) ∪ OPin(3) we have, taking account of the results in 4.3.4 and the Cartan-Dieudonné Theorem (see Appendix).

Theorem. *Let $c \in$ Pin(3) and let $\chi(c)$ be its associated orthogonal transformation in $\mathbf{R}^3$. Then $\chi(c)$ is a rotation if $c \in$ Spin(3) while $\chi(c)$ is an anti-rotation if $c \in$ OPin(3).*

Conversely, for each $T \in$ O(3) there exists a $c \in$ Pin(3) such that $T = \chi(c)$. If moreover $T \in$ SO(3) then there exist unit vectors $\underline{\omega}, \underline{\nu}$ such that $T = \chi(s)$ with $s = \underline{\omega}\,\underline{\nu}$.

Finally, if $T \in$ O(3)\ SO(3), then there exist at most three unit vectors $\underline{\omega}, \underline{\mu}, \underline{\nu}$ such that $T = \chi(u)$ with $u = \underline{\omega}\,\underline{\mu}\,\underline{\nu}$.

The representation χ : Pin(3) $\to$ O(3) just described is called the *vectorial representation of* O(3), its restriction to Spin(3) being the *vectorial representation of* SO(3).

4.3.6 Spin(3) and SO(3) revisited

Let $s \in$ Spin(3) and let $\underline{\omega}, \underline{\nu} \in S^2$ be such that $s = \underline{\omega}\,\underline{\nu}$ (see also 4.3.4, Theorem 2). Then, as $\underline{\nu}\,\underline{\omega} = \overline{\underline{\omega}\,\underline{\nu}}$, we have that

$$\chi(s)(\underline{x}) = s\underline{x}\bar{s}, \quad \underline{x} \in \mathbf{R}^3$$

is the rotation in $\mathbf{R}^3$ determined by s.

Now we wish to characterize this rotation geometrically ; i.e. what are the axis and angle of that rotation. Obviously, if $\underline{\omega}$ and $\underline{\nu}$ are linearly dependent, i.e. $\underline{\omega} = \pm\nu$, then $\chi(s)$ is nothing else but the identity operator. So from now on suppose that $\underline{\omega}$ and $\underline{\nu}$ are linearly independent. As $\chi(s)$ is the composition of a reflection in $\underline{\nu}^\perp$ and $\underline{\omega}^\perp, \chi(s)$ is a rotation about an axis carried by the line $\underline{\omega}^\perp \cap \underline{\nu}^\perp$. Hence, putting $w = \underline{\nu}\times\underline{\omega}$ and taking account of

$$\begin{aligned} \underline{\nu}\times\underline{\omega} &= -\underline{\nu} \wedge \underline{\omega}\, e_{123} \\ &= \underline{\omega} \wedge \underline{\nu}\, e_{123} \end{aligned}$$

we obtain that $\chi(s)$ is a rotation about the vector $\underline{w} = \underline{\omega} \wedge \underline{\nu}\, e_{123}$.

As to the angle of rotation, observe that

$$s = \underline{\omega}\,\underline{\nu} = \underline{\omega}.\underline{\nu} + \underline{\omega} \wedge \underline{\nu}$$

so that, by putting $\underline{X} = \frac{\underline{\omega}\wedge\underline{\nu}}{|\underline{\omega}\wedge\underline{\nu}|}$,

$$s = \underline{\omega}.\underline{\nu} + |\underline{\omega} \wedge \underline{\nu}|\underline{X}.$$

Now let $\theta \in [0, \pi]$ be the angle between $\underline{\nu}$ and $\underline{\omega}$; then

$$s = -\cos\theta + \underline{X}\sin\theta.$$

Using 4.3.2 Proposition 1, we may decompose each vector $\underline{x} \in \mathbf{R}^3$ into

$$\begin{aligned} \underline{x} &= (\underline{x}.\underline{X})\underline{X}^{-1} + (\underline{x} \wedge \underline{X})\underline{X}^{-1} \\ &= x_W + x_{W^\perp} \end{aligned}$$

where W is the plane (through 0) spanned by $\underline{\omega}$ and $\underline{\nu}$ and $\underline{X}^{-1} = -\underline{X}$.

Now, as $\underline{x}_{W^\perp} = \alpha \underline{X} e_{123}$ for some non zero scalar α, we have that $\underline{x}_{W^\perp}$ and $\underline{X}$ commute while, as $\underline{x}_W = \beta\underline{\omega} + \gamma\underline{\nu}$ for some $\beta, \gamma \in \mathbf{R}$, we have that $\underline{x}_W$ and $\underline{X}$ anticommute. Hence

$$\begin{aligned}
\chi(s)(\underline{x}) &= (-\cos\theta + \underline{X}\sin\theta)(\underline{x}_W + \underline{x}_{W^\perp})(-\cos\theta - \underline{X}\sin\theta) \\
&= (-\cos\theta + \underline{X}\sin\theta)\underline{x}_W(-\cos\theta - \underline{X}\sin\theta) \\
&\quad +(-\cos\theta + \underline{X}\sin\theta)\underline{x}_{W^\perp}(-\cos\theta - \underline{X}\sin\theta) \\
&= (\cos 2\theta - \underline{X}\sin 2\theta)\underline{x}_W + \underline{x}_{W^\perp} \\
&= (\cos(-2\theta) + \underline{X}\sin(-2\theta))\underline{x}_W + \underline{x}_{W^\perp}.
\end{aligned}$$

These calculations confirm that $\chi(s)$ is a rotation about the axis $\underline{w} = \underline{\omega} \wedge \underline{\nu} e_{123}$ and this through the angle -2θ. Indeed, it suffices to examine how $\underline{v}$ is transformed. We have

$$\begin{aligned}
\chi(s)(\underline{\nu}) &= (\cos(-2\theta) + \underline{X}\sin(-2\theta))\underline{\nu} \\
&= \underline{\nu}\cos(-2\theta) + \sin(-2\theta)\underline{X}\,\underline{\nu} \\
&= \underline{\nu}(\cos 2\theta + \sin 2\theta\underline{X}) \\
&= \underline{\nu}\cos 2\theta + \sin 2\theta\underline{\nu}\,\underline{X} \\
&= \underline{\nu}\cos 2\theta + \sin 2\theta\underline{\nu}.\underline{X} \\
&= \nu\cos(-2\theta) + \sin(-2\theta)\underline{\nu} \times \frac{(\underline{\nu} \times \underline{\omega})}{|\underline{\nu} \times \underline{\omega}|}.
\end{aligned}$$

4.4 The Kustaanheimo-Stiefel transformation

Let $q = a_0 + a_{23}e_{23} + a_{31}e_{31} + a_{12}e_{12}$ be an arbitrary element of $\mathbf{R}_{0,3}^+ \cong \mathbf{H}$ and consider the element $T(q) = qe_3\overline{q}$. Then $T(q) \in \mathbf{R}^3$ (see also 4.2). Hence, identifying $\mathbf{R}_{0,3}^+$ with $\mathbf{R}^4$, we have that

$$T : \mathbf{R}^4 \to \mathbf{R}^3$$

with

$$T : q \to \underline{x} = T(q) = qe_3\overline{q}$$

is a (quadratic) transformation from $\mathbf{R}^4$ into $\mathbf{R}^3$. It is called the *Kustaanheimo-Stiefel transformation.*

Let us now have a closer look at those $p, q \in \mathbf{R}^4$ for which $T(p) = T(q)$. To this end first observe that $T(p) = T(q)$ implies that $|p| = |q|$ whence we may restrict to

$p, q \in \mathbf{R}^4$ having unit length or to $\mathrm{Sp}(1) \cong \mathrm{Spin}(3)$. Now take $s, t \in \mathrm{Spin}(3)$ and put $u = \bar{s}t$. Then the condition $s e_3 \bar{s} = t e_3 \bar{t}$ implies that $u e_3 \bar{u} = e_3$. This means that $u \in \mathrm{Spin}(3)$, considered as a rotation in $\mathbf{R}^3$, leaves e_3 fixed ; hence $u \in \mathrm{Spin}(2)$.

As Spin(3) acts transitively on S^2 and its stabilizer subgroup fixing e_3 is Spin(2), we may consider the homogeneous space

$$S^2 = \mathrm{Spin}(3)/\mathrm{Spin}(2) = S^3/S^1.$$

The following link can now be made with the famous Hopf fibration (see e.g. [32]). To this end, first recall the identifications

$$\begin{aligned} \mathrm{Spin}(3) &\cong S^3 \\ &\cong \{q \in \mathbf{H} : |q| = 1\} \\ &\cong \left\{ \begin{pmatrix} z & w \\ -\overline{w} & \overline{z} \end{pmatrix} : z, w \in \mathbf{C}, z\overline{z} + w\overline{w} = 1 \right\} \end{aligned}$$

where $z\overline{z} + w\overline{w} = 1$ implies that we may write $z, w \in \mathbf{C}$ as $z = \cos\alpha e^{i\varphi}$, $w = \sin\alpha e^{-i\psi}$, $0 \leq \alpha \leq \pi/2$, $\varphi, \psi \in \mathbf{R}$.

Next, as $u \in$ Spin(2) is of the form $u = \cos\theta/2 + \sin\theta/2\, e_{12}$, $\theta \in \mathbf{R}$, we obtain, by substituting e_{12} by its matrix representation $\begin{pmatrix} i & \\ & -i \end{pmatrix}$, that u may be written as

$$u = \begin{pmatrix} e^{i\theta/2} & \\ & e^{-i\theta/2} \end{pmatrix}.$$

Hence, for $s, t \in$ Spin(3), the requirement $T(s) = T(t)$, or equivalently $t = su$ for some $u \in$ Spin(2), gives rise to, after identifying t and s, respectively, with

$$\begin{pmatrix} z' & w' \\ -\overline{w}' & \overline{z}' \end{pmatrix} \quad \text{and} \quad \begin{pmatrix} z & w \\ -\overline{w} & \overline{z} \end{pmatrix},$$

$$\begin{pmatrix} z' & w' \\ -\overline{w}' & \overline{z}' \end{pmatrix} = \begin{pmatrix} z & w \\ -\overline{w} & \overline{z} \end{pmatrix} \begin{pmatrix} e^{i\theta/2} & \\ & e^{-i\theta/2} \end{pmatrix}$$

or

$$z' = z\, e^{i\theta/2} = \cos\alpha\, e^{i(\varphi+\theta/2)}$$

and

$$w' = w\, e^{-i\theta/2} = \sin\alpha\, e^{-i(\psi+\theta/2)}.$$

5 The Clifford algebra $\mathbf{R}_{0,4}$

In this section we construct the Clifford algebra $\mathbf{R}_{0,4}$, i.e. a real linear associative algebra of maximal dimension containing $\mathbf{R}$ and $\mathbf{R}^4$ such that for each vector $\underline{x} \in \mathbf{R}^4$, $\underline{x}^2 = -|\underline{x}|^2$.

5.1 *Definitions*

If $e = (\underline{e}_1, \underline{e}_2, \underline{e}_3, \underline{e}_4)$ is an orthonormal basis of $\mathbf{R}^4$ then the above requirement implies

$$\underline{e}_j^2 = -1, \quad j = 1, \ldots, 4$$

and

$$\underline{e}_i \underline{e}_j + \underline{e}_j \underline{e}_i = 0, \quad i \neq j.$$

A basis of $\mathbf{R}_{0,4}$ then consists of the elements e_A with $A \subset \{1, \ldots, 4\}$ ordered in a natural way and, if $A = \{\alpha_1, \ldots, \alpha_h\}$,

$$e_A = \underline{e}_{\alpha_1} \underline{e}_{\alpha_2} \cdots \underline{e}_{\alpha_h}.$$

For $A = \emptyset$ we put

$$e_\emptyset = 1,$$

the identity in $\mathbf{R}_{0,4}$. So $\mathbf{R}_{0,4}$ has dimension $2^4 = 16$. For $k = 0, 1, \ldots, 4$, we put

$$\mathbf{R}_{0,4}^k = \text{span}_{\,\mathbf{R}}\{e_A : ^{\#} A = k\}$$

and call it the space of k-vectors.

Any element $a \in \mathbf{R}_{0,4}$ may thus be written as

$$a = \sum_{k=0}^{4} [a]_k$$

with $[a]_k$ the projection of a onto $\mathbf{R}_{0,4}^k$. The spaces $\mathbf{R}$ and $\mathbf{R}^4$ are identified with, respectively, $\mathbf{R}1 = \mathbf{R}_{0,4}^0$ and $\mathbf{R}_{0,4}^1$.

So for $a \in \mathbf{R}_{0,4}$ we have that

$[a]_0 = a_0 1 \in \mathbf{R}$ is a scalar ;

$[a]_1 = a_1 e_1 + \ldots + a_4 e_4$ is a 1-vector or vector ;

$[a]_2 = a_{23}e_{23} + a_{31}e_{31} + a_{12}e_{12} + a_{14}e_{14} + a_{24}e_{24} + a_{34}e_{34}$ is a 2-vector or bi-vector ;

$[a]_3 = a_{234}e_{234} + a_{341}e_{341} + a_{412}e_{412} + a_{123}e_{123}$ is a 3-vector or a tri-vector ;

$[a]_4 = a_{1234}e_{1234}$ is a 4-vector or pseudoscalar.

We introduce the following involutions on $\mathbf{R}_{0,4}$. Putting

$$\hat{e}_A = (-1)^{\#A} e_A$$

and extending the operator $\hat{}$ on $\mathbf{R}_{0,4}$ by linearity, it is called the *main-involution.* Calling

$$\mathbf{R}_{0,4}^+ = \operatorname{span}_{\mathbf{R}}\{e_A :^{\#} A \text{ is even}\}$$

and

$$\mathbf{R}_{0,4}^- = \operatorname{span}_{\mathbf{R}}\{e_A :^{\#} A \text{ is odd}\}$$

we have that

$$\mathbf{R}_{0,4} = \mathbf{R}_{0,4}^+ \oplus \mathbf{R}_{0,4}^-$$

with $\hat{a} = a$, $a \in \mathbf{R}_{0,4}^+$ and $\hat{a} = -a$, $a \in \mathbf{R}_{0,4}^-$. The subalgebra $\mathbf{R}_{0,4}^+$ of $\mathbf{R}_{0,4}$ is called the *even subalgebra.*

Next, if $A = \{\alpha_1, \ldots, \alpha_h\}$, we put for the basic element $e_A, \tilde{e}_A = e_{\alpha_h} e_{\alpha_{h-1}} \cdots e_{\alpha_1}$ and extend the operator $\tilde{}$ by linearity to $\mathbf{R}_{0,4}$. It is called the *reversion* and it is an anti-involution on $\mathbf{R}_{0,4}$.

Finally, putting for all $a \in \mathbf{R}_{0,4}$,

$$\overline{a} = (\tilde{a})\hat{} = (\hat{a})\tilde{},$$

the operator $\bar{}$ is also an anti-involution on $\mathbf{R}_{0,4}$, called *conjugation.*

By means of the conjugation, a norm may be defined on $\mathbf{R}_{0,4}$ by

$$|a|^2 = [a\overline{a}]_0 = [\overline{a}a]_0 = \sum_A a_A^2.$$

Notice that for vectors $\underline{x}, \underline{y} \in \mathbf{R}^4$ we have :

$$\underline{x}\,\underline{y} = \underline{x}.\underline{y} + \underline{x} \wedge \underline{y}$$

where

$$\underline{x}.\underline{y} = -\sum_{j=1}^{4} x_j y_j = - < \underline{x}, \underline{y} >$$

and

$$\underline{x} \wedge \underline{y} = \sum_{i<j} (x_i y_j - x_j y_i) e_{ij}.$$

It is clear that

$$\underline{x}.\underline{y} = \frac{1}{2}(\underline{x}\,\underline{y} + \underline{y}\,\underline{x}) = \left[\underline{x}\,\underline{y}\right]_0$$

and

$$\underline{x} \wedge \underline{y} = \frac{1}{2}(\underline{x}\,\underline{y} - \underline{y}\,\underline{x}) = \left[\underline{x}\,\underline{y}\right]_2;$$

these expressions are called, respectively, the "inner" and "outer" product between vectors.

A bivector of the form $\underline{x} \wedge \underline{y}$ with $\underline{x}$ and $\underline{y}$ vectors is called a 2-blade. Notice that not every bivector is a 2-blade. Indeed, for a general bivector b we have that

$$b^2 = \left[b^2\right]_0 + \left[b^2\right]_2 + \left[b^2\right]_4$$

with

$$\begin{aligned} \left[b^2\right]_0 &= -\sum_{i<j} b_{ij}^2; \\ \left[b^2\right]_2 &= 0; \\ \left[b^2\right]_4 &= (b_{23}b_{14} + b_{31}b_{24} + b_{12}b_{34})e_{1234}. \end{aligned}$$

On the other hand, if $b = \underline{x} \wedge \underline{y}$ is a 2-blade we obtain, by decomposing $\mathbf{R}^4$ into

$$\mathbf{R}^4 = \mathbf{R}\underline{x} \oplus \underline{x}^\perp$$

and $\underline{y}$ into

$$\underline{y} = \underline{y}_{\|} + \underline{y}_{\perp},$$

that

$$b = \underline{x} \wedge \underline{y} = \underline{x} \wedge \underline{y}_{\perp} = \underline{x}\,\underline{y}_{\perp}.$$

Consequently

$$b^2 = \underline{x}\,\underline{y}_{\perp}\underline{x}\,\underline{y}_{\perp} = -\underline{x}\,\underline{y}_{\perp}\underline{x} = -|\underline{x}|^2|\underline{y}_{\perp}|^2 = -|\underline{x} \wedge \underline{y}|^2,$$

and so 2-blades satisfy the Plücker relation

$$b_{23}b_{14} + b_{31}b_{24} + b_{12}b_{34} = 0$$

or simply

$$\left[b^2\right]_4 = 0.$$

Notice however that every bivector b can be written as

$$b = \underline{x} \wedge \underline{y} + \underline{u} \wedge \underline{v}$$

where $\underline{x}, \underline{y}, \underline{u}, \underline{v}$ are vectors such that the planes $\text{span}_{\mathbf{R}}\{\underline{x}, \underline{y}\}$ and $\text{span}_{\mathbf{R}}\{\underline{u}, \underline{v}\}$ are orthogonal. Finally observe that the 2-blade $b = \underline{x} \wedge \underline{y}$ determines the oriented 2-dimensional volume element in $\mathbf{R}^4$.

Next consider vectors $\underline{x}, \underline{y}, \underline{z} \in \mathbf{R}^4$. Then we put

$$\underline{x} \wedge \underline{y} \wedge \underline{z} = \left[\underline{x}\, \underline{y}\, \underline{z}\right]_3 \in \mathbf{R}^3_{0,4}.$$

The element $\underline{x} \wedge \underline{y} \wedge \underline{z}$ clearly determines the oriented 3-dimensional volume element in $\mathbf{R}^4$; it is called a 3-blade. Notice that every trivector is a 3-blade.

Finally, for four vectors $\underline{x}, \underline{y}, \underline{z}, \underline{u} \in \mathbf{R}^4$ we put

$$\underline{x} \wedge \underline{y} \wedge \underline{z} \wedge \underline{u} = \left[\underline{x}\, \underline{y}\, \underline{z}\, \underline{u}\right]_4 \in \mathbf{R}^4_{0,4}.$$

The pseudo-scalar $\underline{x} \wedge \underline{y} \wedge \underline{z} \wedge \underline{u}$ clearly determines the oriented 4-dimensional volume in $\mathbf{R}^4$; it is called a 4-blade. For further properties concerning k-blades and the "outer" product between vectors we refer to [36] and I.1.7.1.

5.2 Algebraic structure of $\mathbf{R}_{0,4}$

It is easy to check that the center Z of $\mathbf{R}_{0,4}$ equals $\mathbf{R}1$. Furthermore the even subalgebra $\mathbf{R}^+_{0,4}$ of $\mathbf{R}_{0,4}$ is isomorphic to $\mathbf{R}_{0,3}$. Indeed, putting

$$\varepsilon_1 = e_{14}, \quad \varepsilon_2 = e_{24} \quad \text{and} \quad \varepsilon_3 = e_{34}$$

we have that $\{\varepsilon_j : j = 1, 2, 3\}$ generates $\mathbf{R}^+_{0,4}$ with

$$\varepsilon_j^2 = -1$$

$$\varepsilon_j \varepsilon_k + \varepsilon_k \varepsilon_j = 0.$$

Taking account of the results obtained in the foregoing section we thus have

$$\mathbf{R}^+_{0,4} \cong \mathbf{R}_{0,3} \cong \mathbf{H} \oplus \mathbf{H}$$

where for $a \in \mathbf{R}^+_{0,4}$,

$$a = \omega_+ a_+ + \omega_- a_-$$

with

$$\begin{aligned} \omega_\pm &= \frac{1}{2}(1 \pm \varepsilon_{123}) \\ &= \frac{1}{2}(1 \pm e_{1234}) \end{aligned}$$

and

$$\begin{aligned} a_\pm \in \mathbf{R}_{0,3}^+ &= \operatorname{span}\{1, \varepsilon_{23}, \varepsilon_{31}, \varepsilon_{12}\} \\ &= \operatorname{span}\{1, e_{23}, e_{31}, e_{12}\} \\ &\cong \mathbf{H}. \end{aligned}$$

Identifying the element $a = (a_+, a_-)$ with the 2×2-matrix $\begin{pmatrix} a_+ & \\ & a_- \end{pmatrix}$ over $\mathbf{H}$ we thus have that $\mathbf{R}_{0,4}^+ \cong \mathbf{H} \oplus \mathbf{H}$ is isomorphic to the algebra of 2×2 matrices over $\mathbf{H}$ of the form $\begin{pmatrix} q_1 & \\ & q_2 \end{pmatrix}$, $q_j \in \mathbf{H}$.

The elements ω_+ and ω_- are represented respectively by $\begin{pmatrix} 1 & 0 \\ 0 & 0 \end{pmatrix}$ and $\begin{pmatrix} 0 & 0 \\ 0 & 1 \end{pmatrix}$ whence $\varepsilon_{123} = \omega_+ - \omega_-$ is represented by $\begin{pmatrix} 1 & \\ & -1 \end{pmatrix}$. Furthermore, as $a \in \mathbf{R}_{0,3}^+$ if and only if $a_+ = a_-$, we obtain that $\mathbf{R}_{0,3}^+ \cong \mathbf{H}$ is represented by the algebra of matrices of the form $\begin{pmatrix} q & \\ & q \end{pmatrix}$, $q \in \mathbf{H}$.

Recalling that $\mathbf{R}_{0,3} = \mathbf{R}_{0,3}^+ \oplus \varepsilon_{123}\mathbf{R}_{0,3}^+$ we find that

$$\begin{aligned} \mathbf{R}_{0,3} &\cong \mathbf{H} \oplus \varepsilon_{123}\mathbf{H} \\ &\cong \mathbf{H}I \oplus \mathbf{H}I' \end{aligned}$$

with

$$I = \begin{pmatrix} 1 & \\ & 1 \end{pmatrix} \quad \text{and} \quad I' = \begin{pmatrix} 1 & \\ & -1 \end{pmatrix}.$$

Returning to $\mathbf{R}_{0,4}$ we have that, as

$$\mathbf{R}_{0,4} = \mathbf{R}_{0,4}^+ \oplus e_4 \mathbf{R}_{0,4}^+$$

each element $a \in \mathbf{R}_{0,4}$ may be written as

$$a = \alpha_1 + e_4 \alpha_2$$

with $\alpha_1, \alpha_2 \in \mathbf{R}_{0,4}^+ \cong \mathbf{R}_{0,3}$ or

$$\alpha_1 = a_1 + \varepsilon_{123} a_2$$

and

$$\alpha_2 = a_3 + \varepsilon_{123} a_4,$$

$$a_1, \ldots, a_4 \in \mathbf{H} \cong \mathbf{R}_{0,3}^+.$$

But, observing that in $\mathbf{R}_{0,4}$, $\mathbf{R}_{0,3}^+$ commutes as well with ε_{123} as with e_4, we obtain that

$$\mathbf{R}_{0,4} \cong \mathbf{H} \otimes_{\mathbf{R}} \operatorname{span}_{\mathbf{R}}\{1, \varepsilon_{123}, e_4, e_4\varepsilon_{123}\}$$

where $\operatorname{span}_{\mathbf{R}}\{1, \varepsilon_{123}, e_4, e_4\varepsilon_{123}\}$ is a subalgebra of $\mathbf{R}_{0,4}$ which is isomorphic to $\mathbf{R}(2)$, the algebra of 2×2 matrices over $\mathbf{R}$.

Indeed, putting $\eta_1 = \varepsilon_{123}$, $\eta_2 = e_4$ we have that $\eta_1^2 = 1$, $\eta_2^2 = -1$, $\eta_1\eta_2 + \eta_2\eta_1 = 0$. So representing η_1 by $\begin{pmatrix} 1 & \\ & -1 \end{pmatrix}$ and η_2 by $\begin{pmatrix} & 1 \\ -1 & \end{pmatrix}$, the algebra $\operatorname{span}_{\mathbf{R}}\{1, \eta_1, \eta_2, \eta_1\eta_2\}$ generated by $\{\eta_1, \eta_2\}$ is isomorphic to the one generated by $\left\{\begin{pmatrix} 1 & \\ & -1 \end{pmatrix}, \begin{pmatrix} & 1 \\ -1 & \end{pmatrix}\right\}$, i.e. $\mathbf{R}(2)$. So we finally obtain that

$$\begin{aligned} \mathbf{R}_{0,4} &\cong \mathbf{H} \otimes_{\mathbf{R}} \operatorname{span}_{\mathbf{R}}\{1, \varepsilon_{123}, e_4, e_4\varepsilon_{123}\} \\ &\cong \mathbf{H} \otimes_{\mathbf{R}} \mathbf{R}(2) \\ &\cong \mathbf{H}(2). \end{aligned}$$

where $\mathbf{H}(2)$ is the (real) algebra of 2×2 matrices over $\mathbf{H}$.

Now let us have a closer look at the identification made above, i.e.

$$\begin{aligned} a &= a_1 \begin{pmatrix} 1 & \\ & 1 \end{pmatrix} + a_2 \begin{pmatrix} 1 & \\ & -1 \end{pmatrix} + a_3 \begin{pmatrix} & 1 \\ -1 & \end{pmatrix} + a_4 \begin{pmatrix} & 1 \\ 1 & \end{pmatrix} \\ &= \begin{pmatrix} a_1 + a_2 & a_3 + a_4 \\ -a_3 + a_4 & a_1 - a_2 \end{pmatrix} \end{aligned}$$

with $a_j \in \mathbf{H}$. Defining on $\mathbf{H}(2)$ the anti-involution

$$\begin{pmatrix} q_1 & q_2 \\ q_3 & q_4 \end{pmatrix} \longrightarrow \overline{\begin{pmatrix} q_1 & q_2 \\ q_3 & q_4 \end{pmatrix}} = \begin{pmatrix} \overline{q}_1 & \overline{q}_3 \\ \overline{q}_2 & \overline{q}_4 \end{pmatrix}$$

where $\overline{q}_j$ is the conjugate of q_j in $\mathbf{H}$, we obtain for $\overline{a}$ that

$$\overline{a} = \begin{pmatrix} 1 & \\ & 1 \end{pmatrix} \overline{a}_1 + \begin{pmatrix} 1 & \\ & -1 \end{pmatrix} \overline{a}_2 + \begin{pmatrix} & -1 \\ 1 & \end{pmatrix} \overline{a}_3 + \begin{pmatrix} & 1 \\ 1 & \end{pmatrix} \overline{a}_4$$

and, as is easily seen, this corresponds exactly to taking conjugation in $\mathbf{R}_{0,4}$ as defined in the foregoing subsection.

5.3 *The group Spin(4)*

Embedding $\mathbf{R}^4$ in $\mathbf{R}_{0,4}$ by identifying $\underline{x} = (x_1, x_2, x_3, x_4) \in \mathbf{R}^4$ with $\underline{x} = \sum_{j=1}^4 e_j x_j$ we have that the unit sphere S^3 of $\mathbf{R}^4$ consists of the set of elements $\underline{\omega} = \sum_{j=1}^4 e_j \omega_j$ with $|\omega| = 1$. Now defining

$$\text{Spin}(4) = \{\prod_{j=1}^{2k} \underline{\omega} : \underline{\omega}_j \in S^3, \quad k \in \mathbf{N}\}$$

we obtain a subgroup of $\mathbf{R}_{0,4}$, called the Spingroup. Obviously, as for each pair $\underline{\omega}_1, \underline{\omega}_2 \in S^3$, $\underline{\omega}_1\underline{\omega}_2 \in \mathbf{R}_{0,4}^+$,

$$\text{Spin}(4) \subset \mathbf{R}_{0,4}^+ \cong \mathbf{R}_{0,3} \cong \mathbf{H} \oplus \mathbf{H}.$$

Moreover, putting $s = \prod_{j=1}^{2k} \underline{\omega}_j \in$ Spin(4), we have that $\bar{s}s = 1$.

Using the results of the foregoing subsection we thus have that, by identifying each $s \in$ Spin(4) with $\begin{pmatrix} q_1 & \\ & q_2 \end{pmatrix}$, $q_j \in \mathbf{H}$, the relation $\bar{s}s = 1$ gives rise to

$$\begin{pmatrix} \bar{q}_1 q_1 & \\ & \bar{q}_2 q_2 \end{pmatrix} = \begin{pmatrix} 1 & \\ & 1 \end{pmatrix},$$

i.e.

$$\text{Spin}(4) \subset \text{Sp}(1) \times \text{Sp}(1) \cong \text{Spin}(3) \times \text{Spin}(3).$$

There is even more - the proof of which is omitted here - namely (see e.g. [63], [23])

$$\text{Spin}(4) \cong \text{Spin}(3) \times \text{Spin}(3).$$

5.4 *The algebra* H *and rotations in* $\mathbf{R}^4$

In this subsection we show how quaternions may be used to represent rotations in $\mathbf{R}^4$. To this end let us first give some supplementary properties of quaternions.

Lemma 5.4.1. *For each $q \in \mathbf{H}$ there exists $b \in Ve\mathbf{H}$ such that $qb \in Ve\mathbf{H}$.*

Proof. Choose $b \in Ve\mathbf{H}$ such that $b.Ve\, q = 0$, i.e. $b \perp Ve\, q \in \mathbf{R}^3$.

then

$$\begin{aligned} qb &= (Re\, q)b + (Ve\, q)b \\ &= (Re\, q)b + (Ve\, q) \times b \end{aligned}$$

indeed belongs to $Ve\mathbf{H}$. ■

Lemma 5.4.2. *Let $q \in \mathbf{H}$ with $|q| = 1$. Then there exist non-zero $a, b \in \mathbf{H}$ such that $q = aba^{-1}b^{-1}$.*

Proof. By virtue of Lemma 5.4.1, there exists $b \in Ve\mathbf{H}$ such that $qb \in Ve\mathbf{H}$. As $|q| = 1$, $|qb| = |b|$, whence there exists a rotation T in $\mathbf{R}^3$ such that $T(b) = qb$. By 4.2.1, there ought to exist $a \in \mathbf{H}$ with $|a| = 1$ such that $T(\underline{x}) = T_a(\underline{x}) = a\underline{x}\bar{a}$ for all $\underline{x} \in Ve\mathbf{H}$. In particular, $qb = ab\bar{a} = aba^{-1}$ and so $q = aba^{-1}b^{-1}$. ■

Henceforth we identify $\mathbf{R}^4$ with $\mathbf{H}$ and write $x = (x_0, x_1, x_2, x_3) \in \mathbf{R}^4$ as

$$x = x_0\mathbf{1} + x_1\mathbf{i} + x_2\mathbf{j} + x_3\mathbf{k}.$$

Theorem 5.4.3. *Let $q, r \in \mathbf{H}$ with $|q| = |r| = 1$. Then the maps Φ_q, $\Psi_r : \mathbf{R}^4 \to \mathbf{R}^4$ given by $\Phi_q(x) = qx$ and $\Psi_r(x) = xr$ define rotations in $\mathbf{R}^4$.*

Proof. Let us look at the case Φ_q, the proof for the case Ψ_r being similar. Clearly Φ_q is linear and as for each $x \in \mathbf{R}^4$,

$$\begin{aligned} |\Phi_q(x)|^2 &= \Phi_q(x)\overline{\Phi_q(x)} \\ &= (qx)(\overline{qx}) \\ &= qx\overline{x}\,\overline{q} \\ &= |x|^2 \end{aligned}$$

we have that Φ_q is an orthogonal transformation. Finally, as by virtue of Lemma 5.4.2, there exist $a, b \in \mathbf{H}\backslash\{0\}$ such that $q = aba^{-1}b^{-1}$, we have that $\det_{\mathbf{R}}(\Phi_q) = 1$ or Φ_q is a rotation. ■

Theorem 5.4.4

(i) Let $q, r \in \mathbf{H}$ with $|q| = |r| = 1$. Then $\Theta_{q,r} : \mathbf{R}^4 \to \mathbf{R}^4$ given by $\Theta_{q,r} = \Phi_q \circ \Phi_{\bar{r}} = \Phi_{\bar{r}} \circ \Phi_q$ is a rotation in $\mathbf{R}^4$.

(ii) Let T be a rotation in $\mathbf{R}^4$. Then there exist $q, r \in \mathbf{H}$ with $|q| = |r| = 1$ such that $T = \Theta_{q,r}$.

Proof.

(i) follows immediately from Theorem 5.4.1.

(ii) Let T be a rotation in $\mathbf{R}^4$ and put $s = T(1)$. Then $|s| = 1$ and the rotation $\Phi_{\bar{s}} \circ T$ leaves 1, and hence $Re\mathbf{H}$ invariant. Consequently, for each $x \in \mathbf{R}^4$,

$$\Phi_{\bar{s}} \circ T(x) = \Phi_{\bar{s}} \circ T(Re\, x) + \Phi_{\bar{s}} \circ T(Ve\, x)$$

$$= \quad Re\, x + \Phi_{\bar{s}} \circ T(Ve\, x)$$

and clearly $\Phi_{\bar{s}} \circ T$, restricted to $\mathbf{R}^3 \cong Ve\mathbf{H}$, is a rotation in $\mathbf{R}^3$. Hence, by means of 4.2.1, there exists $r \in \mathbf{H}$ with $|r| = 1$ such that for all $x \in \mathbf{H}$,

$$\Phi_{\bar{s}} \circ T(Ve\, x) = r(Ve\, x)\bar{r}.$$

But then, for all $x \in \mathbf{H}$,

$$\begin{aligned} rx\bar{r} &= r(Re\, x)\bar{r} + r(Ve\, x)\bar{r} \\ &= \Phi_{\bar{s}} \circ T(x) \end{aligned}$$

and so

$$\begin{aligned} T(x) &= \Phi_s(rx\bar{r}) \\ &= srx\bar{r} \\ &= qx\bar{r} \end{aligned}$$

with $q = sr$.

This indeed proves that $T = \Theta_{q,r}$ for some $q, r \in \mathbf{H}$ with $|q| = |r| = 1$. ∎

Notice that, since any antirotation in $\mathbf{R}^4$ is the composition of any fixed antirotation with a rotation and moreover conjugation in $\mathbf{H}$ is an antirotation, any antirotation in $\mathbf{R}^4$ may be represented by quaternions too.

Appendix : Euclidean m-space

This appendix essentially aims to recall some basic notations, terminology and results concerning the Euclidean m-space $\mathbf{R}^m$ which will be currently used throughout Chapter 0.

A.1. The oriented space $\mathbf{R}^m$

As usual $\mathbf{R}^m$ stands for the set of elements $\underline{x} = (x_1, \ldots, x_m)$, $x_j \in \mathbf{R}$, $j = 1, \ldots, m$, called *vectors*, on which addition and multiplication with real scalars are defined in the usual way by

$$\begin{aligned} \underline{x} + \underline{y} &= (x_1 + y_1, \ldots, x_m + y_m) \\ \lambda \underline{x} &= (\lambda x_1, \ldots, \lambda x_m) \end{aligned}$$

for all $\underline{x} = (x_1, \ldots, x_m)$, $\underline{y} = (y_1, \ldots, y_m) \in \mathbf{R}^m$ and $\lambda \in \mathbf{R}$.

In the m-dimensional real vector space $\mathbf{R}^m$, 1-dimensional, 2-dimensional and $(m-1)$-dimensional subspaces are called, respectively, *lines, planes* and *hyperplanes* (through the origin). Let $e = (\underline{e}_1, \underline{e}_2, \ldots, \underline{e}_m)$ and $e' = (\underline{e}'_1, \underline{e}'_2, \ldots, \underline{e}'_m)$ be bases of $\mathbf{R}^m$ and let $Q = (Q_{ij})$ be the transition matrix from e to e', i.e.

$$e' = eQ$$

or

$$\underline{e}'_j = \sum_{i=1}^{m} Q_{ij} \underline{e}_j, \quad j = 1, \ldots, m.$$

Then e and e' are said to determine the *same orientation* of $\mathbf{R}^m$ if $\det Q > 0$.

It is clear that this relation gives rise to an equivalence relation on the set of all bases of $\mathbf{R}^m$, whence a partition of it into two disjoint classes is obtained. Bases belonging to the same class are called *like-oriented* ; bases belonging to different classes are said to be *oppositely oriented.*

The vector space $\mathbf{R}^m$ is called *oriented* if the bases belonging to one of the classes are called *positive*, those belonging to the other class then being called *negative.* As the change of bases is completely governed by the group $GL(m; \mathbf{R})$ of invertible linear transformations in $\mathbf{R}^m$, we say that $T \in GL(m; \mathbf{R})$ is *orientation preserving* if $\det T > 0$ where $\det T = \det A$, A being the matrix representation of T relative to any basis of $\mathbf{R}^m$.

A.2. The scalar product - Angles-Orthogonality

The *scalar product* between two vectors $\underline{x} = (x_1, x_2, \ldots, x_m)$ and $\underline{y} = (y_1, \ldots, y_m)$ in $\mathbf{R}^m$ is denoted by $< \underline{x}, \underline{y} >$, i.e.

$$< \underline{x}, \underline{y} >= \sum_{j=1}^{m} x_j y_j ;$$

$|\underline{x}| = \sqrt{< \underline{x}, \underline{x} >}$ stands for the *length* of the vector $\underline{x}$.

The space $\mathbf{R}^m$ provided with its scalar product is called Euclidean m-space. A *unit vector* $\underline{u}$ is any vector of length 1, i.e. $|\underline{u}| = 1$. Given nonzero vectors $\underline{x}, \underline{y}$ in $\mathbf{R}^3$, the *angle* between $\underline{x}$ and $\underline{y}$ is that angle θ for which

$$\cos\theta = \frac{< \underline{x}, \underline{y} >}{|\underline{x}||\underline{y}|}, \quad \theta \in [0, \pi] .$$

Two nonzero vectors $\underline{x}$ and $\underline{y}$ are said to be *orthogonal* or *perpendicular*, notation $\underline{x} \perp \underline{y}$, if $< \underline{x}, \underline{y} >= 0$ or equivalently, if $\theta = \pi/2$.

If W is a subspace of $\mathbf{R}^m$ then its *orthogonal complement* is denoted by $W^\perp$, i.e.

$$W^\perp = \{\underline{y} \in \mathbf{R}^m :< \underline{x}, \underline{y} >= 0 \quad \text{for all} \quad \underline{x} \in W\}.$$

Using the direct sum decomposition $\mathbf{R}^m = W \oplus W^\perp$, each vector $\underline{x} \in \mathbf{R}^m$ admits a unique splitting

$$\underline{x} = \underline{x}_W + \underline{x}_{W^\perp}$$

the vectors $\underline{x}_W \in W$ and $\underline{x}_{W^\perp} \in W^\perp$ being called respectively the *orthogonal projections* of $\underline{x}$ on W and $W^\perp$.

The orthogonal projection $\underline{x}_W$ of $\underline{x}$ on W is also characterized as being the (unique) vector in W whose distance to $\underline{x}$ is minimal, i.e.

$$|\underline{x} - \underline{x}_W| = \inf_{\underline{y} \in W} |\underline{x} - \underline{y}|,$$

or as being the unique vector $\underline{x}_W \in W$ such that $\underline{x} - \underline{x}_W \perp W$. If $\underline{u} = \{\underline{u}_1, \ldots, \underline{u}_k\}$ is an *orthonormal basis* of W, i.e. $|\underline{u}_j| = 1, j = 1, \ldots, k$, and $\underline{u}_i \perp \underline{u}_j$, $i \neq j$, then $\underline{x}_W$ is also given by

$$\underline{x}_W = \sum_{j=1}^{k} < \underline{x}, \underline{u}_j > \underline{u}_j .$$

The map $\rho_W : \mathbf{R}^m \to \mathbf{R}^m$ given by

$$\rho_W(\underline{x}) = \underline{x}_W - \underline{x}_{W^\perp}$$

is called the *reflection* of $\mathbf{R}^m$ in W.

Of particular importance in what follows is the case where W is the orthogonal complement of a nonzero vector $\underline{x}$, i.e. $W = \underline{x}^\perp$. We then have that $W^\perp = \text{span}\{\underline{x}\} = \mathbf{R}\underline{x}$ whence any vector $\underline{y} \in \mathbf{R}^m$ admits a unique (orthogonal) decomposition

$$\underline{y} = \underline{y}_{\|} + \underline{y}_{\perp}$$

where $\underline{y}_{\|} \in \mathbf{R}\underline{x}$ and $\underline{y}_{\perp} \in \underline{x}^\perp$ are the orthogonal projections of $\underline{y}$ on respectively $\mathbf{R}\underline{x}$ and $\underline{x}^\perp$. We thus have that $\underline{y}_{\|} = \lambda\underline{x}$ is parallel to $\underline{x}$ and $\underline{y}_{\perp} = \underline{y} - \underline{y}_{\|}$ is perpendicular to $\underline{x}$. In fact the appropriate scalar λ is given by

$$\lambda = \frac{< \underline{y}, \underline{x} >}{|\underline{y}|^2}$$

while, choosing an orthonormal basis $\{\underline{u}_1, \ldots, \underline{u}_{m-1}\}$ of $\underline{x}^\perp$, we have that

$$\underline{y}_{\perp} = < \underline{y}, \underline{u}_1 > \underline{u}_1 + \ldots + < \underline{y}, \underline{u}_{m-1} > \underline{u}_{m-1}.$$

Finally, denoting henceforth the reflection of $\mathbf{R}^m$ in $\underline{x}^\perp$ by $\rho_{\underline{x}}$ and the image of $\underline{y}$ under $\rho_{\underline{x}}$ by $\underline{y}_{\rho}$, we have that

$$\underline{y}_{\rho} = -y_{\|} + \underline{y}_{\perp}.$$

As $\underline{x}^\perp$ is a hyperplane (through 0) in $\mathbf{R}^m$, the reflection $\rho_{\underline{x}}$ is also said to be a *hyperplane reflection* of $\mathbf{R}^m$.

A.3. Orthogonal transformations

A linear transformation T from $\mathbf{R}^m$ to $\mathbf{R}^m$ is *orthogonal* if for all $\underline{x}, \underline{y} \in \mathbf{R}^m$,

$$< T\underline{x}, T\underline{y} > = < \underline{x}, \underline{y} > .$$

Now let T be any linear transformation in $\mathbf{R}^m$ and let A be its matrix representation relative to a basis of $\mathbf{R}^m$ then T is orthogonal if and only if its matrix A relative to any orthonormal basis of $\mathbf{R}^m$ satisfies the relation $A^tA = AA^t = I_m$, A^t and I_m being respectively the *transpose* of A and the $(m \times m)$-*identity* matrix.

A matrix $A \in \mathbf{R}^{m\times m}$ for which $A^tA = AA^t = I_m$ is said to be *orthogonal.* The set of orthogonal transformations in $\mathbf{R}^m$ forms a group under the composition ; it is denoted by $\mathrm{O}(m)$ and called the *orthogonal group* in $\mathbf{R}^m$. If $T \in \mathrm{O}(m)$ we have that det $T = \pm 1$. The set of those $T \in \mathrm{O}(m)$ for which det $T = 1$ forms a subgroup of $\mathrm{O}(m)$, called the *special orthogonal group* and denoted by $\mathrm{SO}(m)$. An orthogonal transformation T belonging to respectively $\mathrm{SO}(m)$ or $\mathrm{O}(m)\setminus \mathrm{SO}(m)$ is called a *rotation* or an *antirotation.*

It thus follows that a rotation is an orthogonal tranformation which preserves the orientation. As is readily seen, a reflection of $\mathbf{R}^m$ in a hyperplane $\underline{x}^\perp(0 \neq \underline{x} \in \mathbf{R}^m)$ is an antirotation. Moreover it may be proved that each $T \in \mathrm{O}(m)$ is the composition of at most m reflections in hyperplanes. Finally, any rotation is the composition of an even number and any antirotation is the composition of an odd number of reflections in hyperplanes. This is the Euclidean version of the so-called *Cartan-Dieudonné Theorem* (see e.g. [51]).

Chapter I

Clifford Algebras and Spinor Spaces

The aim of this chapter is to gather some basic results concerning real and complex Clifford algebras. All material covered is classical, exception made of the approach given in §§4.7 - 4.8 to the explicit realization of spinor space and a Hermitian structure on it.

In section 1 some fundamental notions and properties are recalled for real Clifford algebras. Almost no proofs are given and the reader eager to see them worked out is referred to [63] and [22].

Is section 2 realizations of real Clifford algebras as matrix algebras over **R**, **C** or **H** are studied. Our approach has been much inspired by the one given in [21].

Following mainly the ideas worked out in [1], complex Clifford algebras and their realizations as matrix algebras are studied in section 3.

In section 4 we show how spinor spaces may be used to represent the real groups $\text{Pin}(p,q)$ and $\text{Spin}(p,q)$ which are subgroups of the real Clifford algebra $\mathbf{R}_{p,q}$. If moreover the spinor space under consideration carries a real or a quaternionic structure, then we also discuss how real or quaternionic representations of these groups may be obtained. Part of the material presented here may be found in the recently published book [13], excepted - as we already mentioned - the subsections 4.7 and 4.8 in which an explicit realization is obtained for the spinor space S_{2m} inside the complex Clifford algebra $\mathbf{C}_{2m}$. By constructing a primitive idempotent I, generated by mutually commuting idempotents $I_j = \frac{1}{2}(1 + ie_{2j-1}e_{2j})$ where $\{e_{2j-1}, e_{2j} : j = 1, \ldots, m\}$ is an orthonormal basis for $\mathbf{R}^{0,2m}$, we prove that $\mathbf{C}_{2m}I = (\wedge W)I$. Here $W = \text{Span}_{\mathbf{C}}\{f_j = \frac{1}{2}(e_{2j-1} + ie_{2j}) : j = 1, \ldots, m\}$ is a maximally isotropic subspace of $\mathbf{C}^{2m}$. In such a way, the spinor space is realized both as a minimal left ideal in $\mathbf{C}_{2m}$ and as a Grassmann algebra. In doing this we have been much inspired, on the one hand, by the idea of constructing primitive idempotents in Clifford algebras (see [54] and [5]) and, on the other hand, by the realization of the γ-matrices presented in [62] and [16].

The Spin representation of $\mathbf{C}_{2m}$ is then obtained by left multiplication inside the Clifford algebra itself. Notice that an approach similar to ours was already worked out in [17]. Furthermore an inner product is defined explicitly by means of the vacuum expectation value. Finally we also introduce the Pin and Spin groups for the space $\mathbf{C}^n$ and we show how in the even dimensional case the set IG_n of all maximal isotropic subspaces in $\mathbf{C}^{2n}$ can be identified with the projective space $\mathbf{P}\ (S^{\text{pure}})$ of pure spinors.

1 Real Clifford algebras : general theory

In this section an introduction is given to real Clifford algebras $\mathbf{R}_{p,q}$ and their underlying subgroups $\Gamma(p,q)$, $\text{Pin}(p,q)$ and $\text{Spin}(p,q)$.

1.1 Real orthogonal spaces

Let X be a finite dimensional real vector space with $\dim X = n$ and let $\mathcal{B}$ be a non-degenerate symmetric bilinear form on X, i.e. $\mathcal{B} : X \times X \to \mathbf{R}$ is a function satisfying the properties

(B.1) *(Bilinearity condition)* For all $\lambda \in \mathbf{R}$, $v, v', w, w' \in X$

$$\mathcal{B}(\lambda v + v', w) = \lambda \mathcal{B}(v,w) + \mathcal{B}(v', w)$$

and

$$(v, \lambda w + w') = \lambda \mathcal{B}(v,w) + \mathcal{B}(v,w').$$

(B.2) *(Symmetry condition)* For all $v, w \in X$ $\quad \mathcal{B}(v,w) = \mathcal{B}(w,v)$.

(B.3) *(Non-degeneracy condition)* For each non-zero $v \in X$ there exists a $w \in X$ such that $\mathcal{B}(v,w) \neq 0$.

Then the pair $(X, \mathcal{B})$ is said to be a *non-degenerate n-dimensional real orthogonal space.*

Now let $e = (e_1, \ldots, e_n)$ be an arbitrary ordered basis for X and put for each $(i,j) \in \{1, \ldots, n\} \times \{1, \ldots, n\}$,

$$a_{ij} = \mathcal{B}(e_i, e_j).$$

Then the $n \times n$ real matrix $A = (a_{ij})$ is called the *matrix of $\mathcal{B}$ relative to e*. It describes the action of $\mathcal{B}$ in the following way.

If for $v, w \in X$, $x = [v]_e$ and $y = [w]_e$ stand for the coordinate matrices of v and w relative to e, then $\mathcal{B}(v,w) = x^t A y$, x^t being the transpose of x.

Moreover the symmetry and non-degeneracy conditions of $\mathcal{B}$ respectively imply that A is symmetric and non-singular. Furthermore there exist an ordered basis $e^* = (e_1^*, \ldots, e_n^*)$ for X and $p, q \in \mathbf{N}$ with $p + q = n$ such that

(1) $\mathcal{B}(e_i^*, e_i^*) = 1,\ i = 1, \ldots, p$

(2) $\mathcal{B}(e_i^*, e_i^*) = -1,\ i = p+1, \ldots, p+q$

(3) $\mathcal{B}(e_i^*, e_j^*) = 0,\ i \neq j$

By virtue of (1) – (3) we say that the basis e^* *diagonalizes* the bilinear form $\mathcal{B}$. As a final property we have the so-called *Law of Inertia* or *Sylvester's Theorem*.

(4) The number p of basis vectors e_i^* for which $\mathcal{B}(e_i^*, e_i^*) = 1$ is independent of the choice of the basis e^* diagonalizing $\mathcal{B}$.

A basis e^* for X satisfying (1) – (3) is called an *orthonormal basis* for X. By the choice of such a basis e^*, the function

$$\Phi_{e^*} : X \to \mathbf{R}^{p+q}$$

with

$$\Phi_{e^*}(v) = [v]_{e^*}$$

is an isomorphism of the real orthogonal space $(X, \mathcal{B})$ into the real space $\mathbf{R}^{p+q}$ provided with the non-degenerate symmetric bilinear form $B : \mathbf{R}^{p+q} \to \mathbf{R}$ given by

$$B(x, y) = \sum_{i=1}^{p} x_i y_i - \sum_{i=p+1}^{p+q} x_i y_i.$$

The non-degenerate real orthogonal space $(\mathbf{R}^{p+q}, B)$ will henceforth be denoted by $\mathbf{R}^{p,q}$ and $(X, \mathcal{B})$ is said to be of *type* (p, q) ; its corresponding orthonormal basis will be denoted by e.

The set of linear maps $T : X \to X$ preserving the bilinear form $\mathcal{B}$, i.e. for which $\mathcal{B}(Tv, Tw) = \mathcal{B}(v, w)$, for all $v, w \in X$, forms a group under the operation of composition, called the *pseudo-orthogonal group* associated with the real orthogonal space $(X, \mathcal{B})$; it is denoted by $\mathrm{O}(X)$ and its elements are called *orthogonal maps*.

For any finite p, q and n the groups $\mathrm{O}(\mathbf{R}^{p,q})$ and $\mathrm{O}(\mathbf{R}^{n,0})$ will also be denoted respectively by $\mathrm{O}(p, q)$ and $\mathrm{O}(n)$.

Notice that $\mathrm{O}(p, q)$ may be considered as consisting of those invertible real $n \times n$ matrices Q satisfying $Q^t A Q = A$ where $A = (a_{ij})$ is the matrix of B relative to the standard orthonormal basis for $\mathbf{R}^{p,q}$.

The subgroup of $\mathrm{O}(p, q)$ consisting of those Q having determinant 1 is called the group of *special orthogonal transformations* or *rotations* of $\mathbf{R}^{p,q}$; it is denoted by $\mathrm{SO}(p, q)$.

In the case of $\mathbf{R}^{n,0}$ we also write $\mathrm{SO}(n)$ for $\mathrm{SO}(n; 0)$. The element of $\mathrm{O}(p, q) \backslash \mathrm{SO}(p, q)$ are called *antirotations*.

Notice that for any p, q, the groups $\mathrm{O}(p, q)$ and $\mathrm{O}(q, p)$ are isomorphic, as are the groups $\mathrm{SO}(p, q)$ and $\mathrm{SO}(q, p)$.

Let again $(X, \mathcal{B})$ be a non-degenerate n-dimensional real orthogonal space, let W be a subspace of X and $W^\perp$ be its *orthogonal annihilator*, i.e. $W^\perp$ is the subspace of X given by

$$W^\perp = \{u \in X : \mathcal{B}(u,v) = 0 \text{ for all } v \in W\}$$

Then $X = W \oplus W^\perp$ if and only if W is itself non-degenerate, i.e. if $\mathcal{B}_W$ denotes the restriction of $\mathcal{B}$ to W, then $(W, \mathcal{B}_W)$ is a non-degenerate real orthogonal space.

In particular, for an element $a \in X$, $X = \mathbf{R}a \oplus (\mathbf{R}a)^\perp$ if and only if a is *invertible*, i.e. $\mathcal{B}(a,a) \neq 0$. In this case $H_a = (\mathbf{R}a)^\perp$ is an $(n-1)$-dimensional subspace of X; we call it the *hyperplane associated with the invertible element* $a \in X$. Moreover, as each $v \in X$ then admits a unique decomposition $v = \lambda a + b$ where $\lambda \in \mathbf{R}$ and $b \in H_a$, the map $\rho_a : X \to X$ defined by $\rho_a(v) = -\lambda a + b$ is orthogonal ; it is called the *reflection in the hyperplane* H_a.

The famous *Cartan-Dieudonné Theorem* tells us that any $T \in \mathrm{O}(X)$ is expressible as the composition of a finite number of hyperplane reflections of X, the number of reflections composing T being even when T is a rotation and odd when T is an antirotation.

1.2 Real Clifford algebras : definitions

1.2.1 Let $(X, \mathcal{B})$ be a non-degenerate n-dimensional real orthogonal space and let $\mathcal{A}$ be a real associative algebra with identity 1 such that

(C.1) $\mathcal{A}$ contains copies of $\mathbf{R}$ and X as linear subspaces.

(C.2) For all $v \in X$, $v^2 = \mathcal{B}(v,v)$.

(C.3) $\mathcal{A}$ is generated as a ring by the copies of $\mathbf{R}$ and X, or equivalently, as a real algebra by $\{1\}$ and X.

Then $\mathcal{A}$ is said to be a real *Clifford algebra* for $(X, \mathcal{B})$ and it is denoted by $\mathcal{A} = \mathcal{C}(X)$.

1.2.2 Elementary examples

(1) $\mathbf{R}$ is a Clifford algebra for $\mathbf{R}^{0,0}$ and $\mathbf{R}^{1,0}$

(2) $\mathbf{C}$ is a Clifford algebra for $\mathbf{R}^{0,1}$

(3) $\mathbf{H}$ (the algebra of real quaternions) is a Clifford algebra for $\mathbf{R}^{0,2}$, the space $\mathbf{R}^{0,2}$ being identified with the subspace span$\{i,j\}$ of $\mathbf{H}$.

(4) $\mathbf{H}$ is also a Clifford algebra for $\mathbf{R}^{0,3}$, the space $\mathbf{R}^{0,3}$ being identified now with the space of pure quaternions span$\{i,j,k\}$.

From the examples above it already follows that different real orthogonal spaces may give rise to the same Clifford algebra. This phenomenon will be discussed further in section 1.4.

1.2.3 Let $\mathcal{C}(X)$ be a Clifford algebra for $(X, \mathcal{B})$ and let W be a (non-degenerate) subspace of X. Then clearly the subalgebra of $\mathcal{C}(X)$ generated by the copies of $\mathbf{R}$ and W is a Clifford algebra for $(W, \mathcal{B}_W)$, denoted by $\mathcal{C}(W)$.

Moreover, if $e = \{e_1, \ldots, e_n\}$ is an orthonormal basis for $(X, \mathcal{B})$, the latter being of type (p, q), then the requirement (C.2) implies that

$$(1)\quad e_i^2 = \mathcal{B}(e_i, e_i) = \begin{cases} 1 & , \quad i = 1, \ldots, p \\ -1 & , \quad i = p+1, \ldots, p+q \end{cases}$$

$$(2)\quad e_i e_j + e_j e_i = 0, i \neq j.$$

Indeed, (1) follows directly from (C.2) while (2) is obtained by observing that for all $i \neq j$, on the one hand

$$\begin{aligned} (e_i + e_j)^2 &= \mathcal{B}(e_i + e_j, e_i + e_j) \\ &= \mathcal{B}(e_i, e_i) + \mathcal{B}(e_j, e_j) \\ &= e_i^2 + e_j^2 \end{aligned}$$

while on the other hand

$$(e_i + e_j)^2 = e_i^2 + e_j^2 + e_i e_j + e_j e_i.$$

This prompts the following definitions and results.

Let S be a subset of a real associative algebra $\mathcal{A}$ with identity. Then S is said to be a *non-degenerate orthonormal subset* of $\mathcal{A}$ if

(1) S is linearly independent.

(2) $ab + ba = 0$ for all $a, b \in S$, i.e. S consists of mutually anticommuting elements of $\mathcal{A}$.

(3) $a^2 = 1$ or -1 for all $a \in S$.

If S contains $r + s$ elements, r and s being the number of elements having square $+1$ or -1 respectively, then S is said to be of *type* (r, s).

If Y denotes the linear hull of S then it may be proved that there exists a (unique) symmetric bilinear form $\mathcal{B}^\star$ on Y such that

(1) $a^2 = \mathcal{B}^\star(a, a)$ for all $a \in S$.

(2) The orthogonal space $(Y, \mathcal{B}^\star)$ is isomorphic to $\mathbf{R}^{r,s}$.

(3) If S generates $\mathcal{A}$, then $\mathcal{A}$ is a Clifford algebra for $(Y, \mathcal{B}^\star)$.

For example, consider in $\mathbf{R}(2)$ - the algebra of 2×2 real matrices - the subset S given by

$$S = \left\{ \begin{pmatrix} 0 & 1 \\ 1 & 0 \end{pmatrix}, \begin{pmatrix} 1 & 0 \\ 0 & -1 \end{pmatrix} \right\}.$$

Then S is a non-degenerate orthonormal subset of $\mathbf{R}(2)$ of type $(2,0)$ which generates $\mathbf{R}(2)$. Hence $\mathbf{R}(2)$ is a Clifford algebra for $\mathbf{R}^{2,0}$.

1.3 A basis for the Clifford algebra $C(X)$

Let $C(X)$ be a Clifford algebra for the real orthogonal space $(X, \mathcal{B})$ of type (p,q) and let $e = \{e_1, \ldots, e_p, e_{p+1}, \ldots, e_{p+q}\}$ be an orthogonal basis for $(X, \mathcal{B})$.

Then as we saw

$$e_i^2 = \begin{cases} 1, & i = 1, \ldots, p \\ -1, & i = p+1, \ldots, p+q \end{cases} \tag{1.1}$$

and

$$e_i e_j + e_j e_i = 0, \quad i \neq j.$$

Hence for any two vectors $v, w \in X$ we have that

$$vw + wv = 2\mathcal{B}(v, w)$$

and this since

$$(v+w)^2 = \mathcal{B}(v+w, v+w).$$

It follows from (1.1) that among all the products of two basis vectors e_i and e_j, it suffices to consider those products $e_i e_j$ for which $i < j$.

More generally if $\beta_1, \ldots, \beta_s$ are s distinct natural numbers with $1 \leq \beta_i \leq n$, $i = 1, \ldots, s$, then we have that

$$e_{\beta_1} e_{\beta_2} \ldots e_{\beta_s} = (-1)^\sigma e_{\alpha_1} e_{\alpha_2} \ldots e_{\alpha_s}$$

where $1 \leq \alpha_1 < \alpha_2 < \ldots < \alpha_s \leq n$ and σ is the number of inversions in the permutation

$$\pi = \begin{pmatrix} \alpha_1 & \alpha_2 & \ldots & \alpha_s \\ \beta_1 & \beta_2 & \ldots & \beta_s \end{pmatrix}.$$

Finally consider an arbitrary product $e_{\gamma_1} e_{\gamma_2} \ldots e_{\gamma_r}$ of the basic vectors $e_{\gamma_j} (1 \leq \gamma_j \leq n)$ and assume that the basic vector e_i occurs repeatedly in it; then after ordering the factors according to increasing indices, e_i will appear at most to the first power since $e_i^2 \in \mathbf{R}$. It thus follows in particular that every product $e_{\gamma_1} e_{\gamma_2} \ldots e_{\gamma_r}$ can be reduced to one containing at most n factors.

Putting for each subset $A = \{\alpha_1, \ldots, \alpha_h\}$ of $N = \{1, \ldots, n\}$ with $1 \leq \alpha, < \alpha_2 < \ldots < \alpha_h \leq n$, $e_A = e_{\alpha_1} e_{\alpha_2} \ldots e_{\alpha_h}$ and defining $e_\phi = 1$,, the preceding observations imply that the family

$$(e_A : A = \{\alpha_1, \ldots, \alpha_h\}, 1 \leq \alpha_1 < \alpha_2 < \ldots < \alpha_h \leq n) \tag{1.2}$$

is generating for $\mathcal{C}(X)$ whence $\dim \mathcal{C}(X) \leq 2^n$.

Now we investigate under which conditions the family (1.2) is a basis for $\mathcal{C}(X)$, i.e. when does $\sum_{A \subset N} \lambda_A e_A = 0$ imply that all coefficients λ_A are zero.

To that end notice first that, as for each ordered $B = \{\beta_1, \ldots, \beta_r\} \subset N = \{1, \ldots, n\}$, e_B is invertible,

$$\sum_{A \subset N} \lambda_A e_A = 0 \Leftrightarrow (\sum_{A \subset N} \lambda_A e_A) e_B^{-1} = 0,$$

thus making λ_B the coefficient of 1.

It thus suffices to prove that

$$\sum_{A \subset N} \lambda_A e_A = 0 \Longrightarrow \lambda_\phi = 0. \tag{1.3}$$

As for $j = 1, \ldots, n$ and $A \subset \{1, \ldots, n\}$, e_j either commutes or anticommutes with e_A, we obtain for each $j = 1, \ldots, n$

$$\sum_{A \subset N} \lambda_A e_A = 0 \Longrightarrow \sum_{A \subset N} \lambda_A e_j e_A e_j^{-1} = \sum_{A \subset N} \varepsilon_{A,j} \lambda_A e_A = 0 \tag{1.4}$$

where

$$\varepsilon_{A,j} = \begin{cases} 1 & \text{if } e_j \text{ commutes with } e_A, \\ -1 & \text{if } e_j \text{ anticommutes with } e_A. \end{cases}$$

It follows from (1.3) and (1.4) that

$$\sum_{A \subset N} \lambda_A e_A = 0 \tag{1.5}$$

with the summation taken over all $A \subset \{1, \ldots, n\}$ for which e_A commutes with e_j. But, as $j = 1, \ldots, n$ has been taken arbitrarily, we finally obtain

$$\sum_{A \subset N} \lambda_A e_A = 0 \tag{1.6}$$

with the summation taken over those $A \subset \{1, \ldots, n\}$ for which e_A commutes with all e_j, $j = 1, \ldots, n$.

Now suppose that n is *even*. Then it is easily checked that there is exactly one element e_A satisfying this condition, namely $e_A = e_\phi = 1$.

In that case (1.6) thus reduces to $\lambda_\phi 1 = 0$ implying that $\lambda_\phi = 0$, whence the family (1.2) is also linearly independent or it is a basis for $\mathcal{C}(X)$. Consequently for n even, $\dim \mathcal{C}(X) = 2^n$.

Now suppose that n is *odd* and that $e_N = e_1 e_2 \ldots e_n \notin \mathbf{R}$. Then it is easily checked that e_ϕ and e_N are the only elements which commute with each $e_j, j = 1, \ldots, n$.

In that case (1.6) thus reduces to

$$\lambda_\phi 1 + \lambda_N e_N = 0$$

If $\lambda_N \neq 0$ we find that $e_N \in \mathbf{R}$, a contradiction. Hence λ_N ought to be zero and so again $\lambda_\phi = 0$ and (1.2) is a basis for $\mathcal{C}(X)$ or $\dim \mathcal{C}(X) = 2^n$.

Finally suppose that n is *odd* and that $e_N \in \mathbf{R}$.

If $(X, \mathcal{B})$ is of type (p,q) with $p+q = n = 2k+1$, we find that

$$(e_N)^2 = (-1)^{k(2k+1)+q}.$$

But, since $e_N \in \mathbf{R}$, $(e_N)^2$ is positive, implying that $e_N^2 = 1$.

Consequently $k(2k+1)+q$ or, equivalently, $k+q$ is divisible by 2 or, equivalently, $2k+2q = p+3q-1$ is divisible by 4 whence $p-q-1$ is divisible by 4.
Conversely, if $p-q-1$ is divisible by 4 then n is odd.

We thus obtain that in this case $e_N = \pm 1$ and that for each $A \subset N$ with $^\# A$ odd, $e_A = \pm e_{N \setminus A}$.

But, as from the reasoning made in the case n even it follows that the subfamily $(e_A : {}^\# A \text{ even})$ of the family (1.2) is always linearly independent, we obtain that in the case n odd and $e_N \in \mathbf{R}$, $e_N = \pm 1$ and $\{e_A : {}^\# A \text{ even}\}$ is a maximally free subset of the family (1.2), or $\dim \mathcal{C}(X) = 2^{n-1}$.

Summarizing we have

Theorem. *Let $\mathcal{C}(X)$ be a Clifford algebra for the n-dimensional non-degenerate real orthogonal space $(X, \mathcal{B})$ of type (p,q). Furthermore let $e = (e_1, \ldots, e_p, e_{p+1}, \ldots, e_{p+q})$ be an orthogonal basis for $(X, \mathcal{B})$. Then we have*

(i) If n is even, $\dim \mathcal{C}(X) = 2^n$ *and $\{e_A : A \subset N\}$ is a basis for $\mathcal{C}(X)$.*

(ii) If n is odd and $e_N \notin \mathbf{R}$, then $\dim \mathcal{C}(X) = 2^n$ *and $\{e_A : A \subset N\}$ is a basis for $\mathcal{C}(X)$.*

(iii) If n is odd and $e_N \in \mathbf{R}$, then $e_N = \pm 1$ and $p-q-1 \equiv 0$ ((mod 4)). In that case $\dim \mathcal{C}(X) = 2^{n-1}$ *and $\{e_A : {}^\# A \text{ is even}\}$ is a basis for $\mathcal{C}(X)$.*

1.4 Universal and non-universal Clifford algebras

In the previous subsection we saw that if $\mathcal{C}(X)$ is a Clifford algebra for $(X, \mathcal{B})$ then either $\dim \mathcal{C}(X) = 2^n$ or $\dim \mathcal{C}(X) = 2^{n-1}$.

If $\dim \mathcal{C}(X) = 2^n$ we call it a *universal Clifford algebra* for $(X, \mathcal{B})$; if $\dim \mathcal{C}(X) = 2^{n-1}$, $\mathcal{C}(X)$ is said to be a *non-universal Clifford algebra* for$(X, \mathcal{B})$.

As was shown by Chevalley in [17], every non-degenerate finite dimensional real orthogonal space $(X, \mathcal{B})$ has a universal Clifford algebra.

In the case of the real orthogonal space $\mathbf{R}^{p,q}$, its universal Clifford algebra will be denoted by $\mathbf{R}_{p,q}$.

Notice that, if $(X, \mathcal{B})$ is of type (p,q), then the isomorphism Φ_e between the real orthogonal spaces $(X, \mathcal{B})$ and $(\mathbf{R}^{p,q}, \langle\cdot,\cdot\rangle)$, $e = (e_1, \ldots, e_p, e_{p+1}, \ldots, e_{p+q})$ being an orthonormal basis for $(X, \mathcal{B})$, extends naturally to an isomorphism between their respective universal Clifford algebras.

Of course, by virtue of the section 1.2.3., a real associative algebra $\mathcal{A}$ with identity containing a non-degenerate orthonormal subset S of type (p,q) which generates $\mathcal{A}$, is isomorphic to a Clifford algebra $\mathcal{C}(\mathbf{R}^{p,q})$ for $\mathbf{R}^{p,q}$. In order to verify whether or not $\mathcal{A}$ is a universal Clifford algebra for $\mathbf{R}^{p,q}$ it thus suffices, by virtue of §1.3 Theorem to look at $\dim \mathcal{A}$. For example we saw in 1.2.3. that $\mathbf{R}(2)$ is a Clifford algebra for $\mathbf{R}^{2,0}$. As $\mathbf{R}(2)$ has dimension 2^2, it is isomorphic to the universal Clifford algebra $\mathbf{R}_{2,0}$ for $\mathbf{R}^{2,0}$.

That non-universal Clifford algebras do occur is illustrated by the following example.

Take the space $\mathbf{R}^{0,3}$. Then, as we saw in 1.2.2 (4), $\mathbf{H}$ is a Clifford for $\mathbf{R}^{0,3}$, the latter space being identified with the subspace of $\mathbf{H}$ spanned by $\{i, j, k\}$. Here $n = 3$ is odd and $ijk = -1 \in \mathbf{R}$. The condition $p - q - 1 \equiv 0 \pmod 4$ is clearly satisfied and the set $\{1, ij, ik, jk\} = \{1, k, -j, i\}$ is a basis for $\mathbf{H}$.

The isomorphism between the universal Clifford algebras for $(X, \mathcal{B})$ and $(\mathbf{R}^{p,q}, \langle\cdot,\cdot\rangle)$ suggests that in practice we restrict ourselves to the study of $\mathbf{R}_{p,q}$. This will be systematically done in the sequel.

1.5 The existence of universal Clifford algebras

Let $(X, \mathcal{B})$ be a non-degenerate n-dimensional real orthogonal space of type (p,q). Furthermore, let $\mathcal{Q} : X \to \mathbf{R}$ be the quadratic form associated with $\mathcal{B}$, i.e. for all $v \in X$

$$\mathcal{Q}(v) = \mathcal{B}(v, v),$$

and let $e = (e_1, \ldots e_p, e_{p+1}, \ldots, e_{p+q})$ be an orthonormal basis for $(X, \mathcal{B})$.

Then a universal Clifford algebra for $(X, \mathcal{B})$ may be realized as follows.

Let $\otimes X$ be the tensor algebra over X and let $I(\mathcal{Q})$ be the two-sided ideal in $\otimes X$ generated by elements of the form

$$v \otimes v - \mathcal{Q}(v) \quad v \in X.$$

Then the quotient algebra

$$\mathcal{C}(\mathcal{Q}) = \otimes X / I(\mathcal{Q})$$

is a universal Clifford algebra for $(X, \mathcal{B})$.

As we already mentioned, this realization of universal Clifford algebras is due to Chevalley (see [17]). The universal property of that algebra $\mathcal{C}(\mathcal{Q})$ is illustrated by the following

Theorem. *Let F be a real associative algebra with identity and let $u : X \to F$ be a linear map such that $(u(v))^2 = \mathcal{Q}(v)$ for all $v \in X$. Then there exists a unique homomorphism $\tilde{u}$ from $\mathcal{C}(\mathcal{Q})$ into F such that $u = \tilde{u} o \rho$ where $\rho : \otimes X \to \mathcal{C}(\mathcal{Q})$ is the canonical homomorphism.*

Hence the following diagram is commutative :

$$\begin{array}{ccccc} & & \mathcal{C}(\mathcal{Q}) & & \\ & \rho \nearrow & & \searrow \tilde{u} & \\ X & & \underset{u}{\longrightarrow} & & F \end{array}$$

1.6 Alternative approaches to Clifford algebras

In this subsection we sketch briefly some other approaches to the realization of the universal Clifford algebra $\mathbf{R}_{p,q}$.

1.6.1 The differential form realization

Let E be a finite dimensional real vector space and call $\Lambda = \sum_{p=0}^{n} \Lambda^p$ the algebra of differential forms over E (or Grassmann algebra associated with E).

Then $\Lambda^o = \mathbf{R}, \Lambda^1 = E'$ (the dual of E). Moreover Λ is a vector space having dimension 2^n. If furthermore E carries a metric g, then a metric $\bar{g}$ on E' may be defined as follows.

Let for each $x \in E$, $\mathcal{T}_x$ be the linear functional on E given by

$$\mathcal{T}_x(y) = g(x, y), \quad y \in E.$$

Then $\Phi : E \to E'$ with $\Phi(x) = \mathbf{T}_x, x \in E$, is an isomorphism.

Now define $\bar{g}$ on E' by means of

$$g(x, y) = \bar{g}(\mathcal{T}_x, \mathcal{T}_y), \quad x, y \in E.$$

Then $\bar{g}$ may be extended step by step as follows. For $u \in E'$, $\sigma \in \Lambda^k$ and $\tau \in \Lambda$, put

$$\bar{g}(u, \sigma \wedge \tau) = \bar{g}(u, \sigma) \wedge \tau + (-1)^k \sigma \wedge \bar{g}(u, \sigma).$$

In such a way, for a k-form $\sigma_1 \wedge \sigma_2 \wedge \ldots \wedge \sigma_k \in \Lambda^k, \sigma_j \in E'$, we find that for $u \in E'$

$$\bar{g}(u, \sigma_1 \wedge \ldots \wedge \sigma_k) = \sum_{j=1}^{k} (-1)^{j+1} \bar{g}(u, \sigma_j) \sigma_1 \wedge \ldots \wedge [\sigma_j] \wedge \ldots \wedge \sigma_k$$

where $[\sigma_j]$ means that σ_j does not occur. Notice that $\bar{g}(u,\cdot)$ thus maps Λ^k into Λ^{k-1}. Now define the *Clifford product* $u\sigma$ between $u \in E'$ and $\sigma \in \Lambda$ by $u\sigma = u\wedge\sigma + \bar{g}(u,\sigma)$ and extend it to the whole of Λ by imposing linearity and associativity. If the space (E, g) has dimension n and the metric g is such that given a basis $\{\sigma^1, \ldots, \sigma^n\}$ of 1-forms,

$$g^{ii} = 1 \quad \text{for } i = 1\ldots, p$$

$$g^{ii} = -1 \quad \text{for } i = p+1, \ldots, p+q$$

$$g^{ij} = 0 \quad \text{for } i \neq j$$

where $n = p + q$, then it may be shown that Λ, endowed with the Clifford product is isomorphic to the universal Clifford algebra for (E, g).

Let us give more explicit formulas for calculating the Clifford product — also called "vee-product" and denoted sometimes by $\vee$ — between basis r- and s-forms. As will be seen, this calculation is carried out in two steps.

Let the basis r- and s-forms be given by respectively $\sigma^{i_1}\wedge\ldots\wedge\sigma^{i_r}$ and $\sigma^{j_1}\wedge\ldots\wedge\sigma^{j_s}$. Then

(1) If the sets of indices $\{i_1, \ldots, i_r\}$ and $\{j_1, \ldots, j_s\}$ have k elements in common, say $\{v_1, \ldots, v_k\} = \{i_1, \ldots, i_r\}\cap\{j_1, \ldots, j_s\}$, consider the permutations π_r and π_s given by

$$\pi_r = \begin{pmatrix} i_1 & \ldots & i_k & i_{k+1} & \ldots & i_r \\ \lambda_1 & \ldots & \lambda_{r-k} & v_1 & \ldots & v_k \end{pmatrix}$$

$$\pi_s = \begin{pmatrix} j_1 & \ldots & j_k & j_{k+1} & \ldots & j_1 \\ v_k & \ldots & v_1 & \mu_1 & \ldots & \mu_{s-k} \end{pmatrix}$$

(2) Contract between the identical indices.

Then

$$(\sigma^{i_1}\wedge\ldots\wedge\sigma^{i_r})\vee(\sigma^{j_1}\wedge\ldots\wedge\sigma^{j_s})$$
$$= \ (-1)^{\pi_r}(-1)^{\pi_s}g^{v_1v_1}\ldots g^{v_kv_k}\sigma^{\lambda_1}\wedge\ldots\wedge\sigma^{\lambda_{r-k}}\wedge\sigma^{\mu_1}\wedge\ldots\wedge\sigma^{\mu_{s-k}},$$

where $(-1)^{\pi_r}$ and $(-1)^{\pi_s}$ are the signs of π_r and π_s.

Let us work out a few examples in the case of a 4-dimensional space with metric g having diagonal entries 1 or -1 (all off-diagonal entries are zero) :

$$
\begin{aligned}
\sigma^i \vee \sigma^j &= g^{ij} + \sigma^i \wedge \sigma^j \\
\sigma^i \vee (\sigma^j \wedge \sigma^k) &= g^{ij}\sigma^k - g^{ik}\sigma^j + g^{jk}\sigma^i + \sigma^i \wedge \sigma^j \wedge \sigma^k; \\
(\sigma^i \vee \sigma^j) \vee (\sigma^k \vee \sigma^l) &= g^{ij}g^{kl} + g^{il}g^{jk} - g^{ik}g^{jl} \\
&\quad + g^{ij}\sigma^k \wedge \sigma^l - g^{ik}\sigma^j \wedge \sigma^l + g^{il}\sigma^j \wedge^k \\
&\quad + g^{kl}\sigma^i \wedge \sigma^j + g^{ik}\sigma^i \wedge \sigma^l - g^{jl}\sigma^i \wedge \sigma^k \\
&\quad + \sigma^i \wedge \sigma^j \wedge \sigma^k \wedge \sigma^l.
\end{aligned}
$$

Notice that the realization of the Clifford algebra $\mathbf{R}_{p,q}$ by means of the algebra of exterior differential forms on an n-dimensional vector space with metric g is originally due to Kähler (see [46] and [47], see also [67]).

1.6.2 Matrix algebra realizations

When $n = p + q$ is rather small (say $n \leq 4$) it is often very handy to realize $\mathbf{R}_{p,q}$ directly by exhibiting a set of generators $\gamma^\mu (\mu = 1, \ldots, n)$ for a matrix algebra over $\mathbf{R}, \mathbf{C}$ or $\mathbf{H}$ satisfying

$$
\begin{aligned}
&(\gamma^\mu)^2 = I_n, && \mu = 1, \ldots, p \\
&(\gamma^\mu)^2 = -I_n, && \mu = p+1, \ldots, p+q \\
&\gamma^\mu\gamma^\nu + \gamma^\nu\gamma^\mu = 0, && \mu \neq \nu,
\end{aligned}
$$

I_n being the identity matrix.

Let us illustrate this procedure in the important case of the 4-dimensional space $\mathbf{R}^{3,1}$.

Consider the real 4×4-matrices

$$
\gamma^1 = \begin{pmatrix} 1 & & & \\ & -1 & & \\ & & -1 & \\ & & & 1 \end{pmatrix}; \gamma^2 = \begin{pmatrix} 0 & 1 & & \\ 1 & 0 & & \\ & & 0 & 1 \\ & & 1 & 0 \end{pmatrix}
$$

$$
\gamma^3 = \begin{pmatrix} & & 1 & 0 \\ & & 0 & -1 \\ 1 & 0 & & \\ 0 & -1 & & \end{pmatrix}; \gamma^4 = \begin{pmatrix} 0 & 1 & & \\ -1 & 0 & & \\ & & 0 & 1 \\ & & -1 & 0 \end{pmatrix}.
$$

Then

$$
\begin{aligned}
(\gamma^1)^2 &= (\gamma^2)^2 = (\gamma^3)^2 = I_4; \\
(\gamma^4)^2 &= -I_4;
\end{aligned}
$$

$$\gamma^\mu\gamma^\nu + \gamma^\nu\gamma^\mu = 0, \quad \mu \neq \nu.$$

Hence $S = \{\gamma^{(1)}, \gamma^{(2)}, \gamma^{(3)}, \gamma^{(4)}\}$ is a non-degenerate orthogonal subset of $\mathbf{R}(4)$, its linear hull being isomorphic to $\mathbf{R}^{3,1}$. As S generates $\mathbf{R}(4)$, the algebra of real 4×4 matrices $\mathbf{R}(4)$ is isomorphic to $\mathbf{R}_{3,1}$.

The set S is called the set of *Majorana matrices* (see also 4.5.1).

The classification of real Clifford algebras worked out in section 2 indicates how any Clifford algebra $\mathbf{R}^{p,q}$ may be realized as a matrix algebra over $\mathbf{R}, \mathbf{C}, \mathbf{H}, {}^2\mathbf{R}$ or ${}^2\mathbf{H}$.

1.7 k-Vectors and involutions on $\mathbf{R}_{p,q}$*. The even subalgebra* $\mathbf{R}^+_{p,q}$ *of* $\mathbf{R}_{p,q}$

We introduce three involutions on $\mathbf{R}_{p,q}$ which extend either the identity or minus the identity operator on $\mathbf{R}^{p,q}$. If $e = (e_1, \ldots, e_p, e_{p+1}, \ldots, e_{p+q})$ is an orthonormal basis for $\mathbf{R}^{p,q}$ then throughout this subsection $\{e_A : A \subset \{1, \ldots, n\}\}$ stands for the basis of $\mathbf{R}_{p,q}$ built up by means of e (see also section 1.3).

1.7.1 k-vectors

The linear subspace of $\mathbf{R}_{p,q}$ spanned by the $\binom{n}{k}$ basis vectors e_A satisfying $\#A = k$, is denoted by $\mathbf{R}^k_{p,q}$; its elements are called *k-vectors*. In particular we have that

(1) $\mathbf{R}^0_{p,q}$ is the 1-dimensional subspace of 0-vectors. A basis of $\mathbf{R}^0_{p,q}$ is given by $\{1\}$; its elements are therefore also called *scalars*.

(2) $\mathbf{R}^1_{p,q}$ is the n-dimensional subspace of 1-vectors. It may be identified with $\mathbf{R}^{p,q}$ and its elements are therefore also called *vectors*.

(3) $\mathbf{R}^2_{p,q}$ is the $\binom{n}{2}$-dimensional subspace of 2-vectors which are also called *bivectors*.

(4) $\mathbf{R}^n_{p,q}$ is the 1-dimensional subspace of n-vectors, also called *pseudo-scalars*. A basis of $\mathbf{R}^n_{p,q}$ is given by $\{e_N\}$. The basis element e_N is often called *pseudo-unit*. By definition we thus have that

$$\mathbf{R}_{p,q} = \sum_{k=0}^{n} \oplus \mathbf{R}^k_{p,q}.$$

Denoting by $[a]_k$ the projection of $a \in \mathbf{R}_{p,q}$ into $\mathbf{R}^k_{p,q}$ we obtain that

$$a = \sum_{k=0}^{n} [a]_k \tag{1.7}$$

For $a \in \mathbf{R}_{p,q}$, $[a]_k$ is called its *k-vector part*. Notice that in Clifford algebra it becomes very natural to multiply vectors (i.e. 1-vectors) and more generally r- and s-vectors.

Let us first take a closer look to the product of two 1-vectors.

For all $v, w \in \mathbf{R}^1_{p,q}$ we find that their product vw can be split up into a symmetric and an anti-symmetric part :

$$vw = \frac{1}{2}(vw + wv) + \frac{1}{2}(vw - wv)$$

or

$$vw = v \cdot w + v \wedge w.$$

The "inner product"

$$v \cdot w = \frac{1}{2}(vw + wv) \tag{1.8}$$

is symmetric and represents a scalar ; it vanishes whenever the two vectors are orthogonal (i.e. $v \cdot w = 0$).

Notice that for $v, w \in \mathbf{R}^1_{p,q}$

$$v \cdot w = B(v, w).$$

The "outer product"

$$v \wedge w = \frac{1}{2}(vw - wv) \tag{1.9}$$

is anti-symmetric and is a 2-vector ; it vanishes whenever the two vectors are collinear.

Notice that for the basic vectors of $\mathbf{R}^{p,q}$,

$$e_i e_j = e_i \wedge e_j \ , \ i \neq j$$

and

$$e_i \cdot e_i = e_i^2 = B(e_i, e_i).$$

The notions of "inner" or "outer" products of 1-vectors may be extended in a straightforward way to the case of r- and s-vectors and this in the following way.

First remark that using the notation $[a]_k$ for the k-vector part of $a \in \mathbf{R}_{p,q}$, we obtain for 1-vectors v and w that $v \cdot w = [vw]_0$ and $v \wedge w = [vw]_2$.

Now let $u^{(r)} \in \mathbf{R}^r_{p,q}$ and $v^{(s)} \in \mathbf{R}^s_{p,q}$; then their "inner product" $u^{(r)} \cdot v^{(s)}$ and "outer product" $u^{(r)} \wedge v^{(s)}$ is given by

$$u^{(r)} \cdot v^{(s)} = \begin{cases} [u^{(r)} v^{(s)}]_{|r-s|} & \text{if} \quad r, s > 0 \\ 0 & \text{if} \quad r = 0 \ \mathit{or} \ s = 0 \end{cases}$$

and

$$u^{(r)} \wedge v^{(s)} = [u^{(r)}v^{(s)}]_{r+s}.$$

Consequently the "inner product" between r- and s-vectors is an $|r-s|$-vector or zero while their "outer product" is an $(r+s)$-vector.

More generally, if $a = \sum_{k=0}^{n}[a]_k$ and $b = \sum_{l=0}^{n}[b]_l$ are arbitrary elements in $\mathbf{R}^{p,q}$ decomposed following (1.7), we put

$$a \cdot b = \sum_{k,l}[a]_k \cdot [b]_l$$

and

$$a \wedge b = \sum_{k,l}[a]_k \wedge [b]_l.$$

In particular notice that for a 1-vector u and an r-vector $v^{(r)}$, their product in $\mathbf{R}_{p,q}$ and their "inner" and "outer" products are related by the following fundamental formulas

$$\begin{aligned} u \cdot v^{(r)} &= [uv^{(r)}]_{r-1} = \frac{1}{2}(uv^{(r)} - (-1)^r v^{(r)}u) \\ u \wedge v^{(r)} &= [uv^{(r)}]_{r+1} = \frac{1}{2}(uv^{(r)} + (-1)^r v^{(r)}u) \end{aligned}$$

$$uv^{(r)} = u \cdot v^{(r)} + u \wedge v^{(r)}.$$

For further properties concerning inner and outer products between multivectors we refer to [36]. Referring also to subsection 1.6.1. we thus see how universal Clifford algebras and exterior algebras are intimately related.

1.7.2 The main involution on $\mathbf{R}_{p,q}$. The even subalgebra $\mathbf{R}_{p,q}^{+}$ of $\mathbf{R}_{p,q}$

Let $k \in \{0, 1, \ldots, n\}$ be fixed and define on $\mathbf{R}_{p,q}^{k}$ the operator $u \to \hat{u}$ given by $\hat{u} = (-1)^k u$, i.e.

$$\hat{u} = u \Leftrightarrow k \equiv 0 \pmod 2$$

and

$$\hat{u} = -u \Leftrightarrow k \equiv 1 \pmod 2.$$

Furthermore extend this operator by linearity to $\mathbf{R}_{p,q}$, i.e. put for any $a = \sum_{k=0}^{n}[a]_k \in \mathbf{R}_{p,q}$,

$$\hat{a} = \sum_{k=0}^{n} (-1)^k [a]_k = \sum_{k=0}^{n} [a]_k^{\wedge}.$$

Then the map $a \to \hat{a}$ clearly induces a grading (mod 2) on $\mathbf{R}_{p,q}$, namely

$$\mathbf{R}_{p,q} = \mathbf{R}_{p,q}^{+} \oplus \mathbf{R}_{p,q}^{-}$$

with

$$\mathbf{R}_{p,q}^{+} = \sum_{k \text{ even}} \oplus \mathbf{R}_{p,q}^{k}$$

and

$$\mathbf{R}_{p,q}^{-} = \sum_{k \text{ odd}} \oplus \mathbf{R}_{p,q}^{k}.$$

Obviously

$$\begin{array}{lll} \mathbf{R}_{p,q}^{+} \; \mathbf{R}_{p,q}^{+} & \subset & \mathbf{R}_{p,q}^{+} \\ \mathbf{R}_{p,q}^{+} \; \mathbf{R}_{p,q}^{-} & \subset & \mathbf{R}_{p,q}^{-} \\ \mathbf{R}_{p,q}^{-} \; \mathbf{R}_{p,q}^{+} & \subset & \mathbf{R}_{p,q}^{-} \\ \mathbf{R}_{p,q}^{-} \; \mathbf{R}_{p,q}^{-} & \subset & \mathbf{R}_{p,q}^{+}. \end{array}$$

It follows that the map $a \to \hat{a}$ on $\mathbf{R}_{p,q}$ is an isomorphism satisfying $\hat{\hat{a}} = a$ for all $a \in \mathbf{R}_{p,q}$, i.e. it is an involution on $\mathbf{R}_{p,q}$. We thus have that for all $a, b \in \mathbf{R}_{p,q}$ and $\lambda \in \mathbf{R}$

$$\begin{aligned} (\lambda a + b)\hat{} &= \lambda \hat{a} + \hat{b} \\ (ab)\hat{} &= \hat{a}\hat{b} \\ \hat{\hat{a}} &= a. \end{aligned}$$

This involution is called the *main involution* or *inversion* on $\mathbf{R}_{p,q}$. Notice that it extends the orthogonal transformation $x \to -x$ of $\mathbf{R}^{p,q}$ to $\mathbf{R}_{p,q}$.

Its fixed subalgebra $\mathbf{R}_{p,q}^{+}$ is called the *even subalgebra* of $\mathbf{R}_{p,q}$.

Theorem. *$\mathbf{R}_{p,q}^{+}$ is isomorphic to the universal Clifford algebra for the $(n-1)$-dimensional real orthogonal space $\mathbf{R}^{p,q-1}$ or $\mathbf{R}^{q,p-1}$.*

Proof. Indeed, let $e = \{e_1, \ldots, e_p, e_{p+1}, \ldots, e_{p+q}\}$ be an orthonormal basis for $\mathbf{R}^{p,q}$, take $j \in \{1, \ldots, n\}$ fixed and put $f_l = e_j e_l$, $l = 1, \ldots, j-1$, $j+1, \ldots, n$. Then these $n-1$ elements (in fact bi-vectors) form an orthonormal subset of $\mathbf{R}_{p,q}^{+}$ generating $\mathbf{R}_{p,q}^{+}$.

If $e_j^2 = 1$ (i.e. $j \in \{1, \ldots, p\}$ then

$$(1)\ f_k^2 = -1\ ,\ k = 1, \ldots, p; k \neq j$$

$$(2)\ f_k^2 = 1\ ,\ k = p+1, \ldots, n$$

$$(3)\ f_k f_l + f_l f_k = 0\ ,\ l \neq k,$$

whence $\mathbf{R}_{p,q}^+$ is isomorphic to $\mathbf{R}_{q,p-1}$.

If $e_j^2 = -1$ (i.e. $j \in \{p+1, \ldots, p+q\}$) then

$$(1)\ f_k^2 = 1\ ,\ k = 1, \ldots, p$$

$$(2)\ f_k^2 = -1\ ,\ k = p+1, \ldots, n; k \neq j$$

$$(3)\ f_k f_l + f_l f_k = 0\ ,\ l \neq k,$$

whence $\mathbf{R}_{p,q}^+$ is isomorphic to $\mathbf{R}_{p,q-1}$. ∎

A straightforward corollary of the previous result is that for all $p, q \in \mathbf{N}, \mathbf{R}_{q,p-1}$ and $\mathbf{R}_{p,q-1}$ are isomorphic.

Finally notice that in the case where j has been taken in $\{p+1, \ldots, n\}$ (i.e. $e_j^2 = -1$), then an isomorphism Φ_j between $\mathbf{R}_{p,q-1}$ and $\mathbf{R}_{p,q}^+$ may be explicitly constructed as follows.

Consider the orthonormal basis $\{e_1, \ldots, e_p, e_{p+1}, \ldots, e_{j-1}, e_{j+1}, \ldots, e_n\}$ of $\mathbf{R}^{p,q-1}$ and define $\Phi_j : \mathbf{R}_{p,q-1} \to \mathbf{R}_{p,q}^+$ by

$$\Phi_j(u) = \Phi_j(u_0 + u_1) = u_0 + e_j u_1,$$

where the splitting of $u \in \mathbf{R}_{p,q-1}$ into $u_o + u_1$ corresponds to the decomposition $\mathbf{R}_{p,q-1} = \mathbf{R}_{p,q-1}^+ \oplus \mathbf{R}_{p,q-1}^-$, i.e. $u_0 \in \mathbf{R}_{p,q-1}^+$, $u_1 \in \mathbf{R}_{p,q-1}^-$.

Then one may easily check that Φ_j is indeed an isomorphism.

1.7.3 The reversion

Let $k \in \{0, 1, \ldots, n\}$ be fixed and define on $R_{p,q}^k$ the operator $u \to \tilde{u}$ given by

$$\tilde{u} = u \quad if \quad k \equiv 0, 1 \pmod 4$$

$$\tilde{u} = -u \quad if \quad k \equiv 2, 3 \pmod 4.$$

Furthermore extend this operator by linearity to $\mathbf{R}_{p,q}$, i.e. put for any $a = \sum_{k=0}^n [a]_k \in \mathbf{R}_{p,q}$,

$$\tilde{a} = \sum_{k=0}^{n} [a]_k^{\sim}.$$

Then it may be easily checked that the map $a \to \tilde{a}$ defines an anti-involution on $\mathbf{R}_{p,q}$, i.e. for all $a, b \in \mathbf{R}_{p,q}$ and $\lambda \in \mathbf{R}$,

$$\begin{aligned}(\lambda a + b)\tilde{} &= \lambda\tilde{a} + \tilde{b}\\ (ab)\tilde{} &= \tilde{b}\,\tilde{a}\\ \tilde{\tilde{a}} &= a\end{aligned}$$

It is called the *reversion* on $\mathbf{R}_{p,q}$, the reason for this being that for any product of a finite number of elements of $\mathbf{R}^{p,q}$, say $\prod_{j=1}^{s} x_j$, $x_j \in \mathbf{R}^{p,q}$, we have that

$$(x_1 x_2 \ldots x_s)\tilde{} = x_s x_{s-1} \ldots x_1,$$

i.e. the reversion of such a product is the product of the same elements but now taken in the reverse order.

In particular we obtain that for any basis element $e_A = e_{\alpha_1} \ldots e_{\alpha_s}$ of $\mathbf{R}_{p,q}$,

$$\tilde{e}_A = (-1)^{\frac{s(s-1)}{2}} e_A \text{ with } s = \#A.$$

Finally notice that by means of the reversion the orthogonal transformation $x \to x$ of $\mathbf{R}^{p,q}$ is extended to an anti-involution of $\mathbf{R}_{p,q}$.

1.7.4 The conjugation

As is readily seen the inversion and reversion are commutative operators on $\mathbf{R}_{p,q}$, i.e. for all $a \in \mathbf{R}_{p,q}$

$$(\hat{a})\tilde{} = (\tilde{a})\hat{}.$$

The composition of these two operations is called the *conjugation* on $\mathbf{R}_{p,q}$ and it is denoted by $a \to \alpha(a)$, $a \in \mathbf{R}_{p,q}$.

So, by definition, for all $a \in \mathbf{R}_{p,q}$

$$\alpha(a) = (\hat{a})\tilde{} = (\tilde{a})\hat{},$$

which implies that the conjugation is an anti-involution on $\mathbf{R}_{p,q}$ i.e. for all $a, b \in \mathbf{R}_{p,q}$ and $\lambda \in \mathbf{R}$,

$$\begin{aligned}\alpha(\lambda a + b) &= \lambda\alpha(a) + \alpha(b)\\ \alpha(ab) &= \alpha(b)\alpha(a)\\ \alpha(\alpha(a)) &= a.\end{aligned}$$

In particular, as for any basis element $e_A = e_{\alpha_1} \ldots e_{\alpha_s}$ of $\mathbf{R}_{p,q}$,

$$\hat{e}_A = (-1)^s e_A$$

and

$$\tilde{e}_A = (-1)^{\frac{s(s-1)}{2}} e_A,$$

we find that

$$\alpha(e_A) = (-1)^{\frac{s(s+1)}{2}} e_A.$$

1.8 The centers Z and Z^+ of $\mathbf{R}_{p,q}$ and $\mathbf{R}_{p,q}^+$

Without mentioning it explicitly we already determined in section 1.3 the center of a Clifford algebra $\mathcal{C}(X)$ for a real orthogonal space $(X, \mathcal{B})$.

Using the fact that $\mathbf{R}_{p,q}^+$ is isomorphic to $\mathbf{R}_{q,p-1}$ or $\mathbf{R}_{p,q-1}$ (see section 1.7.2) the center Z^+ of $\mathbf{R}_{p,q}^+$ is isomorphic to the center of $\mathbf{R}_{p,q-1}$ (or the center of $\mathbf{R}_{q,p-1}$).

1.8.1 The center Z of $\mathbf{R}_{p,q}$

(1) The case n even

We saw in section 1.3 that, if $p+q=n$ even, the only basis element e_A of $\mathbf{R}_{p,q}$ which commutes with each basis element e_j of $\mathbf{R}^{p,q}$ is the identity element 1. Hence we have that the center Z of $\mathbf{R}_{p,q}$ is isomorphic to the algebra of real numbers, i.e.

$$Z = \mathbf{R}_{p,q}^0 \cong \mathbf{R} \Leftrightarrow n \quad \text{is even} .$$

(2) The case n odd

Again in section 1.3 we saw that, if $p+q=n$ is odd, the only basis elements e_A of $\mathbf{R}_{p,q}$ which commute with each basis element e_j of $\mathbf{R}_{p,q}$ are the identity element 1 and the pseudo-unit e_N. Hence we have that the center Z of $\mathbf{R}_{p,q}$ is given by $\mathbf{R}_{p,q}^0 \oplus \mathbf{R}_{p,q}^n$, i.e.

$$Z \cong \mathbf{R} \oplus \mathbf{R}e_N \Leftrightarrow n \quad \text{is odd} .$$

A further description of Z is obtained as follows.
As for $n = p+q$ odd,

$$e_N^2 = (-1)^{\frac{p-q-1}{2}}$$

we have that

$$e_N^2 = 1 \Leftrightarrow p - q \equiv 1 \pmod 4$$

and

$$e_N^2 = -1 \Leftrightarrow p - q \equiv 3 \pmod 4.$$

Hence for $n = p + q$ odd,

$$Z \cong {}^2\mathbf{R} = \mathbf{R} \oplus \mathbf{R} \Leftrightarrow p - q \equiv 1 \pmod 4$$

while

$$Z \cong \mathbf{C} \Leftrightarrow p - q \equiv 3 \pmod 4.$$

1.8.2 The center Z^+ of $\mathbf{R}_{p,q}^+$

As $\mathbf{R}_{p,q}^+ \cong \mathbf{R}_{p,q-1} \cong \mathbf{R}_{q,p-1}$, its center Z^+ is given by :

(1) The case n even

If $n = p + q$ is even, then $p + q - 1$ is odd whence

$$Z^+ \cong \mathbf{R} \oplus \mathbf{R}e_N \;\; \Leftrightarrow n \text{is even}$$

Moreover for $n = p + q$ even

$$Z^+ \cong {}^2\mathbf{R} = \mathbf{R} \oplus \mathbf{R} \Leftrightarrow p - q \equiv 2 \pmod 4$$

while

$$Z^+ \cong \mathbf{C} \Leftrightarrow p - q \equiv 0 \pmod 4.$$

(2) The case n odd

If $n = p + q$ is odd then $p + q - 1$ is even whence

$$Z^+ \cong \mathbf{R} \Leftrightarrow n \quad \text{is odd.}$$

1.9 The structure of $\mathbf{R}_{p,q}$ *and* $\mathbf{R}_{p,q}^+$

In order to characterize proper two-sided ideals of $\mathbf{R}_{p,q}$ or $\mathbf{R}_{p,q}^+$, it suffices to find out the existence of non trivial central idempotent elements, i.e. elements $z \in Z$ or $z \in Z^+$ such that $z^2 = z$ and $z \neq 0$ and 1.

There are no such candidates when $Z \cong \mathbf{R}$ or $Z^+ \cong \mathbf{R}$.

However, when $Z \cong \mathbf{R} \oplus \mathbf{R}e_N$ or $Z^+ \cong \mathbf{R} \oplus \mathbf{R}e_N$ and moreover $e_N^2 = 1$, then the elements

$$u = \frac{1 + e_N}{2} \quad \text{and} \quad v = \frac{1 - e_N}{2}$$

are the only non-trivial central idempotents in respectively Z or Z^+.

Furthermore in that case $uv = vu = 0$, i.e. they are mutually orthogonal elements. As for $n = p+q$ even,

$$e_N^2 = (-1)^{\frac{p-q}{2}}$$

while for $n = p+q$ odd,

$$e_N^2 = (-1)^{\frac{p-q-1}{2}} \tag{1.10}$$

this prompts the following results.

1.9.1 The case $n = p+q$ even

(1) $Z \cong \mathbf{R}$ whence $\mathbf{R}_{p,q}$ is simple (i.e. it contains no proper two-sided ideals)

(2) $Z^+ \cong \mathbf{R} \oplus \mathbf{R}e_N$ whence

$(2)_1$ $\mathbf{R}_{p,q}^+$ is simple $\Leftrightarrow p - q \equiv 2 \pmod 4$)

$(2)_2$ $\mathbf{R}_{p,q}^+$ is semi-simple $\Leftrightarrow p - q \equiv 0 \pmod 4$. i.e. $\mathbf{R}_{p,q}^+$ is the direct sum of two mutually annihilating simple ideals, namely

$$\mathbf{R}_{p,q}^+ = \frac{1+e_N}{2}\mathbf{R}_{p,q}^+ \oplus \frac{1-e_N}{2}\mathbf{R}_{p,q}^+.$$

1.9.2 The case $n = p+q$ odd

(1) $Z^+ \cong \mathbf{R}$ whence $\mathbf{R}_{p,q}^+$ is simple.

(2) $Z \cong \mathbf{R} \oplus \mathbf{R}e_N$ whence

$(2)_1$ $\mathbf{R}_{p,q}$ is simple $\Leftrightarrow p - q \equiv 3 \pmod 4$)

$(2)_2$ $\mathbf{R}_{p,q}$ is semi-simple $\Leftrightarrow p - q \equiv 1 \pmod 4$ i.e. $\mathbf{R}_{p,q}$ is the direct sum of two mutually annihilating simple ideals namely

$$\mathbf{R}_{p,q} = \frac{1+e_N}{2}\mathbf{R}_{p,q} \oplus \frac{1-e_N}{2}\mathbf{R}_{p,q}.$$

1.9.3 **Remarks**

(1) The results in the subsections 1.9.1. and 1.9.2. show that the roles of $\mathbf{R}_{p,q}$ and $\mathbf{R}_{p,q}^+$ are somewhat interchanged by going from the case n even to the case n odd. Even more, one could think that the classification of the Clifford algebras $\mathbf{R}_{p,q}$ depends only on $(p-q) \pmod 4$. As will be seen in section 2, the classification of $\mathbf{R}_{p,q}$ is completely described by the so-called periodicity theorem which is given in terms of $(p-q) \pmod 8$.

(2) Whenever $\mathbf{R}_{p,q}$ is simple, the famous Wedderburn Theorem tells us that $\mathbf{R}_{p,q}$ is isomorphic to a full matrix algebra. Again we refer to subsection 2 for further information.

1.10 Subgroups of $\mathbf{R}_{p,q}$

Clifford algebras will now be employed to construct two-fold covering groups for the orthogonal groups and for the rotation groups in particular. Let us recall that the orthogonal group and special orthogonal group for $\mathbf{R}^{p,q}$ are denoted respectively by $\mathrm{O}(p,q)$ and $\mathrm{SO}(p,q)$.

1.10.1 The Clifford group

One of the main objectives in constructing a Clifford algebra for $\mathbf{R}^{p,q}$ is to enable multiplication among vectors and this in such a way that for each $x \in \mathbf{R}^{p,q}$

$$x^2 = B(x,x).$$

Using the main involution this requirement can also be expressed as

$$\begin{aligned} x^2 &= -x\hat{x} \\ &= -\hat{x}x \end{aligned}$$

for all $x \in \mathbf{R}^{p,q}$.

Now let $s \in \mathbf{R}_{p,q}$ be such that it is invertible and that for all $x \in \mathbf{R}^{p,q}$

$$sx\hat{s}^{-1} \in \mathbf{R}^{p,q}$$

and consider the associated linear transformation $\chi(s) : \mathbf{R}^{p,q} \to \mathbf{R}^{p,q}$ given by $\chi(s)x = sx\hat{s}^{-1}$.

Then clearly $\chi(s) \in \mathrm{O}(p,q)$ since

$$\begin{aligned} (\chi(s)x)^2 &= B(\chi(s)x, \chi(s)x >) \\ &= -(\chi(s)x)\hat{}\ \chi(s)x \\ &= -(sx\hat{s}^{-1})\hat{}\ s\,x\,\hat{s}^{-1} \\ &= -\hat{s}\hat{x}s^{-1}s\,x\,\hat{s}^{-1} \\ &= -\hat{x}x \\ &= B(x,x). \end{aligned}$$

Here we used the fact that $\hat{x}x \in \mathbf{R}$ and hence $\hat{x}x \in Z$.

Now $\chi(s)$ being an orthogonal transformation on $\mathbf{R}^{p,q}$, for each $y \in \mathbf{R}^{p,q}$ there aught to exist an $x \in \mathbf{R}^{p,q}$ such that $y = sx\hat{s}^{-1}$ or $s^{-1}y\hat{s} = s^{-1}y((s^{-1})\hat{\ })^{-1} = x \in \mathbf{R}^{p,q}$.

Consequently, if $s \in \mathbf{R}_{p,q}$ is invertible and satisfies $sx\hat{s}^{-1} \in \mathbf{R}^{p,q}$ for all $x \in \mathbf{R}^{p,q}$, the same is true for s^{-1}.

Putting

$$\Gamma(p,q) = \{s \in \mathbf{R}_{p,q} : sx\hat{s}^{-1} \in \mathbf{R}^{p,q} \text{ for all } x \in \mathbf{R}^{p,q}\}$$

we thus obtain easily that $\Gamma(p,q)$ is a group under the multiplication in $\mathbf{R}_{p,q}$. It is called the *Clifford group*.

Now we prove that each invertible $s \in \mathbf{R}^{p,q}$ (i.e. for which $B(s,s) \neq 0$) belongs to $\Gamma(p,q)$.

Indeed, suppose that $s \in \mathbf{R}^{p,q}$ is invertible and let $x \in \mathbf{R}^{p,q}$ be taken arbitrarily. Then, as x admits a unique decomposition in $\mathbf{R}^{p,q}$ given by

$$x = \lambda s + t$$

where $\lambda \in \mathbf{R}$ and $t \in \{s\}^{\perp} = \{u \in \mathbf{R}^{p,q} : B(u,s) = 0\}$, we get :

$$\begin{aligned} \chi(s)x &= s(\lambda s + t)\hat{s}^{-1} \\ &= \lambda s^2\hat{s}^{-1} + st\hat{s}^{-1} \\ &= -\lambda s + t. \end{aligned}$$

Here we used the fact that $\hat{s} = -s$ and $st + ts = 2B(s,t) = 0$.

So $s \in \Gamma(p,q)$ and at the same time it is shown that the action of $\chi(s) \in \mathrm{O}(p,q)$ is nothing else but the reflection in the hyperplane H_s (see also section 1.1). Hence, by means of the Cartan-Dieudonné Theorem, any $T \in \mathrm{O}(p,q)$ is the composition of a finite number (say k) of reflections in hyperplanes H_{s_i}, $s_i \in \mathbf{R}^{p,q}$ being invertible, i.e.

$$T = \textstyle\prod_{i=1}^{k} \chi(s_i).$$

But, as $\Gamma(p,q)$ is a group and for any two elements $u, v \in \Gamma(p,q)$, $\chi(uv) = \chi(u)\chi(v)$, we find that

$$T = \chi(\textstyle\prod_{i=1}^{k} s_i)$$

whence

$$\chi : \Gamma(p,q) \to \mathrm{O}(p,q)$$

defines a representation of the Clifford group $\Gamma(p,q)$, the so-called *vectorial representation*.

Now it is easy to check that if $a \in \mathbf{R}_{p,q}$ is such that for each $x \in \mathbf{R}_{p,q}$, $ax = x\hat{a}$, then $a \in \mathbf{R}$.

Consequently ker χ, the kernel of the representation χ, is given by

$$\ker \chi = \mathbf{R} \setminus \{0\} = \mathbf{R}_0$$

and so

$$\Gamma(p,q)/\mathbf{R}_0 \cong \mathrm{O}(p,q).$$

At the same time we obtain that each $s \in \Gamma(p,q)$ is a finite product of invertible elements $s_i \in \mathbf{R}^{p,q}$, say $s = \prod_{i=1}^{l} s_i$, i.e.

$$\Gamma(p,q) = \{\textstyle\prod_{i=1}^{l} s_i : l \in \mathbf{N}, s_i \in \mathbf{R}^{p,q},\ B(s_i,s_i) \neq 0\}.$$

Finally, as any rotation or anti-rotation on $\mathbf{R}^{p,q}$ is, respectively, the product of an even or odd number of reflections in hyperplanes, we obtain

(1) $s \in \Gamma(p,q)$ induces a rotation $\Leftrightarrow s \in \Gamma^+(p,q) = \Gamma(p,q) \cap \mathbf{R}^+_{p,q}$.
As for a general $u \in \mathbf{R}^+_{p,q}$, $\hat{u} = u$, the rotation induced by $s \in \Gamma^+(p,q)$ is thus given by $x \to sxs^{-1}$.

(2) $s \in \Gamma(p,q)$ induces an antirotation $\Leftrightarrow s \in \Gamma^-(p,q) = \Gamma(p,q) \cap \mathbf{R}^-_{p,q}$.
As for a general $u \in \mathbf{R}^-_{p,q}$, $\hat{u} = -u$, the antirotation induced by $s \in \Gamma^-(p,q)$ is thus given by $x \to -sxs^{-1}$.

The subgroup $\Gamma^+(p,q)$ of $\Gamma(p,q)$ is called the *even Clifford group*. Clearly

$$\Gamma(p,q) = \Gamma^+(p,q) \cup \Gamma^-(p,q) \text{ with } \Gamma^+(p,q) \cap \Gamma^-(p,q) = \phi.$$

Calling χ^+ the restriction of χ to $\Gamma^+(p,q)$ we have that

$$\chi^+ : \Gamma^+(p,q) \to \mathrm{SO}(p,q)$$

defines a representation of $\Gamma^+(p,q)$, its kernel being

$$\ker \chi^+ = \mathbf{R}_0.$$

Hence

$$\Gamma^+(p,q)/\mathbf{R}_0 \cong \mathrm{SO}(p,q).$$

1.10.2 The Pin and Spin groups

Now we introduce a norm to extract out of the Clifford group the two-fold covering group of respectively the orthogonal group and the rotation group.

Define on $\Gamma(p,q)$ the *spinorial norm* $\mathcal{N}$ by

$$\mathcal{N}(s) = \alpha(s)s, \quad s \in \Gamma(p,q).$$

Then for each invertible $s \in \mathbf{R}^{p,q}$, $\mathcal{N}(s) = \alpha(s)s \in \mathbf{R}_0$ so that, $\Gamma(p,q)$ being generated by such elements, $\mathcal{N}(s) \in \mathbf{R}_0$ for all $s \in \Gamma(p,q)$. Moreover it is easily checked that $|\mathcal{N}|$ is a group homomorphism from $\Gamma(p,q)$ onto the multiplicative group $\mathbf{R}_0^+$, where for $s \in \Gamma(p,q)$, $|\mathcal{N}|(s) = |\mathcal{N}(s)|$.

Hence

$$\mathrm{Pin}(p,q) = \{s \in \Gamma(p,q) : |\mathcal{N}(s)| = 1\}$$

and

$$\mathrm{Spin}(p,q) = \{s \in \Gamma^+(p,q) : |\mathcal{N}(s)| = 1\}$$

are normal subgroups of, respectively, $\Gamma(p,q)$ and $\Gamma^+(p,q)$. They are called the *Pin and Spin groups* for the real orthogonal space $\mathbf{R}^{p,q}$.

As is readily seen,

$$\mathrm{Pin}(p,q) \cong \Gamma(p,q)/\mathbf{R}_0^+$$

and

$$\mathrm{Spin}(p,q) \cong \Gamma^+(p,q)/\mathbf{R}_0^+.$$

Notice that, as $e_i \in \mathrm{Pin}(p,q)$ for each $i = 1,\ldots,p+q$, and $e_ie_j \in \mathrm{Spin}(p,q)$, $i < j$, $i,j = 1,2,\ldots,p+q$, $\mathrm{Pin}(p,q)$ and $\mathrm{Spin}(p,q)$ generate, respectively, $\mathbf{R}_{p,q}$ and $\mathbf{R}_{p,q}^+$.

Now define

$$\tilde{\chi} : \mathrm{Pin}(p,q) \to \mathrm{O}(p,q)$$

and

$$\tilde{\chi}^+ : \mathrm{Spin}(p,q) \to \mathrm{SO}(p,q)$$

by

$$\tilde{\chi}(s\mathbf{R}_0^+) = \chi(s), s \in \Gamma(p,q)$$

and

$$\tilde{\chi}^+(s\mathbf{R}_0^+) = \chi^+(s), s \in \Gamma^+(p,q)$$

where χ and χ^+ are the group homomorphisms given in the foregoing subsection.

Then ker$\tilde{\chi} \cong \mathbf{Z}_2 = \{1, -1\}$ and ker$\tilde{\chi}^+ \cong \mathbf{Z}_2 = \{1, -1\}$ whence Pin $(p,q)/\mathbf{Z}_2 \cong$ O(p,q) and Spin $(p,q)/\mathbf{Z}_2 \cong$ SO(p,q).

This implies that Pin (p,q) and Spin (p,q) are two-fold covering groups of, respectively, the orthogonal group O(p,q) and the rotation group SO(p,q). In other words, we have the exact sequences

$$1 \to \mathbf{Z}_2 \to \text{Pin }(p,q) \to \text{O}(p,q) \to 1$$

and

$$1 \to \mathbf{Z}_2 \to \text{Spin }(p,q) \to \text{SO}(p,q) \to 1.$$

Notice that, although in general Pin (p,q) and Pin (q,p) are not identical, Spin (p,q) and Spin (q,p) are isomorphic for any p and q and this since $\mathbf{R}^+_{p,q}$ and $\mathbf{R}^+_{q,p}$ are isomorphic.

Finally the subgroup Spin $_+(p,q)$ of Spin (p,q) is defined by

$$\text{Spin }_+(p,q) = \{s \in \text{Spin }(p,q) : \mathcal{N}(s) = 1\}$$

Its image in SO(p,q) under $\tilde{\chi}^+$ is then a subgroup of SO(p,q), called the *proper Lorentz group* of $\mathbf{R}^{p,q}$ and denoted by SO$_+(p,q)$.

Spin$_+(p,q)$ is thus a two-fold covering group of SO$_+(p,q)$, which in its turn is the identity component of SO(p,q).

Let us illustrate this in the case of the four dimensional real orthogonal space $\mathbf{R}^{1,3}$, also called *space-time*.

The orthogonal group O$(1,3)$ is called the *Lorentz group*, its elements being called *Lorentz transformations*. The proper Lorentz group SO$_+(1,3)$ is also called the *Lorentz rotation group* and Spin$_+(1,3)$ is said to be the *Spin 1/2-representation* of SO$_+(1,3)$.

1.10.3 The Pin and Spin groups for $\mathbf{R}^{0,n}$

Consider the real orthogonal space $\mathbf{R}^{0,n} (n \geq 1)$ and its universal Clifford algebra $\mathbf{R}_{0,n}$. Then, if $e = \{e_1, \ldots, e_n\}$ is an orthonormal basis for $\mathbf{R}^{0,n}$, for each vector $x = \sum_{j=1}^n x_j e_j \in \mathbf{R}^{0,n}$ we have that $x^2 = B(x,x) = -\sum_{j=1}^n x_j^2$, so that any non-zero vector in $\mathbf{R}^{0,n}$ is invertible, its inverse being given by $\frac{\alpha(x)}{x^2}$.

Hence the unit sphere S^{n-1} in $\mathbf{R}^n$ may be identified with the set of those $s \in \mathbf{R}^{0,n}$ such that $\alpha(s)s = \mathcal{N}(s) = 1$. Denoting the Pin and Spin groups for $\mathbf{R}^{0,n}$ by Pin(n) and Spin(n), it is thus easily obtained that

$$\text{Pin }(n) = \{\textstyle\prod_{i=1}^k s_i \ : \ k \in \mathbf{N}, \ s_i \in S^{n-1}\}$$

and

$$\mathrm{Spin}\,(n) = \{\Pi_{i=1}^{2k} s_i \ : \ k \in \mathbf{N},\ s_i \in S^{n-1}\}.$$

Clearly, as for $n = 1$, $\mathbf{R}^{0,1} = \mathbf{R}e_1$ with $e_1^2 = -1$, we have that

$$\mathrm{Spin}\,(1) = \{1, -1\} \cong \mathbf{Z}_2.$$

Characterizations of Spin (n) in the cases $n = 2, 3, 4$ were treated in Chapter 0.

1.10.4 Lie algebras of Spin groups

In the preceding subsections we saw how the groups $\Gamma(p,q)$, $\mathrm{Pin}(p,q)$, $\mathrm{Spin}_+(p,q)$ and $\mathrm{Spin}(p,q)$ arise as subgroups of the Clifford algebra $\mathbf{R}_{p,q}$. Hence their associated Lie algebras can be realized as subsets of $\mathbf{R}_{p,q}$. More precisely, it may be proved that the Lie algebra of the Clifford group $\Gamma(p,q)$ is the sub Lie algebra of $\mathbf{R}_{p,q}$ consisting of the scalars and bivectors, while the Lie algebra of the groups $\mathrm{Pin}(p,q)$, $\mathrm{Spin}(p,q)$ and $\mathrm{Spin}_+(p,q)$ is the bivector space $\mathbf{R}^2_{p,q}$ of $\mathbf{R}_{p,q}$.

Let us indicate this in the case of the Lie algebras Lie $(\Gamma(p,q))$ and Lie $(\mathrm{Spin}_+(p,q))$.

To this end first remark that for all $x, y, z \in \mathbf{R}^{p,q}$, the commutator of xy and z, namely $[xy, z] = xyz - zxy$, again belongs to $\mathbf{R}^{p,q}$.

Indeed, as

$$\begin{aligned} yz + zy &= 2y \cdot z \\ xz + zx &= 2x \cdot z \end{aligned}$$

we get

$$xyz - zxy = 2xy \cdot z - 2yz \cdot x \in \mathbf{R}^{p,q}.$$

Hence for all $j, k \in \mathbf{N}$ with $j + k = l$ we obtain by induction that for $x, y, z \in \mathbf{R}^{p,q}$

$$\sum_{\substack{j,k \\ j+k=l}} (-1)^k \frac{1}{j!k!} (xy)^j z (xy)^k \in \mathbf{R}^{p,q}.$$

Consequently, as $xy \in \mathbf{R}^0_{p,q} \oplus \mathbf{R}^2_{p,q}$, $(xy)\hat{} = xy$ and so $(\exp(xy))\hat{} = \exp(xy)$, whence for all $t \in \mathbf{R}$ and $x, y, z \in \mathbf{R}^{p,q}$, $\exp(txy)z((\exp(txy))\hat{})^{-1} = \exp(txy)z\exp(-txy) \in \mathbf{R}^{p,q}$.

Hence $\exp(txy) \in \Gamma(p,q)$ which implies that $xy \in \mathrm{Lie}\,(\Gamma(p,q))$.

As to $\mathrm{Lie}\,\mathrm{Spin}_+(p,q)$, notice that for each $a \in \mathbf{R}_{p,q}$,

$$\alpha(\exp(a)) = \exp\,\alpha(a)$$

so that, if $\exp(a) \in \Gamma(p,q)$,

$$\mathcal{N}(\exp(a)) = \exp\,\alpha(a)\exp(a).$$

Consequently, for $x, y \in \mathbf{R}^{p,q}$

$$\exp(xy) \in \mathrm{Spin}_+(p,q)$$

$$\begin{aligned}
&\Leftrightarrow \ \mathcal{N}(\exp(xy)) = 1 \\
&\Leftrightarrow \ \exp(\alpha(x)y)\exp(xy) = 1 \\
&\Leftrightarrow \ \exp(yx + xy) = 1 \\
&\Leftrightarrow \ xy + yx = 0 \\
&\Leftrightarrow \ x \cdot y = 0.
\end{aligned}$$

Therefore each element of $\mathrm{Lie}(\mathrm{Spin}_+(p,q))$ can be written as a linear combination of products xy where $x, y \in \mathbf{R}^{p,q}$ with x orthogonal to y. But for such x and y, $xy \in \mathbf{R}^2_{p,q}$ whence

$$\mathrm{Lie}(\mathrm{Spin}_+(p,q)) = \mathbf{R}^2_{p,q}.$$

Analogously

$$\mathrm{Lie}(\mathrm{Spin}(p,q)) = \mathbf{R}^2_{p,q};$$

$$\mathrm{Lie}(\mathrm{Pin}(p,q)) = \mathbf{R}^2_{p,q}.$$

2 Classification of the real Clifford algebras $\mathbf{R}_{p,q}$

The aim of this section is to give matrix algebras which are isomorphic to the algebras $\mathbf{R}_{p,q}$ up to $n = p + q < 8$. Using the modulo 8 periodicity theorem of real Clifford algebras a complete table of matrix algebras representing $\mathbf{R}_{p,q}$ for all $p, q \in \mathbf{N}$ may then be constructed.

2.1 A reduction theorem

In this subsection a theorem will be proved which will enable us to build "high dimensional" Clifford algebras out of "small dimensional" ones.

Let X be a finite dimensional real vector space endowed with a non-degenerate symmetric bilinear form $\mathcal{B}$; let Y be an *even* dimensional subspace of X such that the restriction $\mathcal{B}_Y$ of $\mathcal{B}$ to Y is non-degenerate too. Furthermore let $Y^\perp$ be the orthogonal complement of Y in X and let $\mathcal{C}(X,\mathcal{B})$, $\mathcal{C}(Y,\mathcal{B}_Y)$ and $\mathcal{C}(Y^\perp, \varepsilon_Y^2\mathcal{B}_{Y^\perp})$ stand for the universal Clifford algebras generated respectively by the orthogonal spaces $(X,\mathcal{B})$, $(Y,\mathcal{B}_Y)$ and $(Y^\perp, \varepsilon_Y^2\mathcal{B}_{Y^\perp})$ where ε_Y and $\varepsilon_Y^2\mathcal{B}_{Y^\perp}$ are defined as follows. If $\dim Y = t$ and $\{e_1, \ldots, e_t\}$ is an orthonormal basis for $(Y,\mathcal{B}_Y)$, then ε_Y stands for the element

$$\varepsilon_Y = e_1 e_2 \ldots e_t.$$

Moreover $\mathcal{B}_{Y^\perp}$ denotes the restriction of $\mathcal{B}$ to $Y^\perp$ whence $\varepsilon_Y^2\mathcal{B}_{Y^\perp}$ defines a non-degenerate symmetric bilinear form on $Y^\perp$.

Theorem 2.1.1. (Dimensional Reduction theorem) *Under the previous assumptions we have*

$$\mathcal{C}(X,\mathcal{B}) \cong \mathcal{C}(Y,\mathcal{B}_Y) \otimes \mathcal{C}(Y^\perp, \varepsilon_Y^2\mathcal{B}_{Y^\perp})$$

Proof. Suppose $(Y,\mathcal{B}_Y)$ to be of type $\mathbf{R}^{p_1,q_1}(p_1 + q_1 = t$ even) and $(Y^\perp, \varepsilon_Y^2\mathcal{B}_{Y^\perp})$ to be of type $\mathbf{R}^{p_2,q_2}(p_2 + q_2 = n - t,\ n = \dim X)$. Furthermore let $\{e_1, \ldots, e_t\}$ be an orthonormal basis for $(Y,\mathcal{B}_Y)$ and $\{e_{t+1}, \ldots, e_n\}$ be an orthonormal basis for $(Y^\perp, \varepsilon_Y^2\mathcal{B}_{Y^\perp})$, i.e.

$$e_\alpha e_{\alpha'} + e_{\alpha'} e_\alpha = 2\mathcal{B}(e_{\alpha'}, e_{\alpha'})$$

and

$$e_\beta e_{\beta'} + e_{\beta'} e_\beta = 2\varepsilon_Y^2\mathcal{B}(e_\beta, e_{\beta'})$$

for all $\alpha, \alpha' = 1, \ldots, t$ and $\beta, \beta' = t+1, \ldots, n$.

Then define the following elements in $\mathcal{C}(Y,\mathcal{B}_Y) \otimes \mathcal{C}(Y_\perp, \varepsilon_Y^2\mathcal{B}_{Y^\perp})$:

$$\begin{aligned} \gamma_\alpha &= e_\alpha \otimes 1 \quad , \quad \alpha = 1, \ldots, t; \\ \gamma_\beta &= \varepsilon_Y \otimes e_\beta \quad , \quad \beta = t+1, \ldots, n, \end{aligned}$$

where 1 is the identity element in $\mathbf{R}_{p_2,q_2}$.

Obviously

$$\gamma_\alpha\gamma_{\alpha'} + \gamma_{\alpha'}\gamma_\alpha = 2\mathcal{B}(e_\alpha, e_{\alpha'})$$

and

$$\gamma_\beta\gamma_{\beta'} + \gamma_{\beta'}\gamma_\beta = 2\mathcal{B}(e_\beta, e_{\beta'})$$

for all $\alpha, \alpha' = 1, \ldots, t$ and $\beta, \beta' = t+1, \ldots, n$.

But, as $\dim Y = p_1 + q_1 = t$ is even, ε_Y anticommutes with $e_\alpha, \alpha = 1, \ldots, t$, so that

$$\gamma_\alpha\gamma_\beta + \gamma_\beta\gamma_\alpha = 0$$

for all $\alpha = 1, \ldots, t$ and $\beta = t+1, \ldots, n$. As $\dim(\mathcal{C}(Y, \mathcal{B}_Y) \otimes \mathcal{C}(Y^\perp, \varepsilon_Y^2\mathcal{B}_{Y^\perp})) = 2^n$ and the elements $\gamma_\alpha, \gamma_\beta, \alpha = 1, \ldots, t, \beta = t+1, \ldots, n$ form a set of generators satisfying the basic relations with respect to the bilinear form $\mathcal{B}$ we obtain the desired isomorphism between $\mathcal{C}(Y, \mathcal{B}_Y) \otimes \mathcal{C}(Y^\perp, \varepsilon_Y^2\mathcal{B}_{Y^\perp})$ and $\mathcal{C}(X, \mathcal{B})$. ■

Corollary 2.1.2. (Recurrence Relation). *Let $\mathbf{R}_{p_1,q_1}$ and $\mathbf{R}_{p_2,q_2}$ be the universal Clifford algebras for the real orthogonal spaces $\mathbf{R}^{p_1,q_1}$ and $\mathbf{R}^{p_2,q_2}$ where p_1+q_1 is even. Furthermore let ε be the pseudo-unit of $\mathbf{R}_{p_1,q_1}$. Then we have*

(i) For $\varepsilon^2 = 1$ ($\Leftrightarrow$ $p_1 - q_1 \equiv 0 (mod\ 4)$), $\mathbf{R}_{p_1,q_1} \otimes \mathbf{R}_{p_2,q_2} \cong \mathbf{R}_{p_1+p_2,q_1+q_2}$.

(ii) For $\varepsilon^2 = -1$ ($\Leftrightarrow$ $p_1 - q_1 \equiv 2 (mod\ 4)$), $\mathbf{R}_{p_1,q_1} \otimes \mathbf{R}_{p_2,q_2} \cong \mathbf{R}_{p_1+q_2,q_1+p_2}$.

Corollary 2.1.3. *Let $p_1, q_1 \in \mathbf{N}$ be such that $p_1 + q_1 = 2$ and let $(p,q) \in \mathbf{N} \times \mathbf{N}$. Then we have :*

(i) If $(p_1, q_1) = (2, 0)$, whence $\varepsilon^2 = -1$, then $\mathbf{R}_{2,0} \otimes \mathbf{R}_{p,q} \cong \mathbf{R}_{q+2,p}$.

(ii) $(p_1, q_1) = (0, 2)$, whence $\varepsilon^2 = -1$, then $\mathbf{R}_{0,2} \otimes \mathbf{R}_{p,q} \cong \mathbf{R}_{q,p+2}$.

(iii) If $(p_1, q_1) = (1, 1)$, whence $\varepsilon^2 = 1$, then $\mathbf{R}_{1,1} \otimes \mathbf{R}_{p,q} \cong \mathbf{R}_{p+1,q+1}$.

2.2 Low dimensional cases

In this subsection we derive matrix representations for the Clifford algebras $\mathbf{R}_{p,q}$ where $p + q \leq 2$.

2.2.1 The Clifford algebras $\mathbf{R}_{0,0}$; $\mathbf{R}_{0,1}$; $\mathbf{R}_{1,0}$

By definition $\mathbf{R}_{0,0} = \mathbf{R}$, its basis being given by the identity element $e_\phi = 1$.
As $\mathbf{R}_{0,1}$ is generated by 1 and e_1 with $e_1^2 = -1$, $\mathbf{R}_{0,1}$ is isomorphic to $\mathbf{C}$.
$\mathbf{R}_{1,0}$ is generated by 1 and e_1 with $e_1^2 = 1$ so that it is isomorphic to ${}^2\mathbf{R} = \mathbf{R} \oplus \mathbf{R}$.

2.2.2 The Clifford algebras $\mathbf{R}_{2,0}$; $\mathbf{R}_{1,1}$; $\mathbf{R}_{0,2}$

Consider the Pauli-matrices

$$\sigma^0 = \begin{pmatrix} 1 & 0 \\ 0 & 1 \end{pmatrix} \quad ; \quad \sigma^1 = \begin{pmatrix} 0 & 1 \\ 1 & 0 \end{pmatrix} \quad ;$$

$$i\sigma^2 = \begin{pmatrix} 0 & 1 \\ -1 & 0 \end{pmatrix} \quad ; \quad \sigma^3 = \begin{pmatrix} 1 & 0 \\ 0 & -1 \end{pmatrix}.$$

(1) The case $\mathbf{R}_{2,0}$

Take $\gamma^1 = \sigma^3$, $\gamma^2 = \sigma^1$; then

$$(\gamma^1)^2 = (\gamma^2)^2 = \begin{pmatrix} 1 & 0 \\ 0 & 1 \end{pmatrix}$$

and

$$\gamma^1\gamma^2 + \gamma^2\gamma^1 = \begin{pmatrix} 0 & 0 \\ 0 & 0 \end{pmatrix}.$$

Hence the matrices γ^1 and γ^2 generate a 2^2-dimensional real associative algebra $\mathcal{A}$ which is isomorphic to $\mathbf{R}_{2,0}$.
As $\mathcal{A} \subset \mathbf{R}(2)$, the algebra of 2×2 real matrices,

$$\mathbf{R}_{2,0} \cong \mathbf{R}(2).$$

(2) The case $\mathbf{R}_{1,1}$

Choose $\gamma^1 = \sigma^1$, $\gamma^2 = i\sigma^2$; then

$$(\gamma^1)^2 = \begin{pmatrix} 1 & 0 \\ 0 & 1 \end{pmatrix} ; (\gamma^2)^2 = \begin{pmatrix} -1 & 0 \\ 0 & -1 \end{pmatrix} ;$$

$$\gamma^1\gamma^2 + \gamma^2\gamma^1 = \begin{pmatrix} 0 & 0 \\ 0 & 0 \end{pmatrix}.$$

Hence the matrices γ^1, γ^2 generate a 2^2-dimensional real associative algebra $\mathcal{A}$ which is isomorphic to $\mathbf{R}_{1,1}$. As $\mathcal{A} \subset \mathbf{R}(2)$, we have that

$$\mathbf{R}_{1,1} \cong \mathbf{R}(2).$$

(3) The case $\mathbf{R}_{0,2}$

Choose $\gamma^1 = i\sigma^1$, $\gamma^2 = i\sigma^2$; then

$$(\gamma^1)^2 = (\gamma^2)^2 = \begin{pmatrix} -1 & 0 \\ 0 & -1 \end{pmatrix},$$

$$\gamma^1\gamma^2 + \gamma^2\gamma^1 = \begin{pmatrix} 0 & 0 \\ 0 & 0 \end{pmatrix}.$$

Hence the matrices γ^1 , γ^2 generate a 2^2-dimensional real associative algebra $\mathcal{A}$ which is isomorphic to $\mathbf{R}_{0,2}$.
As an arbitrary element $a \in \mathcal{A}$ may be written as

$$a = a_0 I_4 + a_1\gamma^1 + a_2\gamma^2 + a_{12}\gamma^1\gamma^2$$

where $a_0, \ldots, a_{12} \in \mathbf{R}$ and

$$(\gamma^1)^2 = (\gamma^2)^2 = (\gamma^1\gamma^2)^2 = -I_4,$$

it is clear that $\mathcal{A}$ is isomorphic to $\mathbf{H}$, the algebra of real quaternions and so

$$\mathbf{R}_{0,2} \cong \mathbf{H}.$$

2.3 Some tensor products of algebras

In the final classification of $\mathbf{R}_{p,q}$ we shall frequently make use of the following well known tensor products of algebras :

$$\begin{aligned} \mathbf{R}(n) \otimes_{\mathbf{R}} \mathbf{R}(m) &\cong \mathbf{R}(nm) \\ \mathbf{R}(n) \otimes_{\mathbf{R}} \mathbf{C} &\cong \mathbf{C}(n) \\ \mathbf{R}(n) \otimes_{\mathbf{R}} \mathbf{H} &\cong \mathbf{H}(n). \end{aligned} \tag{2.1}$$

As usual, when K is $\mathbf{R}$, $\mathbf{C}$ or $\mathbf{H}$, $K(n)$ stands for the (real) algebra of $n \times n$ matrices over K.

Moreover we have that

$$\begin{aligned} \mathbf{C} \otimes_{\mathbf{R}} \mathbf{C} &\cong \mathbf{C} \oplus \mathbf{C} = {}^2\mathbf{C} \\ \mathbf{H} \otimes_{\mathbf{R}} \mathbf{C} &\cong \mathbf{C}(2) \\ \mathbf{H} \otimes_{\mathbf{R}} \mathbf{H} &\cong \mathbf{R}(4) \end{aligned} \tag{2.2}$$

Notice that by means of (2.1) and (2.2), any (finite) tensor product of matrix algebras over $\mathbf{R}$, $\mathbf{C}$ or $\mathbf{H}$ may be expressed as a sum of matrix algebras over $\mathbf{R}$, $\mathbf{C}$ or $\mathbf{H}$. Examples of such tensor products are worked out in 2.4.2.

2.4 The periodicity theorem

In this subsection it is shown how the characterization of $\mathbf{R}_{p,q}$ as a matrix algebra over $\mathbf{R}$, $\mathbf{C}$ or $\mathbf{H}$ is completely determined by the value of $p-q \pmod 8$.

To this end we prove

Theorem (The periodicity theorem) *For all p, q in* $\mathbf{N}$

$$\mathbf{R}_{p,q+8} \cong \mathbf{R}_{p,q} \otimes \mathbf{R}(16) \cong \mathbf{R}_{p+8,q}.$$

Proof. Let us prove the isomorphism $\mathbf{R}_{p,q+8} \cong \mathbf{R}_{p,q} \otimes \mathbf{R}(16)$, the other case being similar.

Using repeatedly Corollary 2.1.3., we get :

$$\begin{aligned}\mathbf{R}_{p,q+8} &\cong \mathbf{R}_{0,2} \otimes \mathbf{R}_{q+6,p} \\ &\cong \mathbf{R}_{0,2} \otimes \mathbf{R}_{2,0} \otimes \mathbf{R}_{p,q+4} \\ &\cong \mathbf{R}_{0,2} \otimes \mathbf{R}_{2,0} \otimes \mathbf{R}_{0,2} \otimes \mathbf{R}_{q+2,p} \\ &\cong \mathbf{R}_{0,2} \otimes \mathbf{R}_{2,0} \otimes \mathbf{R}_{0,2} \otimes \mathbf{R}_{2,0} \otimes \mathbf{R}_{p,q}\end{aligned}$$

But, as $\mathbf{R}_{0,2} \cong \mathbf{H}$ and $\mathbf{R}_{2,0} \cong \mathbf{R}(2)$, we obtain the desired result using the appropriate identities out of (2.1) and (2.2). ■

Theorem 2.4.1. tells us that all $\mathbf{R}_{p,q}$ may be constructed provided we know them for $p+q \leq 8$. The latter cases are given in the following table.

$p+q$ \ $q-p$	-8	-7	-6	-5	-4	-3	-2	-1	0	1	2	3	4	5	6	7	8
0									**R**								
1								^{2}**R**		**C**							
2							**R**(2)		**R**(2)		**H**						
3						**C**(2)		^{2}**R**(2)		**C**(2)		^{2}**H**					
4					**H**(2)		**R**(4)		**R**(4)		**H**(2)		**H**(2)				
5				^{2}**H**(2)		**C**(4)		^{2}**R**(4)		**C**(4)		^{2}**H**(2)		**C**(4)			
6			**H**(4)		**H**(4)		**R**(8)		**R**(8)		**H**(4)		**H**(4)		**R**(8)		
7		**C**(8)		^{2}**H**(4)		**C**(8)		^{2}**R**(8)		**C**(8)		^{2}**H**(4)		**C**(8)		^{2}**R**(8)	
8	**R**(16)		**H**(8)		**H**(8)		**R**(16)		**R**(16)		**H**(8)		**H**(8)		**R**(16)		**R**(16)

We wish to point out that

(1) The horizontal lines corresponding to $p+q \leq 2$ were obtained in 2.2.

(2) The horizontal lines corresponding to $3 \leq p+q \leq 4$ may be constructed by considering tensor products of the form $\mathbf{R}_{0,2} \otimes \mathbf{R}_{p,q}$ and $\mathbf{R}_{0,2} \otimes \mathbf{R}_{p,q}$ with $p+q \leq 2$ and using appropriate identities out of (2.1) and (2.2).

(3) The horizontal lines corresponding to $5 \leq p+q \leq 8$ are obtained successively by taking in a suitable way tensor products of the form

$$\begin{aligned} \mathbf{R}_{1,1} \otimes \mathbf{R}_{p,q} &\cong \mathbf{R}_{p+1,q+1} \\ \mathbf{R}_{0,2} \otimes \mathbf{R}_{s,0} &\cong \mathbf{R}_{0,s+2} \\ \mathbf{R}_{2,0} \otimes \mathbf{R}_{0,s} &\cong \mathbf{R}_{s+2,0} \end{aligned}$$

and identities out of (2.1) and (2.2).

Let us illustrate this procedure by showing how the line corresponding to $p+q=7$ may be obtained from the line corresponding to $p+q=5$. As is seen in the table

$$\begin{aligned} \mathbf{R}_{5,0} &\cong {}^2\mathbf{H}(2) \\ \mathbf{R}_{4,1} &\cong \mathbf{C}(4) \\ \mathbf{R}_{3,2} &\cong {}^2\mathbf{R}(4) \\ \mathbf{R}_{2,3} &\cong \mathbf{C}(4) \\ \mathbf{R}_{1,4} &\cong {}^2\mathbf{H}(2) \\ \mathbf{R}_{0,5} &\cong \mathbf{C}(4). \end{aligned}$$

Hence, in virtue of $\mathbf{R}_{1,1} \otimes \mathbf{R}_{p,q} \cong \mathbf{R}_{p+1,q+1}, p+q=5$, and $\mathbf{R}_{1,1} \cong \mathbf{R}(2)$, we obtain :

$$\begin{aligned} \mathbf{R}_{6,1} &\cong {}^2\mathbf{H}(2) \otimes \mathbf{R}(2) \\ &\cong \mathbf{H} \otimes {}^2\mathbf{R}(2) \otimes \mathbf{R}(2) \\ &\cong \mathbf{H} \otimes {}^2\mathbf{R}(4) \\ &\cong {}^2\mathbf{H}(4); \end{aligned}$$

$$\begin{aligned} \mathbf{R}_{5,2} &\cong \mathbf{C}(4) \otimes \mathbf{R}(2) \\ &\cong \mathbf{R}(4) \otimes \mathbf{R}(2) \otimes \mathbf{C} \\ &\cong \mathbf{C}(8); \end{aligned}$$

$$\begin{aligned}\mathbf{R}_{4,3} &\cong {}^2\mathbf{R}(4)\otimes\mathbf{R}(2)\\ &\cong {}^2\mathbf{R}(8);\end{aligned}$$

$$\begin{aligned}\mathbf{R}_{3,4} &\cong \mathbf{C}(4)\otimes\mathbf{R}(2)\\ &\cong \mathbf{C}(8);\end{aligned}$$

$$\begin{aligned}\mathbf{R}_{2,5} &\cong {}^2\mathbf{H}(2)\otimes\mathbf{R}(2)\\ &\cong \mathbf{H}\otimes {}^2\mathbf{R}(2)\otimes\mathbf{R}(2)\\ &\cong {}^2\mathbf{H}(4);\end{aligned}$$

$$\begin{aligned}\mathbf{R}_{1,6} &\cong \mathbf{C}(4)\otimes\mathbf{R}(2)\\ &\cong \mathbf{C}(8).\end{aligned}$$

Finally

$$\begin{aligned}\mathbf{R}_{0,7} &\cong \mathbf{R}_{0,2}\otimes\mathbf{R}_{5,0}\\ &\cong \mathbf{H}\otimes {}^2\mathbf{H}(2)\\ &\cong \mathbf{H}\otimes\mathbf{H}\otimes {}^2\mathbf{R}(2)\\ &\cong \mathbf{R}(4)\otimes {}^2\mathbf{R}(2)\\ &\cong {}^2\mathbf{R}(8)\end{aligned}$$

and

$$\begin{aligned}\mathbf{R}_{7,0} &\cong \mathbf{R}_{2,0}\otimes\mathbf{R}_{0,5}\\ &\cong \mathbf{R}(2)\otimes\mathbf{C}(4)\\ &\cong \mathbf{C}(8).\end{aligned}$$

3 Complex Clifford algebras

In this section we introduce the complex Clifford algebra $\mathbf{C}_n$ and give its full classification in terms of matrix algebras over $\mathbf{C}$.

3.1 Chevalley's definition revisited

In 1.5 we described how to define the universal Clifford algebra $\mathcal{C}(X)$ associated with the real n-dimensional non-degenerate orthogonal space $(X, \mathcal{B})$.

In fact it was shown by Chevalley (see [17]) that any n-dimensional vector space V over a field F with characteristic zero and provided with a non-degenerate symmetric bilinear form $\mathcal{B}$ admits a universal Clifford algebra $\mathcal{C}(V)$ given by

$$\mathcal{C}(V) = \otimes V / I(\mathcal{Q}).$$

Here $I(\mathcal{Q})$ is the two-sided ideal in $\otimes V$ generated by elements of the form

$$v \otimes v - \mathcal{Q}(v), \quad v \in V,$$

$\mathcal{Q}$ being the quadratic form associated with $\mathcal{B}$.

Again, if $\{e_1, \ldots, e_n\}$ is an orthonormal basis of $(V, \mathcal{B})$, then the set of elements

$$e_A = e_{\alpha_1} e_{\alpha_2} \ldots e_{\alpha_s}, \qquad 1 \leq \alpha_1 < \ldots < \alpha_s \leq n,$$

forms a basis for $\mathcal{C}(V)$, whence $\mathcal{C}(V)$ is a linear associative algebra over F having dimension 2^n.

3.2 Complex Clifford algebras

3.2.1 When the field F is chosen to be $\mathbf{C}$, then each n-dimensional complex vector space V provided with a non-degenerate symmetry bilinear form $\mathcal{B}$ is isomorphic to the space $\mathbf{C}^n$ equipped with the bilinear form

$$B(z, z') = -\sum_{j=1}^{n} z_j z_j',$$

i.e. one can always find an orthonormal basis $\{e_1, \ldots, e_n\}$ for $(V, \mathcal{B})$ such that

$$\mathcal{B}(e_i, e_j) = -\delta_{ij}, \quad i, j = 1, \ldots, n.$$

In the sequel it is tacitly understood that such a basis has been chosen for $(V, \mathcal{B})$.

Referring to Chevalley's construction we thus obtain the existence of a universal Clifford algebra $\mathcal{C}(V)$ for $(V, \mathcal{B})$.

3.2.2 Now let $(X, \mathcal{B})$ be an n-dimensional non-degenerate real orthogonal space of type (p, q) and consider its complexification

$$X_{\mathbf{C}} = X \otimes_{\mathbf{R}} \mathbf{C}$$

provided with the complexified non-degenerate symmetric bilinear form $\mathcal{B}_{\mathbf{C}}$ of $\mathcal{B}$. Then the universal Clifford algebra for $(X_{\mathbf{C}}, \mathcal{B}_{\mathbf{C}})$ is denoted by $\mathbf{C}_{p,q}$.

Of course, for any two pairs (p, q) and (p', q') with $p + q = p' + q' = n$, we have that

$$\mathbf{C}_{p,q} \cong \mathbf{C}_{p'q'}.$$

In the sequel we write $\mathbf{C}^n$ for the complexification of the real orthogonal space of type $(0, n)$ and $\mathbf{C}_n$ for its associated Clifford algebra $\mathbf{C}_{0,n}$.

Moreover the orthonormal basis $\{e_j \otimes 1 : j = 1, \ldots, n\}$ of the complex quadratic space $\mathbf{C}^n$ will still be denoted by $\{e_1, \ldots, e_n\}$.

3.2.3. Let $\mathbf{R}_{p,q}$ be the universal Clifford algebra for $\mathbf{R}^{p,q}$ and consider the tensor product

$$\tilde{\mathbf{C}}_{p,q} = \mathbf{R}_{p,q} \otimes_{\mathbf{R}} \mathbf{C}.$$

Then clearly $\tilde{\mathbf{C}}_{p,q}$ is isomorphic to $\mathbf{C}_{p,q}$; it is called the *complexified Clifford algebra* of $\mathbf{R}_{p,q}$.

In $\tilde{\mathbf{C}}_{p,q}$ we consider the element

$$\begin{aligned} \theta &= i^{\nu} e_1 e_2 \ldots e_n \\ &= i^{\nu} e_N \end{aligned}$$

where $\nu \in \mathbf{N}$ is chosen in such a way that $\theta^2 = 1$.

The element θ is called the *chirality operator*.

Henceforth we only consider $\tilde{\mathbf{C}}_{0,n}$ and denote it by $\tilde{\mathbf{C}}_n$.

3.2.4. Summarizing we have that for any $(V, \mathcal{B})$

$$\mathcal{C}(V) \simeq \mathbf{C}_n \simeq \tilde{\mathbf{C}}_n.$$

3.3 Involutions on $\mathbf{C}_n$ *- The even subalgebra* $\mathbf{C}_n^+$ *of* $\mathbf{C}_n$

3.3.1 Involutions

Quite in the same way as we defined on $\mathbf{R}_{p,q}$, the maps $a \to \hat{a}$, $a \to \tilde{a}$ and $a \to \alpha(a)$ may be defined on $\mathbf{C}_n$. They are still called *inversion, reversion* and *conjugation* on $\mathbf{C}_n$.

The anti-involution on $\mathbf{C}_n \simeq \tilde{\mathbf{C}}_{p,q}$ which is obtained by taking the tensor product of the conjugation α on $\mathbf{R}_{p,q}$ and the classical complex conjugation on $\mathbf{C}$ is called the *bar-map* on $\mathbf{C}_n$ and denoted by $a \to \bar{a}$.

Notice that the restriction of the bar-map to the real subalgebra $\mathbf{R}_{p,q}$ of $\mathbf{C}_n$ is just the conjugation α.

3.3.2 k-vectors in $\mathbf{C}_n$ - The algebra $\mathbf{C}_n^+$

Call for $k = 0, 1, \ldots, n$ fixed,

$$\mathbf{C}_n^k = \{ \sum_{\# A = k} a_A e_A : a_A \in \mathbf{C}\}$$

the set of *k-vectors* in $\mathbf{C}_n$.

Then

$$\mathbf{C}_n = \sum_{k=0}^{n} \oplus \mathbf{C}_n^k,$$

so that an arbitrary element $a \in \mathbf{C}_n$ admits the decomposition

$$a = \sum_{k=0}^{n} [a]_k,$$

$[a]_k$ being the projection of a onto $\mathbf{C}_n^k$.

Furthermore put

$$\mathbf{C}_n^+ = \sum_{k\ \mathit{even}} \mathbf{C}_n^k$$

and

$$\mathbf{C}_n^- = \sum_{k\ \mathit{odd}} \mathbf{C}_n^k.$$

Then $\mathbf{C}_n^+$ is called the *even subalgebra* of $\mathbf{C}_n$.

Theorem. $\mathbf{C}_n^+ \cong \mathbf{C}_{n-1}$.

Proof. Take the standard orthogonal basis $\{e_1, \ldots, e_n\}$ of $\mathbf{C}^n$ and put $e_j' = e_j e_n$, $j = 1, \ldots, n-1$.

Then the set $\{e_1', \ldots, e_{n-1}'\}$ generates a Clifford algebra which is isomorphic to $\mathbf{C}_{n-1}$ while the set $\{e_1 e_n, \ldots, e_{n-1} e_n\}$ generates $\mathbf{C}_n^+$. ∎

Finally notice that by putting for each $a, b \in \mathbf{C}_n$,

$$(a, b) = [a\bar{b}]_0,$$

an inner product is defined on the complex space $\mathbf{C}_n$.

For any $a = \sum_A a_A e_A$, its norm is thus given by

$$|a|^2 = (a,a) = \sum_A |a_A|^2.$$

In order to turn it into an algebra-norm we should take

$$|a|_0^2 = 2^n |a|^2.$$

3.4 Structure of $\mathbf{C}_n$

It is immediately clear that the center Z of $\mathbf{C}_n$ is given by

$$Z = \mathbf{C} \quad \text{if} \quad n \quad \text{is even}$$

and

$$Z = \mathbf{C} \oplus \mathbf{C}e_N \quad \text{if} \quad n \quad \text{is odd.}$$

Moreover, considering the chirality operator

$$\theta = e_N \quad \text{if} \quad e_N^2 = 1$$

and

$$\theta = i\, e_N \quad \text{if} \quad e_N^2 = -1$$

we find that the elements $\frac{1\pm\theta}{2}$ are primitive idempotents which project $\mathbf{C}_n^+$ on simple components when n is even and $\mathbf{C}_n$ on simple components when n is odd.

So we may summarize as follows.

For $n = 2m$ *even* we have :

(1) $\mathbf{C}_n$ is simple

(2) $Z = \mathbf{C}$

(3) $\mathbf{C}_n^+ = \dfrac{1-\theta}{2}\mathbf{C}_n^+ \oplus \dfrac{1+\theta}{2}\mathbf{C}_n^+$,

i.e. $\mathbf{C}_n^+$ is the direct sum of two mutually annihilating simple ideals $\frac{1-\theta}{2}\mathbf{C}_n^+$ and $\frac{1+\theta}{2}\mathbf{C}_n^+$.

For $n = 2m+1$ *odd* we have :

(1) $\mathbf{C}_n$ is not simple and $\mathbf{C}_n = \frac{1-\theta}{2}\mathbf{C}_n \oplus \frac{1+\theta}{2}\mathbf{C}_n$,

i.e. $\mathbf{C}_n$ is the direct sum of two mutually annihilating simple ideals $\frac{1-\theta}{2}\mathbf{C}_n$ and $\frac{1+\theta}{2}\mathbf{C}_n$.

(2) $Z = \mathbf{C} \oplus \mathbf{C}e_N$

(3) $\mathbf{C}_n^+$ is simple.

As to the center Z^+ of $\mathbf{C}_n^+$ we clearly have

$$Z^+ = \mathbf{C} \oplus \mathbf{C}e_N \Leftrightarrow n \quad \text{is even}$$

and

$$Z^+ = \mathbf{C} \qquad\qquad \Leftrightarrow n \quad \text{is odd.}$$

3.5 Classification of complex Clifford algebras

Let $n \in \mathbf{N}$ be given and let $\{e_1, \ldots, e_n\}$ be the standard orthonormal basis for the complex quadratic space $\mathbf{C}^n$.

Theorem.

(i) For $n = 2m$, $\mathbf{C}_n \cong \mathbf{C}(2^m)$

(ii) For $n = 2m + 1$, $\mathbf{C}_n \cong \mathbf{C}(2^m) \oplus \mathbf{C}(2^m)$.

Proof. Let $\{e_1, \ldots, e_n\}$ and $\{e_1', e_2'\}$ be orthonormal bases for $\mathbf{C}^n$ and $\mathbf{C}^2$ respectively and consider in $\mathbf{C}_n \otimes_{\mathbf{C}} \mathbf{C}_2$ the $n+2$ elements

$$\begin{aligned} e_1'' &= 1 \otimes e_1' \\ e_2'' &= 1 \otimes e_2' \\ e_{j+2}'' &= i\, e_j \otimes e_1' e_2', \qquad j = 1, \ldots, n. \end{aligned}$$

Then they satisfy the relations

$$\begin{aligned} e_k''^2 &= -1, \quad k = 1, \ldots, n+2 \\ e_r'' e_s'' + e_s'' e_r'' &= 0, \qquad r \neq s. \end{aligned}$$

As moreover $\prod_{k=1}^{n+2} e_k'' \neq \pm 1 \otimes 1$, they generate the universal complex Clifford algebra $\mathbf{C}_{n+2}$. But, as by construction $\mathbf{C}_{n+2} \subset \mathbf{C}_n \otimes_{\mathbf{C}} \mathbf{C}_2$ and $\dim \mathbf{C}_{n+2} = \dim \mathbf{C}_n \otimes_{\mathbf{C}} \mathbf{C}_2 = 2^{n+2}$, we obtain that

$$\mathbf{C}_{n+2} \cong \mathbf{C}_n \otimes_{\mathbf{C}} \mathbf{C}_2. \tag{3.1}$$

Obviously, for $n = 1$,

$$\begin{aligned} \mathbf{C}_1 &= \mathbf{C}1 \oplus \mathbf{C}e_1 \\ &\simeq \mathbf{C} \oplus \mathbf{C} \end{aligned} \tag{3.2}$$

while for $n = 2$, by defining

$$\psi : \mathbf{C}^2 \to \mathbf{C}(2)$$

with

$$\psi(e_1) = \begin{pmatrix} i & 0 \\ 0 & -i \end{pmatrix} = i\sigma^3$$

$$\psi(e_2) = \begin{pmatrix} 0 & 1 \\ -1 & 0 \end{pmatrix} = i\sigma^2,$$

σ^2, σ^3 being Pauli-matrices, it is clear that ψ may be extended to an algebra isomorphism between $\mathbf{C}_2$ and $\mathbf{C}(2)$, i.e.

$$\mathbf{C}_2 \cong \mathbf{C}(2). \tag{3.3}$$

By means of (3.1), (3.2) and (3.3) we get by induction that

$$\mathbf{C}_n \cong \mathbf{C}(2^m) \quad \text{for} \quad n = 2m$$

and

$$\mathbf{C}_n \cong \mathbf{C}(2^m) \oplus \mathbf{C}(2^m) \quad \text{for} \quad n = 2m + 1. \quad \blacksquare$$

Remarks.

(1) We saw in §3.4 that $\mathbf{C}_n$ is simple when $n = 2m$ and $\mathbf{C}_n$ is semi-simple when $n = 2m + 1$. This is in full accordance with the Theorem just proved

(2) Let $n \in \mathbf{N}$ be given ($n \geq 1$).
We saw that $\mathbf{C}_{n-1}$ and $\mathbf{C}_n^+$ are isomorphic (see 3.3.2). We thus obtain that $\mathbf{C}_n^+$ is a full matrix algebra when n is odd, while it is a direct sum of two full matrix algebras for n even.

4 Spinor spaces

The main purpose of this section is to describe how representations of the real groups $\mathrm{Pin}(p,q)$ and $\mathrm{Spin}(p,q)$ may be derived from representations of complex Clifford algebras on appropriate finite dimensional complex vector spaces V, called spinor spaces. In this way complex representations of the groups $\mathrm{Pin}(p,q)$ and $\mathrm{Spin}(p,q)$ are investigated.

However, if the spinor space V under consideration allows a real or quaternionic structure, then real or quaternionic representations of these groups are also discussed.

Furthermore a realization is obtained of the spinor space as a minimal left ideal inside the Clifford algebra $\mathbf{C}_{2m}$; this leads to the construction of the γ-matrices.

Notice that our realization of the spinor space brings together two ideas : the first one which considers the spinor space as a minimal left ideal of $\mathbf{C}_{2m}$, the second one which realizes the spinor space as a Grassmann algebra, the latter construction being known to physicists as the Fock space for free fermions.

Finally the existence of Hermitian structures on a spinor space is investigated.

4.1 Real and quaternionic structures on complex vector spaces

In this subsection the notion of a real or quaternionic structure on a finite dimensional complex vector space V is introduced.

4.1.1 The conjugate space of a complex vector space

Let V be a complex vector space and let $V^\star$ stand for the complex vector space defined as follows :

(1) $V = V^\star$ (in the sense of sets of vectors)

(2) For each $\lambda \in \mathbf{C}$ and $v^\star \in V^\star$, the vector $\lambda v^\star$ in $V^\star$ is equal to the vector $\bar{\lambda} v$ in V.

Denoting by * the identity map from $V \to V^\star$ we have that for all $\lambda \in \mathbf{C}$ and $v^\star \in V^\star$,

$$\lambda v^\star = (\bar{\lambda} v)^\star$$

whence the *-map : $V \to V^\star$ is anti-linear.

Moreover $(V^\star)^\star = V$.

The space $V^\star$ is called the *conjugate space* of V.

If V and W are complex vector spaces and $f : V \to W$ is a linear map, then its *conjugate map* $f^\star : V^\star \to W^\star$ is defined by

$$f^\star(v^\star) = (f(v))^\star, \quad v \in V.$$

Clearly $f^\star$ is a linear map from $V^\star$ into $W^\star$.

4.1.2 Real structures on a complex vector space

Let V be a complex vector space. A complex linear map $c : V \to V^\star$ is called a *real structure on* V if and only if $c^\star c = \mathbf{1}$, $\mathbf{1}$ being the identity operator on V.

If c is a real structure on V, then we associate with it the map $\tilde{c} : V \to V$ given by

$$\tilde{c}(v) = (c(v))^\star = c^\star(v^\star).$$

Clearly $\tilde{c}$ is anti-linear on V and $\tilde{c}^2 = \mathbf{1}$, which means that $\tilde{c}$ may be regarded upon as being an "*anti-involution*" on V.

A real structure on V gives rise to a splitting of V into its real and imaginary parts (with respect to c), denoted respectively by $Re\, V$ and $Im\, V$, and given by

$$\begin{aligned} Re\, V &= \{v \in V : \tilde{c}v = v\}, \\ Im\, V &= \{v \in V ; \tilde{c}v = -v\}. \end{aligned}$$

So $Re\, V$ and $Im\, V$ are the eigenspaces of $\tilde{c}$ whence V, considered as a real vector space, admits the decomposition

$$V = Re\, V \oplus Im\, V;$$

here $Re\, V$ and $Im\, V$ are real subspaces of V.

We thus saw that a real structure c on V implies the existence of an "anti-involution" $\tilde{c}$ on V.

The converse is also true.

Indeed, suppose that $\tilde{c}$ is an "anti-involution" on V, i.e. $\tilde{c} : V \to V$ is an anti-linear map satisfying $\tilde{c}^2 = \mathbf{1}$.

Then clearly the map $c : V \to V^\star$ defined by

$$c(v) = (\tilde{c}(v))^\star, \quad v \in V,$$

is a real structure on V.

Let us illustrate this by starting from a real (finite-dimensional) vector space W.

Taking the complexification $V_{\mathbf{C}}$ of V i.e. $V_{\mathbf{C}} = V \otimes_{\mathbf{R}} \mathbf{C}$, it is well known that $V_{\mathbf{C}}$ may be identified with (see [58])

$$V_{\mathbf{C}} = \{x + iy : x, y \in V\}.$$

Now define $\tilde{c} : V_{\mathbf{C}} \to V_{\mathbf{C}}$ by

$$\tilde{c}(x + iy) = x - iy.$$

Then clearly $\tilde{c}$ is an anti-involution on $V_{\mathbf{C}}$.

4.1.3 Quaternionic structures on a complex vector space

Let again V be a complex vector space. Then a complex linear map $c : V \to V^\star$ such that $c^\star c = -\mathbf{1}$ is said to be a *quaternionic structure* on V.

Once we have a quaternionic structure c on V, it becomes clear from what follows how V may be turned into a vector space over $\mathbf{H}$.

Define the quaternionic units i, j and k as being the operators on V given by

$$\begin{aligned} i(v) &= iv \\ j(v) &= (c(v))^\star \\ k(v) &= i(j(v)). \end{aligned}$$

Then clearly $i^2 = -\mathbf{1}$.

Moreover $j^2 = -\mathbf{1}$ and $ij + ji = 0$.

Indeed,

$$\begin{aligned} j^2(v) &= j(j(v)) \\ &= c(c(v)^\star)^\star \\ &= c^\star(c(v)) \\ &= -\mathbf{1}(v) \end{aligned}$$

while for all $v \in V$,

$$(ij + ji)(v) = i(c(v))^\star + (c(iv))^\star$$

so that, as

$$\begin{aligned} (c(iv))^\star &= c^\star(iv)^\star \\ &= c^\star(-iv^\star) \\ &= -ic^\star(v^\star) \\ &= -i(c(v))^\star, \end{aligned}$$

$$(ij + ji)(v) = 0.$$

Note that, as j is the composition of c and the *-map, j is anti-linear on V.

Finally, from the properties of i and j, we may derive easily that k is anti-linear on V with $k^2 = -\mathbf{1}$ and $ki + ik = kj + jk = 0$.

Consequently the set of operators on V of the form $x_0\mathbf{1} + x_1 i + x_2 j + x_3 k$, $x_0, \ldots, x_3 \in \mathbf{R}$, determines a real associative algebra which is isomorphic to $\mathbf{H}$.

If for each $\lambda = x_0\mathbf{1} + x_1 i + x_2 j + x_3 k \in \mathbf{H}$ and $v \in V$, we put

$$\lambda v = (x_0\mathbf{1} + x_1 i + x_2 j + x_3 k)(v),$$

then clearly V becomes a (left) vector space over $\mathbf{H}$.

4.1.4 Some relations between vector spaces over different fields

If V is a vector space over $\mathbf{C}$, then obviously it may be considered as a vector space over $\mathbf{R}$, the latter being denoted by $V_{\mathbf{R}}$.

Clearly, if V is finite dimensional over $\mathbf{C}$, say $\dim_{\mathbf{C}} V = n$, then $V_{\mathbf{R}}$ is finite dimensional over $\mathbf{R}$ with $\dim_{\mathbf{R}} V_{\mathbf{R}} = 2n$.

If V is a real vector space, then its complexification $V_{\mathbf{C}}$ is given by $V_{\mathbf{C}} = V \otimes_{\mathbf{R}} \mathbf{C}$.

If V is finite dimensional over $\mathbf{R}$, say $\dim_{\mathbf{R}} V = n$, then $V_{\mathbf{C}}$ is finite dimensional over $\mathbf{C}$ with $\dim_{\mathbf{C}} V_{\mathbf{C}} = n$.

If V is a finite dimensional vector space over $\mathbf{C}$, then the following hold :

$$\begin{aligned} (V_{\mathbf{R}})_{\mathbf{C}} &\cong V \oplus V^\star \\ V_{\mathbf{R}}^\star &\cong V_{\mathbf{R}} \\ (V^\star)^\star &\cong V. \end{aligned} \tag{4.1}$$

If V is a finite dimensional vector space over $\mathbf{R}$ then

$$(V_{\mathbf{C}})_{\mathbf{R}} \cong V \oplus V. \tag{4.2}$$

As usual the symbol " $\cong$" denotes that the vector spaces under consideration are isomorphic.

4.2 *Representations of algebras and groups*

In this subsection we recall some basic facts about representations of algebras and groups on finite dimensional vector spaces V.

4.2.1 Representations of algebras

In the sequel A will stand for a (finite dimensional) linear associative algebra over the field k, V will be a (finite dimensional) vector space over the field K and $\Phi : A \to \mathrm{End}_K(V)$ will be a representation of A on V.

Here the following three cases will be considered : $k = \mathbf{R}$ and $K = \mathbf{R}$; $k = \mathbf{R}$ and $K = \mathbf{C}$; $k = \mathbf{C}$ and $K = \mathbf{C}$.

Let us recall that a representation Φ of A on V is a k-linear homomorphism of A into $\mathrm{End}_K(V)$, the space V being then called on A-*module* (more precisely on A-module over K) or even a representation of A.

An A-module V is said to be *irreducible* (over K) if it contains no proper subspace W (over K) invariant under all $\Phi(a)$, $a \in A$. In this case we also say that the family $\Phi(A)$ of linear transformations on V *acts irreducibly* on V. If V_1 and V_2 are A-modules (over the same field K) and Φ_1, Φ_2 are corresponding representations of A, then a K-linear map $F : V_1 \to V_2$ is called *intertwining* for Φ_1 and Φ_2 if the following diagram commutes for all $a \in A$:

$$\begin{array}{ccc} & F & \\ V_1 & \longrightarrow & V_2 \\ \Phi_1(a) \downarrow & & \downarrow \Phi_2(a) \\ V_1 & \longrightarrow & V_2 \\ & F & \end{array}$$

i.e. for all $a \in A$ and $v \in V_1$,

$$F(\Phi_1(a)(v)) = \Phi_2(a)(F(v)).$$

The A-modules V_1 and V_2 are said to be *equivalent* if there exists an intertwining map $F : V_1 \to V_2$ which is an isomorphism between V_1 and V_2. We then write $V_1 \cong V_2$.

Now let A be a *real* algebra. Then an A-module V can be a real or a complex vector space.

Suppose first that V is a real vector space and let $\Phi : A \to \operatorname{End}_{\mathbf{R}}(V)$ be a representation of A on V.

Then, by considering the complexification $V_{\mathbf{C}}$ of V, a representation $\tilde{\Phi}$ of A on $V_{\mathbf{C}}$ may be obtained as follows.

Define for each $v = x + iy \in V_{\mathbf{C}}$ and $a \in A$,

$$\tilde{\Phi}(a)(v) = \Phi(a)(x) + i\Phi(a)(y).$$

Then clearly $\tilde{\Phi}(a) \in \operatorname{End}_{\mathbf{C}}(V_{\mathbf{C}})$ for all $a \in A$, whence $V_{\mathbf{C}}$ is an A-module (over $\mathbf{C}$).

Next suppose that V is a complex vector space and that $\Phi : A \to \operatorname{End}_{\mathbf{C}}(V)$ is a representation of A on V. Then by taking $V_{\mathbf{R}}$, i.e. by considering V as a real vector space, $\Phi : A \to \operatorname{End}_{\mathbf{R}}(V_{\mathbf{R}})$ becomes a representation of A on $V_{\mathbf{R}}$.

Finally, suppose again that V is a complex vector space and that $\Phi : A \to \operatorname{End}_{\mathbf{C}}(V)$ is a representation of A on V. Then its *complexification* $\Phi_{\mathbf{C}}$ is obtained as follows.

Consider the complexification $A_{\mathbf{C}} = A \otimes_{\mathbf{R}} \mathbf{C}$ of the real algebra A. Then for an arbitrary element $\sum_{(j)} a_j \otimes \lambda_j$ in $A_{\mathbf{C}}$, where $a_j \in A$ and $\lambda_j \in \mathbf{C}$, we put

$$\Phi_{\mathbf{C}}\Big(\sum_{(j)} a_j \otimes \lambda_j\Big) = \sum_{(j)} \lambda_j \Phi(a_j).$$

Clearly $\Phi_{\mathbf{C}}$ is a representation of $A_{\mathbf{C}}$ on V.

Notice that, when V is a complex irreducible A-module, then $V_{\mathbf{R}}$ is a real A-module which may, but need not be, irreducible.

If V_1 and V_2 are equivalent complex A-modules, then clearly $V_{1,\mathbf{R}}$ and $V_{2,\mathbf{R}}$ are equivalent real A-modules, the converse not being true in general as is shown by the following counter example.

Take $A = \mathbf{C}$ and consider A as a real algebra.

Moreover take $V_1 = V_2 = \mathbf{C}$. Then defining $\Phi_j : A \to \mathrm{End}_{\mathbf{C}}(V_j)$, $j = 1, 2$, by

$$\Phi_1(a)(v) = a\,v,$$

$$\Phi_2(a)(v) = \bar{a}\,v,$$

we have that Φ_1 and Φ_2 are representations of A on $\mathbf{C} = V_1 = V_2$.

Moreover the real A-modules $V_{1,\mathbf{R}}$ and $V_{2,\mathbf{R}}$ are equivalent, the intertwining map $F : V_{1,\mathbf{R}} \to V_{2,\mathbf{R}}$ being given by $F(v) = \bar{v}$.

But clearly there exists no intertwining map for Φ_1 and Φ_2.

Finally suppose that A is a real or complex algebra and that V is a complex A-module associated to the representation $\Phi : A \to \mathrm{End}_{\mathbf{C}}(V)$.

Then the "*conjugate representation*" $\Phi^\star : A \to \mathrm{End}_{\mathbf{C}}(V^\star)$ is defined by

$$\Phi^\star(a) = (\Phi(a))^\star, \quad a \in A,$$

i.e. for each $a \in A$ and $v^\star \in V^\star$,

$$\Phi^\star(a)(v^\star) = (\Phi(a))^\star(v^\star).$$

Notice that, when A is a complex algebra, then the "conjugate representation" $\Phi^\star$ of A on $V^\star$ is anti-linear since for each $a \in A$ and $\lambda \in \mathbf{C}$,

$$\begin{aligned} \Phi^\star(\lambda a) &= (\Phi(\lambda a))^\star \\ &= (\lambda \Phi(a))^\star \\ &= \bar{\lambda}\Phi^\star(a). \end{aligned}$$

Now suppose that A is a real algebra, V is a complex vector space and Φ is a representation of A on V. Then by considering the complexification $\Phi_{\mathbf{C}}$ of Φ we have that the restrictions of $\Phi_{\mathbf{C}}$ and $\Phi^\star_{\mathbf{C}}$ to $A \subset A \otimes_{\mathbf{R}} \mathbf{C}$ are representations of A on V and $V^\star$ respectively. Denoting these representations by $\Phi_{\mathbf{C}}|A$ and $\Phi^\star_{\mathbf{C}}|A$ we have for their complexifications $(\Phi_{\mathbf{C}}|A)_{\mathbf{C}}$ and $(\Phi^\star_{\mathbf{C}}|A)_{\mathbf{C}}$ that

$$(\Phi_{\mathbf{C}}|A)_{\mathbf{C}} = \Phi_{\mathbf{C}}$$

while

$$(\Phi^\star_{\mathbf{C}}|A)_{\mathbf{C}} \neq \Phi^\star_{\mathbf{C}}.$$

The first relation is trivial, while the second one follows immediately from the observation that $\Phi^\star_{\mathbf{C}}$ is anti-linear and $(\Phi^\star_{\mathbf{C}}|A)_{\mathbf{C}}$ is linear.

Obviously the relations (4.1) and (4.2) remain valid for A-modules.

We end up this subsection by recalling Schur's Lemma (see e.g. [58])

Theorem. *Let V be a finite-dimensional complex vector space. If a family of transformations $\mathcal{A}$ acts irreducibly on V, then any linear transformation C which commutes with every $T \in \mathcal{A}$ is of the form $C = \lambda \mathbf{1}$ where $\lambda \in C$ and $\mathbf{1}$ is the identity transformation.*

4.2.2 Representations of groups

Let G be a real group and let V be a (finite dimensional) vector space over K, K being either $\mathbf{R}$ or $\mathbf{C}$. Then a representation Φ of G on V is a homomorphism from G into $\mathrm{End}_K(V)$, the space V being then called a *G-module*.

The notations and definitions introduced in the foregoing subsection admit straightforward analogues in the case of G-modules.

In what follows we shall be particularly interested in representations of the spinor groups $\mathrm{Pin}(p,q)$ and $\mathrm{Spin}(p,q)$ introduced in 1.10.2.

4.3 Dirac and Weyl spinors

In this section irreducible representations of the complex Clifford algebra $\mathbf{C}_n$ are discussed.

For the results mentioned we refer the reader to [44].

As the structure of such $\mathbf{C}_n$-modules depends heavily on the parity of n, we shall discuss the even and odd dimensional cases separately.

4.3.1 The even-dimensional case

Putting $n = 2m$, we saw in Theorem 3.5 that $\mathbf{C}_{2m}$ is isomorphic to the full matrix algebra $\mathbf{C}(2^m)$ whence there exists a unique (up to equivalence) irreducible (complex) $\mathbf{C}_{2m}$-module, denoted henceforth by S_{2m}. As is commonly accepted, S_{2m} is called the *space of Dirac spinors.*

As realizations of S_{2m} we mention :

(1) The space $\mathbf{C}^{2^m}$ of column vectors, the action of $\mathbf{C}_{2m}$ being matrix multiplication (after identification of $\mathbf{C}_{2m}$ with $\mathbf{C}(2^m)$).

(2) Any minimal left ideal of $\mathbf{C}_{2m}$, the action of $\mathbf{C}_{2m}$ being left multiplication.

(3) The Fock space $(\wedge W)I$ where W is a maximal isotropic subspace of $\mathbf{C}^{2m}$ and I is a primitive idempotent in $\mathbf{C}_{2m}$. For further details we refer to 4.7.

4.3.2 The odd dimensional case

Putting $n = 2m-1$, we saw in Theorem 3.5 that $\mathbf{C}_{2m-1}$ is isomorphic to $\mathbf{C}(2^{m-1}) \oplus \mathbf{C}(2^{m-1})$.

The unique irreducible representations of the first (resp. second) factor will be denoted by S_{2m}^+ (resp. S_{2m}^-) and, as is commonly accepted, the spaces $S_{2m}^\pm$ are called the *spaces of Weyl spinors.* Alternative names of Weyl spinors are *half-spinors* (see [17]), *semi-spinors* (see [22]) or *reduced spinors* (see [62]).

The spaces of Weyl spinors $S_{2m}^\pm$ are most of all realized as "half-spaces" of the space of Dirac spinors S_{2m}. We mention :

(1) As $\mathbf{C}_{2m-1} \cong \mathbf{C}(2^{m-1}) \oplus \mathbf{C}(2^{m-1}) \cong \mathbf{C}_{2m}^{+}$, a Weyl spinor may be considered as being a half-column of a $\mathbf{C}^{2^m}$-column vector representing a Dirac spinor.

(2) As $\mathbf{C}_{2m-1} \cong \mathbf{C}_{2m}^{+}$, we have that, when considering the chirality operator $\theta \in \mathbf{C}_{2m}^{+}$ (see 3.4), $\theta^2 = \mathbf{1}$, so that we may define $S_{2m}^{\pm}$ as the eigenspaces of θ acting on S_{2m}, i.e.

$$S_{2m}^{\pm} = \{v \in S_{2m} : \theta v = \pm v\}.$$

(3) If $(\wedge W)I$ is the Fock space representation of $\mathbf{C}_{2m}$, then S_{2m}^{+} (resp. S_{2m}^{-}) is realized as $(\wedge^{\text{even}} W)I$ (resp. $(\wedge^{\text{odd}} W)I$) (see again 4.7).

It should be noted that the distinction between S_{2m}^{+} and S_{2m}^{-} is, of course, conventional.

That the spaces S_{2m}^{+} and S_{2m}^{-} of Weyl spinors are indeed unequivalent representations of $\mathbf{C}_{2m-1}$ follows immediately from the fact that the chirality operator $\theta \in \mathbf{C}_{2m}^{+}$ is represented by $\pm\mathbf{1}$ on $S_{2m}^{\pm}$.

4.3.3 Abstract realizations of $\mathbf{C}_{2m}$ and $\mathbf{C}_{2m-1}$

Let S_{2m} be a space of Dirac spinors and define again $S_{2m}^{\pm}$ as the eigenspaces of the chirality operator θ acting on S_{2m}.

As each element $a^{+} \in \mathbf{C}_{2m}^{+}$ commutes with θ (see also §3.4) it is clear that the action of a^{+} on $S_{2m} = S_{2m}^{+} \oplus S_{2m}^{-}$ may be described by a (2×2)-block matrix of the form

$$a^{+} = \left(\begin{array}{c|c} \alpha & 0 \\ \hline 0 & \delta \end{array}\right)$$

where $\alpha : S_{2m}^{+} \to S_{2m}^{+}$ and $\delta : S_{2m}^{-} \to S_{2m}^{-}$.

Analogously, as each element $a^{-} \in \mathbf{C}_{2m}^{-}$ anti-commutes with θ, its action on S_{2m} is given by a (2×2)-block matrix

$$a^{-} = \left(\begin{array}{c|c} 0 & \beta \\ \hline \gamma & 0 \end{array}\right)$$

where $\beta : S_{2m}^{-} \to S_{2m}^{+}$ and $\gamma : S_{2m}^{+} \to S_{2m}^{-}$.

Hence, since each $a \in \mathbf{C}_{2m}$ admits the decomposition $a = a^{+} + a^{-}$ with $a^{\pm} \in \mathbf{C}_{2m}^{\pm}$, a may be represented by

$$a = \left(\begin{array}{c|c} \alpha & \beta \\ \hline \gamma & \delta \end{array}\right).$$

Now denoting an orthonormal basis of $\mathbf{C}^{2m}$ by

$$e = \{e_0, e_1, \ldots, e_{2m-1}\}$$

we have that, as $e_i\theta = -\theta e_i, \quad i = 0, \dots, 2m-1$,

$$e_i = \left(\begin{array}{c|c} 0 & e_i^+ \\ \hline e_i^- & 0 \end{array}\right)$$

with $e_i^\pm : S_{2m}^\mp \to S_{2m}^\pm$.

Furthermore, defining for each $i = 1, \dots, 2m-1$,

$$\epsilon_i = -e_0 e_i,$$

the set $\{\epsilon_i \cdot i = 1, \dots, 2m-1\}$ is generating $\mathbf{C}_{2m}^+ \cong \mathbf{C}_{2m-1}$ with

$$\epsilon_i = \left(\begin{array}{c|c} -e_0^+ e_i^- & \\ \hline & -e_0^- e_i^+ \end{array}\right).$$

Obviously, $\epsilon_i : S_{2m}^\pm \to S_{2m}^\pm$ where

$$\epsilon_i | S_{2m}^+ = -e_0^+ e_i^- \quad \text{and} \quad \epsilon_i | S_{2m}^- = -e_0^- e_i^+ .$$

In such a way (abstract) realizations are given of $\mathbf{C}_{2m}$ and $\mathbf{C}_{2m-1} \cong \mathbf{C}_{2m}^+$.

4.4 Dirac and Weyl spinors revisited : Complex representations of Pin(p, q) and Spin(p, q)

In the foregoing subsection we defined the spaces of Dirac and Weyl spinors as representation spaces of the complex Clifford algebras $\mathbf{C}_n$.

As their name indicates, spinors are mainly used as irreducible complex representations of the real groups $\text{Pin}(p, q)$ and $\text{Spin}(p, q)$ lying in the real algebra $\mathbf{R}_{p,q}$. This is possible since $\mathbf{R}_{p,q}$ is a (real) subalgebra of $\mathbf{C}_{p+q}$ whence any representation of $\mathbf{C}_{p+q}$ yields a representation of $\text{Pin}(p, q)$ and $\text{Spin}(p, q)$.

Of course the parity of $n = p + q$ will play a fundamental role in finding out irreducible complex representations of $\text{Pin}(p, q)$ and $\text{Spin}(p, q)$.

4.4.1 The even dimensional case

Putting $n = p + q = 2m$, we have that the space of Dirac spinors S_{2m} forms an irreducible complex representation of $\text{Pin}(p, q)$ and this since $\text{Pin}(p, q)$ generates $\mathbf{R}_{p,q}$.

The space of Dirac Spinors S_{2m} is therefore called the *basic spinor representation* of $\text{Pin}(p, q)$.

As $\text{Spin}(p, q) \subset \mathbf{R}_{p,q}^+ \subset \mathbf{C}_{p+q}^+ \cong \mathbf{C}_{2m-1}$ and $\text{Spin}(p, q)$ is generating $\mathbf{R}_{p,q}^+$, the spaces of Weyl spinors $S_{2m}^\pm$ are unequivalent irreducible (complex) $\text{Spin}(p, q)$-modules.

They are called *basic spinor representations* of $\text{Spin}(p, q)$.

4.4.2 The odd dimensional case

Putting $n = p + q = 2m - 1$ we have that, as $\mathrm{Pin}(p, q) \subset \mathbf{R}_{p,q} \subset \mathbf{C}_{p+q} \cong \mathbf{C}^+_{2m}$, the spaces of Weyl spinors $S^{\pm}_{2m}$ are unequivalent irreducible (complex) representations of $\mathrm{Pin}(p, q)$.

Furthermore, since $\mathrm{Spin}(p, q) \subset \mathbf{R}^+_{p,q} \subset \mathbf{C}^+_{p+q} \cong \mathbf{C}_{2m-2}$, we have three possibilities to obtain irreducible complex representations of $\mathrm{Spin}(p, q)$, namely S^+_{2m}, S^-_{2m} and S_{2m-2}.

As $\mathbf{C}^+_{p+q}$ is simple and hence admits only one irreducible complex representation (up to equivalence), we have that in this case $S^+_{2m} \simeq S^-_{2m} \simeq S_{2m-2}$.

4.4.3 Summary

The results of 4.4.1. and 4.4.2. may be summarized in the following table :

Parity	Signature	Basic $\mathrm{Spin}(p, q)$-repr.	Basic $\mathrm{Pin}(p, q)$-repr.
even	$p + q = 2m$	S^+_{2m}, S^-_{2m}	S_{2m}
odd	$p + q = 2m - 1$	$S_{2m-2} \simeq S^+_{2m} \simeq S^-_{2m}$	S^+_{2m}, S^-_{2m}

The above table tells us that the roles of the (complex) irreducible $\mathrm{Pin}(p, q)$ and $\mathrm{Spin}(p, q)$ representations are interchanged when shifting the dimension by 1.

For this reason we restrict ourselves in the sequel to the even dimensional case, i.e. $p + q = 2m$.

Let us recall that we obtained the basic spinor representations of $\mathrm{Pin}(p, q)$ and $\mathrm{Spin}(p, q)$ by considering the respective chains of inclusions

$$\mathrm{Pin}(p, q) \subset \mathbf{R}_{p,q} \subset \mathbf{C}_{p+q}$$

and

$$\mathrm{Spin}(p, q) \subset \mathbf{R}^+_{p,q} \subset \mathbf{C}^+_{p+q} \cong \mathbf{C}_{p+q-1}$$

and using the property that $\mathrm{Pin}(p, q)$ generates $\mathbf{R}_{p,q}$, respectively $\mathrm{Spin}(p, q)$ generates $\mathbf{R}^+_{p,q}$ (see also 1.10.2).

It should be noted that there are an infinite number of other irreducible complex G-modules ($G = \mathrm{Pin}(p, q)$ or $G = \mathrm{Spin}(p, q)$) which can however not be extended to $\mathbf{C}_{p+q}$-modules.

4.5 Real and quaternionic representations of Pin(p, q) and Spin(p, q) - Majorana spinors

The aim of this subsection is to investigate under which conditions the spaces of Dirac and Weyl spinors admit real or quaternionic structures and to determine the possibility of obtaining in such cases real or quaternionic representations of $\mathrm{Pin}(p, q)$ and $\mathrm{Spin}(p, q)$.

We start with

Definition. *Let V be an irreducible complex (finite-dimensional) G-module, G being a real group. Then a* $\mathbf{C}$*-linear map $c : V \to V^\star$ is said to be a real (resp. quaternionic) structure on V if*

(i) $c^\star c = \mathbf{1}$ in the real case or $c^\star c = \text{-}\mathbf{1}$ in the quaternionic case

(ii) c is an intertwining map between the complex G-modules V and $V^\star$.

So the problem here not only consists of determining a real or complex structure c on V in the sense of section 4.1, but to check also whether or not c is an intertwining map between V and $V^\star$.

Notice that if c is a real structure on the irreducible complex representation V of G, then $Re\, V$ and $Im\, V$ are equivalent irreducible real representations of G, whence $V_{\mathbf{R}} = Re\, V \oplus Im\, V$ becomes a reducible real representation of G.

In the case of finding out irreducible representations of G, we thus have two possibilities for the space V, namely

(1) V does not admit a real structure.

Then $V_{\mathbf{R}}$ is an irreducible real representation of G ;

(2) There exists a real structure on V.

Then $Re\, V$ and $Im\, V$ are equivalent irreducible real representations of G.

As we already pointed out, V stands for the space of Dirac or Weyl spinors, while G is either $\text{Pin}(p,q)$ or $\text{Spin}(p,q)$. Moreover, as we agreed upon, we assume that $p+q=2m$ is even.

4.5.1 Real and quaternionic structures on the space of Dirac spinors - Majorana spinors

Before we discuss the existence of real or quaternionic structures on the $\text{Pin}(p,q)$-module S_{2m}, we wish to point out some remarks concerning representations of $\mathbf{R}_{p,q}$ into $\text{End}(S_{2m})$ and $\text{End}(S^\star_{2m})$.

As $\mathbf{R}_{p,q}$ is a real subalgebra of $\mathbf{C}_{2m}$ with $\mathbf{C}_{2m} = \mathbf{R}_{p,q} \oplus_{\mathbf{R}} \mathbf{C}$ and S_{2m} is a representation space for $\mathbf{C}_{2m}$ we have that, when calling $\gamma : \mathbf{C}_{2m} \to \text{End}_{\mathbf{C}}(S_{2m})$ such a representation, $\gamma|\mathbf{R}_{p,q}$ and $\gamma^\star|\mathbf{R}_{p,q}$ are representations of $\mathbf{R}_{p,q}$ on S_{2m} and $S^\star_{2m}$ respectively. Hence we can consider their complexifications $(\gamma|\mathbf{R}_{p,q})_{\mathbf{C}} = \gamma$ and $(\gamma^\star|\mathbf{R}_{p,q})_{\mathbf{C}}$ which are representations of $\mathbf{C}_{2m}$ on S_{2m} and $S^\star_{2m}$ respectively.

Notice that $(\gamma^\star|\mathbf{R}_{p,q})_{\mathbf{C}}$ may of course depend upon p and q.

We have

Theorem. *Consider the $\mathbf{R}_{p,q}$-modules S_{2m} and $S^\star_{2m}$. Then there exists an intertwining map $c : S_{2m} \to S^\star_{2m}$ such that either $c^\star c = \mathbf{1}$ or $c^\star c = -\mathbf{1}$ and both cases cannot occur simultaneously.*

Moreover if $\eta = q - p$, we have that $c^\star c = \mathbf{1}$ for $\eta = 0$ or 6 (mod 8) while $c^\star c = -\mathbf{1}$ for $\eta = 2$ or 4 (mod 8).

Proof. As γ and $(\gamma^\star | \mathbf{R}_{p,q})_{\mathbf{C}}$ are both complex irreducible representations of $\mathbf{C}_{2m}$, there ought to exist an intertwining map c for them.

That c is unique (up to a non-zero constant multiple) may be proved as follows.

Suppose c' is also an intertwining map for γ and $(\gamma^\star | \mathbf{R}_{p,q})_{\mathbf{C}}$. Then $c^{-1}c'$ is an intertwining map for γ on S_{2m} whence, S_{2m} being irreducible, Schur's Lemma yields the existence of $\lambda \in \mathbf{C}\backslash\{0\}$ such that $c^{-1}c' = \lambda\mathbf{1}$.

Consequently $c' = \lambda c$.

Now consider the intertwining map $c^\star c : S_{2m} \to S_{2m}$. Then again by Schur's Lemma there ought to exist $\mu \in \mathbf{C}\backslash\{0\}$ such that $c^\star c = \mu\mathbf{1}$.

We claim that $\mu \in \mathbf{R}\backslash\{0\}$.

Indeed, on the one hand we have for all $v \in V$ that

$$\begin{aligned} c(\mu v) &= c(c^\star c(v)) \\ &= (c^\star c(c(v))^\star)^\star \\ &= (\mu(c(v))^\star)^\star \\ &= \bar{\mu} c(v) \end{aligned}$$

while on the other hand

$$c(\mu v) = \mu c(v).$$

Hence $\mu = \bar{\mu}$ or $\mu \in \mathbf{R}$.

Multiplying c by $|\mu|^{-1/2}$ we obtain that $c^\star c = \pm\,\mathbf{1}$ and clearly both cases cannot occur simultaneously.

In order to find out when $c^\star c = \mathbf{1}$ or $c^\star c = -\mathbf{1}$, it is sufficient to check under which conditions $\mathbf{R}_{p,q}$ may be realized as a matrix algebra over $\mathbf{R}$ or $\mathbf{H}$.

Looking at the table of classification for $\mathbf{R}_{p,q}$ (see 2.4.2) we have the following possibilities.

(1) If $\eta = 0$ or 6 (mod 8), then $\mathbf{R}_{p,q} \cong \mathbf{R}(2^m) \subset \mathbf{C}(2^m) \cong \mathbf{C}_{p+q}$, whence $\mathbf{R}_{p,q}$ acts on $V = \mathbf{C}^{2^m}$. Denoting an arbitrary $v \in V$ by $v = (v_1, v_2, \ldots, v_{2^m})$, a real structure c on V is given by

$$c(v_1, \ldots, v_{2^m}) = (\bar{v}_1, \ldots, \bar{v}_{2^m})^\star,$$

i.e. c is given by complex conjugation.

(2) If $\eta = 2$ or 4 (mod 8), then $\mathbf{R}_{p,q} \cong \mathbf{H}(2^{m-1})$ whence $\mathbf{R}_{p,q}$ acts on the (real) space $\mathbf{H}^{2^{m-1}}$ of (real) dimension 4.2^{m-1}.

Now we turn this space into a complex vector space by defining for each $v \in \mathbf{H}^{2^{m-1}}$ and $\lambda = (\alpha, \beta) \in \mathbf{C}$,

$$\lambda v = \alpha v + \beta v i$$

where the multiplication vi is performed in $\mathbf{H}$.

In such a way $\mathbf{H}^{2^{m-1}}$ becomes a complex vector space V with $\dim_{\mathbf{C}} V = 2.2^{m-1} = 2^m$.

Moreover, if $a \in \mathbf{H}(2^{m-1})$, $\lambda \in \mathbf{C}$ and $v \in \mathbf{H}^{2^{m-1}}$,

$$a(\lambda v) = \lambda a(v)$$

whence a admits a representation as a complex matrix.

Consequently there exists an embedding $\Phi : \mathbf{H}(2^{m-1}) \to \mathbf{C}_{2m} \cong \text{End}\,(V)$ and this since $\mathbf{C}_{2m} \cong \mathbf{C}(2^m)$ and $V \cong \mathbf{C}^{2^m}$.

The question now arises whether or not there exists a linear map $c : V \to V^\star$ such that

$$c^\star c = -\mathbf{1}$$

and for all $a \in \mathbf{H}(2^{m-1})$,

$$c \circ \Phi(a) = \Phi(a)^\star \circ c.$$

We claim that $c =^\star \circ J$, with $Jv = vj$ for all $v \in \mathbf{H}^{2^{m-1}}$, satisfies these requirements.

Indeed, first note that for all $\lambda \in \mathbf{C}$,

$$\begin{aligned} J(\lambda v) &= v\lambda j \\ &= vj\bar{\lambda} = \bar{\lambda} J(v) \end{aligned}$$

whence

$$\begin{aligned} c(\lambda v) &= (J(\lambda v))^\star \\ &= (\bar{\lambda} J(v))^\star \\ &= \lambda (J(v))^\star \\ &= \lambda c(v) \end{aligned}$$

and so c is linear.

Moreover

$$\begin{aligned} c^\star c(v) &= c^\star((J(v))^\star) \\ &= (c(J(v)))^\star \\ &= ((J(J(v)))^\star)^\star \\ &= J^2(v) \\ &= -v \end{aligned}$$

and so c is a quaternionic structure.

Finally for all $a \in \mathbf{H}(2^{m-1})$,

$$\begin{aligned} c\Phi(a)(v) &= c(av) \\ &= (J(av))^\star \\ &= (avj)^\star \\ &= (\Phi(a)(vj))^\star \\ &= \Phi(a)^\star((vj)^\star) \\ &= \Phi(a)^\star(c(v)) \end{aligned}$$

and so c is an intertwining map. ∎

We have thus obtained that in the case $\eta = 0$ or 6 (mod 8), although the space of Dirac spinors S_{2m} is irreducible as a Pin(p,q)-module over $\mathbf{C}$, it is reducible as a Pin(p,q)-module over $\mathbf{R}$. Indeed, it splits into two (equivalent) irreducible pieces, namely the eigenspaces $Re\, S_{2m}$ and $Im\, S_{2m}$ of $\tilde{c}$ (see also 4.1.1). Either of these real Pin(p,q)-modules is usually called the *space of Majorana spinors.*

It means that (in even dimensions) Majorana spinors exist for $\eta = 0$ or 6 (mod 8) only.

4.5.2 Real and quaternionic structures on the space of Weyl spinors - Majorana-Weyl spinors

Let us consider the Spin(p,q)-modules S_{2m}^+ and S_{2m}^-. Then we have

Theorem. *Putting $\eta = q - p$, we have*

(i) If $\eta = 0$ (mod 8) *then there exist real structures on the spaces $S_{2m}^{\pm}$.*

(ii) If $\eta = 4$ (mod 8) *then there exist quaternionic structures on the spaces $S_{2m}^{\pm}$.*

(iii) In the cases $\eta = 2$ or 6 (mod 8) *neither real nor quaternionic structures exist on $S_{2m}^{\pm}$.*

Proof. As we saw in Theorem 4.5.1, there exists an intertwining map $c : S_{2m} \to S_{2m}^\star$ such that $c^\star c = \mathbf{1}$ if $\eta = 0$ or 6 (mod 8) and $c^\star c = -1$ if $\eta = 2$ or 4 (mod 8).

Now Weyl spinor spaces can be realized as eigenspaces of the chirality operator θ (see 4.3.2), i.e.

$$S_{2m}^{\pm} = \{v \in S_{2m} : \theta v = \pm v\}.$$

Here

$$\theta = i^{\nu} e_N$$

with $\nu = \frac{q-p}{2} = \frac{\eta}{2}$.

We claim that

$$\theta^{\star} c = (-1)^{\nu} c\theta.$$

Indeed, considering S_{2m} as being realized by a minimal left ideal of $\mathbf{C}_{2m}$, we have that θ acts on S_{2m} by left multiplication.

Since c is an intertwining map we have that

$$e_N^{\star} c = c e_N$$

whence

$$\begin{aligned} i^{\nu} e_N^{\star} c &= i^{\nu} c e_N \\ &= c\theta. \end{aligned}$$

Consequently, as for each $\lambda \in \mathbf{C}$ and $f : V \to V$ linear, $(\lambda f)^{\star} = \bar{\lambda} f^{\star}$, we obtain

$$((-i)^{\nu} e_N)^{\star} c = c\theta$$

or

$$\begin{aligned} \theta^{\star} c &= (-1)^{\nu} c\theta. \\ &= (-1)^{\frac{\eta}{2}} c\theta \end{aligned}$$

So for $\eta = 0$ or 4 (mod 8), c preserves the spaces $S^{\pm}_{2m}$ (i.e. it maps $S^{\pm}_{2m}$ into its respective conjugate space), thus proving the first two assertions of the theorem.

As for $\eta = 2$ or 6 (mod 8), c interchanges the spaces $S^{\pm}_{2m}$ (i.e. it maps $S^{\pm}_{2m}$ into the respective conjugate spaces $(S^{\mp}_{2m})^{\star}$, this c cannot be a candidate for obtaining a possible intertwining map.

Suppose that in the case $\nu = 2$ or 6 (mod 8) there would exist a real or quaternionic structure c' on $S^{\pm}_{2m}$. To fix the ideas, assume that c' is a real structure on $S^{\pm}_{2m}$.

As the map c described above maps S^{+}_{2m} onto $(S^{-}_{2m})^{\star}$ and S^{-}_{2m} onto $(S^{+}_{2m})^{\star}$, we get that the composition $c^{-1}c'$ is an intertwining isomorphism between S^{+}_{2m} and S^{-}_{2m} which is impossible since they are unequivalent $\mathrm{Spin}(p,q)$-modules.

This proves the assertion (3). ∎

We have thus obtained that in the case $\nu = 0$ (mod 8), although the space of Weyl spinors $S^{\pm}_{2m}$ is irreducible as a $\mathrm{Spin}(p,q)$-module over $\mathbf{C}$, it is reducible as a $\mathrm{Spin}(p,q)$-module over $\mathbf{R}$. Indeed, it splits into two (equivalent) irreducible pieces, namely the eigenspaces $Re\, S^{\pm}_{2m}$ and $Im\, S^{\pm}_{2m}$ of $\tilde{c}$ (see again 4.1.1). Either of these real $\mathrm{Spin}(p,q)$-modules is usually called the *space of Majorana-Weyl spinors.*

It means that (in even dimensions) Majorana-Weyl spinors exist for $\nu = 0$ (mod 8) only.

4.6 *Hermitian structures on spinor spaces*

In this section it will be shown in which way a non-degenerate Hermitian form may be defined on the spinor space S_{2m}.

A realization of such a form will be discussed in section 4.8.

4.6.1 The canonical isomorphism between $(V')^\star$ and $(V^\star)'$-Hermitian maps

Let again V be a finite dimensional complex vector space and let V' be its dual space. Then it is meaningful to consider the spaces $(V')^\star$ and $(V^\star)'$ and, of course, these spaces are isomorphic.

An explicit isomorphism ι_V between $(V^\star)'$ and $(V')^\star$ will now be constructed as follows.

Recalling that

$$(V^\star)' = \{\mathcal{T} : V^\star \to \mathbf{C} : \mathcal{T} \quad \text{is} \quad \mathbf{C}\text{-linear}\}$$

and

$$(V')^\star = \{\mathcal{U}^\star : \mathcal{U} : V \to \mathbf{C} \quad \text{is} \quad \mathbf{C}\text{-linear}\}$$

we define $\iota_V : (V^\star)' \to (V')^\star$ by putting for all $\mathcal{T} \in (V^\star)'$ and $w \in V$

$$(\iota_V(\mathcal{T}))^\star(w) = \overline{\mathcal{T}(w^\star)}$$

or

$$\iota_V(\mathcal{T})(w^\star) = (\overline{\mathcal{T}(w^\star)})^\star \tag{4.3}$$

It is understood that in the sequel we do not distinguish between the spaces $(V')^\star$ and $(V^\star)'$, the identification being made through the *canonical isomorphism* ι_V.

Moreover we shall frequently use the notations

$$(V')^\star = V'^\star$$

and

$$(V^\star)' = V^{\star\prime}$$

This identification also leads in a natural way to the following considerations.

Let U and V be (finite dimensional) complex vector spaces and let $H : U \to V$ be a linear map. Then we may consider the linear maps $(H^\star)'$ and $(H')^\star$ which are, respectively, the dual map of the conjugate map $H^\star$ of H and the conjugate map of the dual map H' of H.

As the diagram

$$
\begin{array}{ccccc}
 & & (H^{\star\prime}) & & \\
 & (V^{\star})' & \rightarrow & (U^{\star})' & \\
\iota_V & \downarrow & & \downarrow & \iota_U \\
 & (V')^{\star} & \rightarrow & (U')^{\star} & \\
 & & (H')^{\star} & &
\end{array}
$$

is commutative, we may identify $(H^\star)'$ and $(H')^\star$ through the relation

$$(H')^\star \circ \iota_V = \iota_U \circ (H^\star)'$$

In the sequel we use the notations

$$(H')^\star = H'^\star$$

and

$$(H^\star)' = H^{\star\prime}$$

Now let $H : V \to V'^\star$ be a linear map and let $H^\star : V^\star \to (V'^\star)^\star = V'$ be its conjugate map. Then the dual map $H^{\star\prime}$ of $H^\star$ acts from V into $V^{\star\prime}$, where we of course identified $(V')'$ and V in a classical way by associating with each element $\mathbf{T} \in (V')'$ the (unique) element $v \in V$ such that $\mathbf{T} = \mathbf{T}_v$ with $\mathbf{T}_v(\mathcal{T}) = \mathbf{T}(v)$ for all $\mathcal{T} \in V'$.

So by definition we have that for $\mathbf{T} = \mathbf{T}_v \in (V')'$,

$$H^{\star\prime}(v) = H^{\star\prime}(\mathbf{T}_v) = \mathcal{U}^\star \in V^{\star\prime} \tag{4.4}$$

is given by

$$\mathcal{U}^\star = \mathbf{T}_v \circ H^\star,$$

i.e. for all $w^\star \in V^\star$,

$$
\begin{aligned}
\mathcal{U}^\star(w^\star) &= (\mathbf{T}_v \circ H^\star)(w^\star) \\
&= (H^\star(w^\star))(v) \\
&= (H(w))^\star(v).
\end{aligned}
$$

Consequently $\iota_V \circ (H^\star)'$ is a linear map from V into $(V')^\star$.

This prompts the following

Definition. *A linear map $H : V \to V'$ is said to be Hermitian if and only if $\iota_V \circ H^{\star\prime} = H$.*

4.6.2 A Hermitian structure on S_{2m}

Let $p, q \in \mathbf{N}$ be such that $p + q = 2m$ and consider the spinor space S_{2m} as a representation space of $\mathbf{R}_{p,q} \subset \mathbf{C}_{2m}$. Then, starting from a representation $\gamma : \mathbf{C}_{2m} \to$ End (S_{2m}) and making use of the conjugation α on $\mathbf{C}_{2m}$ (see also 3.3.1), we may define for each $a \in \mathbf{C}_{2m}$ the operator $\gamma_\alpha(a) \in$ End (S_{2m}) in the following way :

$$\gamma_\alpha(a) = \gamma(\alpha(a)).$$

Putting for $a \in \mathbf{C}_{2m}$,

$$\gamma'_\alpha(a) = (\gamma_\alpha(a))'$$

and

$$\gamma'^{\star}_\alpha(a) = ((\gamma_\alpha(a))')^{\star}$$

we have that $(\gamma'^{\star}_\alpha | \mathbf{R}_{p,q})_{\mathbf{C}}$ is a representation of $\mathbf{C}_{2m}$ on $S'^{\star}_{2m}$.

Theorem 1. *Let $p + q = 2m$ and let γ be a representation of $\mathbf{C}_{2m} \cong \mathbf{R}_{p,q} \otimes_{\mathbf{R}} \mathbf{C}$ on $S_{2m} = S$. Then there exists a non-trivial Hermitian intertwining map $H : S \to S'^{\star}$ between the representations γ and $(\gamma'^{\star}_\alpha | \mathbf{R}_{p,q})_{\mathbf{C}}$. Moreover this map H is unique up to a scalar multiple $\lambda \in \mathbf{R}\backslash\{0\}$.*

Proof. As γ and $(\gamma'^{\star}_\alpha | \mathbf{R}_{p,q})_{\mathbf{C}}$ are both (irreducible) representations of $\mathbf{C}_{2m}$, there ought to exist an intertwining map $T : S \to S'^{\star}$.

Consequently, for all $a \in \mathbf{R}_{p,q}$, the following diagram is commutative :

$$\begin{array}{ccc} & T & \\ S & \longrightarrow & S'^{\star} \\ \gamma(a) \downarrow & & \downarrow \gamma'^{\star}_\alpha(a) \\ S & \longrightarrow & S'^{\star} \\ & T & \end{array}$$

Moreover the map T is unique up to a scalar multiple $\lambda \in \mathbf{C}\backslash\{0\}$.

Indeed, let U also be such a map, then $T^{-1} \circ U$ is intertwining for γ on S whence, by virtue of Schur's Lemma, there exists a $\lambda \in \mathbf{C}\backslash\{0\}$ such that $T^{-1} \circ U = \lambda 1$ on S or $U = \lambda T$.

Furthermore, the above diagram implies that

$$\begin{array}{ccc} & T^{\star} & \\ S^{\star} & \longrightarrow & S' \\ \gamma^{\star}(a) \downarrow & & \downarrow \gamma'_\alpha(a) \\ S^{\star} & \longrightarrow & S' \\ & T^{\star} & \end{array}$$

and

$$\begin{array}{ccc} & T^{\star\prime} & \\ S & \longrightarrow & S^{\star\prime} \\ \gamma_\alpha(a)\ \downarrow & & \downarrow\ \gamma^{\star\prime}_{\alpha}(a) \\ S & \longrightarrow & S^{\star\prime} \\ & T^{\star\prime} & \end{array}$$

are commutative diagrams too.

Here we of course identified S'' with S while $\gamma^{\star\prime}_{\alpha}(a)$ stands for $((\gamma_\alpha(a))^\star)'$.

Applying the canonical isomorphism ι_S between $S^{\star\prime}$ and $S'^\star$ we may thus consider $T^{\star\prime}$ as being an intertwining map between S and $S'^\star$, whence there ought to exist $\beta \in \mathbf{C}\backslash\{0\}$ such that

$$T^{\star\prime} = \beta T. \tag{4.5}$$

Now we claim that there exists $\mu \in \mathbf{C}\backslash\{0\}$ with $|\mu| = 1$ such that $H = \mu T$ is Hermitian.

To this end we first prove that $|\beta| = 1$.

Indeed, on the one hand, taking conjugates in (4.5), we obtain that

$$T' = \bar{\beta} T^\star \tag{4.6}$$

while on the other hand, taking duals in (4.6), we get :

$$T = \bar{\beta} T^{\star\prime}.$$

Consequently

$$T^{\star\prime} = \beta T = \beta\bar{\beta} T^{\star\prime}$$

or $|\beta| = 1$.

Now take $\mu \in \mathbf{C}$ such that $\mu^2 = \beta$ and put

$$H = \mu T.$$

Then, as $|\mu| = 1$,

$$\begin{aligned} H^\star = \bar{\mu} T^{\star\prime} &= \bar{\mu}\beta T \\ &= \frac{\bar{\mu}}{\mu}\beta(\mu T) \\ &= H. \end{aligned}$$

Obviously H is then a Hermitian intertwining map between S qnd $S'^{\star}$ (see also Remark 1 below) and the maps of the form λH, $\lambda \in \mathbf{R}\backslash\{0\}$, are the only ones satisfying the same requirements. ∎

Remarks.

(1) In the foregoing proof we concluded that from $H^{\star\prime} = H$ it followed that H was Hermitian. This is not in accordance with the definition we gave of a Hermitian map. The main reason for this (erroneous) conclusion is that at a certain stage of the proof we considered $T^{\star\prime}$ as being an intertwining map between S and $S'^{\star}$. A correct statement would have been "$\iota_S \circ T^{\star\prime}$ is an intertwining map between S and $S'^{\star}$". But this would have led us to some technical subtleties which we wanted to omit.

(2) Note that the Hermitian map H obtained in the foregoing Theorem 1 depends on the conjugation α. From the reasoning made it is clear that if we replace α by any other anti-automorphism ν on $\mathbf{C}_{2m}$, the existence of a corresponding Hermitian map between S and $S'^{\star}$ may be shown.

In the sequel H_α will stand for the Hermitian map associated with α.
Putting for all $v, w \in S$

$$(v, w) = (H_\alpha(w))^{\star}(v) \tag{4.7}$$

then clearly $(\ ,\ ) : S \times S \to \mathbf{C}$ is a sesqui-linear form on S.

Theorem 2. *The sesqui-linear form $(\ ,\ )$ on S given by (4.7) is non-degenerate and Hermitian.*

Proof. Let us first show that $(\ .\ )$ is a Hermitian form.

As H_α is Hermitian, i.e. $\iota_S \circ H^{\star\prime}_{\alpha}$, we have by the definition itself of the identification map ι_S and of the action of $H^{\star\prime}$ on S that for all $v, w \in S$,

$$\begin{aligned}
(w, v) &= (H_\alpha(V))^{\star}(w) \\
&= (\iota_S(H^{\star\prime}_{\alpha}(v)))^{\star}(w) \\
&= \overline{(H^{\star\prime}_{\alpha}(v))w^{\star}} \\
&= \overline{(H^{\star}_{\alpha}(w^{\star}))(v)} \\
&= \overline{(H_\alpha(w))^{\star}(v)} \\
&= \overline{(v, w)}.
\end{aligned}$$

Now assume that $w \in S$ is such that for all $v \in S$, $(v, w) = 0$. Then $(H(w))^{\star}(v) = 0$ for all $v \in S$, whence $(H(w))^{\star} = 0$ or $H(w) = 0$.

As H is an isomorphism, $w = 0$. ■

We end up this subsection by considering the important case $\mathbf{R}_{0,2m}$ (i.e. $p = 0, q = 2m$), which will play an essential role in developing a function theory for harmonic spinors (see Chapter II).

We have

Theorem 3. *Let γ be a representation of $\mathbf{C}_{2m} \cong \mathbf{R}_{0,2m} \otimes_{\mathbf{R}} \mathbf{C}$ on the space of Dirac spinors S_{2m} and let $H_\alpha : S \to S'^*$ be the corresponding Hermitian intertwining map. Then its associated non-degenerate Hermitian form (,) determines an inner product on S_{2m}.*

Proof. See section 4.8.3, where a proof is given using the so-called Fock space as a realization of S_{2m}. ■

Remark. The foregoing discussions all deal with the case $p + q = 2m$ even, i.e. the case of Dirac spinors. When $p+q = 2m-1$ is odd, then of course one may investigate the existence of forms on the spaces $S^{\pm}_{2m}$ of Weyl spinors.

It turns out that the restriction of the form (,) obtained in Theorem 1 to the space of Weyl spinors is non-degenerate for p even and trivial for p odd (see [13]).

As in the following chapters we shall only be involved with the case $\mathbf{R}_{o,q}$, we thus have that the form (,) is positive, as well on S_{2m} as on $S^{\pm}_{2m}$.

4.7 A realization of spinor space : the Fock space

In section 3.5 we proved that $\mathbf{C}_{2m}$ is a full matrix algebra $\mathbf{C}(2^m)$ and that $\mathbf{C}_{2m+1}$ is isomorphic to $\mathbf{C}(2^m) \oplus \mathbf{C}(2^m)$.

The aim of this section is to construct an explicit representation of $\mathbf{C}_{2m}$ and this by realizing the spinor space S_{2m} as a minimal left ideal of $\mathbf{C}_{2m}$. The representation of $\mathbf{C}_{2m}$ is then obtained by left multiplication.

An explicit representation of $\mathbf{C}_{2m-1}$ is also briefly discussed.

4.7.1 Complex Structures

In this subsection we recall some results concerning complex structures. For further details the reader is referred to [58].

Let X be an n-dimensional real vector space. Then a linear transformation J on X is said to be a *complex structure* on X if $J^2 = -1$.

Once a complex structure J on X is given, it turns out that X may be transformed into a complex vector space and this in the following way. Retaining the addition of vectors in X, put for each $\lambda = a + ib \in \mathbf{C}$ and $v \in X$,

$$\lambda v = av + bJ(v).$$

Then clearly X becomes a vector space over $\mathbf{C}$, denoted henceforth by X_J.

Notice that in X_J, $iv = J(v)$ for all $v \in X$.

It is also straightforward to verify that X_J is finite-dimensional and that, if $\dim X_J = m$ and $\{v_1, \ldots, v_m\}$ is a basis for X_J, then $\{v_1, \ldots, v_m, Jv_1, \ldots, Jv_m\}$ is a basis for X, whence $n = 2m$ or X is *even dimensional.*

Let us now discuss the relationship between complex structures on a real finite-dimensional vector space X and the complexification $X_{\mathbf{C}} = X \otimes_{\mathbf{R}} \mathbf{C}$ of X.

As is well known, any vector $v \in X_{\mathbf{C}}$ may be written in a unique way as

$$v = v' + iv'' \tag{4.8}$$

where $v', v'' \in X$; its conjugate $\bar{v}$ is then defined by

$$\bar{v} = v' - iv''. \tag{4.9}$$

Now suppose that X is an even dimensional real vector space with $\dim X = 2m$ and that J is a complex structure on X.

Then calling

$$W = \{v - iJ(v) : v \in X\} \tag{4.10}$$

and introducing its *conjugate* $\bar{W}$ with

$$\bar{W} = \{v + iJ(v) : v \in X\},$$

we have that

$$X_{\mathbf{C}} = W \oplus \bar{W},$$

W and $\bar{W}$ being m-dimensional subspaces of $X_{\mathbf{C}}$.

Moreover, by extending J to a complex-linear map $J_{\mathbf{C}}$ on $X_{\mathbf{C}}$, we have that $J_{\mathbf{C}}^2 = -\mathbf{1}$ on $X_{\mathbf{C}}$ and that W and $\bar{W}$ are precisely the eigenspaces of $J_{\mathbf{C}}$ corresponding to respectively the eigenvalues i and $-i$ of $J_{\mathbf{C}}$.

But the converse also holds, i.e. suppose that $X_{\mathbf{C}}$ is the sum of a subspace W and its conjugate $\bar{W}$, then it may be shown that X carries a complex structure J such that W and $\bar{W}$ are precisely the eigenspaces of $J_{\mathbf{C}}$ corresponding to, respectively, the eigenvalues i and $-i$.

Thus we have that the correspondence between the set of all complex structures J on an (even) dimensional real vector space X and the set of all subspaces W of $X_{\mathbf{C}}$ for which $X_{\mathbf{C}} = W \oplus \bar{W}$ is one-to-one.

Finally let us discuss the relationship between bilinear forms $\mathcal{B}$ of type $(0, 2m)$ on a $2m$-dimensional real vector space X, complex structures on X and maximally isotropic subspaces of $(X_{\mathbf{C}}, \mathcal{B}_{\mathbf{C}})$.

To this end, first suppose that J is a complex structure on the $2m$-dimensional real vector space X, that W is the m-dimensional subspace of $X_{\mathbf{C}}$ given by (4.10) and that $\{f_1, \ldots, f_m\}$ is a basis of W. Then, following (4.8), for each $j = 1, \ldots, m$, there exist unique vectors $e_{2j-1}, e_{2j} \in X$ such that

$$f_j = e_{2j-1} + ie_{2j}$$

and, by virtue of (4.10),

$$Je_{2j-1} = -e_{2j}$$

and

$$Je_{2j} = e_{2j-1}$$

Obviously, $\{\bar{f}_j \; : \; j = 1, \ldots, m\}$ and $\{e_{2j-1}, e_{2j} \; : \; j = 1, \ldots, m\}$ are then bases of respectively $\bar{W}$ and X. Moreover, as we already observed, W and $\bar{W}$ are the eigenspaces of $J_{\mathbf{C}}$ corresponding to respectively the eigenvalues i and $-i$.

Now defining for all $p, q \in \{1, \ldots, m\}$,

$$\mathcal{B}(e_p, e_q) = -\delta_{p,q}$$

and extending $\mathcal{B}$ bilinearly to $X \times X$, we have that $(X, \mathcal{B})$ is a non-degenerate real orthogonal space of type $(0, 2m)$ and that J preserves the bilinear form $\mathcal{B}$ i.e. $J \in \mathrm{SO}(2m)$.

Moreover, providing $X_{\mathbf{C}}$ with the complexified non-degenerate symmetric bilinear form $\mathcal{B}_{\mathbf{C}}$ of $\mathcal{B}$ we have that W is a maximally isotropic subspace of $(X_{\mathbf{C}}, \mathcal{B}_{\mathbf{C}})$, i.e. $\mathcal{B}_{\mathbf{C}}|W = 0$ and for each subspace V of $X_{\mathbf{C}}$ on which $\mathcal{B}_{\mathbf{C}}$ vanishes identically, $\dim V \leq \dim W = m$.

Indeed, as $J_{\mathbf{C}}$ preserves the bilinear form $\mathcal{B}_{\mathbf{C}}$ and as for each $w \in W$, $J_{\mathbf{C}}(w) = iw$, we have that for all $v_1, v_2 \in W$,

$$\begin{aligned} \mathcal{B}_{\mathbf{C}}(v_1, v_2) &= \mathcal{B}_{\mathbf{C}}(J_{\mathbf{C}}v_1, J_{\mathbf{C}}v_2) \\ &= \mathcal{B}_{\mathbf{C}}(iv_1, iv_2) \\ &= -\mathcal{B}_{\mathbf{C}}(v_1, v_2), \end{aligned}$$

whence $\mathcal{B}_{\mathbf{C}}(v_1, v_2) = 0$.

Moreover, as each maximally isotropic subspace of the $2m$-dimensional space $(X_{\mathbf{C}}, \mathcal{B}_{\mathbf{C}})$ has exactly dimension m, the assertion is proved.

Now we claim that the converse is also true, i.e. given a $2m$-dimensional non-degenerate real orthogonal space $(X, \mathcal{B})$ of type $(0, 2m)$ and a maximally isotropic subspace W of $(X_{\mathbf{C}}, \mathcal{B}_{\mathbf{C}})$, then there exists a complex structure J on X with $J \in \mathrm{SO}(2m)$ such that W is the eigenspace of the complex linear extension $J_{\mathbf{C}}$ of J corresponding to the eigenvalue i.

First of all observe that if J is a complex structure on $(X, \mathcal{B})$, then $\det J = 1$ whence $J \in \mathrm{SO}(2m)$.

Now define the negative definite sesquilinear form h on $X_{\mathbf{C}}$ by

$$h(v, w) = \mathcal{B}_{\mathbf{C}}(v, \bar{w}), \quad v, w \in X_{\mathbf{C}}$$

where $\overline{w}$ is the conjugate of w (see(4.9)) and let $\{f_1, \ldots, f_m\}$ be an orthonormal basis of the maximally isotropic subspace W with respect to h, i.e.

$$h(f_j, f_k) = -\delta_{jk}.$$

Following (4.8), put for each $j = 1, \ldots, m$,

$$f_j = \frac{1}{\sqrt{2}}(e_{2j-1} + ie_{2j})$$

with e_{2j-1}, $e_{2j} \in X$.

Then $\{e_{2j-1}, e_{2j} : j - 1, \ldots, m\}$ is an orthonormal basis for $(X, \mathcal{B})$.

Indeed, for $j \neq k$ we obtain from

$$0 = \mathcal{B}_{\mathbf{C}}(f_j, f_k)$$

and

$$0 = h(f_j, f_k)$$

that

$$\begin{aligned} \mathcal{B}(e_{2j-1}, e_{2k-1}) &= 0 \\ \mathcal{B}(e_{2j}, e_{2k-1}) &= 0 \\ \mathcal{B}(e_{2j-1}, e_{2k}) &= 0 \\ \mathcal{B}(e_{2j}, e_{2k}) &= 0, \end{aligned}$$

while for $j = k$, we get from

$$0 = \mathcal{B}_{\mathbf{C}}(f_j, f_j)$$

and

$$-1 = h(f_j, f_j)$$

that

$$\mathcal{B}(e_{2j-1}, e_{2j}) = 0$$

and

$$\mathcal{B}(e_{2j}, e_{2j}) = \mathcal{B}(e_{2j-1}, e_{2j-1}) = -1.$$

Now putting for each $j = 1, \ldots, m$,

$$J(e_{2j}) = e_{2j-1}$$

and

$$J(e_{2j-1}) = -e_{2j}$$

and extending J linearly to X, it is easily seen that J is a complex structure on X.

Thus we have that the correspondence between the set of all complex structures J on a $2m$-dimensional real orthogonal space $(X, \mathcal{B})$ of type $(0, 2m)$ and the set of all maximally isotropic subspaces W of $(X_{\mathbf{C}}, \mathcal{B}_{\mathbf{C}})$ is one-to-one.

4.7.2 Complex structures on $\mathbb{R}^{0,2m}$ and corresponding minimal left ideals in $\mathbb{C}_{2m}$

Consider the real orthogonal space $\mathbb{R}^{0,2m}$, its complexification $\mathbb{C}^{2m}$ and the associated complex Clifford algebra $\mathbb{C}_{2m}$. Then we have that $\mathbb{C}^{2m}$ is a subspace of $\mathbb{C}_{2m}$.

Now let $\{e_{2j-1}, e_{2j} : j = 1, \ldots, m\}$ be an orthonormal basis for $\mathbb{C}^{2m}$ and define for each $j = 1, \ldots, m$, the elements f_j and $\bar{f}_j$ by

$$\begin{aligned} f_j &= \tfrac{1}{2}(e_{2j-1} + ie_{2j}) \\ \text{and} \quad & \\ \bar{f}_j &= -\tfrac{1}{2}(e_{2j-1} - ie_{2j}), \end{aligned} \tag{4.11}$$

i.e. $\bar{f}_j$ is obtained from f_j by taking the bar-map in $\mathbb{C}_{2m}$ (see also 3.3)

Notice that from the relations (4.8) we obviously obtain that

$$\begin{aligned} e_{2j-1} &= f_j - \bar{f}_j \\ \text{and} \quad & \\ e_{2j} &= \tfrac{1}{i}(f_j + \bar{f}_j) \end{aligned} \tag{4.12}$$

Clearly $W = \text{span}_{\mathbb{C}}\{f_1, \ldots, f_m\}$ and $\bar{W} = \text{span}_{\mathbb{C}}\{\bar{f}_1, \ldots, \bar{f}_m\}$ are maximally isotropic subspaces of $\mathbb{C}^{2m}$ and $\mathbb{C}^{2m} = W \oplus \bar{W}$.

Hence W determines a complex structure J on $\mathbb{R}^{0,2m}$ which preserves the bilinear form, i.e. $J^2 = -1$ and $J \in \mathrm{SO}(2m)$.

Notice also that in $\mathbb{C}_{2m}$,

$$\begin{aligned} f_j f_k + f_k f_j &= 0; \\ \bar{f}_j \bar{f}_k + \bar{f}_k \bar{f}_j &= 0; \\ \bar{f}_j f_k + f_k \bar{f}_j &= 0 \quad \text{if} \quad k \neq j; \\ \bar{f}_j f_j &= \frac{1}{2}(1 - ie_{2j-1}e_{2j}). \end{aligned} \tag{4.13}$$

From the first two relations of (4.13) it follows that the subalgebras of $\mathbb{C}_{2m}$ generated by W and $\bar{W}$ are isomorphic to, respectively, ΛW and $\Lambda \bar{W}$, the Grassmann algebras determined by W and $\bar{W}$.

But there is more, namely ΛW and $\Lambda \bar{W}$ are subalgebras of $\mathbb{C}_{2m}$. Let us indicate this in the case of ΛW.

Indeed, by (1.9) and the first relation of (4.13)

$$\begin{aligned} f_j \wedge f_k &= \frac{1}{2}(f_j f_k - f_k f_j) \\ &= f_j f_k \end{aligned}$$

whence by iteration, for any $A = \{\alpha_1, \ldots, \alpha_h\} \subset \{1, \ldots, m\}$ where $1 \leq \alpha_1 < \alpha_2 < \ldots < \alpha_h \leq m$,

$$f_A = f_{\alpha_1} \wedge f_{\alpha_2} \wedge \ldots \wedge f_{\alpha_h}$$

$$= f_{\alpha_1} f_{\alpha_2} \cdots f_{\alpha_h}. \tag{4.14}$$

This relation clearly implies that $\wedge W$ is a subalgebra of $\mathbf{C}_{2m}$.

Putting for $j = 1, \ldots, m$,

$$i_j = -e_{2j-1}e_{2j}$$

and

$$I_j = \frac{1}{2}(1 + i\, i_j) = \bar{f}_j f_j,$$

we have that the set $\{I_1, \ldots, I_m\}$ consists of mutually commuting idempotent elements in $\mathbf{C}_{2m}$.

Next, defining $I \in \mathbf{C}_{2m}$ by

$$I = I_1 I_2 \ldots I_m,$$

we claim that I is a primitive idempotent in $\mathbf{C}_{2m}$.

This will follow from

Theorem. $\mathbf{C}_{2m} I = (\wedge W) I$.

Proof. First remark that for all $j = 1, \ldots, m$, $\bar{f}_j I = 0$.

Next, let again $A = \{\alpha_1, \ldots, \alpha_h\} \subset \{1, \ldots, m\}$ be ordered in such a way that $1 \leq \alpha_1 < \alpha_2 < \ldots < \alpha_h \leq m$.

Then, by means of (4.13), the following possibilities occur for $\bar{f}_j f_A I$.

If $j \notin A$, then

$$\bar{f}_j f_A I = (-1)^h f_A \bar{f}_j I = 0,$$

while, if $j \in A$, say $j = \alpha_t$,

$$\begin{aligned} \bar{f}_j f_A I &= (-1)^{t+1} f_{A'} \bar{f}_j f_j I \\ &= (-1)^{t+1} f_{A'} I \end{aligned}$$

where $A' = A \backslash \{\alpha_t\}$.

This implies that, if $a \in \mathbf{C}_{2m}$ is an element generated by the sets $\{f_1, \ldots, f_m\}$ and $\{\bar{f}_1, \ldots, \bar{f}_m\}$, then there exists $b \in \wedge W$ such that

$$aI = bI.$$

But, as $\mathbf{C}^{2m} = W \oplus \bar{W}$, any element of $\mathbf{C}_{2m}$ is generated by the sets $\{f_1, \ldots, f_m\}$ and $\{\bar{f}_1, \ldots, \bar{f}_m\}$, whence $\mathbf{C}_{2m} I \subset (\wedge W) I$.

As moreover $\wedge W \subset \mathbf{C}_{2m}$, we finally obtain that $\mathbf{C}_{2m} I = (\wedge W) I$. ■

Remark. From the foregoing proof we easily obtain the following important property concerning the multiplication of $I\bar{f}_B$ and $f_A I$ where $A, B \subset \{1, \ldots, m\}$, namely :

$$(I\bar{f}_B)(f_A I) = \delta_{A,B} I. \tag{4.15}$$

Let us recall once more that $^-$ stands for the bar-map on $\mathbf{C}_{2m}$ introduced in section 3.3.

Corollary. *I is a primitive idempotent in $\mathbf{C}_{2m}$ and so $(\wedge W)I = \mathbf{C}_{2m}I$ is a minimal left ideal in $\mathbf{C}_{2m}$.*

Proof. As a complex vector space, $(\wedge W)I$ has dimension 2^m, whence by virtue of the foregoing theorem, the left ideal $\mathbf{C}_{2m}I$ of $\mathbf{C}_{2m}$ has also dimension 2^m, which is in fact the dimension of a minimal left ideal in $\mathbf{C}_{2m}$ since $\mathbf{C}_{2m} \cong \mathbf{C}(2^m)$.

Hence I and $\mathbf{C}_{2m}I$ must be respectively a primitive idempotent and a minimal left ideal in $\mathbf{C}_{2m}$. ■

4.7.3 Fock space and γ-matrices

Since $(\wedge W)I = \mathbf{C}_{2m}I$ is a minimal left ideal of $\mathbf{C}_{2m}$, this space may be considered as a space S_{2m} of Dirac spinors where the representation of $\mathbf{C}_{2m}$ on S_{2m} is realized by left multiplication.

To be more precise, define for any $a \in \mathbf{C}_{2m}$, $\gamma(a) : (\wedge W)I \to (\wedge W)I$ by

$$\gamma(a)(bI) = abI.$$

Then $\gamma(a) \in \operatorname{End}_{\mathbf{C}}((\wedge W)I)$.

We have

Theorem 1. *For $j = 1, \ldots, m$,*

(i) $\gamma(f_j)f_A I = (f_j \wedge f_A)I$

(ii) $\gamma(\bar{f}_j)f_A I = \bar{f}_j(f_A)I$

where $\bar{f}_j(f_A) = 0$ if $j \notin A$ and $\bar{f}_j(f_A) = (-1)^{t+1} f_{A'}$ if $j = \alpha_t$ with $A' = A \backslash \{\alpha_t\}$.

Corollary 1. *For all $j, k = 1, \ldots, m$, we have in $\operatorname{End}_{\mathbf{C}}((\wedge W)I)$ that*

(i) $\gamma(f_j)\gamma(f_k) + \gamma(f_k)\gamma(f_j) = 0$

(ii) $\gamma(\bar{f}_j)\gamma(\bar{f}_k) + \gamma(\bar{f}_k)\gamma(\bar{f}_j) = 0$

(iii) $\gamma(f_j)\gamma(\bar{f}_k) + \gamma(\bar{f}_k)\gamma(f_j) = \delta_{jk}\mathbf{1}$.

As the relations (4.11) may suggest, we now put

$$\begin{aligned} \gamma_{2j-1} &= \gamma(e_{2j-1}) &= \gamma(f_j) - \gamma(\bar{f}_j) \\ \text{and} \\ \gamma_{2j} &= \gamma(e_{2j}) &= \tfrac{1}{i}[\gamma(f_j) + \gamma(\bar{f}_j)]. \end{aligned} \tag{4.16}$$

Since, by virtue of Corollary 1, the set $\{\gamma(f_j), \gamma(\bar{f}_j) : j = 1, \ldots, m\}$ is a linearly independent subset of $\text{End}_{\mathbf{C}}((\wedge W)I)$, the same may be proved for the set $\{\gamma_{2j-1}, \gamma_{2j} : j = 1, \ldots, m\}$.

Moreover for all $r, s = 1, \ldots, 2m$,

$$\gamma_r\gamma_s + \gamma_s\gamma_r = -2\delta_{rs}\mathbf{1}$$

whence the set $\{\gamma_1, \ldots, \gamma_{2m}\}$ generates a subalgebra $\mathcal{A}$ of $\text{End}_{\mathbf{C}}((\wedge W)I)$ which is isomorphic to $\mathbf{C}_{2m}$.

But, as $\dim \mathcal{A} = 2^{2m} = \dim \text{End}_{\mathbf{C}}((\wedge W)I)$, we have that

$$\mathcal{A} = \text{End}\,\mathbf{C}((\wedge W)I) \cong \mathbf{C}_{2m}$$

and so $(\wedge W)I$ is a representation space for $\mathbf{C}_{2m}$ over $\mathbf{C}$ or a space of Dirac spinors.

The matrices $\gamma_r, r = 1, \ldots, 2m$, are called *$\gamma$-matrices* and the realization $(\wedge W)I$ of the space of Dirac spinors is called the *Fock space.* The representation $\gamma : \mathbf{C}_{2m} \to \text{End}_{\mathbf{C}}((\wedge W)I)$ is called the *basic spinor representation* .

4.7.4 A realization of Weyl spinors

As we saw in the foregoing subsection, a realization of the space of Dirac spinors is obtained by taking $S_{2m} = (\wedge W)I$.

In 4.3.2. we saw that the space of Weyl spinors could be considered as being the eigenspaces of the chirality operator θ acting on S_{2m}, namely

$$S_{2m}^{\pm} = \{v \in S_{2m} : \theta v = \pm v\}.$$

Let us recall that, as in general $\theta = i^{\nu} e_N$ with $\nu = \frac{q-p}{2}$, we obtain for the case $\mathbf{C}_{2m}$ under consideration (i.e. $p = 0, q = 2m$) that $\theta = i^m e_N$.

Now we claim that, when S_{2m} is realized by $(\wedge W)I$, then S_{2m}^+ and S_{2m}^- are realized by, respectively, $(\wedge^{\text{even}} W)I$ and $(\wedge^{\text{odd}} W)I$.

In order to prove this result, we first make the following observations.

Obviously

$$\mathbf{C}_{2m} I = \mathbf{C}_{2m}^+ I \oplus \mathbf{C}_{2m}^- I$$

and

$$(\wedge W)I = (\wedge^{\text{even}} W)I \oplus (\wedge^{\text{odd}} W)I.$$

Using the relations (4.11)-(4.14) as well as the properties

$$\bar{f}_k I = 0$$

and

$$\bar{f}_k f_k I = I,$$

valid for each $k = 1, \ldots, m$, we get in a straightforward way

Lemma 1.

(i) $(\wedge^{\text{even}} W)I = \mathbf{C}_{2m}^{+} I$.

(ii) $(\wedge^{\text{odd}} W)I = \mathbf{C}_{2m}^{-} I$.

Furthermore we have

Lemma 2.

(i) $\theta|(\wedge^{\text{even}} W)I = \mathbf{1}$

(ii) $\theta|(\wedge^{\text{odd}} W)I = -\mathbf{1}$.

Proof. First observe that for each $k = 1, \ldots, m$, $e_{2k-1}e_{2k}I_k = (-i)I_k$ whence

$$e_N I = (-i)^m I.$$

Consequently, as $\theta = i^m e_N$, $\theta I = I$.

Taking an arbitrary basis element $f_A I$ of $(\wedge W)I$, we thus get

$$\theta(f_A I) = f_A \theta I = f_A I \quad \text{if } {}^{\#}A \text{ is even}$$

and

$$\theta(f_A I) = -f_A \theta I = -f_A I \quad \text{if } {}^{\#}A \text{ is odd.} \quad \blacksquare$$

From Lemma 1 and Lemma 2 we now obtain the desired realization of $S_{2m}^{\pm}$.

Theorem 1.

(i) $S_{2m}^{+} = (\wedge^{\text{even}} W)I \subset \mathbf{C}_{2m}^{+}$

(ii) $S_{2m}^{-} = (\wedge^{\text{odd}} W)I \subset \mathbf{C}_{2m}^{-}$.

Proof. It follows from Lemma 1 that $(\wedge^{\text{even}} W)I$ is a subspace of S_{2m}^{+}. But as it has the same dimension as S_{2m}^{+}, both spaces should coincide and so (i) is proved.

In a same way we get (ii). $\blacksquare$

4.7.5 The γ-matrices revisited

It is clear from the definition of the γ-matrices that for each $r = 1, \ldots, 2m$,

$$\gamma_r : S_{2m}^{+} \to S_{2m}^{-}$$

and

$$\gamma_r : S_{2m}^{-} \to S_{2m}^{+}.$$

Ordering the basis elements $f_A I$ of S_{2m} in such a way that first the basis elements of S_{2m}^{+} are taken, followed by those of S_{2m}^{-}, we thus obtain for any γ-matrix that, with respect to such an ordered basis, it has the form (see also 4.3.3)

$$\left(\begin{array}{c|c} 0 & \star \\ \hline \star & 0 \end{array}\right).$$

Let us illustrate this in the cases $m = 1$ and $m = 2$.

Example 1 : $m = 1$

Take the orthonormal basis $\{e_1, e_2\}$ of $\mathbf{R}^{0,2}$ and construct $W = \mathrm{span}_{\mathbf{C}}\{f_1\}$ with $f_1 = \frac{1}{2}(e_1 + ie_2)$.

Then $i_1 = -e_1e_2$, $I = I_1 = \frac{1}{2}(1 + ii_1) = \bar{f}_1 f_1$, and $S_2 = (\wedge W)I = (\mathbf{C} \oplus \mathbf{C}f_1)I = S_2^{+} \oplus S_2^{-}$ with $S_2^{+} = \mathbf{C}I$ and $S_2^{-} = \mathbf{C}f_1 I$.

A straightforward calculation yields

$$\gamma_1 = \begin{pmatrix} & -1 \\ 1 & \end{pmatrix} \quad \text{and} \quad \gamma_2 = \begin{pmatrix} & -i \\ -i & \end{pmatrix},$$

which are known as *Pauli-matrices*.

Example 2 : $m = 2$

Consider the orthonormal basis $\{e_1, e_2, e_3, e_4\}$ of $\mathbf{R}^{0,4}$ and construct $W = \mathrm{span}_{\mathbf{C}}\{f_1, f_2\}$ with $f_1 = \frac{1}{2}(e_1 + ie_2)$ and $f_2 = \frac{1}{2}(e_3 + ie_4)$.

Then $i_1 = -e_1e_2$, $i_2 = -e_3e_4$, $I_1 = \frac{1}{2}(1 + ii_1)$, $I_2 = \frac{1}{2}(1 + ii_2)$ and $I = I_1 I_2$.

So $S_4 = (\wedge W)I = S_4^{+} \oplus S_4^{-}$ with $S_4^{+} = (\mathrm{span}_{\mathbf{C}}\{1, f_1 \wedge f_2\})I$ and $S_4^{-} = (\mathrm{span}_{\mathbf{C}}\{f_1, f_2\})I$.

Direct calculations lead to

$$\gamma_1 = \left(\begin{array}{cc|cc} & & -1 & \\ & & & 1 \\ \hline 1 & & & \\ & -1 & & \end{array}\right) \quad ; \quad \gamma_2 = \left(\begin{array}{cc|cc} & & -i & \\ & & & -i \\ \hline -i & & & \\ & -i & & \end{array}\right)$$

$$\gamma_3 = \left(\begin{array}{cc|cc} & & & -1 \\ & & -1 & \\ \hline & 1 & & \\ 1 & & & \end{array}\right) \quad ; \quad \gamma_4 = \left(\begin{array}{cc|cc} & & & -i \\ & & i & \\ \hline & i & & \\ -i & & & \end{array}\right)$$

which are known as (a Euclidean version of) *Dirac matrices.*

Remark. As we already pointed out in 4.3.3, a realization of $\mathbf{C}_{2m-1} \cong \mathbf{C}_{2m}^{+}$ is obtained by (2×2)-block matrices of the form

$$\left(\begin{array}{c|c} \star & 0 \\ \hline 0 & \star \end{array}\right).$$

Let us illustrate this in the case $n = 2m - 1 = 3$ or $m = 2$.

By taking the basis e_i, $i = 0, 1, 2, 3$, which generates $\mathbf{C}_4$ and by putting $\varepsilon_j = -e_0 e_j$, $j = 1, 2, 3$, we get from Example 2 that

$$\varepsilon_1 \Leftrightarrow \left(\begin{array}{cc|cc} -i & & & \\ & i & & \\ \hline & & i & \\ & & & -i \end{array}\right) \quad ;$$

$$\varepsilon_2 \Leftrightarrow \left(\begin{array}{cc|cc} & 1 & & \\ -1 & & & \\ \hline & & & 1 \\ & & -1 & \end{array}\right) \quad ;$$

$$\varepsilon_3 \Leftrightarrow \left(\begin{array}{cc|cc} & i & & \\ i & & & \\ \hline & & & i \\ & & i & \end{array}\right).$$

4.8 A Hermitian structure on the Fock space

In section 4.6 we saw how by means of a Hermitian map H between S_{2m} and $S_{2m}^{\prime\star}$, a non-degenerate Hermitian form could be defined on S_{2m}.

The aim of this section is to give a realization of H and its corresponding Hermitian form in the case when S_{2m} is the Fock space $(\wedge W)I$ introduced in the foregoing subsection.

Let us briefly recall that $(\wedge W)I = \mathbf{C}_{2m} I \subset \mathbf{C}_{2m}$ so that, by taking the bar-map, $\overline{(\wedge W)I} = I(\wedge \bar{W}) = I\mathbf{C}_{2m} \subset \mathbf{C}_{2m}$.

In what follows we use the notations S and $\bar{S}$ for $(\wedge W)I$ and $I(\wedge \bar{W})$.

4.8.1 The vacuum expectation value $<>$

If we look again how $\mathbf{C}_{2m}$ was realized as $\mathrm{End}(S)$ by means of left multiplication on S, we find in particular that for each $j = 1, \ldots, m$,

$$f_j(1I) = f_j I$$

and

$$\bar{f}_j(1I) = 0.$$

Hence the elements I, f_j and $\bar{f}_j$ may be regarded upon as being respectively a vacuum vector and creation and annihilation operators.

We have

Theorem 1. $I\mathbf{C}_{2m}I \cong \mathbf{C}I$.

Proof. First observe that, as $\bar{\mathbf{C}}_{2m}\mathbf{C}_{2m} = \mathbf{C}_{2m}$, $I(\wedge\bar{W})(\wedge W)I = I\mathbf{C}_{2m}I$.

Moreover, as was already pointed out in (4.15) for all $A, B \subset \{1, \ldots, m\}$, $(I\bar{f}_B)(f_A I)$ $\delta_{A,B}I$. ■

Corollary. *For each $a \in \mathbf{C}_{2m}$, there exists a unique scalar, denoted by $< a >$, such that $IaI =< a > I$.*

Definition. *The function $<>$: $\mathbf{C}_{2m} \to \mathbf{C}$ is called the vacuum expectation value.*

Remarks.

(1) It is clear that $<>$ is a linear functional on $\mathbf{C}_{2m}$.

(2) The relation (4.15) implies that for any pair of basic elements $f_A, f_B \in \wedge W$,
$< \bar{f}_B f_A >= \delta_{AB}$. (4.17)

Theorem 2. *For all $a \in \mathbf{C}_{2m}$, $< \bar{a} >= \overline{< a >}$*

Proof. Take $A, B \subset \{1, \ldots, m\}$. Then by (4.17)

$$\delta_{AB} = \overline{< \bar{f}_B f_A >} =< \bar{f}_A f_B >=< \overline{\bar{f}_B f_A} >$$

and for each $\lambda \in \mathbf{C}$,

$$\begin{aligned} \overline{< \lambda \bar{f}_B f_A >} =< \bar{\lambda} \bar{f}_A f_B > &= \bar{\lambda} < \bar{f}_B f_A > \\ &= \overline{< \lambda \bar{f}_B f_A >}. \end{aligned}$$

The two foregoing relations already imply that for each element $a \in \mathbf{C}_{2m}$ of the form $a = \sum_{A,B} \mu_{AB} \bar{f}_B f_A$, $< \bar{a} >= \overline{< a >}$.

But, as $\mathbf{C}_{2m}$ is generated by $f_j, \bar{f}_k, j, k = 1, \ldots, m$, the proof is done. ■

4.8.2 The dual S' of S

The aim of this subsection is to show how the dual space of the spinor space S may also be identified with a subspace of $\mathbf{C}_{2m}$.

Indeed, we have

Theorem 1. $S' \cong \bar{S}$.

Proof. For any element $I\bar{t} \in \bar{S}$ define $\mathcal{T}_{I\bar{t}} : S \to \mathbf{C}$ by $\mathcal{T}_{I\bar{t}}(sI) = <\bar{t}s>$.

Then clearly $\mathcal{T}_{I\bar{t}} \in S'$.

Moreover, as for all $A, B \subset \{1, \ldots, m\}$,

$$\mathcal{T}_{I\bar{f}_B}(f_A I) = \delta_{AB},$$

we obtain that $\{\mathcal{T}_{I\bar{f}_B} : B \subset \{1, \ldots, m\}\}$ is the dual basis of the basis $\{f_A I : A \subset \{1, \ldots, m\}\}$ for S. ■

Remark. From the foregoing proof it follows that to each $\mathcal{T} \in S'$ there corresponds a unique $t \in \Lambda W$ such that $\mathcal{T} = \mathcal{T}_{I\bar{t}}$.

Henceforth we write $\mathcal{T} = I\bar{t}$.

Let us also recall that the action of $\mathcal{T}$ on $sI \in S$ is given by

$$\mathcal{T}(sI) = \mathcal{T}_{I\bar{t}}(sI) = <\bar{t}s> . \tag{4.18}$$

As for any (finite dimensional) vector space V,

$$\text{End}(V) \cong V \otimes V',$$

we obtain by the foregoing Theorem 1 that

$$\begin{aligned} \mathbf{C}_{2m} &\cong S \otimes S' \\ &\cong S \otimes \bar{S} \end{aligned}$$

We even have more, namely

Theorem 2. $\mathbf{C}_{2m} = S\bar{S}$

Proof. First note that in a classical way $S\bar{S}$ stands for the subalgebra of $\mathbf{C}_{2m}$ generated by elements of the form $sI\bar{t}$ with $s, t \in \Lambda W$.

Now consider for $A, B \subset \{1, \ldots, m\}$ the elements $f_A I \bar{f}_B \in S\bar{S}$. As their action as linear operators on S is given by left multiplication, we find that for any $C \subset \{1, \ldots, m\}$,

$$(f_A I f_B)(f_C I) = \delta_{BC} f_A I.$$

Ordering the subsets $C \subset \{1, \ldots, m\}$ in some way and calling $\mathcal{B}$ the ordered basis of S consisting of the elements $f_C I$, we thus obtain that the matrix representation $[f_A I \bar{f}_B]_{\mathcal{B}}$ of $f_A I \bar{f}_B$ with respect to $\mathcal{B}$ contains one element 1, the other elements all being zero.

Hence $\{[f_A I \bar{f}_B]_{\mathcal{B}} : A, B \subset \{1, \ldots, m\}\}$ determines the standard basis of $\mathbf{C}(2^m)$, which indeed implies that $\mathbf{C}_{2m} = S\bar{S}$. ■

Corollary 1. $\mathbf{C}_{2m} = \overline{S\bar{S}}$.

4.8.3 An inner product on the Fock space

Throughout this subsection we shall make constant use of the fact that as well S as S' are lying inside $\mathbf{C}_{2m}$.

As we saw in 4.6.2, by means of the Hermitian map $H_\alpha : S \to S'^\star$, we may define a non-degenerate Hermitian form $(\ ,\ )$ on S by putting for all $v, w \in S$,

$$(v, w) = (H_\alpha(w))^\star(v).$$

Let us now have a closer look at the realizations of H_α and $(\ ,\ )$ in the case of $\mathbf{R}_{0,2m}$.

Of crucial importance in the definition of H_α was the representation $(\gamma_\alpha'^\star|\mathbf{R}_{0,2m})_\mathbf{C}$ of $\mathbf{C}_{2m}$ on $S'^\star$ associated with the representation γ of $\mathbf{C}_{2m}$ on S.

Here, of course, γ is the basic spinor representation of $\mathbf{C}_{2m}$ obtained in 4.7.3.

Now let $\mathcal{T} = I\bar{t} \in S'$. Then for all $v = sI \in S$ and $a \in \mathbf{R}_{0,2m}$,

$$\gamma_\alpha'^\star(a)(\mathcal{T}^\star)(v^\star) = (\mathcal{T}(\gamma(\alpha(a))(v)))^\star.$$

But, since the action of $\gamma(\alpha(a))$ on v is given by left multiplication in $\mathbf{C}_{2m}$ and that of $\mathcal{T}$ on S is described by (4.18), we get :

$$\gamma_\alpha'^\star(a)(\mathcal{T}^\star)(v^\star) = (< \bar{t}\alpha(a)s >)^\star.$$

Furthermore, as $a \in \mathbf{R}_{0,2m}$ and so $\alpha(a) = \bar{a}$, we obtain :

$$\begin{aligned} \gamma_\alpha'^\star(a)(\mathcal{T}^\star)(v^\star) &= (< \bar{t}\bar{a}s >)^\star \\ &= (< \overline{at}s >)^\star. \end{aligned}$$

By the convention made, this means that for $\mathcal{T} = I\bar{t} \in S'$,

$$\gamma_\alpha'^\star(\mathcal{T}^\star) = (I\overline{at})^\star. \tag{4.19}$$

Now define $\tilde{H} : S \to S'^\star = \bar{S}^\star$ by

$$\tilde{H}(u) = \bar{u}^\star \quad , \quad u = tI \in S. \tag{4.20}$$

Clearly we then have that for each $v = sI \in S$,

$$(\tilde{H}(u))^\star(v) = < \bar{t}s > .$$

Theorem 1. *Let $\tilde{H} : S \to S'^\star$ be defined as in (4.20). Then*

(i) $\tilde{H} = H_\alpha$

(ii) The form $(\ ,\ ) : S \times S \to \mathbf{C}$ given by
$(v, w) = (\tilde{H}(v))^\star(w)$
is an inner product on S.

Proof. In order to prove (i), we show that $\tilde{H}$ is a Hermitian intertwining map between S and $S'^{\star}$.

From the definition of $\tilde{H}$ it immediately follows that $\tilde{H}$ is an isomorphism.

Moreover, for each $a \in \mathbf{R}_{0,2m}$ and $u = tI \in S$ we have that

$$\tilde{H}(\gamma(a)u) = \gamma'^{\star}_{\alpha}(a)(\tilde{H}(u)).$$

Indeed, on the one hand, since $\gamma(a)u = atI$,

$$\tilde{H}(\gamma(a)u) = (I\bar{a}t)^{*}$$

while on the other hand, by virtue of (4.19)

$$\begin{aligned} \gamma'^{\star}_{\alpha}(a)(\tilde{H}(u)) &= \gamma'^{\star}_{\alpha}(a)(\bar{u}^{\star}) \\ &= (I\bar{a}t)^{\star}. \end{aligned}$$

We thus proved that $\tilde{H}$ is an intertwining map.

Furthermore, we show that $\tilde{H}$ is Hermitian, i.e. $\iota_S \circ \tilde{H}^{\star\prime} = \tilde{H}$.

Indeed, take arbitrary elements $v = sI$ and $w = tI$ in S. Then

$$(\tilde{H}(v))^{\star}(w) = < \bar{s}t >$$

and, using (4.3) and (4.4),

$$\begin{aligned} ((\iota_S(\tilde{H}^{\star\prime}(v)))^{\star}(w) &= \overline{\tilde{H}^{\star\prime}(v)(w^{\star})} \\ &= \overline{\tilde{H}^{\star}(w^{\star}(v)} \\ &= \overline{(\tilde{H}(w))^{\star}(v)} \\ &= \overline{< \bar{t}s >}, \end{aligned}$$

whence, by virtue of 4.8.1 Theorem 2,

$$((\iota_S \circ \tilde{H}^{\star\prime})(v))^{\star}(w) = < \bar{s}t > .$$

In view of 4.6.2 Theorem 1, this implies that $\tilde{H} = H_{\alpha}$, whence, by means of 4.6.2 Theorem 2, the form (,) on S given by $(v, w) = (\tilde{H}(v))^{\star}(w)$ is non-degenerate and Hermitian.

To prove that it is positive, it suffices to remark that for any two basis elements $f_A I$ and $f_B I$ of S,

$$\begin{aligned} (f_A I, f_B I) &= (\tilde{H}(f_A I))^{\star}(f_B I) \\ &= < \bar{f}_A f_B > \\ &= \delta_{AB}. \quad \blacksquare \end{aligned}$$

Remarks.

(1) The foregoing Theorem tells us that, if we put for any two elements $v = sI$ and $w = tI$ in S

$$(v, w) = < \bar{s}t >,$$

then (,) is an inner product on S.

(2) Now considering γ as a representation of $\mathbf{C}_{2m}$ on the inner product space S, we have that for all $a \in \mathbf{C}_{2m}$,

$$(\gamma(a))^{\dagger} = \gamma(\bar{a}).$$

Indeed, for all $v = sI$ and $w = tI$ in S,

$$\begin{aligned}(\gamma(a)v, w) &= < \overline{a\bar{s}}t > \\ &= (v, \gamma(\bar{a})w).\end{aligned}$$

In particular, for $a \in \mathbf{R}_{0,2m}$,

$$\gamma(a)^{\dagger} = \gamma(\alpha(a)).$$

Consequently, for all $j = 1, \ldots, 2m$,

$$\gamma(e_j)^{\dagger} = -\gamma(e_j)$$

or $\gamma(e_j)$ is anti-Hermitian.

(3) For any $v, w \in S$ and $a \in \mathbf{R}_{0,2m}$,

$$\begin{aligned}(\gamma(a)v, \gamma(a)w) &= (v, \gamma(\alpha a)(\gamma(a)w)) \\ &= (v, \gamma(\alpha(a)a)(w)).\end{aligned}$$

Calling for each $a \in \mathbf{R}_{0,2m}$,

$$N(a) = \alpha(a)a$$

and G the subset of $\mathbf{R}_{0,2m}$ consisting of those elements $a \in \mathbf{R}_{0,2m}$ for which $N(a) \in \{1, -1\}$, we have that G is a (multiplicative) subgroup of $\mathbf{R}_{0,2m}$.

It is called the *Wall group* for the pair $(\mathbf{R}_{0,2m}, \alpha)$.

We so obtain that for all $a \in G$,

$$(\gamma(a)v, \gamma(a)w) = N(a)(v, w).$$

For a more general description of Wall groups, we refer to [13].

4.9 The complex Spin group and pure spinors

Just as the real Clifford algebra $\mathbf{R}_{p,q}$ contains the subgroups $\mathrm{Pin}(p, q)$ and $\mathrm{Spin}(p, q)$ which are covering groups of, respectively, the real orthogonal group $O(p, q)$ and the group of rotations $SO(p, q)$, the complex Clifford algebra $\mathbf{C}_n$ contains subgroups of the multiplicative group of units in $\mathbf{C}_n$ which are covering spaces for, respectively, the groups $O(n, \mathbf{C})$ and $SO(n, \mathbf{C})$ corresponding to the standard bilinear form B on $\mathbf{C}^n$.

4.9.1 The groups $\mathrm{Pin}(n, \mathbf{C})$ and $\mathrm{Spin}(n, \mathbf{C})$

The *Clifford group* $\Gamma(n, \mathbf{C})$ is a multiplicative subgroup in $\mathbf{C}_n$ defined by

$$\Gamma(n,\mathbf{C}) = \{g \in \mathbf{C}_n : g \text{ is a unit and } gz\hat{g}^{-1} \in \mathbf{C}^n \text{ for all } z \in \mathbf{C}^n\}.$$

Each vector $v \in \mathbf{C}^n$ with $B(v,v) \neq 0$ belongs to the Clifford group and the Clifford group is generated by such vectors.

The linear map $\chi(g) : \mathbf{C}^n \to \mathbf{C}^n$ with $\chi(g)z = gz\hat{g}^{-1}$ preserves the form B whence the map $\chi : \Gamma(n,\mathbf{C}) \to \mathrm{O}(n,\mathbf{C})$ is a representation of the Clifford group on $\mathbf{C}^n$ by orthogonal matrices. The kernel of the homomorphism χ is the multiplicative group $\mathbf{C}\setminus\{0\}$ of complex numbers.

Defining the spinorial norm $\mathcal{N}$ on $\Gamma(n,\mathbf{C})$ by

$$\mathcal{N}(g) = \alpha(g)g,$$

and observing that $\mathcal{N}(g)$ is a homomorphism of $\Gamma(n,\mathbf{C})$ into the multiplicative group $\mathbf{C}\setminus\{0\}$, we introduce the groups

$$\mathrm{Pin}(n,\mathbf{C}) = \{g \in \Gamma(n,\mathbf{C}) : \mathcal{N}(g) = \pm 1\}$$

and

$$\mathrm{Spin}(n,\mathbf{C}) = \{g \in \Gamma(n,\mathbf{C}) \cap \mathbf{C}_n^+ : \mathcal{N}(g) = \pm 1\}.$$

These groups are not connected and we denote by $\mathrm{Pin}_\pm(n,\mathbf{C})$ and $\mathrm{Spin}_\pm(n,\mathbf{C})$ their components mapped to ± 1. So the groups $\mathrm{Pin}_+(n,\mathbf{C})$ and $\mathrm{Spin}_+(n,\mathbf{C})$ are the connected components containing the identity of, respectively, $\mathrm{Pin}(n,\mathbf{C})$ and $\mathrm{Spin}(n,\mathbf{C})$.

Restricting the homomorphism χ to $\mathrm{Pin}(n,\mathbf{C})$ and $\mathrm{Spin}(n,\mathbf{C})$, we obtain the exact sequences

$$1 \longrightarrow \{\pm 1\} \longrightarrow \mathrm{Pin}_+(n,\mathbf{C}) \xrightarrow{\chi} \mathrm{O}(n,\mathbf{C}) \to 1$$

and

$$1 \to \{\pm 1\} \longrightarrow \mathrm{Spin}_+(n,\mathbf{C}) \xrightarrow{\chi} \mathrm{SO}(n,\mathbf{C}) \longrightarrow 1.$$

Indeed, the only elements in $\mathbf{C}\setminus\{0\}$ with norm equal to 1 are ± 1. It can be proved that the group $\mathrm{Spin}_+(n,\mathbf{C})$ is the universal covering group of the group $\mathrm{SO}(n,\mathbf{C})$. Similarly, the kernel of the map χ restricted to $\mathrm{Pin}_-(n,\mathbf{C})$, resp. $\mathrm{Spin}_-(n,\mathbf{C})$, is $\{\pm i\}$ and the map χ restricted to these sets is again a double covering of $\mathrm{O}(n,\mathbf{C})$, resp. $\mathrm{SO}(n,\mathbf{C})$.

Each element of the group $\mathrm{Pin}_+(n,\mathbf{C})$ is a finite product of vectors $v \in \mathbf{C}^n$ with $B(v,v) = \pm 1$, while the elements in $\mathrm{Spin}_+(n,\mathbf{C})$ can be written as a product of an even number of such vectors.

Let us discuss now the map χ and its inverse in more details. On the one hand, if $g \in \mathrm{Spin}_+(n,\mathbf{C})$ is given and if $\{e_1,\ldots,e_n\}$ is the canonical basis of $\mathbf{C}^n$, then the columns of the matrix $\chi(g) \in \mathrm{SO}(n,\mathbf{C})$ are the components of the vectors ge_ig^{-1}, with respect to the canonical basis, $i = 1,\ldots,n$. Note also that due to the fact that $\mathrm{Spin}_+(n,\mathbf{C}) \subset \mathbf{C}_{2n}^+$, we have $\hat{g} = g$.

On the other hand, it is more difficult to describe the preimage $g = \chi^{-1}(\mathcal{A})$ of a matrix $\mathcal{A} \in \mathrm{SO}(n, \mathbf{C})$. We shall compute now some examples in the even dimensional case which will be used in Ch. V.; a general discussion can be found e.g. in [68].

Let $\{e_1, \ldots, e_{2n}\}$ be the canonical basis for $\mathbf{C}^{2n}$ giving rise to the corresponding isotropic basis $f_j, \bar{f}_j, j = 1, \ldots, n$ (see 4.7.2.) The direct sum of the spaces $W = \mathrm{span}\{f_j, j = 1, \ldots, n\}$ and $\overline{W} = \mathrm{span}\{\bar{f}_j, j = 1, \ldots, n\}$ is then the whole space $\mathbf{C}^{2n}$. Recall also that it was shown in 4.7.2 that the products $I_j = \bar{f}_j f_j, j = 1, \ldots, n$, form a commuting set of idempotents. Let us furthermore write matrices $\mathcal{A} \in \mathrm{SO}(2n, \mathbf{C})$ in block form with respect to the splitting $\mathbf{C}^{2n} = W \oplus \overline{W}$.

Example 4.9.1. Let $\mathcal{A} = \begin{pmatrix} A & 0 \\ 0 & (A^t)^{-1} \end{pmatrix} \in \mathrm{SO}(2n, \mathbf{C})$ be such that the matrix $A \in \mathbf{C}(n)$ is a diagonal matrix of the form

$$A = \mathrm{diag}(1 + a_1, \ldots, 1 + a_n),\ a_i \in \mathbf{C},\ a_i \neq -1.$$

Then its preimage in the Spin group may be constructed as follows.

First notice that the element $g = \Pi_{j=1}^n (1 + a_j) I_j$ is invertible. Indeed, choosing for each $j = 1, \ldots, n$, $b_j \in \mathbf{C}$ such that $(1 + a_j)(1 + b_j) = 1$, the inverse element of g is given by $g^{-1} = \Pi_1^n (1 + b_j) I_j$. Moreover, using elementary computations we may easily prove that g belongs to the Clifford group and that the corresponding orthogonal matrix is exactly the matrix $\mathcal{A}$. Clearly the norm of g is given by $\mathcal{N}(g) = \Pi_1^n (1 + a_j)$ and so the preimage of $\mathcal{A}$ is given by $\{\pm g / \sqrt{\mathcal{N}(g)}\}$.

Example 4.9.2 Let $\mathcal{A} = \begin{pmatrix} I_n & A \\ 0 & I_n \end{pmatrix} \in \mathrm{SO}(2n, \mathbf{C})$ whence A is antisymmetric. Then the preimage $\pm g = \chi^{-1}(\mathcal{A}) \in \mathrm{Spin}(2n, \mathbf{C})$ is given by

$$g = \exp\{\frac{1}{2} \sum_{i,j=1}^{n} A_{ij} \bar{f}_i \bar{f}_j\}.$$

Indeed, let us first recall the following standard identity, which may be easily proved by induction. Putting $[a, b] = ab - ba$, we have for all $a, b \in \mathbf{C}_{2n}$ that

$$e^a b = (b + [a, b] + 1/2! [a, [a, b]] + \ldots) e^a.$$

Taking $a = \frac{1}{2} \sum_{i,j=1}^n A_{ij} \bar{f}_i \bar{f}_j$ we get by straightforward computations that $[a, f_k] = \sum_i A_{ik} \bar{f}_i$ and $[a, \bar{f}_k] = 0$. Hence for each $v \in \mathbf{C}^{2n}$ we have $[a, [a, v]] = 0$. Consequently, denoting the action of the linear map $\mathcal{A}$ on a vector v by $A(v)$, we have that

$$e^a v_1 e^{-a} = v_1 + A(v_1), v_1 \in W,\ e^a v_2 e^{-a} = v_2, v_2 \in \overline{W}$$

and so the induced rotation is described by the matrix $\mathcal{A}$.

Moreover, $\mathcal{N}(g)$ being a complex number, we get $\mathcal{N}(g) = 1$ since

$$\mathcal{N}(g)I = I\alpha(g)gI = I.$$

4.9.2 Pure spinors

As was shown in 4.7.1 the set of all complex structures J on a given $2n$-dimensional real orthogonal space can be identified with the set IG_n of all maximal isotropic subspaces in the complexification $\mathbf{C}^{2n}$ (see also V.3.2). We now give a third interpretation of the same set by identifying it with the space of pure spinors.

Let us recall the realizations of, respectively, the spinor space S_{2n} by $(\Lambda W)I$, W being the span of $f_1, \ldots f_n$, and the space of half spinors $S^+ = S_{2n}^+$ by $(\Lambda^{\text{even}} W)I$ (see 4.7).

For each nontrivial spinor $s \in S^+$ we define the set $W_s = \{v \in \mathbf{C}^{2n} : v\,s = 0\}$. Having two vectors $v, v' \in W_s$, then $0 = (vv' + v'v)s = 2(v, v')s$ whence $(v, v') = 0$. So the space W_s is an isotropic subspace and, as we know already, the maximal dimension of an isotropic subspace is n.

We define then the *space of pure spinors* S^{pure} by

$$S^{\text{pure}} = \{s \in S^+ : \; \dim W_s = n\}.$$

As a nonzero multiple of a pure spinor is again a pure spinor and the corresponding isotropic subspaces clearly coincide, we have defined a map from the projective space $\mathbf{P}(S^{\text{pure}})$ of pure spinors into the space IG_n of all maximal isotropic subspaces.

For example, taking the element $I \in S^+$, it is clear that it is a pure spinor and that the corresponding isotropic subspace is exactly the span W of $f_1, \ldots, f_n$.

Moreover, the group $\mathrm{O}(2n, \mathbf{C})$ is acting transitively on IG_n and the connected components of IG_n are given by the image of W under the action of $\mathrm{SO}(2n, \mathbf{C})$ and $\mathrm{O}(2n, \mathbf{C}) \setminus \mathrm{SO}(2n, \mathbf{C})$, respectively. Denoting the corresponding components by IG_n^+ and IG_n^-, then clearly IG_n^+ is the orbit of W under the action of $\mathrm{SO}(2n, \mathbf{C})$ while the isotropic subspaces W_s, $s \in S^+$, always belong to IG_n^+.

To construct the inverse map, take a subspace $W' \in IG_n^+$ and a matrix $\mathcal{A} \in \mathrm{SO}(2n, \mathbf{C})$ with columns $f_1', \ldots, f_{2n}'$ such that W' is the span of $f_1', \ldots, f_n'$. Then there exists an element $g \in \mathrm{Spin}(2n, \mathbf{C})$ such that $\chi(g) = \mathcal{A}$. As for a spinor $s \in S^+$ given by $s = gI$ we have that

$$f_i' s = g f_i g^{-1} g I = 0, \; i = 1, \ldots, n,$$

s is a pure spinor corresponding to the subspace W'.

So the map described establishes a bijection of the space IG_n^+ with the projective space $\mathbf{P}(S^{\text{pure}})$ of pure spinors.

Chapter II

Monogenic functions

This chapter is entirely devoted to fundamental concepts concerning nullsolutions of the Dirac operator.

Section 0 discusses how realizations of the (massless) Dirac and Weyl equations in $\mathbf{R}^{m+1}$ may be obtained by using appropriate complex representations of, respectively the groups $\mathrm{Pin}(m+1)$ and $\mathrm{Spin}(m+1)$. Moreover some basic properties of nullsolutions to these equations, already available in [10], are briefly recalled. All results are formulated in the case of the Dirac equation and it is indicated how, by performing some elementary notational changes, one may obtain similar results in the case of the Weyl equation.

Section 1 studies the spaces $\mathcal{M}^+(k;V)$ and $\mathcal{M}^-(k;V)$ of so-called spherical monogenics of degree k.

In the case of the plane - as is well known - the space $\mathcal{H}(k)$ of spherical harmonics of degree k decomposes into $\mathcal{H}(k) = \mathcal{O}^+(k) \oplus \mathcal{O}^-(k)$ where $\mathcal{O}^\pm(k) = \mathrm{span}_{\mathbf{C}}\{e^{\pm ik\theta} : \theta \in \mathbf{R}\}$ are exactly the eigenspaces of the Cauchy-Riemann operator restricted to the unit circle S^1. The main result of this section describes how in the higher dimensional case an analogous decomposition of the space $\mathcal{H}(k;V)$ of V-valued spherical harmonics of degree k may be obtained, namely $\mathcal{H}(k;V) = \mathcal{M}^+(k;V) \oplus \mathcal{M}^-(k-1;V)$ where $\mathcal{M}^\pm(k;V)$ are eigenspaces of the spherical Dirac operator. This result gives rise to a lot of fundamental properties about the decomposition into spherical monogenics of functions living on the unit sphere S^m. Among others we cite the orthogonal decomposition of $L_2(S^m;V)$ into spaces of spherical monogenics (Theorem 1.3.4), the Laurent and Taylor series for monogenic functions (§1.6), the Fischer decomposition of homogeneous V-valued polynomials of degree k (Theorem 1.10.1) and the decomposition $L_2(S^m;V) = \mathcal{M}_2(S^m;V) \oplus \mathcal{M}_{0,2}(S^m;V)$ obtained in 1.8.3 where $\mathcal{M}_2(S^m;V)$ generalizes the classical Hardy space $\mathfrak{H}_2$. The latter decomposition then leads to the definition of the Hilbert-Riesz transform on $L_2(S^m;V)$, which in its turn may be used in factorizing square roots of differential operators related to the Casimir operators $C_B(H)$ and $C_B(L)$ (§1.11).

We end up this section by showing how complex analysis in the plane naturally fits into Clifford analysis by taking $m = 1$.

It should be noticed that, up to subsection 1.9, all results remain valid for null solutions of the Cauchy-Riemann operator as well while those appearing in §§1.10 -

1.12 only hold in the case of the Dirac operator.

In section 2 we study nullsolutions of the Dirac operator in open domains Ω of $\mathbf{R}^m = \mathbf{R}^{m_1} \times \mathbf{R}^{m_2} \times \ldots \times \mathbf{R}^{m_n}$ which are invariant under the compact subgroup $G = \text{Spin}(m_1) \times \text{Spin}(m_2) \times \ldots \times \text{Spin}(m_n)$ of Spin(m). This leads to the notion of polyaxially monogenic functions and related Laurent equations. The special cases $n = 2$ (biaxially monogenics), $m_1 = 1$ (axially monogenics) and $m_1 = m_2 = \ldots = m_n = 2$ (toroidally monogenics) are treated in more detail.

In section 3 we touch upon the notion of separately monogenic functions and show how the theory of holomorphic functions of several complex variables fits into it.

In section 4 a method is worked out to construct orthonormal bases for the space of inner spherical monogenics of degree k.

In section 5 generalized Taylor series are first studied for functions which are monogenic in an open subset $\tilde{\Omega}$ of $\mathbf{R}^m$ where $\mathbf{R}^m$ is split into $\mathbf{R}^m = \mathbf{R}^p \times \mathbf{R}^q$ and Ω is SO(p)-invariant. The construction of such Taylor series relies upon a generalized Cauchy-Kovalevska Theorem for functions of the form $P_k(\underline{x})A(\underline{y})$ where $P_k \in M^+(p;k;C)$ and A is analytic in $\tilde{\Omega} \cap \mathbf{R}^q$. Finally, using the duality between, respectively, the space $M_r(K;C)$ of functions right monogenic on K and the space $M_{\ell,0}(\mathbf{R}^m \backslash K;C)$ of functions left monogenic outside K and vanishing at infinity, $K \subset \mathbf{R}^q$ being compact, a generalized Laurent series is obtained for functions $f \in M_{\ell,0}(\mathbf{R}^m \backslash K;C)$. The decompositions thus obtained will play a fundamental role when studying residue theory in Chapter IV.

Notice that, in contrast with section 1, the sections 2 - 5 deal with monogenic functions in $\mathbf{R}^m$; moreover all results obtained there are valid only in the case of the Dirac operator.

0 The Dirac and Weyl equations : Elementary Function Theory

In this section some basic properties are recalled of solutions of the (massless) Dirac and Weyl equations in $(m+1)$-dimensional Euclidean space. For detailed proofs we refer the reader to [10].

At present let us focus our attention to the case of the **Dirac** equation which looks like

$$\partial_x f = 0. \tag{0.1}$$

Classically S-valued solutions of (0.1) are studied, S being a spinor space ; in [10] a systematic study of C-valued solutions of (0.1) was undertaken, C being a Clifford algebra.

However, both approaches are related to each other in the following way.

If f is a C-valued solution then, by writing C as a sum of minimal left ideals, f may be decomposed into a sum of S-valued solutions. Moreover, multiplying a

$\mathcal{C}$-valued solution from the right by a constant spinor $s \in S$, an S-valued solution fs is thus obtained. So the reader should realize that, whenever in a formula $\mathcal{C}$-valued solutions are occurring, this formula may immediately be turned into a corresponding one valid for S-valued solutions by multiplying it from the right with some constant spinor s.

Nevertheless notice that by using the Clifford algebra instead of the spinor language in writing out formulas we may benefit of some considerable advantages, namely

(i) A lot of these formulas are not valid when only taking spinor-valued solutions into consideration.

(ii) The space of $\mathcal{C}$-valued solutions is a right $\mathcal{C}$-module, a property which is often useful in itself (see e.g. [52] §7).

Using appropriate realizations of the spinor space S, the Dirac equation can also be written in the form (see e.g. 0.1.3)

$$f\partial_x = 0 \tag{0.2}$$

and again it is meaningful too to study $\mathcal{C}$-valued solutions of (0.2) (see [10]).

Obviously the theory of functions related to (0.2) is similar to the one for (0.1).

However, as will be seen when developing the function theory, some important $\mathcal{C}$-valued functions - especially various kernel functions - satisfy both (0.1) and (0.2) and hence they are not spinor-valued. Let us recall that $\mathcal{C}$ can be identified (in the even dimensional case) with the space End (S) of all endomorphisms of S.

In order to avoid tedious repetitions in formulating theorems we present in subsection 0.1.1 a multiplication table in which all possibilities for the value-spaces are considered at once.

0.1 The Dirac and Weyl equations in $\mathbf{R}^{m+1}$

In this subsection we first give a general setting for the Dirac and Weyl equations in $\mathbf{R}^{m+1}$ and then describe how, by using the realization of the spinor spaces inside an appropriate Clifford algebra, we may obtain realizations of these equations.

0.1.1 Notations

Let $m \in \mathbf{N}$ be fixed with $m \geq 1$.

An arbitrary element $x \in \mathbf{R}^{m+1}$ is written as

$$x = (x_0, x_1, \ldots, x_m) = (x_0, \underline{x})$$

with $\underline{x} = (x_1, \ldots, x_m) \in \mathbf{R}^m$.

For realizations of $\mathbf{R}^{m+1}$ as (real) subspaces of an appropriate (complex) Clifford algebra, we already refer the reader to 0.1.3, Remark (4).

Introducing polar coordinates in $\mathbf{R}^{m+1}$ we write $x \in \mathbf{R}^{m+1}$ as $x = r\xi$ where $r = |x|$ and $\xi \in S^m$, S^m of course denoting the unit sphere in $\mathbf{R}^{m+1}$. We thus have that $\xi = \frac{x}{|x|}$.

The area of S^m is denoted by A_{m+1} and, as is well known,

$$A_{m+1} = \frac{2\pi^{(m+1)/2}}{\Gamma(\frac{m+1}{2})} = (m+1)V_{m+1},$$

V_{m+1} being the volume of the $(m+1)$-dimensional unit ball.

Finally we put

$$\mathbf{R}_0^{m+1} = \mathbf{R}^{m+1}\backslash\{0\}.$$

Now let $\Omega \subset \mathbf{R}^{m+1}$ be open (and non empty). Then in a classical way we denote by

(1) $C_k(\Omega)(k \geq 0)$ the space of k-times continuously differentiable C-valued functions in Ω.

(2) $\mathcal{E}(\Omega)$ the space of C_∞-functions in Ω.

(3) $\mathcal{A}(\Omega)$ the space of analytic functions in Ω.

(4) $\mathcal{D}(\Omega)$ the space of C_∞-functions in Ω having compact support contained in Ω.

If U is an $(m+1)$-dimensional measurable subset of $\mathbf{R}^{m+1}$ and $1 \leq p \leq +\infty$, then

(5) $L_p(U)$ stands for the Banach space of p-integrable functions on U $(1 \leq p < +\infty)$ or the Banach space of essentially bounded measurable functions on $U(p = \infty)$.

Let $\mathcal{F}$ be one of the spaces (1) - (5) and let V be a (complex) subspace of $\mathcal{C}$. Then we may consider the associated space $\mathcal{F} \otimes_{\mathbf{C}} V$ of V-valued functions. If moreover V is a left or right ideal of $\mathcal{C}$, then it is clear that $\mathcal{F} \otimes_{\mathbf{C}} V$ may also be considered as a left or a right module over $\mathcal{C}$.

In the cases (1) - (5), the associated spaces $\mathcal{F} \otimes_{\mathbf{C}} V$ are henceforth denoted by respectively

$$C_k(\Omega; V);\ \mathcal{E}(\Omega; V);\ \mathcal{A}(\Omega; V);\ \mathcal{D}(\Omega; V);\ L_p(U; V).$$

In the same way, replacing Ω or U by S^m we introduce the spaces

$$C_k(S^m; V);\ \mathcal{E}(S^m; V);\ \mathcal{A}(S^m; V);\ L_p(S^m; V).$$

Throughout the book V will be one of the following spaces (see also I. 4.7)

(i) S, the space of Dirac or Weyl spinors realized in $\mathcal{C}$.

(ii) $\overline{S}$, the dual space of the spinor space S, realized too in $\mathcal{C}$.

(iii) $\mathbf{C}I$, to be identified with the field of scalars $\mathbf{C}$.

(iv) The Clifford algebra $\mathcal{C}$.

As in quite a lot of formulas which describe the solutions of (0.1) or (0.2), pointwise multiplication between $\mathcal{F}\otimes_{\mathbf{C}}V$-valued functions is involved, $\mathcal{F}$ and V depending upon the functions under consideration, it is absolutely necessary to specify when such a multiplication is defined.

To this end, first notice that if for the functions f, g and h the products $(fg)h$ and $f(gh)$ exist, then $(fg)h = f(gh)$.

Indeed, as these products are realized pointwisely in the Clifford algebra $\mathcal{C}$ and associativity is valid for the multiplication in $\mathcal{C}$, the assertion follows immediately.

The only thing left now is to investigate when the multiplication between such functions is meaningful. The answer is given by the (4×4)-multiplication table below, where the symbol $\star$ indicates that the multiplication under consideration is not defined.

f \ g	$\mathbf{C}I$	S	$\overline{S}$	$\mathcal{C}$
$\mathbf{C}I$	$\mathbf{C}I$	$\mathbf{C}I$	$\overline{S}$	$\overline{S}$
$\overline{S}$	$\mathbf{C}I$	$\mathbf{C}I$	$\star$	$\overline{S}$
S	S	$\star$	$\mathcal{C}$	$\star$
$\mathcal{C}$	S	S	$\star$	$\mathcal{C}$

Notice that once the multiplication between three elements taken in a certain order and by using brackets is meaningful, it also makes sense when replacing the brackets and the result remains the same, i.e. "associativity" holds for the multiplication.

Let us illustrate this table in the case when $m+1=2k$ is even.

By the convention made and taking account of the results obtained in I 4.7, we take V equal to $S = (\wedge W)I$, $\overline{S} = I(\wedge\overline{W})$, $\mathbf{C}I$ or $\mathcal{C} = \mathbf{C}_{2k}$. Furthermore we choose $\mathcal{F} = L_2(S^m)$.

On $L_2(S^m; V)$ we then define an inner product and norm by

$$(f,g) = \int_{S^M} \overline{f(\omega)}g(\omega)d\Sigma(\omega) \tag{0.3}$$

and

$$\| f \|^2 = 2^{m+1}\int_{S^m} [\overline{f(\omega)}f(\omega)]_0 d\Sigma(\omega). \tag{0.4}$$

Here $d\Sigma(\omega)$ stands for the elementary surface element on S^m.

Whenever the product between $\bar{f}$ and g is meaningful in the sense of the multiplication table, we have that the expressions (0.3) and (0.4) are well defined.

Let us discuss two important cases which will appear frequently in the sequel.

(i) The case $V = \mathcal{C}$

When $V = \mathcal{C}$, the product $\bar{f}g$ is $\mathcal{C}$-valued too and hence (0.3) gives rise to a $\mathcal{C}$-valued inner product. In such a way $L_2(S^m; \mathcal{C})$ becomes a right $\mathcal{C}$- Hilbert module (see also [10])

(ii) The case $V = S$

When $V = S$, the product $\bar{f}g$ is $\mathbf{C}I$-valued and, using the vacuum expectation introduced in I.4.8, we obtain

$$\begin{aligned} (f, g) &= \int_{S^m} I\bar{f}(\omega)g(\omega)I d\Sigma(\omega) \\ &= I\int_{S^m} < \bar{f}(\omega)g(\omega) > d\Sigma(\omega). \end{aligned} \tag{0.5}$$

Furthermore as for all $t, s \in \wedge W$,

$$\begin{aligned} [I\bar{t}sI]_0 &= < \bar{t}s > [I]_0 \\ &= < \bar{t}s > \frac{1}{2^{m+1}}, \end{aligned}$$

we have that

$$\| f \|^2 = \int_{S^m} < \bar{f}(\omega)f(\omega) > d\Sigma(\omega). \tag{0.6}$$

The expression (0.3) thus gives rise to a $\mathbf{C}I$-valued inner product which we may obviously identify with a $\mathbf{C}$-valued inner product. In such a way $L_2(S^m; S)$ becomes a complex Hilbert space.

The need for a multiplication table as given above becomes immediately clear when establishing one of the most fundamental formulas in the theory of monogenic functions, namely Stokes' Theorem (see section 0.2.1)

0.1.2 The Dirac and Weyl equations : general setting

Let $e = (e_0, e_1, \ldots, e_m)$ be a non-degenerate orthonormal set of type $(0, m+1)$, let $\mathcal{C}$ be the complex Clifford algebra generated by it and let $\Omega \subset \mathbf{R}^{m+1}$ be open.

If (ρ, V) is a basic representation of Pin $(m+1)$ or a direct sum of such representations, then the *(massless) Dirac equation* for $f \in C_1(\Omega; V)$ is given by

$$\partial_x f = 0 \quad \text{in} \quad \Omega \tag{0.7}$$

where

$$\partial_x = \sum_{j=0}^{m} \rho(e_j)\partial_{x_j} \tag{0.8}$$

is called the *Dirac operator* in $\mathbf{R}^{m+1}$.

Putting

$$\varepsilon_i = -e_0 e_i = \overline{e}_0 e_i, \quad i = 1, \ldots, m,$$

then the set $\{\varepsilon_1, \ldots, \varepsilon_m\}$ generates Spin $(m+1)$.

Now letting (ρ, V) be a basic Spin $(m+1)$-representation or a direct sum of such representations, the *Weyl equation* for $f \in C_1(\Omega; V)$ is given by

$$D_x f = 0 \tag{0.9}$$

where

$$D_x = \partial_{x_0} + \sum_{i=1}^{m} \rho(\varepsilon_i)\partial_{x_i} \tag{0.10}$$

is called the *Cauchy-Riemann operator* in $\mathbf{R}^{m+1}$.

Of course, any of the equations (0.8) or (0.10) will split into pieces of the type which will now be discussed in more detail.

As in the definition of the Dirac equation a sum of basic representations of the group Pin $(m+1)$ is involved, in what follows it is sufficient to look at one such basic representation of Pin $(m+1)$. By virtue of I. 4.4, we obtain that for $m+1 = 2k$ even, there exists (up to equivalence) only one basic spinor representation, namely S_{2k}, whence in that case we are left with one Dirac equation where $C_1(\Omega; S_{2m})$-solutions are considered. For $m+1 = 2k-1$ odd, there are two unequivalent basic Pin $(m+1)$- representations, namely S_{2k}^+ and S_{2k}^- whence in that case we are left with two possibilities for the Dirac equation, corresponding, respectively, to the study of $C_1(\Omega; S_{2k}^+)$- or $C_1(\Omega; S_{2k}^-)$-solutions.

Similarly, as in the definition of the Weyl equation a sum of basic representations of Spin $(m+1)$ is involved, we may restrict ourselves to looking at one such basic representation. Again by I.4.4 we get that for $m+1 = 2k$ even, two Weyl equations should be studied depending upon the case whether f belongs to $C_1(\Omega, S_{2k}^+)$ or $C_1(\Omega; S_{2k}^-)$, while for $m+1 = 2k-1$ odd, only one Weyl equation appears which corresponds to the study of $C_1(\Omega; S_{2k-2})$-solutions f.

0.1.3 Realizations of the Dirac and Weyl equations

As we already pointed out in the beginning of this section we shall consider two types of solutions to the Dirac and Weyl equations, namely spinor-valued and Clifford algebra-valued solutions.

As to the spinor-valued solutions we shall systematically use the realizations of the spinor spaces S as subspaces of an appropriate Clifford algebra (see I.4). Furthermore the representation of the Pin- and Spin groups are then realized by multiplication inside the Clifford algebra. Finally, we shall indicate which objects from the multiplication table should be taken in the cases under consideration.

As we saw in I.4.4.3, the basic representations of Pin $(m+1)$ and Spin $(m+1)$ depend upon the parity of $m+1$.

I. The case $m+1=2k$ even

We have $\mathcal{C}_{m+1} = \mathbf{C}_{2k}$.

As (equivalent) irreducible representation spaces for Pin $(2k)$ we take the space $S_{2k} = (\wedge W)I = \mathbf{C}_{2k}I$ and its dual $\overline{S}_{2k} = I(\wedge\overline{W}) = I\overline{\mathbf{C}}_{2k}$. Notice that $I \in \mathbf{C}_{2k}^+$.

For S_{2k} and $\overline{S}_{2k}$ valued functions the Dirac equation reads, respectively,

$$(\sum_{j=0}^{m} e_j\partial_{x_j})f = 0 \tag{0.11}$$

and

$$f(\sum_{j=0}^{m} e_j\partial_{x_j}) = 0. \tag{0.12}$$

As (equivalent) irreducible representation spaces for Spin $(2k)$ we take the space $S_{2k}^+ = (\wedge^{\text{even}}W)I = \mathbf{C}_{2k}^+ I$ and its dual $\overline{S}_{2k}^+ = I(\wedge^{\text{even}}\overline{W}) = I\overline{\mathbf{C}}_{2k}^+$.

For S_{2k}^+ and $\overline{S}_{2k}^+$ valued functions the Weyl equation reads, respectively,

$$(\partial_{x_0} + \sum_{i=1}^{m} \varepsilon_i\partial_{x_i})f = 0 \tag{0.13}$$

and

$$f(\partial_{x_0} + \sum_{i=1}^{m} \varepsilon_i\partial_{x_i}) = 0. \tag{0.14}$$

II. The case $m+1=2k-1$ odd

We have $\mathbf{C}_{m+1} \cong \mathbf{C}_{2k}^+$.

As (equivalent) irreducible representation spaces for Pin $(2k-1)$ we take the space $S_{2k}^+ = (\wedge^{\text{even}} W)I = \mathbf{C}_{2k}^+ I$ and its dual $\overline{S}_{2k}^+ = I(\wedge^{\text{even}}) = I\overline{\mathbf{C}}_{2k}^+$. Again $I \in \mathbf{C}_{2k}^+$. For S_{2k}^+ and $\overline{S}_{2k}^+$ valued functions the Dirac equation is then similar to, respectively, (0.11) and (0.12).

As (equivalent) irreducible representation spaces for Spin $(2k-1)$ we take the space $S_{2k}^+ = (\wedge^{\text{even}} W)I = \mathbf{C}_{2k}^+ I$ and $\overline{S}_{2k}^+ = I(\wedge^{\text{even}} \overline{W}) = I\overline{\mathbf{C}}_{2k}^+$.

For S_{2k}^+ and $\overline{S}_{2k}^+$ valued functions the Weyl equation is then similar to, respectively, (0.13) and (0.14).

As to Clifford algebra-valued solutions, the following possibilities are on hand.

In the case of the Dirac equation we take $\mathcal{C} = \mathbf{C}_{m+1}$, i.e. $\mathcal{C} = \mathbf{C}_{2k}$ for $m+1 = 2k$ even and $\mathcal{C} = \mathbf{C}_{2k}^+$ for $m+1 = 2k-1$ odd.

In the case of the Weyl equation we take either $\mathcal{C} = \mathbf{C}_{m+1}$, i.e. $\mathcal{C} = \mathbf{C}_{2k}$ for $m+1 = 2k$ even and $\mathcal{C} = \mathbf{C}_{2k}^+$ for $m+1 = 2k-1$ odd, or $\mathcal{C} = \mathbf{C}_{m+1}^+$ i.e. $\mathcal{C} = \mathbf{C}_{2k}^+ \cong \mathbf{C}_m$ for $m+1 = 2k$ even and $\mathcal{C} = (\mathbf{C}_{2k}^+)^+ \cong \mathbf{C}_m$ for $m+1 = 2k-1$ odd.

In Remark (4) below, we discuss how to obtain such realizations.

Remarks.

(1) We use the notations ∂_x and D_x for the operators

$$\partial_x = \sum_{j=0}^{m} e_j \partial_{x_j}$$

and

$$D_x = \partial_{x_0} + \sum_{i=1}^{m} \varepsilon_i \partial_{x_i}$$

and call them the *Dirac* and *Cauchy-Riemann operators* in $\mathbf{R}^{m+1}$.

Since we use the notation $x = (x_0, \underline{x})$ for an arbitrary element $x \in \mathbf{R}^{m+1}$ with $x_0 \in \mathbf{R}$ and $\underline{x} \in \mathbf{R}^m$, we shall frequently write ∂_x as

$$\partial_x = e_0 \partial_{x_0} + \partial_{\underline{x}}$$

where

$$\partial_{\underline{x}} = \sum_{i=1}^{m} e_i \partial_{x_i}$$

is a realization of the Dirac operator in $\mathbf{R}^m$.

(2) Traditionally there are two possibilities for writing the Weyl equation, namely

$$(\partial_{x_0} + \sum_{i=1}^{n} \varepsilon_i \partial_{x_i}) f = 0$$

or

$$(\partial_{x_0} - \sum_{i=1}^{m} \varepsilon_i \partial_{x_i}) f = 0.$$

It is possible to obtain both cases by choosing suitably the realization of the representation.

(3) Solutions to the equations (0.11) and (0.13) or (0.12) and (0.14) are called, respectively, *left* or *right monogenic functions* in $\Omega \subset \mathbf{R}^{m+1}$. It should be noticed however that commonly spinor valued solutions to these equations are called *harmonic spinors.*

The space of V-valued left monogenic functions in Ω is denoted by $M(\Omega; V)$; its elements are henceforth called *monogenic* for short.

The theory of monogenic functions is now commonly called *Clifford analysis.*

(4) The identification of $\mathbf{R}^{m+1}$ in the Dirac and Weyl cases

(4.1) The case $m+1 = 2k$ even Let $e = (e_0, e_1, \ldots, e_m)$ be an orthonormal basis for $\mathbf{R}^{0,m+1}$.

In the case of the Dirac equation, $x \in \mathbf{R}^{m+1}$ is identified with $x = \sum_{j=0}^{m} x_j e_j$.

As to the case of the Weyl equation, notice that $\varepsilon_i = \overline{e}_0 e_i \in \mathbf{C}_{2k}^{+}$ with $\varepsilon_i^2 = -1$, $i = 1, \ldots, m$, whence the set $\{\varepsilon_1, \ldots, \varepsilon_m\}$ may also be regarded as an orthonormal set of vectors generating a real orthogonal space of type $(0, m)$ and thus as a set of vectors generating $\mathbf{R}^{0,m}$.

We then identify $x \in \mathbf{R}^{m+1}$ with $x_0 + \sum_{i=1}^{m} x_i \varepsilon_i$.

(4.2) The case $m+1 = 2k-1$ odd

Let $\eta = (\eta_1, \eta_2, \ldots, \eta_{2k})$ be an orthonormal basis for $\mathbf{R}^{0,m+2} = \mathbf{R}^{0,2k}$ and put

$$\begin{aligned} e_j &= -\eta_1 \eta_{j+2} \\ &= \overline{\eta}_1 \eta_{j+2}, \quad j = 0, 1, \ldots, 2k-2. \end{aligned}$$

Then the set $e = (e_0, e_1, \ldots, e_m)$ clearly consists of bivectors in $\mathbf{C}_{2k}$ whence they belong to $\mathbf{C}_{2k}^{+}$. However, as $e_j^2 = -1$, $j = 0, \ldots, m$, this set may also be considered as an orthonormal set of generators of a real orthogonal space of type $(0, m+1)$ and thus as a set of vectors generating $\mathbf{R}^{0,m+1}$.

In the case of the Dirac equation we identify $x \in \mathbf{R}^{m+1}$ with $x = \sum_{j=0}^{m} x_j e_j$. As to the case of the Weyl equation, putting for $i = 1, \ldots, m$, $\varepsilon_i = \overline{e}_0 e_i$, we have that ε_i still belongs to $\mathbf{C}_{2k}^{+}$. However, as $\varepsilon_i^2 = -1$, $i = 1, \ldots, m$, the set $\{\varepsilon_1, \ldots, \varepsilon_m\}$ may again be regarded as an orthonormal set of generators of a real orthogonal space of type $(0, m)$ and hence as a set of vectors generating $\mathbf{R}^{0,m}$ whence we identify $x \in \mathbf{R}^{m+1}$ with $x = x_0 + \sum_{i=1}^{m} x_i \varepsilon_i$.

Notice that in both cases the set $\{e_0, e_1, \ldots, e_m\}$ generates the complex Clifford algebra $\mathbf{C}_{m+1}$ while the set $\{\varepsilon_1, \ldots, \varepsilon_m\}$ generates the complex Clifford algebra $\mathbf{C}_m$.

The reader should be conscious of both interpretations of the elements $e_0, e_1, \ldots, e_m$ in the odd dimensional case, respectively, the elements $\varepsilon_1, \ldots, \varepsilon_m$ in the even and odd dimensional cases.

0.1.4 The spherical Dirac and Cauchy-Riemann operators

Let us first of all remark that

$$\partial_x^2 = -\Delta_x\,,\ \overline{\partial}_x\partial_x = \partial_x\overline{\partial}_x = \Delta_x$$

and

$$D_x\overline{D}_x = \overline{D}_xD_x = \Delta_x,$$

Δ_x being the Laplacian in $\mathbf{R}^{m+1}$.

Using polar coordinates $x = r\xi$, with $r = |x|$ and $\xi \in S^m$, we may write ∂_x as

$$\partial_x = \xi\partial_r + \frac{1}{r}\partial_\xi$$

where ∂_ξ is a differential operator depending only on the angular coordinates.

Putting

$$\Gamma_\xi = \overline{\xi}\partial_\xi$$

we have thus defined a first order differential operator acting on S^m, called the *spherical Dirac operator.*

Obviously

$$\partial_x = \xi(\partial_r + \frac{1}{r}\Gamma_\xi).$$

As is well known, the Laplacian Δ_x has the expression

$$\Delta_x = \partial_r^2 + \frac{m}{r}\partial_r + \frac{1}{r^2}\Delta_\xi, \tag{0.15}$$

Δ_ξ being (up to minus sign) the Laplace-Beltrami operator on S^m.

On the other hand, taking account of $[\partial_r, \xi] = 0$ and $[\partial_r, \Gamma_\xi] = 0$,

$$\begin{aligned}
\Delta_x &= -\xi(\partial_r + \frac{1}{r}\Gamma_\xi)\xi(\partial_r + \frac{1}{r}\Gamma_\xi)\\
&= \partial_r^2 + \partial_r\frac{1}{r}\Gamma_\xi - \frac{1}{r}\xi\partial_r\Gamma_\xi\xi - \frac{1}{r^2}\xi\Gamma_\xi\xi\Gamma_\xi\\
&= \partial_r^2 - \frac{1}{r^2}\Gamma_\xi + \frac{1}{r}\Gamma_\xi\partial_r - \frac{1}{r}\xi\Gamma_\xi\xi\partial_r - \frac{1}{r^2}\xi\Gamma_\xi\xi\Gamma_\xi\\
&= \partial_r^2 + \frac{1}{r}\partial_r(\Gamma_\xi - \xi\Gamma_\xi\xi) - \frac{2}{r^2}(\Gamma_\xi + \xi\Gamma_\xi\xi\Gamma_\xi).
\end{aligned}$$

We thus obtain that

$$\Gamma_\xi - \xi \Gamma_\xi \xi = m\mathbf{1}$$

and

$$(\mathbf{1} + \xi \Gamma_\xi \xi)\Gamma_\xi = -\Delta_\xi$$

or

$$\Delta_\xi = ((m-1)\mathbf{1} - \Gamma_\xi)\Gamma_\xi, \tag{0.16}$$

the latter formula thus yielding a factorization of the Laplace-Beltrami operator on S^m in terms of the spherical Dirac operator.

Notice also that, as on the one hand

$$\overline{x}\partial_x = r\overline{\xi}\xi(\partial_r + \frac{1}{r}\Gamma_\xi) = r\partial_r + \Gamma_\xi$$

while on the other hand

$$\begin{aligned}
\overline{x}\partial_x &= (\overline{e}_0 x_0 - \underline{x})(e_0 \partial_{x_0} + \partial_{\underline{x}}) \\
&= x_0 \partial_{x_0} + \overline{e}_0(x_0 \partial_{\underline{x}} - \underline{x}\partial_{x_0}) - \underline{x}\partial_{\underline{x}} \\
&= \sum_{j=0}^{m} x_j \partial_{x_j} - \underline{x} \wedge \partial_{\underline{x}} + \overline{e}_0(x_0 \partial_{\underline{x}} - \underline{x}\partial_{x_0}) \\
&= r\partial_r - \underline{x} \wedge \partial_{\underline{x}} + \overline{e}_0(x_0 \partial_{\underline{x}} - \underline{x}\partial_{x_0})
\end{aligned}$$

we get :

$$\begin{aligned}
\Gamma_\xi &= \overline{e}_0(x_0 \partial_{\underline{x}} - \underline{x}\partial_{x_0}) - \underline{x} \wedge \partial_{\underline{x}} \\
&= \overline{x} \wedge \partial_x. \\
&= -\sum_{i<j} e_{ij}(x_i \partial_{x_j} - x_j \partial_{x_i}.
\end{aligned} \tag{0.17}$$

Hence Γ_ξ has bi-vector coefficients (see also Remark (2) below).

In the case of the Cauchy-Riemann operator analogous computations may be made.

Now putting for $x \in \mathbf{R}^{m+1}$, $x = (x_0, \underline{x}) = x_0 + \sum_{i=1}^{m} x_i \varepsilon_i$, the change into polar coordinates $x = r\omega$ yields :

$$\begin{aligned}
D_x &= \omega\partial_r + \frac{1}{r}\partial_\omega \\
&= \omega(\partial_r + \frac{1}{r}\Gamma_\omega^\star)
\end{aligned}$$

$\Gamma^\star_\omega$ again being a first order differential operator acting on S^m, called the *spherical Cauchy-Riemann operator.*

Calling $\tilde{\Gamma}^\star_\omega = \overline{\partial}_\omega \omega$, i.e. the action of $\tilde{\Gamma}^\star_\omega$ on a function f is given by $\tilde{\Gamma}^\star_\omega f = \overline{\partial}_\omega(\omega f)$, we have :

$$\begin{aligned}
\Delta_x &= \overline{D}_x D_x \\
&= (\partial_r + \frac{1}{r}\tilde{\Gamma}^\star_\omega)\overline{\omega}\omega(\partial_r + \frac{1}{r}\Gamma^\star_\omega) \\
&= (\partial_r + \frac{1}{r}\tilde{\Gamma}^\star_\omega)(\partial_r + \frac{1}{r}\Gamma^\star_\omega) \\
&= \partial_r^2 + \frac{1}{r}(\Gamma^\star_\omega + \tilde{\Gamma}^\star_\omega) + \frac{1}{r^2}(\tilde{\Gamma}^\star_\omega\Gamma^\star_\omega - \Gamma^\star_\omega).
\end{aligned}$$

With (0.15) we obtain

$$\Gamma^\star_\omega + \tilde{\Gamma}^\star_\omega = m\mathbf{1}$$

and

$$\Delta_\omega = ((m-1)\mathbf{1} - \Gamma^\star_\omega)\Gamma^\star_\omega, \tag{0.18}$$

an expression completely similar to (0.16).

Again, as on the one hand

$$\begin{aligned}
\overline{x}D_x &= r\overline{\omega}\omega(\partial_r + \frac{1}{r}\Gamma^\star_\omega) \\
&= r\partial_r + \Gamma^\star_\omega
\end{aligned}$$

while on the other hand

$$\begin{aligned}
\overline{x}D_x &= (x_0 - \underline{x})(\partial_{x_0} + \partial_{\underline{x}}) \\
&= x_0\partial_{x_0} - \underline{x}\partial_{\underline{x}} + x_0\partial_{\underline{x}} - \underline{x}\partial_{x_0} \\
&= \sum_{j=0}^{m} x_j\partial_{x_j} + \sum_{i<j}\varepsilon_i\varepsilon_j(x_j\partial_{x_i} - x_i\partial_{x_j}) + x_0\partial_{\underline{x}} - \underline{x}\partial_{x_0},
\end{aligned}$$

we have that

$$\Gamma^\star_\omega = x_0\partial_{\underline{x}} - \underline{x}\partial_{x_0} + \sum_{i<j}\varepsilon_i\varepsilon_j(x_j\partial_{x_i} - x_i\partial_{x_j}). \tag{0.19}$$

Hence the spherical Cauchy-Riemann operator has vector and bi-vector coefficients, which is clearly in contrast with the case of the spherical Dirac operator (see also Remark (2) below).

Remarks

(1) Looking at the formulas (0.17) and (0.19) we see that, since

$$\begin{aligned} \Gamma_\xi &= \bar{e}_0(x_0\partial_{\underline{x}} - \underline{x}\partial_{x_0}) - \underline{x} \wedge \partial_{\underline{x}} \\ &= \bar{e}_0(x_0\partial_{\underline{x}} - \underline{x}\partial_{x_0}) + \sum_{\substack{i<j \\ i,j=1,\dots,m}} e_i e_j (x_j \partial_{x_i} - x_i \partial_{x_j}) \end{aligned}$$

and

$$\Gamma_\omega^\star = x_0\partial_{\underline{x}} - \underline{x}\partial_{x_0} + \sum_{\substack{i<j \\ i,j=1,\dots,m}} \varepsilon_i \varepsilon_j (x_j \partial_{x_i} - x_i \partial_{x_j})$$

we can formally pass from Γ_ξ to $\Gamma_\omega^\star$ by replacing e_0 by 1 and e_i by ε_i, $i = 1, \dots, m$.

In fact this is also the case in the notation of $x \in \mathbf{R}^{m+1}$, namely $x = x_0 e_0 + \sum_{i=1}^m x_i e_i$ or $x = x_0 + \sum_{i=1}^m x_i \varepsilon_i$ in, respectively, the Dirac and Weyl cases. Moreover notice that, when in the calculations leading to (0.16) and (0.19) we systematically replace formally e_0 by 1 and e_i by ε_i, we arrive at the expressions (0.18) and (0.19).

These remarks are fundamental since they indicate that when we shall establish formulas in the Dirac case, they will be presented in such a way that by applying the formal substitution mentioned above, the analogous formulas for the Weyl case will be obtained straightforwardly. Such procedure enables us to avoid tedious repetitions in formulating theorems and establishing proofs of them.

Of course it will be explicitly mentioned when this procedure does not apply.

(2) When in (0.17) we write Γ_ξ as

$$\Gamma_\xi = \bar{x} \wedge \partial_x \tag{0.20}$$

we are giving a nice illustration of the double interpretation of the elements $e_0, e_1, \dots, e_m$ in the case $m + 1 = 2k - 1$ odd (see also 0.1.3 Remark (4)).

Indeed, since $e_0, e_1, \dots, e_m$ in fact belong to $\mathbf{C}_{2k}^+$, they are bivectors, a property which is essentially needed in studying S_{2k}^+-valued solutions to the equation $\partial_x f = 0$. However, in the formula (0.20), the outer product should not be taken in the sense of the outer product between bivectors (see I.1.7.1) but in the sense of the outer product between vectors, thus interpreting the elements $e_0, e_1, \dots, e_m$ as vectors generating $\mathbf{R}^{0,m+1}$.

The same applies to the expression (0.19) of the spherical Cauchy-Riemann operator $\Gamma_\omega^\star$ when we say that it has vector and bivector coefficients. Indeed, we then interpret the elements $\varepsilon_1, \dots, \varepsilon_m$ as vectors generating $\mathbf{R}^{0,m}$ and not as bivectors belonging to $\mathbf{C}_{m+1}$.

(3) Let us calculate explicitly Γ_ξ and $\Gamma_\omega^\star$ in the case $m = 1$ and see how Remark (1) then applies.

Passing to polar coordinates $x = r\xi$ in the plane we have that, as

$$\begin{pmatrix} \partial_{x_0} \\ \partial_{x_1} \end{pmatrix} = \begin{pmatrix} \cos\theta & -r\sin\theta \\ \sin\theta & r\cos\theta \end{pmatrix} \begin{pmatrix} \partial_r \\ \frac{1}{r^2}\partial_\theta \end{pmatrix},$$

$$\begin{aligned} \partial_x &= e_0\partial_{x_0} + e_1\partial x_1 \\ &= (e_0\cos\theta + e_1\sin\theta)\partial_r + \frac{1}{r}(e_1\cos\theta - e_0\sin\theta)\partial_\theta \\ &= \xi(\partial_r + \frac{1}{r}\overline{e}_0 e_1\partial_\theta) \end{aligned}$$

or

$$\Gamma_\xi = \overline{e}_0 e_1 \partial_\theta.$$

In a similar way we find for $D_x = \partial_{x_1} + \varepsilon_1\partial_{x_1}$, that

$$D_x = \omega(\partial_r + \frac{1}{r}\varepsilon_1\partial_\theta)$$

or

$$\Gamma^\star_\omega = \varepsilon_1\partial_\theta.$$

So indeed, replacing in the expression of Γ_ξ e_0 by 1 and e_1 by ε_1, we get $\Gamma^\star_\omega$.

0.2 Elementary function theory in $\mathbf{R}^{m+1}$

In this section we list some basic properties of solutions to the (massless) **Dirac** equation in $\mathbf{R}^{m+1}$. Analogous formulas for solutions to Weyl's equation are obtained immediately by substituting $e_0 = 1$ and $e_i = \varepsilon_i$, $i = 1, \ldots, m$, in the statements and proofs.

The functions considered are V-valued C_1-functions in some open subset $\Omega \subset \mathbf{R}^{m+1}$ and it is understood that in the statements of theorems all cases of V are allowed as for as the multiplications involved are well defined (see the multiplication table in 0.1.1).

In what follows we shall freely use notions and results from the theory of harmonic functions for which we refer to [29].

0.2.1 Stokes' Formula and Cauchy's Theorems

Let U, V and W be elements of the multiplication table such that $UV \subset W$ and let $\mathcal{E}^\star(\Omega)$ be the space of differential forms in $\Omega \subset \mathbf{R}^{m+1}$. Furthermore, let $\tilde{\omega}, \tilde{\tau} \in \mathcal{E}^\star(\Omega)$ and let $u \in U$, $v \in V$.

Then, taking $\omega = \tilde{\omega} \otimes u \in \mathcal{E}^\star(\Omega) \otimes U$ and $\tau = \tilde{\tau} \otimes v \in \mathcal{E}^\star(\Omega) \otimes V$, we define $\omega \wedge \tau \in \mathcal{E}^\star(\Omega) \otimes W$ by

$$\omega \wedge \tau = (\tilde{\omega} \wedge \tilde{\tau}) \otimes uv.$$

Moreover the action of the exterior differential d on vector valued forms is defined componentwise, i.e.

$$d(\tilde{\omega} \otimes u) = d\tilde{\omega} \otimes u.$$

We thus obtain straightforwardly

Lemma. *If ω is a k-form and τ is an ℓ-form, then*

$$d(\omega \wedge \tau) = d\omega \wedge \tau + (-1)^k \omega \wedge d\tau.$$

Now let $\sum$ be an $(m+1)$-dimensional compact oriented manifold-with-boundary $\partial\Sigma$ and define the oriented $\mathcal{C}$-valued surface element $d\sigma$ on $\partial\Sigma$ by

$$d\sigma = \sum_{j=0}^{m} (-1)^j e_j d\hat{x}_j$$

where for each $j = 0, 1, \ldots, m$,

$$d\hat{x}_j = dx_0 \wedge \ldots \wedge [dx_j] \wedge \ldots \wedge dx_m.$$

Then, if $n = (n_0, n_1, \ldots, n_m)$ - the outwardly pointed unit normal - is identified with $n = \sum_{j=0}^{n} n_j e_j$ and $d\Sigma$ stands for the "classical" surface element, we have that

$$d\sigma = n d\Sigma.$$

Furthermore by

$$dx = dx_0 \wedge dx_1 \wedge \ldots \wedge dx_m$$

the $(m+1)$-dimensional (oriented) volume element is meant.

Theorem. (Stokes' Formula) Stokes' Formula *If $f \in C_1(\Omega; V)$ and $g \in C_1(\Omega; U)$ then for every $\Sigma \subset \Omega$*

$$\int_{\partial\Sigma} g d\sigma f = \int_{\Sigma} [(g\partial_x) f + g(\partial_x f)] dx.$$

Proof. As by assumption $f \in \mathcal{E}^0(\Omega) \otimes V$, $g \in \mathcal{E}^0(\Omega) \otimes U$ and moreover $d\sigma \in \mathcal{E}^{(m)}(\Omega) \otimes \mathcal{C}$, we have, by virtue of Lemma 0.2.1., that

$$\begin{aligned} d(g d\sigma f) &= d((g d\sigma) f) \\ &= d(g d\sigma) f + (-1)^m g d\sigma \wedge df \\ &= dg \wedge d\sigma f + (-1)^m (g d\sigma) \wedge df. \end{aligned}$$

But for each $j = 0, 1, \ldots, m$,

$$dx_j \wedge d\sigma = e_j dx$$

and

$$d\sigma \wedge dx_j = (-1)^m e_j dx,$$

whence

$$dg \wedge d\sigma f = (g\partial_x) f dx$$

and

$$(-1)^m g d\sigma \wedge df = g(\partial_x f) dx. \quad \blacksquare$$

Remark. The main cases in which Stokes' Formula is applicable are the following (see also the multiplication table in 0.1.1)

(i) $V = S$ and $U = S$, then $\overline{S}CS = \mathbf{C}I$;

(ii) $V = S$ and $U = C$, then $CCS = S$;

(iii) $V = C$ and $U = \overline{S}$, then $\overline{S}CC = \overline{S}$;

(iv) $V = C$ and $U = C$, then $CCC = C$.

Corollary 1. *Let $\partial_x f = 0$ and $g\partial_x = 0$. Then for all $\Sigma \subset \Omega$*

$$\int_{\partial\Sigma} g d\sigma f = 0.$$

Corollary 2. *Let $\partial_x f = 0$. Then for all $\Sigma \subset \Omega$*

$$\int_{\partial\Sigma} d\sigma f = 0.$$

Corollary 1 and Corollary 2 are called *Cauchy's Theorems* for monogenic functions.

0.2.2 Cauchy's Integral Formula and its applications

A principal tool in the study of the equation $\Delta_x f = g$ is the fundamental solution $N(x)$ of Δ_x given by (see [29])

$$N(x) = \begin{cases} \frac{1}{(1-m)A_{m+1}|x|^{m-1}} & , \quad m > 1 \\ \frac{1}{2\pi}\log|x| & , \quad m = 1. \end{cases}$$

It satisfies the following properties :

(N1) N is locally integrable in $\mathbf{R}^{m+1}$, i.e. $N \in L_1^{\mathrm{loc}}(\mathbf{R}^{m+1})$

(N2) N is analytic in $\mathbf{R}_0^{m+1}$

(N3) $\Delta_x N = \delta$, i.e. for all $\varphi \in \mathcal{D}(\Omega)$,

$$\begin{aligned} < \Delta_x N, \varphi > &= < \delta, \varphi > \\ &= \varphi(0). \end{aligned}$$

By means of the relation $\partial_x \overline{\partial}_x = \Delta_x$, we immediately obtain that the function $E(x)$ given by

$$\begin{aligned} E(x) &= \overline{\partial}_x(N(x)) \\ &= (N(x))\overline{\partial}_x \\ &= \frac{1}{A_{m+1}} \frac{\overline{x}}{|x|^{m+1}} \end{aligned}$$

satisfies the following properties

(1) E is both a left and right $\mathcal{C}$-valued monogenic function in $\mathbf{R}_0^{m+1}$ with $\lim_{x\to\infty} E(x) = 0$.

(2) $E \in \mathcal{A}(\mathbf{R}_0^{m+1}; \mathcal{C})$.

(3) $E \in L_1^{\mathrm{loc}}(\mathbf{R}^{m+1}; \mathcal{C})$, the bi-$\mathcal{C}$-module of locally integrable $\mathcal{C}$-valued functions in $\mathbf{R}^{m+1}$.

Thus considering E both as a left and right $\mathcal{C}$-distribution we moreover have

(4) $\partial_x E = E\partial_x = \delta$, where δ is the δ-distribution defined on the bi-$\mathcal{C}$-module $\mathcal{D}(\mathbf{R}^{m+1}; \mathcal{C})$.

In view of these properties E is obviously a left and a right fundamental solution of the Dirac operator ∂_x.

Using the one-point compactification $\mathbf{R}^{m+1} \cup \{\infty\}$ of $\mathbf{R}^{m+1}$, we already observed that $\lim_{x\to\infty} E(x) = 0$. The function E is therefore an example of a function which is monogenic in a neighbourhood of infinity and moreover vanishes at infinity.

More generally, for $R > 0$ given we call $M_0(\mathbf{R}^{m+1}\backslash\overline{B}(R); V)$ the space of V-valued functions f which are monogenic in $\mathbf{R}^{m+1}\backslash\overline{B}(R)$ and satisfy $\lim_{x\to\infty} f(x) = 0$.

Theorem 1. **(Cauchy's Integral Formula)** *Let $f \in C_1(\Omega, V)$ and let $\Sigma \subset \Omega$. Then*

$$\int_{\partial\Sigma} E(y-x)d\sigma_y f(y) - \int_\Sigma E(y-x)\partial_y f(y)dy = \begin{cases} f(x), & x \in \overset{\circ}{\Sigma} \\ 0, & x \in \Omega\backslash\Sigma \end{cases}.$$

Proof. First let $x \in \Omega\backslash\Sigma$. Call $\eta = d(x,\Sigma)$ and let $\Omega^\star$ be an open $\eta/2$-neighbourhood of Σ. Then applying Stokes' Formula to $\Sigma \subset \Omega^\star$ we get, by putting $g(y) = E(y-x)$,

$$\begin{aligned} \int_{\partial\Sigma} E(y-x)d\sigma_y f(y) &= \int_\Sigma [(E(y-x)\partial_y)f(y) + E(y-x)(\partial_y f)]dy \\ &= \int_\Sigma E(y-x)(\partial_y f)dy. \end{aligned}$$

Now take $x \in \overset{\circ}{\Sigma}$ fixed and choose $R > 0$ such that $\overline{B}(x,R) \subset \overset{\circ}{\Sigma}$. Then again by Stokes' Formula

$$\int_{\partial(\Sigma\backslash B)} E(y-x)d\sigma_y f(y) = \int_{\Sigma\backslash B} E(y-x)(\partial_y f)dy. \tag{0.21}$$

But, taking $R' > 0$ in such a way that $\Sigma \subset \overset{\circ}{B}(x,R')$, we have that for some constant $C > 0$, independent of R',

$$\int_\Sigma |E(y-x)|dy \le CR',$$

whence, $\partial_y f$ being continuous in Ω, $E(y-x)(\partial_y f)$ is integrable on Σ.

Consequently, by taking the limit for $R \to 0+$ in the right hand side of (0.21), we get

$$\int_\Sigma E(y-x)(\partial_y f)dy.$$

As to the left hand side of (0.21), it can be put into the form

$$\int_{\partial\Sigma} E(y-x)d\sigma_y f(y) - \int_{\partial B(x,R)} E(y-x)d\sigma_y f(y)$$

where, using standard arguments,

$$\lim_{R\to 0+} \int_{\partial B(x,R)} E(y-x)d\sigma_y f(y) = f(x). \quad \blacksquare$$

Corollary. *If $\partial_x f = 0$ in Ω then for each $\Sigma \subset \Omega$,*

$$\int_{\partial\Sigma} E(y-x)d\sigma_y f(y) = \begin{cases} f(x), & x \in \overset{\circ}{\Sigma} \\ 0, & x \in \Omega\backslash\Sigma \end{cases} .$$

Remark. Cauchy's Integral Formula has been established for left monogenic functions. Obviously the reasonings and calculations made above may be applied to obtain a Cauchy Integral Formula for right monogenic functions in Ω, i.e. satisfying $f\partial_x = 0$.

It reads as follows :

Let $f \in C_1(\Omega; V)$ satisfy $f\partial_x = 0$ in Ω. Then for each $\Sigma \in \Omega$,

$$\int_{\partial\Sigma} f(y)d\sigma_y E(y-x) = \begin{cases} f(x), & x \in \overset{\circ}{\Sigma} \\ 0, & x \in \Omega\backslash\Sigma \end{cases} .$$

Notice here that V is either equal to $\overline{S}$ or to C.

The function $E(y-x)(x \neq y,\ x, y \in \mathbf{R}^{m+1})$ appearing in Cauchy's Integral Formula is called the *Cauchy kernel.*

Now let $\Sigma \subset \mathbf{R}^{m+1}$ be fixed and define for each $f \in C_0(\partial\Sigma; V)$ its *Cauchy transform* $\hat{f}$ by

$$\hat{f}(x) = \int_{\partial\Sigma} E(y-x)d\sigma_y f(y). \tag{0.22}$$

Then clearly $\hat{f} \in \mathcal{E}(\mathbf{R}^{m+1}\backslash\partial\Sigma; V)$ and applying the operator ∂_x we get $\partial_x\hat{f} = 0$ in $\mathbf{R}^{m+1}\backslash\Sigma$, whence $\hat{f} \in M(\mathbf{R}^{m+1}\backslash\partial\Sigma; V)$.

Combining this result with Cauchy's Integral Formula for monogenic functions we thus obtain that $M(\Omega; V) \subset \mathcal{E}(\Omega; V)$ so that for all $f \in M(\Omega; V)$,

$$\Delta_x f = \overline{\partial}_x\partial_x f = 0 \quad \text{in} \quad \Omega,$$

i.e f is a harmonic V-valued function in Ω.

Notice that if $\{\psi_1, \ldots, \psi_d\}$ is a basis for the complex vector space V, then for a V-valued C_2-function f in Ω with

$$f(x) = \sum_{k=1}^{d} f_k(x)\psi_k,$$

we have, Δ_x being a "scalar" differential operator :

$$\Delta_x f = 0 \quad \text{in} \quad \Omega \Leftrightarrow \Delta_x f_k = 0 \quad \text{in} \quad \Omega, \quad k = 1, \ldots, d.$$

The foregoing result implies that quite a lot of fundamental properties for harmonic functions will be inherited by monogenic functions. We here list some of them.

First of all notice that, as harmonic functions are real-analytic, each monogenic function in Ω is real-analytic in Ω, i.e. for each $a \in \Omega$ there exists $R(a) > 0$ such that in $\overset{\circ}{B}(a, R(a)) \subset \Omega$, f admits the multiple Taylor series

$$f(x) = \sum_{\alpha} (x-a)^{\alpha} c_{\alpha}.$$

Here for any multi-index $\alpha = (\alpha_0, \alpha_1, \ldots, \alpha_m) \in \mathbf{N}^{m+1}$,

$$(x-a)^{\alpha} = (x_0 - a_0)^{\alpha_0} \ldots (x_m - a_m)^{\alpha_m}$$

and

$$c_{\alpha} = \frac{1}{\alpha!} \partial^{\alpha} f(a)$$

with

$$\alpha! = \alpha_0! \ldots \alpha_m!$$

and

$$\partial^{\alpha} = \frac{\partial^{\alpha_0 + \ldots + \alpha_m}}{\partial_{x_0}^{\alpha_0} \ldots \partial_{x_m}^{\alpha_m}}.$$

Moreover, if $\Omega = \mathring{B}(a, R)$ it may be proved that we may take $R(a) = \frac{1}{\sqrt{2}} R$.

Theorem 2. (Weierstrass' Theorem) *Let $(f_k)_{k \in \mathbf{N}}$ be a sequence in $M(\Omega; V)$ which converges uniformly to f on each compact subset $K \subset \Omega$. Then*

(i) $f \in M(\Omega; V)$

(ii) For each multi-index $\beta = (\beta_0, \beta_1, \ldots, \beta_m) \in \mathbf{N}^{m+1}$, the sequence $(\partial^{\beta} f_k)_{k \in \mathbf{N}}$ converges uniformly on each compact subset $K \subset \Omega$ to $\partial^{\beta} f$.

Notice that Weierstrass' Theorem tells us that $M(\Omega; V)$ is a Fréchet space when provided with the topology of uniform convergence on compacta. Moreover, if $V = \mathcal{C}$, it even becomes a right Fréchet $\mathcal{C}$-module.

Corollary 1. *Let $(f_k)_{k \in \mathbf{N}}$ be a sequence in $M(\Omega; V)$ such that the series $\sum_{k=0}^{\infty} f_k$ converges uniformly on each compact subset $K \subset \Omega$. Then*

(i) Its sum $f \in M(\Omega; V)$.

(ii) For each multi-index $\beta \in \mathbf{N}^{m+1}$, $\partial^{\beta} f = \sum_{k=0}^{\infty} \partial^{\beta} f_k$, the latter series being uniformly convergent on each compact subset $K \subset \Omega$.

Theorem 3. (Mean Value Theorem) *Let $f \in M(\Omega; V)$.*
Then for all $x \in \Omega$ and $R > 0$ such that $\overline{B}(x, R) \subset \Omega$,

$$f(x) = \frac{m+1}{R^{m+1} A_{m+1}} \int_{B(x,R)} f(y) dy.$$

Theorem 4. *Let $(f_k)_{k\in\mathbf{N}}$ be a sequence in $M(\Omega;V)$ which converges to a function f in $L_2(K;V)$ and this for each $K\subset\Omega$ compact, i.e. $(f_k)_{k\in\mathbf{N}}$ is locally L_2-convergent to f. Then f may be monogenically extended to Ω and the sequence $(f_k)_{k\in\mathbf{N}}$ converges uniformly to f on each compact subset $K\subset\Omega$.*

Proof. Take an arbitrary closed ball $\overline{B}(a,r)\subset\Omega$ and let $0<r<R$ be such that the closed ball $\overline{B}(a,R)$ is still contained in Ω. Then by the Mean Value Theorem, for each $x\in\overline{B}(a,R)$, $0<r'<\frac{R-r}{2}$ and $k,\ell\in\mathbf{N}$, we have

$$f_k(x)-f_\ell(x)=\frac{m+1}{A_{m+1}r'^{m+1}}\int_{B(x,r')}(f_k(y)-f_\ell(y))dy.$$

Consequently

$$\sup_{x\in\overline{B}(a,r)}|f_k(x)-f_\ell|\ \le\ \frac{1}{r'^{m+1}}\|f_k-f_\ell\|^2_{L_2(\overline{B}(a,R))}$$

converges to zero if $\inf(k,\ell)\to\infty$.

But this implies that $(f_k)_{k\in\mathbf{N}}$ is a Cauchy sequence in $M(\Omega;V)$ so that indeed f may be monogenically extended to Ω and $(f_k)_{k\in\mathbf{N}}$ converges uniformly on each compact subset $K\subset\Omega$ to f. ■

0.2.3 Morera's and Painlevé' s Theorem

Cauchy's Theorem tells us that if f is monogenic in Ω, then for each compact interval $J=\prod_{j=0}^{m}[a_j,b_j]\subset\Omega$, $\int_{\partial J}d\sigma f=0$.

A converse of this result reads as follows.

Theorem 1. (Morera's Theorem) *Let $f\in C_0(\Omega;V)$ be such that for each compact interval $J\subset\Omega$, $\int_{\partial J}d\sigma f=0$. Then $f\in M(\Omega;V)$.*

By means of Morera's Theorem we get

Theorem 2. (Painlevé' s Theorem) *Let $\underline{\Omega}\subset\mathbf{R}_m$ be open and let $\Omega\subset\mathbf{R}^{m+1}$ be open such that $\Omega\cap\mathbf{R}^m=\underline{\Omega}$. Furthermore suppose that $f\in M(\Omega\backslash\underline{\Omega};V)$ and that $\lim_{x_0\to0+}f(x_0,\underline{x})$, $\lim_{x_0\to0-}f(x_0,\underline{x})$ exist, are equal and determine a continuous function on $\underline{\Omega}$. Then $f\in M(\Omega;V)$.*

For the proofs of Theorem 1 and Theorem 2, we refer to [10].

0.2.4 The Cauchy-Kovalevska Theorem

In this subsection the following problem is treated :

"Given a V-valued function f which is analytic in an open subset $\underline{\Omega}$ of $\mathbf{R}^m$, does there exist a monogenic function f^* in some open neighbourhood Ω of $\underline{\Omega}$ in $\mathbf{R}^{m+1}$ such that $f^*|_{x_0=0} = f$ in $\underline{\Omega}$? "

The answer to this problem yields the so-called *Cauchy-Kovalevska Theorem* for monogenic functions. As will be seen, it enables us to generate monogenic functions.

Let us first recall that if $\underline{\Omega}$ is open in $\mathbf{R}^m$, then an open neighbourhood Ω of $\underline{\Omega}$ in $\mathbf{R}^{m+1}$ is said to be x_0-normal if for each $x \in \Omega$ the line segment $\{x + te_0 : t \in \mathbf{R}\} \cap \Omega$ is connected and contains just one point in $\underline{\Omega}$.

Theorem. *Let $\underline{\Omega} \subset \mathbf{R}^m$ be open and connected and let $f \in \mathcal{A}(\underline{\Omega}; V)$. Then the function f^* given by*

$$f^*(x) = \sum_{k=0}^{\infty} \frac{1}{k!}(-x_0)^k(\overline{e}_0\partial_{\underline{x}})^k f(\underline{x})$$

satisfies $\partial_x f^* = 0$ in an open connected and normal neighbourhood Ω of $\underline{\Omega}$ in $\mathbf{R}^{m+1}$.

Moreover $f^*|_{x_0=0} = f$ in $\underline{\Omega}$.

Proof. As f is analytic on $\underline{\Omega}$, for each $K \subset \underline{\Omega}$ compact there exist positive constants $C(K)$ and λ_K, depending upon K, such that for all multi-indices $\beta \in \mathbf{N}^m$ and $k \in \mathbf{N}$

$$\sup_{\underline{x}\in K} |\partial_{\underline{x}}^k f(\underline{x})| \leq C(K)(2k)!\lambda_K^k.$$

So it is easily seen that the series determining f^* and all its derived series converge normally on each compact subset of $\Omega = \bigcup_{K\subset\underline{\Omega}}(\overset{\circ}{K} \times [-\frac{1}{\lambda_K}, \frac{1}{\lambda_K}[)$, where clearly Ω is an open, connected and normal neighbourhood of $\underline{\Omega}$ in $\mathbf{R}^{m+1}$ with $\Omega \cap \mathbf{R}^m = \underline{\Omega}$.

Hence $f^* \in \mathcal{E}(\Omega; V)$ and, obviously, $f^*(0, \underline{x}) = f$ in $\underline{\Omega}$.

Now notice that an analytic function g in Ω satisfies $\partial_x g = 0$ if and only if $(\partial_{x_0} + \overline{e}_0\partial_{\underline{x}})g = 0$ whence

$$g(x) = \exp(-x_0\overline{e}_0\partial_{\underline{x}})f(\underline{x}) = f^*(x). \quad \blacksquare$$

The foregoing theorem tells us that, given $f \in \mathcal{A}(\underline{\Omega}; V)$, $\underline{\Omega} \subset \mathbf{R}^m$ being open and connected, there exists at least one couple (Ω, f^*) consisting of an open connected and normal neighbourhood Ω of $\underline{\Omega}$ in $\mathbf{R}^{m+1}$ and a monogenic V-valued function f^* in Ω such that $f^*|_{x_0=0} = f$ in $\underline{\Omega}$.

Now call $((\Omega_\alpha, f_\alpha^*))_{\alpha\in A}$ the family of all couples $(\Omega_\alpha, f_\alpha^*)$ satisfying the above property. Then putting

$$\Omega = \bigcup_{\alpha\in A} \Omega_\alpha$$

and

$$f^*(x) = f_\alpha^*(x), \quad x \in \Omega_\alpha,$$

it is clear that the couple $(\Omega, f^\star)$ satisfies the same property.

Obviously, if the family $(\Omega_\alpha, f_\alpha^\star)_{\alpha\in A}$ is ordered by

$$(\Omega_\alpha, f_\alpha^\star) \leq (\Omega_\beta, f_\beta^\star)$$

if

$$\Omega_\alpha \subset \Omega_\beta \quad \text{and} \quad f_\beta^\star | \Omega_\alpha = f_\alpha^\star,$$

then $(\Omega, f^\star)$ is a maximal element and in view of its construction, it is unique.

We call $(\Omega, f^\star)$ the *Cauchy-Kovalevska extension* of $(\underline{\Omega}, f)$ and for convenience we say $f^\star$ is the *CK-extension* of f.

Remarks.

(1) From the proof of the Cauchy-Kovalevska Theorem it follows that in the case of the Dirac operator the CK-extension of $f \in \mathcal{A}(\underline{\Omega}; V)$ is formally given by

$$f^\star(x) = (e^{-x_0\bar{e}_0\partial_{\underline{x}}})f(\underline{x}).$$

Applying the conventions made we thus have that in the case of the Cauchy-Riemann operator the CK-expansion of $f \in \mathcal{A}(\underline{\Omega}; V)$ reads :

$$\begin{aligned} f^\star(x) &= (e^{-x_0\partial_{\underline{x}}})f(\underline{x}) \\ &= \sum_{k=0}^{\infty} \frac{(-1)^k}{k!} x_0^k(\partial_{\underline{x}})^k f(\underline{x}). \end{aligned}$$

(2) Let again f be a V-valued analytic function in $\underline{\Omega} \subset \mathbf{R}^m$ and let (Ω', f') be such that Ω' is an open connected and normal neighbourhood of $\underline{\Omega}$ in $\mathbf{R}^{m+1}$, $f' \in M(\Omega', V)$ and $f'(0, \underline{x}) = f(\underline{x})$. Then clearly f' is the restriction to Ω' of the CK-extension $f^\star$ of f.

(3) Notice that if f is C-valued, then the function $f^\star$ in Theorem 0.2.13 takes its values in $\operatorname{span}_{\mathbf{C}}\{1, e_0e_i : i = 1, \ldots, m\} = \mathbf{C}1 \oplus \sum_{i=1}^m \oplus \mathbf{C}e_0e_i$.

(4) Of course analogous results may be obtained as to right monogenic CK-extensions. Notice however that in that case $V = \overline{S}$, the dual of the spinor space S, or $V = C$.

Examples.

(1) Let $\ell \in \{1, \ldots, m\}$ and consider the analytic function x_ℓ in $\mathbf{R}^m$.

Putting

$$z_\ell = x_\ell - x_0\bar{e}_0e_\ell,$$

it is easily seen that z_ℓ is monogenic in $\mathbf{R}^{m+1}$ and, as $z_\ell|_{x_0=0} = x_\ell$, z_ℓ is the CK-extension of x_ℓ.

It will be shown in 1.5.4 (2) how by means of the variables z_ℓ, the homogeneous monogenic polynomials $V_{\underline{\alpha}}^*(x)$ of degree $|\underline{\alpha}|$, $\underline{\alpha} \in \mathbf{N}^m$, may be constructed.

(2) Let $k \in \mathbf{N}$ and define the function p_k on $\mathbf{R}^m$ $times \mathbf{R}_0^m$ by

$$\begin{aligned} p_k(\underline{x},\underline{t}) &= (<\underline{x},\underline{t}>)^k \\ &= (\sum_{j=1}^m x_j t_j)^k. \end{aligned}$$

Then for each $\underline{t} \in \mathbf{R}_0^M$ fixed, p_k is clearly a C-valued homogeneous polynomial of degree k in $\mathbf{R}^m$.

We claim that its CK-extension p_k^*, which for convenience will still be denoted by p_k, is given by

$$p_k(x,\underline{t}) = (<\underline{x},\underline{t}> - \underline{t} e_0 x_0)^k.$$

Indeed, straightforward calculations show that $\partial_x p_k = 0$ in $\mathbf{R}^{m+1}$ and obviously $p_k(x,\underline{t})|_{x_0=0} = p_k(\underline{x},\underline{t})$.

Notice that for $\underline{t} \in \mathbf{R}_0^m$ fixed, p_k is a homogeneous polynomial of degree k in $\mathbf{R}^{m+1}$ and that, as $p_k(\underline{x},\underline{t})$ is C-valued, $p_k(x,\underline{t})$ is $\text{span}_{\mathbf{C}}\{1, e_0 e_i : i = 1,\ldots,m\}$-valued.

(3) Consider a C-valued plane wave f, i.e. f depends only on the scalar product $<\underline{x},\underline{t}>$, $\underline{x},\underline{t} \in \mathbf{R}^m$. Then in the case of the Cauchy-Riemann operator we have, putting $u = <\underline{x},\underline{t}>$, that

$$f^*(x) = \sum_{k=0}^\infty \frac{(-x_0\underline{t})^k}{k!} \frac{\partial^k f}{\partial u^k}$$

and since

$$(\partial_{\underline{x}})^k f(<\underline{x},\underline{t}>) = \frac{\partial^k f}{\partial u^k} \underline{t}^k.$$

Let us apply this result to the following two examples

(3.1) $f(<\underline{x},\underline{t}>) = <\underline{x},\underline{t}>^j$

We have

$$\begin{aligned} f^*(x) &= \sum_{k=0}^j \frac{(-x_0\underline{t})^k}{k!} j(j-1)\ldots(j-k+1)u^{j-k} \\ &= (<\underline{x},\underline{t}> - x_0\underline{t})^j. \end{aligned}$$

We thus reobtain the expression given in the foregoing example.

(3.2) $f(<\underline{x},\underline{t}>) = e^{i<\underline{x},\underline{t}>}$

We have

$$\begin{aligned} f^*(x) &= \sum_{k=0}^{\infty} \frac{(-x_0\underline{t})^k}{k!} i^k e^{iu} \\ &= e^{i<\underline{x},\underline{t}>} \sum_{k=0}^{\infty} \frac{(-ix_0\underline{t})^k}{k!} \end{aligned}$$

If $k = 2\ell$ we get that

$$\frac{(-ix_0\underline{t})^{(2\ell)}}{(2\ell)!} = \frac{x_0^{2\ell}|\underline{t}|^{2\ell}}{(2\ell)!}$$

whence

$$\sum_{k\,\text{even}} \frac{(-ix_0\underline{t})^k}{k!} = ch(x_0|\underline{t}|).$$

If $k = 2\ell + 1$ we obtain that

$$\begin{aligned} \frac{(-ix_0\underline{t})^{2\ell+1}}{(2\ell+1)!} &= -i\frac{x_0\underline{t}}{(2\ell+1)!} x_0^{2\ell}|\underline{t}|^{2\ell} \\ &= -i\frac{(x_0|\underline{t}|)^{2\ell+1}}{(2\ell+1)!}\frac{\underline{t}}{|\underline{t}|} \end{aligned}$$

whence

$$\sum_{k\,\text{odd}} \frac{(-ix_0\underline{t})^k}{k!} = -i\frac{\underline{t}}{|\underline{t}|} sh(x_0|\underline{t}|).$$

Consequently

$$f^*(x) = e^{i<\underline{x},\underline{t}>}(ch(x_0|\underline{t}|) - i\frac{\underline{t}}{|\underline{t}|} sh(x_0|\underline{t}|))$$

determines a monogenic extension of the kernel function $e^{i<\underline{x},\underline{t}>}$ appearing in the Fourier transform.

0.2.5 The inversion operator I

Let $0 \neq x \in \mathbf{R}^{m+1}$ with $x = \sum_{j=0}^m x_j e_j$. Then its inverse is given by $\frac{\overline{x}}{|x|^2}$ and the bijection $\psi : x \to \frac{\overline{x}}{|x|^2}$ from $\mathbf{R}_0^{m+1}$ to $\mathbf{R}_0^{m+1}$ may be extended to $\mathbf{R}^{m+1} \cup \{\infty\}$ by putting $\psi(0) = \infty$ and $\psi(\infty) = 0$.

Now let $R > 0$ be fixed and consider the open ball $\mathring{B}(R)$ centred at the origin. Then by means of ψ, the set $B^*(R) = \mathring{B}(R)\backslash\{0\}$ is bijectively transformed into $\mathbf{R}^{m+1}\backslash\overline{B}(1/R)$ where $\overline{B}(1/R)$ is the closed ball centred at the origin with radius $1/R$.

As is well known, by using the inversion $x = \frac{x'}{|x'|^2}$ in $\mathbf{R}^{m+1}$, Kelvin's formula for $m \geq 2$ reads

$$\Delta_x u(x) = |x'|^{m+3}\Delta_{x'}[\frac{1}{|x'|^{m-1}}u(\frac{x'}{|x'|^2})].$$

This formula implies that u is harmonic in $\mathring{B}(R)$ if and only if $\frac{1}{|x'|^{m-1}}u(\frac{x'}{|x'|^2})$ is harmonic in $\mathbf{R}^{m+1}\backslash\overline{B}(1/R)$ with $\lim_{x'\to\infty}\frac{1}{|x'|^{m-1}}u(\frac{x'}{|x'|^2}) = 0$.

Now define the *inversion operator* $\mathbf{I}$ on $M(\mathring{B}(R); V)$ by

$$\mathbf{I}f(x) = \frac{\overline{x}}{|x|^{m+1}}f(\frac{\overline{x}}{|x|^2}). \tag{0.23}$$

and put $x = r\xi$ with $r = |x|$, $\xi \in S^m$.

Then we have

Theorem. *Let $R > 0$ be given and let $f \in M(\mathring{B}(R); V)$.*
Then

(i) $\mathbf{I}f \in M_0(\mathbf{R}^{m+1}\backslash\overline{B}(1/R); V)$

(ii) $\mathbf{I}^2 f = f$.

Proof.

(i) Obviously $\mathbf{I}f$ is defined in $\mathbf{R}^{m+1}\backslash\overline{B}(1/R)$ and $\lim_{x\to\infty}\mathbf{I}f(x) = 0$.

In order to prove that $\mathbf{I}f$ is monogenic in $\mathbf{R}^{m+1}\backslash\overline{B}(1/R)$, first observe that for any $f \in C_2(\Omega; V)$, $\Omega \subset \mathbf{R}^{m+1}$ open,

$$\Delta_x(xf) = 2\partial_x f + x\Delta_x f$$

so that

$$\partial_x f = 0 \Leftrightarrow \Delta_x f = 0 \quad \text{and} \quad \Delta_x(xf) = 0. \tag{0.24}$$

From $\partial_x f = 0$ in $\mathring{B}(R)$ we thus obtain on the one hand $\Delta_x f(x) = 0$ and hence $\Delta_x f(\overline{x}) = 0$ in $\mathring{B}(R)$ which implies that $\Delta_x(\frac{1}{r^{m+1}}f(\frac{\overline{x}}{r^2})) = 0$ in $\mathbf{R}^{m+1}\backslash\overline{B}(1/R)$.

On the other hand, from $\partial_x f = 0$ it follows, by virtue of (0.24), that $\Delta_x(xf(x)) = 0$ and hence $\Delta_x(\overline{x}f(\overline{x})) = 0$, which in its turn implies that $\Delta_x(\frac{1}{r^{m-1}}\frac{\overline{x}}{r^2}f(\frac{\overline{x}}{r^2})) = \Delta_x(\frac{\overline{x}}{r^{m+1}}f(\frac{\overline{x}}{r^2})) = 0$ in $\mathbf{R}^{m+1}\backslash\overline{B}(1/R)$.

Putting

$$g(x) = \frac{\overline{x}}{|x|^{m+1}}f(\frac{\overline{x}}{|x|^2}) = \mathbf{I}f(x)$$

we thus have that in $\mathbf{R}^{m+1}\backslash\overline{B}(1/R)$

$$\Delta_x(xg) = 0 \quad \text{and} \quad \Delta_x g = 0.$$

In view of (0.24) this is equivalent to $\partial_x g = 0$ in $\mathbf{R}^{m+1} \backslash \overline{B}(1/R)$.

(ii) A straightforward calculation yields that for each $x \in \mathring{B}(R) \backslash \{0\}$

$$\mathbf{I}(\mathbf{I}f(x)) = f(x). \quad \blacksquare$$

Corollary. *Let* $(f_k)_{k \in \mathbf{N}}$ *be a sequence of monogenic functions in* $\mathring{B}(R)$ *(resp. in* $\mathbf{R}^{m+1} \backslash \overline{B}(1/R)$ *) which converges uniformly on each compact subset* $K \subset \mathring{B}(R)$ *(resp.* $K \subset \mathbf{R}^{m+1} \backslash \overline{B}(1/R)$*) to a function* f.

Then $(\mathbf{I}f_k)_{k \in \mathbf{N}}$ *converges uniformly to* $\mathbf{I}f$ *on each compact subset of* $\mathbf{R}^{m+1} \backslash \overline{B}(1/R)$ *(resp.* $\mathring{B}(R) \backslash \{0\}$*).*

As we have seen in the foregoing theorem, the inversion operator $\mathbf{I}$ maps the space $M(\mathring{B}(R); V)$ into $M_0(\mathbf{R}^{m+1} \backslash \overline{B}(1/R); V)$. In 1.6.5 it will be seen that $\mathbf{I}$ is in fact an isomorphism between these spaces.

1 SPHERICAL MONOGENICS IN $\mathbf{R}^{m+1}$

In this section we study nullsolutions of the **Dirac** operator which are either homogeneous V-valued polynomials of degree k or homogeneous V-valued functions of degree $-(m+k)$ in $\mathbf{R}^{m+1}$. As before V stands for the spinor space S or the Clifford algebra $\mathcal{C}$. These functions are called spherical monogenics of degree k ; they are substantial building blocks in investigating the local behaviour of monogenic functions in open domains Ω of $\mathbf{R}^{m+1}$ which are invariant under $\mathrm{SO}(m+1)$. Whenever in the case $V = S$ a basis of S is needed in order to express formulas in a comprehensive way, we use the notation $\{v_1, \ldots, v_d\}$ for such a basis. Obviously $\{\overline{v}_1, \ldots, \overline{v}_d\}$ then becomes a basis for S.

Notice that most of the results obtained in this section reduce automatically to analogous results for nullsolutions of the Weyl equation and this by simply substituting e_0 by 1 and e_i by ε_i, $i = 1, \ldots, m$ (§§1.1 -1.9). The cases where such automatism does not hold are cited explicitly (§§1.10 - 1.12).

1.1 Definitions

Let $k \in \mathbf{N}$.

(1) Call $M^+(k; V)$ the space of homogeneous monogenic V-valued polynomials of degree k in $\mathbf{R}^{m+1}$. An arbitrary element of it, denoted by P_k, is called an *inner spherical monogenic of order k*.

(2) Call $M^-(k; V)$ the space of homogeneous monogenic functions of degree $-(k+m)$ in $\mathbf{R}_0^{m+1}$ with values in V. An arbitrary element of it, denoted by Q_k, is called an *outer spherical monogenic of order k. outer spherical monogenic of order k*

(3) Call $\mathcal{M}^+(k; V)$ and $\mathcal{M}^-(k; V)$ the spaces consisting of the restrictions to the unit sphere S^m of respectively the elements belonging to $M^+(k; V)$ and $M^-(k; V)$ and put

$$\mathcal{M}(k; V) = \mathcal{M}^+(k; V) + \mathcal{M}^-(k; V).$$

The elements of $\mathcal{M}(k; V)$ are called *spherical monogenics of degree k*. Arbitrary elements of $\mathcal{M}^+(k; V)$ and $\mathcal{M}^-(k; V)$ will still be denoted by P_k and Q_k.

1.2 Remarks

(1) In the notations used for the spaces of spherical monogenics we did not mention the dimension of the Euclidean space we are working in. When explicit reference to this dimension is needed, we shall adopt the notations $M^+(m+1, k; V)$, $\mathcal{M}^+(m+1; k; V)$, $\mathcal{M}^-(m+1); k; V)$ and $\mathcal{M}(m+1; k; V)$.

(2) In the case $V = C$, the spaces $M^+(k;C)$, $M^-(k;C)$, $\mathcal{M}^+(k;C)$ and $\mathcal{M}^-(k;C)$ are right C - modules. In the case $V = S$ they are just complex vector spaces.

(3) The Cauchy-Kovalevska extension theorem (see 0.2.4) tells us that the space $\mathcal{P}(m;k;V)$ of V-valued homogeneous polynomials in $\mathbf{R}^m$ of degree k and $M^+(k;V)$ are isomorphic (as complex vector spaces, in the case of $V = C$ also as right C-modules).

Indeed, take $P \in \mathcal{P}(m;k;V)$ and put

$$P^\star(x) = \sum_{j=0}^{\infty} \frac{1}{j!}(-x_0)^j(\bar{e}_0\partial_{\underline{x}})^j P(\underline{x}).$$

Then $P^\star$ is a homogeneous polynomial of degree k in $\mathbf{R}^{m+1}$ and, being the CK-extension of P, it thus belongs to $M^+(k,V)$. It moreover satisfies $P^\star(0,\underline{x}) = P(\underline{x})$.

Conversely, let $P_k \in M^+(k,V)$; then its restriction $\tilde{P}_k$ to $\mathbf{R}^m$ with $\tilde{P}_k(\underline{x}) = P_k(0,\underline{x})$ clearly belongs to $\mathcal{P}(m;k;V)$. Moreover, by the principle of analytic continuation, its CK-extension $\tilde{P}_k^\star$ coincides with P_k.

Denoting by $CK(P)$ the CK-extension of $P \in \mathcal{P}(m;k;V)$ and by $R(P_k)$ the restriction of $P_k \in M^+(k,V)$ to $\mathbf{R}^m$, we thus proved that CK establishes an isomorphism between $\mathcal{P}(m;k;V)$ and $M^+(k;V)$, where its inverse is given by R.

(4) In the sequel we shall often refer to the following spaces and their respective dimensions.

(4.1) The space $\mathcal{P}(m;k)$ of $\mathbf{C}$-valued homogeneous polynomials of degree k in $(x_1,\ldots,x_m)$. We have

$$\dim \mathcal{P}(m;k) = K(m;k) = \frac{(k+m-1)!}{k!(m-1)!}$$

(4.2) The space $\mathcal{H}(m+1;k)$ of $\mathbf{C}$-valued spherical harmonics of degree k in $\mathbf{R}^{m+1}$. We have

$$\begin{aligned}\dim \mathcal{H}(m+1;k) &= N(m+1;k)\\ &= K(m+1;k) - K(m+1;k-2)\\ &= K(m;k) + K(m;k-1)\end{aligned}$$

As $\mathcal{P}(m;k;V) = \mathcal{P}(m;k) \otimes V$ we have

$$\dim \mathcal{P}(m;k;V) = K(m;k) \cdot \dim V.$$

Furthermore, denoting by $\mathcal{H}(k;V)$ the space of V-valued spherical harmonics of degree k in $\mathbf{R}^{m+1}$ we have, as $\mathcal{H}(k;V) = \mathcal{H}(m+1;k) \otimes V$, that

$$\dim \mathcal{H}(k; V) = N(m+1; k) \cdot \dim V.$$

(5) The inversion $\mathbf{I}$ establishes an isomorphism between the spaces $M^+(k; V)$ and $M^-(k; V)$, whence

$$\begin{aligned} \dim M^-(k; V) &= \dim M^+(k; V) \\ &= \dim V.K(m; k). \end{aligned}$$

Indeed, from 0.2.5 it already follows that for each $P_k \in M^+(k; V)$, $Q_k = \mathbf{I} P_k$ is monogenic in $\mathbf{R}_0^{m+1}$.

Furthermore

$$Q_k(x) = \mathbf{I} P_k(x) = \frac{\bar{x}}{|x|^{m+1}} P_k(\frac{\bar{x}}{|x|^2})$$

is homogeneous of degree $-(k+m)$ whence $\mathbf{I} P_k \in M^-(k; V)$.

Conversely, taking an arbitrary $Q_k \in \mathcal{M}^-(k; V)$ and putting

$$P_k(x) = \mathbf{I} Q_k(x) = \frac{\bar{x}}{|x|^{m+1}} Q_k(\frac{\bar{x}}{|x|^2})$$

we have that P_k is monogenic in $\mathbf{R}^{m+1} \backslash \{0\}$.

But, passing to polar coordinates $x = r\xi$, we find :

$$\lim_{x \to 0} P_k(x) = \lim_{r \to 0} (r^k \bar{\xi} Q_k(\bar{\xi})) = \begin{cases} \bar{\xi} Q_0(\bar{\xi}), & k = 0 \\ 0, & k \neq 0 \end{cases}$$

whence P_k is monogenic in $\mathbf{R}^{m+1}$ (see Theorem 0.2.12).

Finally P_k is homogeneous of degree k and, as clearly $\mathbf{I} P_k = Q_k$, we have that $\mathbf{I}$ is surjective. As $\mathbf{I}$ is obviously linear, the result follows immediately.

(6) For all $k, l \in \mathbf{N}$

$$M^+(k; V) \cap M^-(l; V) = \{0\}. \tag{1.1}$$

Note also that it will make sense to speak of the dimension of the right $\mathcal{C}$-modules $M^+(k; \mathcal{C})$ and $M^-(k; \mathcal{C})$ in the case $V = \mathcal{C}$.

This notion will in fact be a consequence of a more general result concerning right $\mathcal{C}$-modules E, namely :

"Let E be a right $\mathcal{C}$-module having a finite basis $\{F_1, \ldots, F_s\}$, i.e. $\{F_1, \ldots, F_s\}$ is a generating set for E which is moreover right $\mathcal{C}$-linearly independent. Then any other right $\mathcal{C}$-linearly independent subset of E consisting of exactly s elements is again a basis for E".

This property, which may be proved easily, allows us to call $\dim_{\mathcal{C}} E = s$. As we shall see in Theorem 1.5.3. and Theorem 1.5.7. , in the case $V = \mathcal{C}$ we have

$$K(m;k) = \dim_C M^+(k,C) = \dim_C M^-(k,C).$$

(7) It is clear that for each $Q_k \in M^-(k;V)$

$$\lim_{x\to\infty} Q_k(x) = 0$$

whence $M^-(k) \subset M_0(\mathbf{R}^{m+1};V)$.

(8) Finally notice that by the definition itself and (3), arbitrary elements of the spaces $\mathcal{M}^+(k;V)$ and $\mathcal{M}^-(k;V)$ are given respectively by $P_k(\xi)$ and $\bar{\xi}P_k(\bar{\xi})$ with $P_k \in M^+(k)$.

1.3 Γ_ξ acting on $L_2(S^m;V)$ - Projection operators

Consider $L_2(S^m;V)$, the space of V-valued square integrable functions on S^m, provided with the inner product (see also 0.1.1.)

$$(f,g) = \int_{S^m} \bar{f}(\omega)g(\omega)d\Sigma(\omega).$$

Let us recall that in the case $V = S = CI$, we have that for $f = sI$ and $g = tI$,

$$\int_{S^m} \overline{f(\omega)}g(\omega)d\Sigma(\omega) = I\int_{S^m} < \overline{s(\omega)}t(\omega) > d\Sigma(\omega)$$

whence we may consider the inner product to be C-valued and $L_2(S^m;S)$ so becomes a complex Hilbert space.

In the special case where s and t are C-valued and v,w are elements in S we write for $f = sv$ and $g = tw$ that

$$\int_{S^m} \overline{f(\omega)}g(\omega)d\Sigma(\omega) = \bar{v}\int_{S^m} \overline{s(\omega)}t(\omega)d\Sigma(\omega)w.$$

If moreover $\int_{S^m} \bar{s}(\omega)t(\omega)d\Sigma(\omega) = c \in \mathbf{C}$, we write

$$\begin{aligned}\int_{S^m} \overline{f(\omega)}g(\omega)d\Sigma(\omega) &= c(v,w)\\ &= c\bar{v}w\end{aligned}$$

In the case $V = C$ we have that $(\ ,\ )$ is a C-valued inner product which is turned into a complex inner product space by putting for all $f,g \in L_2(S^m;C)$,

$$< f,g > = [(f,g)]_0 = \int_{S^m} [\overline{f(\omega)}g(\omega)]_0 d\Sigma(\omega).$$

We then consider $L_2(S^m; C)$ both as a right Hilbert module over C and as a complex Hilbert space.

A right C-linear operator $T : L_2(S^m; C) \to L_2(S^m; C)$ is said to be self-adjoint if it is densely defined and $T = T^*$ where T^* is defined in such a way that dom $(T) =$ dom T^* and $(T_g, f) = (g, T^* f)$ for all $f, g \in$ dom (T).

This definition of course implies that self-adjointness in $(L_2(S^m; C), (\ ,\ ))$ implies self-adjointness in $(L_2(S^m; C), <\ ,\ >)$. Clearly $\mathcal{M}^+(k; V)$ and $\mathcal{M}^-(k; V)$ are subspaces of $L_2(S^m; V)$. We have

Theorem 1.3.1. *For each $k, l \in \mathbf{N}$, $\mathcal{M}^+(k; V)$ and $\mathcal{M}^-(l; V)$ are orthogonal, i.e. for any $P_k \in \mathcal{M}^+(k; V)$ and $Q_l \in \mathcal{M}^-(l; V)$,*

$$\begin{aligned}(P_k, Q_l) &= \int_{S^m} \bar{P}_k(\omega) Q_l(\omega) d\Sigma(\omega) \\ &= 0.\end{aligned}$$

Consequently the sum $\mathcal{M}(k; V) = \mathcal{M}^+(k; V) \oplus \mathcal{M}^-(k; V)$ is direct.

Proof. As $Q_l \in \mathcal{M}^-(l; V)$ there exists $P_l \in \mathcal{M}^+(l; V)$ such that (see 1.2 (3))

$$Q_l(x) = \mathbf{I} P_l(x) = \frac{\bar{x}}{|x|^{m+1}} P_l\left(\frac{\bar{x}}{|x|^2}\right),\ x \in \mathbf{R}_0^{m+1}. \tag{1.2}$$

But as $\bar{P}_l(\bar{x})$ is right monogenic in $\mathbf{R}^{m+1}$ we obtain by Cauchy's Theorem that

$$\int_{S^m} \bar{P}_l(\bar{\omega}) d\sigma_\omega P_k(\omega) = 0.$$

Using the relation (1.2) we get

$$\begin{aligned}0 &= \int_{S_m} \bar{Q}_l(\omega) \omega d\sigma_\omega P_k(\omega) \\ &= \int_{S^m} \bar{Q}_l(\omega) P_k(\omega) d\Sigma\omega. \ \blacksquare\end{aligned}$$

Now we claim that $\mathcal{M}^+(k; V)$ and $\mathcal{M}^-(k; V)$ are the only eigenspaces of the spherical Dirac operator Γ_ξ acting on $L_2(S^m; V)$. To be more precise we shall prove that if we take dom $\Gamma_\xi = C_\infty(S^m; V)$ then Γ_ξ is essentially self adjoint on $L_2(S^m; V)$ and that $\mathcal{M}^+(k; V)$ and $\mathcal{M}^-(k; V)$ are the only eigenspaces of its closure $\overline{\Gamma}_\xi$.

Theorem 1.3.2.

(i) For each $k \in \mathbf{N}$, $P_k(\xi) \in \mathcal{M}^+(k; V)$ is a V-valued spherical harmonic of order k, i.e.

$$\Delta_\xi P_k(\xi) = (-k)(k + m - 1) P_k(\xi).$$

(ii) For each $l \in \mathbf{N}$, $Q_l(\xi) \in \mathcal{M}^-(l;V)$ is a V-valued spherical harmonic of order $l+1$, i.e.

$$\Delta_\xi Q_l(\xi) = -(l+1)(l+m)Q_l(\xi).$$

(iii) If $k \neq l$ then

$$(P_k, P_l) = (Q_k, Q_l) = 0$$

while for all $k, l \in \mathbf{N}$

$$(P_k, Q_l) = 0.$$

(iv) For all $k \in \mathbf{N}$

$$\mathcal{H}(k;V) = \mathcal{M}^+(k;V) \oplus \mathcal{M}^-(k-1;V),$$

where for $k = 0$ we put by definition $\mathcal{M}^-(-1;V) = \{0\}$.

Proof. Let us recall that (see 0.16)

$$\Delta_\xi = (m-1-\Gamma_\xi)\Gamma_\xi$$

and that ∂_x is expressed in spherical coordinates $x = r\xi$ by

$$\partial_x = \xi(\partial_r + \frac{1}{r}\Gamma_\xi).$$

Now for each $k \in \mathbf{N}$,

$$P_k(x) = r^k P_k(\xi)$$

and

$$(\frac{\partial}{\partial r} + \frac{1}{r}\Gamma_\xi) r^k P_k(\xi) = 0$$

whence

$$\Gamma_\xi P_k(\xi) = (-k)P_k(\xi).$$

But then

$$\Delta_\xi P_k(\xi) = (-k)(k+m-1)P_k(\xi)$$

or

$$P_k(\xi) \in \mathcal{H}(k;V).$$

Analogously, for each $l \in \mathbf{N}$,

$$Q_l(x) = r^{-(l+m)} Q_l(\xi)$$

and

$$\left(\frac{\partial}{\partial r} + \frac{1}{r}\Gamma_\xi\right)\left(r^{-(l+m)} Q_l(\xi)\right) = 0$$

whence

$$\Gamma_\xi Q_l(\xi) = (l+m) Q_l(\xi).$$

But then

$$\Delta_\xi Q_l(\xi) = -(l+1)(l+m) Q_l(\xi)$$

or

$$Q_l(\xi) \in \mathcal{H}(l+1; V).$$

As for any $k \neq l$

$$\mathcal{H}(k; V) \perp \mathcal{H}(l; V),$$

$$(P_k, P_l) = (Q_k, Q_l) = 0, \quad k \neq l.$$

Furthermore, as we have seen in Theorem 1.3.1

$$(P_k, Q_l) = 0 \quad \text{for all} \quad k, l \in \mathbf{N}.$$

Consequently for all $k \in \mathbf{N}$, $k \geq 1$,

$$\mathcal{M}^+(k; V) \oplus \mathcal{M}^-(k-1; V)$$

is an orthogonal direct sum in $\mathcal{H}(k; V)$.

But as for $k \geq 1$

$$\begin{aligned} \dim \mathcal{H}(k; V) &= \dim V.N(m+1; k)(m+1; k) \\ &= \dim V(K(m; k) + K(m; k-1)) \\ &= \dim \mathcal{M}^+(k; V) + \dim \mathcal{M}^-(k-1; V) \end{aligned}$$

we obtain that

$$\mathcal{H}(k; V) = \mathcal{M}^+(k; V) \oplus \mathcal{M}^-(k-1; V). \blacksquare$$

Corollary 1.3.3. *Let $k \in \mathbf{N}$. Then each $S_k \in \mathcal{H}(k; V)$ admits a unique orthogonal decomposition*

$$S_k = \mathbf{P}_k S_k + \mathbf{Q}_{k-1} S_k$$

with $\mathbf{P}_k S_k \in \mathcal{M}^+(k;V)$ *and* $\mathbf{Q}_{k-1} S_k \in \mathcal{M}^-(k-1;V)$ *where*

$$\mathbf{P}_k S_k(\xi) = \frac{k+m-1-\Gamma_\xi}{2k+m-1} S_k(\xi)$$

and

$$\mathbf{Q}_{k-1} S_k(\xi) = \frac{k+\Gamma_\xi}{2k+m-1} S_k(\xi).$$

Remarks. Let S_k be a **C**-valued spherical harmonic of order k. Then, by virtue of the results and remarks in 0.1.4. and Corollary 1.3.3. we have

(1) In the case of the Dirac operator, the functions $\mathbf{P}_k S_k$ and $\mathbf{Q}_{k-1} S_k$ are $\mathbf{C}^0_{m+1} \oplus \mathbf{C}^2_{m+1}$ valued.

(2) In the case of the Cauchy-Riemann operator, the functions $\mathbf{P}_k S_k$ and $\mathbf{Q}_{k-1} S_k$ are $\mathbf{C}^1_{m+1} \oplus \mathbf{C}^2_{m+1}$ valued.

Now let $f \in L_2(S^m;V)$. Then by means of the expansion of each element in $L_2(S^m)$ into spherical harmonics (see [29]) we obtain that

$$f(\xi) = \sum_{k=0}^{\infty} \mathbf{S}_k f(\xi).$$

where for each $k \in N$

$$\mathbf{S}_k f(\xi) = \frac{N(m+1;k)}{A_{m+1}} \int_{S^m} P_{k,m+1}(\omega.\xi) f(\omega) d\Sigma(\omega),$$

$P_{k,m+1}$ being the Legendre polynomial of degree k in $(m+1)$-dimensions (see e.g. [40]).

Of course the operator $\mathbf{S}_k : L_2(S^m;V) \to \mathcal{H}(k;V)$ is nothing else but the orthogonal projection of $L_2(S^m;V)$ onto $\mathcal{H}(k;V)$.

Henceforth we denote for each $k \in \mathbf{N}$,

$$\mathbf{S}(k) = \mathbf{S}_k.$$

By means of Corollary 1.3.3. we may introduce for each $k \in \mathbf{N}$ the operators

$$\mathbf{P}(k) : L_2(S^m;V) \to \mathcal{M}^+(k;V)$$

and

$$\mathbf{Q}(k) : L_2(S^m;V) \to \mathcal{M}^-(k;V)$$

given by

$$\mathbf{P}(k)f = \mathbf{P}_k(\mathbf{S}(k)f)$$

and

$$\mathbf{Q}(k)f = \mathbf{Q}_k(\mathbf{S}(k+1)f).$$

We then have that $\mathbf{P}(k)$ and $\mathbf{Q}(k)$ are orthogonal projections with

$$\mathbf{P}(k) = \frac{k+m-1-\Gamma_\xi}{2k+m-1}\mathbf{S}(k)$$

and

$$\mathbf{Q}(k) = \frac{k+1+\Gamma_\xi}{2k+m+1}\S(k+1)$$

Moreover

$$\mathbf{S}(k) = \mathbf{P}(k) + \mathbf{Q}(k-1).$$

Finally putting for each $k \in \mathbf{N}$

$$\mathbf{\Pi}(k) = \mathbf{P}(k) + \mathbf{Q}(k);$$

then clearly $\mathbf{\Pi}(k)$ is the orthogonal projection of $L_2(S^m;V)$ onto $\mathcal{M}(k;V)$.

We so obtain

Theorem 1.3.4. *Let $f \in L_2(S^m;V)$. Then there exist unique sequences of spherical monogenics $(\mathbf{P}(k)f)_{k\in\mathbf{N}}$ and $(\mathbf{Q}(k)f)_{k\in\mathbf{N}}$ such that*

$$\begin{aligned} f &= \sum_{k=0}^{\infty}(\mathbf{P}(k)f + \mathbf{Q}(k)f) \\ &= \sum_{k=0}^{\infty}\mathbf{\Pi}(k)f. \end{aligned}$$

Moreover

$$\begin{aligned} \|f\|^2 &= \sum_{k=0}^{\infty}\|\mathbf{\Pi}(k)f\|^2 \\ &= \sum_{k=0}^{\infty}(\|\mathbf{P}(k)f\|^2 + \|\mathbf{Q}(k)f\|^2). \end{aligned}$$

Finally, if $g \in L_2(S^m;V)$ admits the expansion

$$g(\omega) = \sum_{k=0}^{\infty}\mathbf{\Pi}(k)g(\omega)$$

then

$$(f,g) = \sum_{k=0}^{\infty}(\mathbf{P}(k)f\,,\,\mathbf{P}(k)g)$$
$$+\sum_{k=0}^{\infty}(\mathbf{Q}(k)f\,,\,\mathbf{Q}(k)g).$$

Notice that the decomposition of $f \in L_2(S^m;V)$ into a sequence of spherical monogenics $(\mathbf{P}(k)f)_{k\in\mathbf{N}}$ and $(\mathbf{Q}(k)f)_{k\in\mathbf{N}}$ yields the following relation between the projection operators $\mathbf{P}(k)$ and $\mathbf{Q}(k)$.

Define the operator $\boldsymbol{\Omega}$ on $L_2(S^m;V)$ by $\boldsymbol{\Omega} f = \underline{\omega} f$. Then clearly

$$\mathbf{Q}(k) = -\boldsymbol{\Omega}\mathbf{P}(k)\boldsymbol{\Omega}.$$

"Par abus de langage" we write for each $f \in L_2(S^m;V)$ that

$$\mathbf{Q}(k)f(\underline{\omega}) = -\underline{\omega}\mathbf{P}(k)(\underline{\omega}f)(\underline{\omega}).$$

To prove that Γ_ξ is an essentially self adjoint operator on $L_2(S^m;V)$ we shall refine the spectral family $(\tilde{E}_\lambda)_{\lambda\in\mathbf{R}}$ appearing in the spectral decomposition

$$\Delta_\xi = \int_{-\infty}^{+\infty} \lambda d\tilde{E}_\lambda$$

of the Laplace-Beltrami operator Δ_ξ on $L_2(S^m;V)$.

The spectral family $(\tilde{E}_\lambda)_{\lambda\in\mathbf{R}}$ is given explicitly as follows. For $f \in L_2(S^m;V)$, f has a unique decomposition

$$f = \sum_{k=0}^{\infty} \mathbf{S}(k)f$$

where $\mathbf{S}(k)f \in \mathcal{H}(k;V)$. $\tilde{E}_\lambda$ is then written explicitly as

$$\tilde{E}_\lambda = \sum_{-k(k+m-1)\leq\lambda} \mathbf{S}(k)f.$$

We now define the family $U_\lambda(\lambda \in \mathbf{R})$ of projection operators on $L_2(S^m;V)$ by

$$U_\lambda f = \sum_{k\geq-\lambda} \mathbf{P}(k)f \quad \text{for}\lambda \leq 0$$

and

$$U_\lambda f = U_0 f + \sum_{k+m\leq\lambda} \mathbf{Q}(k)f \quad \text{for}\lambda > 0.$$

Now we claim that $(U_\lambda)_{\lambda\in\mathbf{R}}$ is a spectral family.

Indeed, from Theorem 1.3.2 it follows immediately that for all $\lambda \in \mathbf{R}$

$$U_\lambda = U_\lambda^2 = U_\lambda^*$$

and that for all $\lambda, \mu \in \mathbf{R}$

$$U_\lambda U_\mu = U_{\min(\lambda,\mu)}.$$

Moreover

$$s - \lim_{\lambda\to-\infty} U_\lambda = \mathbf{0}$$

and

$$s - \lim_{\lambda\to+\infty} U_\lambda = \mathbf{1}.$$

Indeed, if $-t \leq \lambda \leq -t+1$ and $\mu \geq -t(t+m-1)$ with $t > 0$, we have that for f arbitrary

$$U_\lambda f = U_\lambda \tilde{E}_\mu f$$

whence

$$\|U_\lambda f\| \leq \|\tilde{E}_\mu f\|.$$

Taking the limit for $t \to +\infty$ gives $\mu \to -\infty$ and $\lambda \to -\infty$ and so

$$\lim_{\lambda\to-\infty} \|U_\lambda f\| \leq \lim_{\mu\to-\infty} \|\tilde{E}_\mu f\| = 0.$$

If$(t+m-1) < \lambda \leq (t+m)$ and $\mu \geq -t(t+m-1)$ we have that

$$(U_\lambda - \mathbf{1})f - (U_\lambda - \mathbf{1})\tilde{E}_\mu f$$

whence

$$\|(U_\lambda - \mathbf{1})f\| \leq \|\tilde{E}_\mu f\|.$$

Taking the limit for $t \to +\infty$ gives $\lambda \to +\infty$ and $\mu \to -\infty$ and so

$$\lim_{\lambda\to+\infty} \|(U_\lambda - \mathbf{1})f\| \leq \lim_{\mu\to-\infty} \|\tilde{E}_\mu f\| = 0.$$

Moreover the function $\lambda \to U_\lambda$ is clearly continuous from the right.

Since $(U_\lambda)_{\lambda\in\mathbf{R}}$ is a spectral family, the operator

$$\hat{\Gamma} = \int_{-\infty}^{+\infty} \lambda dU_\lambda$$

is selfadjoint on $L_2(S^m; V)$.

Theorem 1.3.5. *Taking* $dom(\Gamma_\xi) = C_\infty(S^m; V)$ *we have that* Γ_ξ *is an essentially selfadjoint operator on* $L_2(S^m; V)$*, its closure being* $\hat{\Gamma}$.

Proof. We first prove that $\hat{\Gamma}$ is an extension of Γ_ξ and afterwards that

$$\mathrm{dom}(\hat{\Gamma}) = \mathrm{dom}(\overline{\Gamma_\xi}).$$

Take $f \in \mathbf{C}_\infty(S^m; V)$ arbitrarily. Then

$$\hat{\Gamma} f = \Gamma_\xi f.$$

Since $\mathbf{C}_\infty(S^m; V) \subset L_2(S^m; V)$, by virtue of Theorem 1.3.4 we can write f as

$$f = \sum_{k=0}^{\infty} (\mathbf{P}(k)f + \mathbf{Q}(k)f).$$

Moreover, since $\Gamma_\xi f \in \mathbf{C}_\infty(S^m; V)$ we obtain similarly that

$$\Gamma_\xi f = \sum_{k=0}^{\infty} \mathbf{P}(k)\Gamma_\xi f + \mathbf{Q}(k)\Gamma_\xi f.$$

We claim that for each $k \in \mathbf{N}$,

$$\mathbf{P}(k)\Gamma_\xi f = -k\mathbf{P}(k)f$$

and

$$\mathbf{Q}(k)\Gamma_\xi f = (k+m)\mathbf{Q}(k)f.$$

To prove e.g. that $\mathbf{P}(k)\Gamma_\xi f = -k\mathbf{P}(k)f$, it is sufficient to show that for each element $P_k \in \mathcal{M}^+(k; V)$

$$(\mathbf{P}(k)\Gamma_\xi f, P_k) = -k(\mathbf{P}(k)f, P_k)$$

since then, as (see Theorem 1.3.2)

$$\begin{aligned} (\mathbf{P}(k)\Gamma_\xi f + k\mathbf{P}(k)f, P_k) &= 0, \\ (\mathbf{P}(k)\Gamma_\xi f + k\mathbf{P}(k)f, P_\ell) &= 0, \quad \ell \neq k, \end{aligned}$$

and

$$(\mathbf{P}(k)\Gamma_\xi f + k\mathbf{P}(k)f, Q_\ell) = 0, \quad \ell \in \mathbf{N},$$

we have that

$$\mathbf{P}(k)\Gamma_\xi f + k\mathbf{P}(k)f = 0.$$

So let $P_k \in \mathcal{M}^+(k;V)$ and put for all $r > 0$

$$F(r\xi) = f(\xi)$$

and

$$G(r\xi) = \Gamma^{-(m+k)}\xi P_k(\xi).$$

Then by virtue of Cauchy's Theorem applied to $\sum = \bar{B}(R_1)\backslash\mathring{B}(R_2)$ with $0 < R_2 < R_1$ we get

$$\int_{\partial\Sigma} \bar{F} d\sigma G = \int_{\Sigma} [(\bar{F}\partial_x)G + \bar{F}(\partial_x G)]dx.$$

On the one hand

$$\begin{aligned}\int_{\partial\Sigma} \bar{F} d\sigma G &= (R_1^{-k} - R_2^{-k}) \int_{S^m} \bar{f}(\xi)\xi(\xi P_k(\xi))d\Sigma(\xi)\\ &= -(R_1^{-k} - R_2^{-k})(f, P_k).\end{aligned}$$

On the other hand, as $\partial_x G = 0$ in $\mathbf{R}_0^{m+1}$ and

$$\begin{aligned}\bar{F}\partial_x = -\overline{\partial_x F} &= \overline{-\xi(\partial_r + \frac{1}{r}\Gamma_\xi)F}\\ &= \overline{-\frac{\xi}{r}\Gamma_\xi f},\end{aligned}$$

$$\begin{aligned}&\int_{\Sigma} [(\bar{F}\partial_x)G + \bar{F}(\partial_x G)]dx\\ &= \int_{\Sigma} \overline{\Gamma_\xi(f)}\frac{\xi}{r} r^{-(m+k)}\xi P_k(\xi)dx\\ &= -\int_{R_2}^{R_1} r^{-(k+1)}dr \int_{S^m} \overline{\Gamma_\xi(f)}P_k(\xi)d\Sigma(\xi).\end{aligned}$$

So for $k \neq 0$ we get

$$-(R_1^{-k} - R_2^{-k})(f, P_k) = \frac{1}{k}(R_1^{-k} - R_2^{-k})(\Gamma_\xi f, P_k)$$

whence

$$-k(f, P_k) = (\Gamma_\xi f, P_k).$$

For $k = 0$ we have

$$0 = \ln\frac{R_2}{R_1}(\Gamma_\xi f, P_0)$$

whence again, as $\ln \frac{R_2}{R_1} \neq 0$,

$$0 = (\Gamma_\xi f, P_0) = 0(f, P_0).$$

By the orthogonality of spherical monogenics (see Theorem 1.3.2) we thus find

$$(\mathbf{P}(k)\Gamma_\xi f, P_k) = -k(\mathbf{P}(k)f, P_k).$$

Analogously we can prove that for each $k \in \mathbf{N}$,

$$\mathbf{Q}(k)\Gamma_\xi f = (k+m)\mathbf{Q}(k)f.$$

We thus obtain that $f \in \text{dom}\,(\hat{\Gamma})$ since

$$\begin{aligned}\int_{-\infty}^{+\infty} \lambda^2 d(f, U_\lambda f) &= \sum_{k=0}^{\infty} k^2 \|\mathbf{P}(k)f\|^2 + (k+m)^2 \|\mathbf{Q}(k)f\|^2 \\ &= \|\Gamma_\xi f\|^2 < +\infty.\end{aligned}$$

Here we relied upon the fact that

$$U_\lambda \mathbf{P}(k)f = \begin{cases} 0 & \text{if } \ k > -\lambda \\ -k\mathbf{P}(k)f & \text{if } \ k \leq -\lambda \end{cases}$$

and

$$U_\lambda \mathbf{Q}(k)f = \begin{cases} 0 & \text{if } \ k+m > \lambda \\ (k+m)\mathbf{Q}(k)f & \text{if } \ k+m \leq \lambda. \end{cases}$$

Putting for each $n \in \mathbf{N}$

$$f_n = \sum_{k=0}^{n} (\mathbf{P}(k)f + \mathbf{Q}(k)f)$$

we have that

$$f_n \to f$$

and

$$\hat{\Gamma} f_n = \sum_{k=0}^{n} (-k\mathbf{P}(k)f + (k+m)\mathbf{Q}(k)f) \quad \to \quad \Gamma_\xi f,$$

whence, $\hat{\Gamma}$ being closed, $f \in \text{dom}(\hat{\Gamma})$ with $\hat{\Gamma} f = \Gamma_\xi f$.

As $\hat{\Gamma}$ is a closed extension of Γ_ξ, we surely have that $\overline{\Gamma}_\xi \subset \hat{\Gamma}$ so that it is sufficient to prove that $\text{dom}(\hat{\Gamma}) \subset \text{dom}(\hat{\Gamma}_\xi)$ in order to conclude that $\overline{\Gamma}_\xi = \hat{\Gamma}$.

To this end take $f \in \text{dom}(\hat{\Gamma})$ and put for each $n \in \mathbf{N}$

$$f_n = \sum_{k=0}^{n}(\mathbf{P}(k)f + \mathbf{Q}(k)f) = U_{n+m}f - U_{-n-1}f.$$

Then clearly $f_n \to f$.

Moreover, as a finite linear combination of elements in $C_\infty(S^m; V)$, $f_n \in \mathrm{dom}(\Gamma_\xi)$ for each $n \in \mathbf{N}$ while $f \in \mathrm{dom}(\hat{\Gamma})$ implies that

$$\lim_{n\to\infty} \Gamma_\xi f_n = \lim_{n\to\infty} \hat{\Gamma} f_n = \sum_{k=0}^{\infty}(-k\mathbf{P}(k)f + (k+m)\mathbf{Q}(k)f)$$

converges to $\hat{\Gamma}f$. Consequently $f \in \mathrm{dom}\overline{\Gamma}_\xi$. ■

Henceforth we denote $\hat{\Gamma}$ by Γ_ξ. The foregoing results clearly yield

Theorem 1.3.6. *$\mathcal{M}^+(k;V)$ and $\mathcal{M}^-(k;V)$ are the only eigenspaces of Γ_ξ acting on $L_2(S^m;V)$. They correspond, respectively, to the eigenvalues $(-k)$ and $(k+m)$.*

Theorem 1.3.6. tells us that Γ_ξ is a self adjoint operator having pure point spectrum $\sigma_p(\Gamma_\xi) = \{-k, k+m : k \in \mathbf{N}\}$.

It thus follows that for any real valued Borel measurable function φ on the real line satisfying $\lim_{\lambda\to\pm\infty} |\varphi(\lambda)| = \infty$,

$$\varphi(\Gamma_\xi) = \int_{-\infty}^{+\infty} \varphi(\lambda)dU_\lambda$$

is also selfadjoint, its spectral family $(\Phi_\lambda)_{\lambda\in\mathbf{R}}$ being given by

$$\Phi_\lambda = \int_{B_\lambda} dU_\mu$$

with

$$B_\lambda = \{\mu : \varphi(\mu) \leq \lambda\}.$$

In particular, if $\varphi(\lambda) = \sum_{j=0}^{n} a_j\lambda^j$ is a polynomial of degree n over $\mathbf{R}$ $(n \geq 1)$, we obtain taking account of

$$\mathrm{dom}(\varphi(\Gamma_\xi)) = \mathrm{dom}(\Gamma_\xi^n) = \mathrm{dom}(\sum_{j=0}^{n} a_j\Gamma_\xi^j) :$$

Theorem 1.3.7. *The operator $\sum_{j=0}^{n} a_j\Gamma_\xi^j$ is selfadjoint in $L_2(S^m;V)$ and has pure point spectrum*

$$\sigma_p(\sum_{j=0}^{n} a_j\Gamma_\xi^j) = \{\lambda : \lambda = \sum_{j=0}^{n} a_j\mu^j, \quad \mu \in \sigma_p(\Gamma_\xi)\}.$$

1.4 An integral formula for inner spherical monogenics

Let $k \in \mathbf{N}$ be fixed. Then we saw in Remark 1.2 (3) that the operator CK determines an isomorphism between $\mathcal{P}(m;k;V)$ and $M^+(k;V)$.

Putting for $R_k \in \mathcal{P}(m;k;V), CK(R_k)(x) = P_k(x)$, we now derive a representation of $P_k(x)$ in terms of so-called *plane waves.*

By means of the Fischer decomposition of $R_k \in \mathcal{P}(m;k;V)$ into solid spherical harmonics (see [29]) we have

$$R_k(\underline{x}) = |\underline{x}|^k \sum_{\ell=0}^{[k/2]} S_{k-2\ell}(\underline{\xi}).$$

The Hecke-Funk Theorem (see [40]) now tells us that

$$\begin{aligned}
&|x|^k \int_{S^{m-1}} (<\underline{\xi},\underline{\omega}>)^k S_{k-2\ell}(\underline{\omega}) d\Sigma(\underline{\omega} \\
&= |\underline{x}|^k A_{m-1} S_{k-2\ell}(\underline{\xi}) \int_{-1}^1 s^k P_{k-2\ell,m}(s)(1-s^2)^{\frac{m-3}{2}} ds \\
&= |\underline{x}|^k S_{k-2\ell}(\underline{\xi}) N_{k-2\ell,k}.
\end{aligned}$$

Here

$$\begin{aligned}
N_{k-2\ell,k} &= A_{m-1} \int_{-1}^1 s^k P_{k-2\ell,m}(s)(1-s^2)^{\frac{m-3}{2}} ds \\
&= A_{m-1}(\frac{1}{2})^{k-2\ell} \frac{\Gamma(\frac{m-1}{2}k!}{\Gamma(k-2\ell+\frac{m+1}{2})(2\ell)!} \int_{-1}^1 s^{2\ell}(1-s^2)^{k-2\ell+\frac{m-3}{2}} ds \\
&= \frac{k!\pi^{\frac{m-1}{2}}\Gamma(\ell+\frac{1}{2})}{(2\ell)!2^{k-2\ell-1}\Gamma(k-\ell+\frac{m}{2})}
\end{aligned}$$

and $P_{k-2\ell,m}$ is the Legendre polynomial of degree $(k-2\ell)$ in m dimensions.

We thus obtain by putting

$$\eta_k(R_k(\underline{x})) = |\underline{x}|^k \sum_{\ell=0}^{[k/2]} \frac{S_{k-2\ell}(\underline{\xi})}{N_{k-2\ell,k}},$$

and $\underline{x} = |\underline{x}|\underline{\xi}$ that, as for each $\underline{\omega} \in S^{m-1}$ and $k \in \mathbf{N}$,

$$\begin{aligned}
(<\underline{x},\underline{\omega}>)^k &= |\underline{x}|^k(<\underline{\xi},\underline{\omega}>)^k, \\
R_k(\underline{x}) &= \int_{S^{m-1}} (<\underline{x},\underline{\omega}>)^k \eta_k(R_k(\underline{\omega})) d\Sigma(\underline{\omega}).
\end{aligned}$$

By virtue of Example (2) in 0.2.4, we thus get the following integral representation for the CK-extension $P_k(x)$ of $R_k(\underline{x})$:

$$P_k(x) = \int_{S^{m-1}} (<\underline{x},\underline{\omega}> - \underline{\omega} e_0 x_0)^k \eta_k(R_k(\underline{\omega})) d\Sigma(\underline{\omega}).$$

1.5 A basis for $M^+(k;V)$ and $M^-(k;V)$

In this section particular bases for $M^+(k;V)$ and $M^-(k;V)$ will be constructed.

Let, as usual, when $\underline{\alpha} = (\underline{\alpha}_1, \ldots, \underline{\alpha}_m) \in \mathbf{N}^m$ is a multiindex, $\underline{x}^\alpha, \partial^{\underline{\alpha}}, \underline{\alpha}!$ and $|\underline{\alpha}|$ stand form respectively,

$$\begin{aligned} \underline{x}^\alpha &= x_1^{\alpha_1} x_2^{\alpha_2} \ldots x_m^{\alpha_m}, \quad \underline{x} \in \mathbf{R}^m; \\ \partial^{\underline{x}} &= \partial_{x_1}^{\alpha_1} \partial_{x_2}^{\alpha_2} \ldots \partial_{x_m}^{\alpha_m}; \\ \underline{\alpha}! &= \alpha_1! \alpha_2! \ldots \alpha_m!; \\ |\underline{\alpha}| &= \sum_{j=1}^{m} \alpha_j. \end{aligned}$$

I. The case $M^+(k;V)$

Definition 1.5.1. *Let $\underline{\alpha} \in \mathbf{N}^m$ with $|\underline{\alpha}| = k$. Then we put*

$$V_{\underline{\alpha}}(x) = CK\left(\frac{1}{\underline{\alpha}!}\underline{x}^{\underline{\alpha}}\right).$$

Remarks 1.5.2.

(1) By the definition itself of the CK-extension, we have that

$$\begin{aligned} V_{\underline{\alpha}}(x) &= \sum_{j=0}^{\infty} \frac{(-1)^j x_0^j}{j!}[(\bar{e}_0 \partial_{\underline{x}})^j \underline{x}^{\underline{\alpha}}] \\ &= \sum_{j=0}^{|\underline{\alpha}|} \frac{(-1)^j x_0^j}{j!}[(\bar{e}_0 \partial_{\underline{x}})^j \underline{x}^{\underline{\alpha}}] \end{aligned}$$

whence $V_{\underline{\alpha}}(x)$ is a homogeneous (left) monogenic polynomial of degree $|\underline{\alpha}| = k$.

(2) As $\underline{x}^{\underline{\alpha}}$ is $\mathbf{R}$-valued, we obtain that $V_{\underline{\alpha}}(x)$ takes its values in

$$\text{span}_{\mathbf{R}}\{1, \bar{e}_0 e_i : i = 1, \ldots, m\}.$$

Moreover, as in the case of the Cauchy-Riemann operator, for all $j = 0, 1, \ldots,$ and $|\underline{\alpha}| = k$,

$$D_{\underline{x}}^j \underline{x}^{\underline{\alpha}} = \underline{x}^{\underline{\alpha}} D_{\underline{x}}^j$$

we see immediately that $V_{\underline{\alpha}}(x)$ is both left and right monogenic then.

As to the case of the Dirac operator, notice that

$$\begin{aligned} \overline{V}_{\underline{\alpha}}(x) &= \sum_{j=0}^{|\alpha|} \frac{(-1)^j x_0^j}{j!}[(\overline{\partial}_{\underline{x}}\overline{\overline{e}}_0)^j \underline{x}^{\underline{\alpha}}] \\ &= \sum_{j=0}^{|\alpha|} \frac{(-1)^j x_0^j}{j!}[\underline{x}^{\underline{\alpha}}(\partial_{\underline{x}}\overline{e}_0)^j] \end{aligned}$$

is the right monogenic CK-extension of $\underline{x}^{\underline{\alpha}}$ or $\overline{V}_{\underline{\alpha}}(x)$ is right monogenic then.

Theorem 1.5.3. *Let* $k \in \mathbf{N}$.

(i) The set $\{V_{\underline{\alpha}} : |\underline{\alpha}| = k\}$ *is a basis for* $M^+(k;C)$.

(ii) If $\{v_1, \dots, v_d\}$ *is a basis for* S, *then the set* $\{V_{\underline{\alpha}} v_j : |\underline{\alpha}| = k; j = 1, \dots, d\}$ *is a basis for* $M^+(k;S)\}$.

Proof.

(i) Since $\{\underline{x}^{\underline{\alpha}} : |\underline{\alpha}| = k\}$ is a basis for $\mathcal{P}(m;k;C)$ and CK is an isomorphism between $\mathcal{P}(m;k;C)$ and $M^+(k;C)$, the desired result follows immediately.

(ii) Is a direct consequence of (i). ■

Remarks 1.5.4.

(1) Following the notations of 0.2.4. Example (2), we put

$$\begin{aligned} z_\ell &= CK(x_\ell) \\ &= x_\ell - x_0 \overline{e}_0 e_\ell. \end{aligned}$$

(2) If $\underline{\alpha} = (\alpha_1, \dots, \alpha_m) \in \mathbf{N}^m$, we shall often use the notation $\underline{\alpha} = (\ell_1, \dots, \ell_k)$ for it where $k = |\underline{\alpha}|$ and $1 \le \ell_1 \le \ell_2 \le \dots \ell_k \le m$. The element $V_{\underline{\alpha}}$ will then be written as $V_{\ell_1 \dots \ell_k}$.

Taking $\underline{\alpha} \in \mathbf{N}^m$ fixed with $|\underline{\alpha}| = k$, and calling $S_{\underline{\alpha}}$ the set of all permutations of the sequence $\underline{\alpha}$, we put

$$V^\star_{\underline{\alpha}}(x) = \frac{1}{k!} \sum_{\pi \in S_{\underline{\alpha}}} z_{\pi(\underline{\alpha})}.$$

We now claim that $V^\star_{\underline{\alpha}}(x) = V_{\underline{\alpha}}(x)$ for all $\underline{\alpha} \in \mathbf{N}^m$.

Indeed, by straightforward calculations we obtain that $\partial_x V^\star_{\underline{\alpha}}(x) = 0$ so that, since moreover $V^\star_{\underline{\alpha}}(x)|_{x_0=0} = \frac{1}{\underline{\alpha}!}\underline{x}^{\underline{\alpha}}$, the desired result follows immediately.

(3) In order to write out any $P^* \in M^+(k;C)$ in terms of the basis $\{V_{\underline{\alpha}} : |\underline{\alpha}| = k\}$, notice first that $P^* = CK(P)$ for some (unique) $P \in \mathcal{P}(m;k;C)$. Putting

$$P(\underline{x}) = \sum_{|\alpha|=k} \frac{1}{\alpha!}\underline{x}^{\underline{\alpha}} a_{\underline{\alpha}}$$

with $a_{\underline{\alpha}} = \partial^{\underline{\alpha}} P$, we thus obtain that

$$P^*(x) = \sum_{|\alpha|=k} V_{\underline{\alpha}}(x) a_{\underline{\alpha}}$$

with

$$a_{\underline{\alpha}} = \partial^{\underline{\alpha}} P^* = \partial^{\underline{\alpha}} P.$$

(4) Notice that for all $\underline{\alpha}, \underline{\beta} \in \mathbf{N}^m$,

$$\partial^{\underline{\beta}} V_{\underline{\alpha}}(x)|_{x=0} = \delta_{\underline{\beta},\underline{\alpha}}.$$

(5) We saw in 0.2.4 Example that for $k \in \mathbf{N}$ given and $\underline{t} \in \mathbf{R}_0^m$ fixed, the function

$$p_k(x, \underline{t}) = (< \underline{x}, \underline{t} > -x_0 \overline{e}_0 \underline{t})^k$$

belongs to $M^+(k;C)$; it may thus be written as a linear combination of the basic polynomials $V_{\underline{\alpha}}, |\underline{\alpha}| = k$.

We claim that

$$p_k(x, \underline{t}) = k! \sum_{|\alpha|=k} V_{\underline{\alpha}}(x) \underline{t}^{\underline{\alpha}}$$

Indeed, both functions are monogenic. Moreover they have the same restriction to $\mathbf{R}^m$, namely

$$k! \sum_{|\underline{\alpha}|=k} \frac{1}{\alpha!} \underline{x}^{\alpha} \underline{t}^{\underline{\alpha}}.$$

In such a way, the function $p_k(x, \underline{t})$ is a generating function for the basic polynomials $V_{\underline{\alpha}}, |\underline{\alpha}| = k$.

Finally notice that, when setting

$$\begin{aligned} p(x, \underline{t}) &= \sum_{\ell=1}^{m} z_\ell t_\ell \\ &= < \underline{x}, \underline{t} > -x_0 \overline{e}_0 \underline{t}, \end{aligned}$$

we have, taking into account Remark (2), that

$$(p(x,\underline{t}))^k = k! \sum_{|\underline{\alpha}|=k} V_{\underline{\alpha}}(x)\underline{t}^{\underline{\alpha}}$$

or

$$p(x,\underline{t})^k = p_k(x,\underline{t}).$$

II. The case $M^-(k;V)$

Definition 1.5.5. *Let $\underline{\alpha} \in \mathbf{N}^m$ with $|\underline{\alpha}| = k$. Then we put*

$$W_{\underline{\alpha}}(x) = (-1)^{|\underline{\alpha}|}\partial^{\underline{\alpha}}E(x)$$

where $E(x) = \frac{1}{A_{m+1}}\frac{\overline{x}}{|x|^{m+1}}$ is the fundamental solution of ∂_x.

Again using the notation $\underline{\alpha} = (\ell_1,\ldots,\ell_k)$ for $\underline{\alpha} = (\alpha_1,\ldots,\alpha_m)$, where $|\alpha| = k$ and $1 \leq \ell_1 \leq_2 \leq \ldots \leq \ell_k \leq m$, we also put $W_{\underline{\alpha}}(x) = W_{\ell_1\ldots\ell_k}(x)$.

Remarks 1.5.6.

(1) By the definition itself we have that $W_{\underline{\alpha}}(x)$ is a homogeneous monogenic function of degree $-(m+k)$ in $\mathbf{R}_0^{m+1}$ or $W_{\underline{\alpha}}(x) \in M^-(k;C)$.

(2) As $E(x)$ is both left and right monogenic and takes its values in

$$\text{span}_{\mathbf{R}}\{e_0, e_1, \ldots, e_m\},$$

the same holds for $W_{\underline{\alpha}}(x)$.

Now we prove some orthogonality relations between the functions $V_{\underline{\alpha}}$ and $W_{\underline{\beta}}$ for $\underline{\alpha}, \underline{\beta} \in \mathbf{N}^m$.

Lemma 1.5.7. *Let $\underline{\alpha}, \underline{\beta} \in \mathbf{N}^m$ and let $R > 0$. Then*

(i) $\int_{\partial B(R)} W_{\underline{\beta}}(u)d\sigma_u V_{\underline{\alpha}}(u) = \delta_{\underline{\beta},\underline{\alpha}}$.

(ii) $\int_{\partial B(R)} W_{\underline{\alpha}}(u)d\sigma_u W_{\underline{\beta}}(u) = 0$.

Proof.

(i) Since $V_{\underline{\alpha}}$ is left monogenic we obtain by Cauchy's formula that for all $x \in \mathring{B}(R)$

$$V_{\underline{\alpha}}(x) = \int_{\partial B(R)} E(u-x)d\sigma_u V_{\underline{\alpha}}(u).$$

Hence for any $\underline{\beta} \in \mathbf{N}^m$ (see also 1.5.4 Remark (4))

$$\begin{aligned}\partial^{\underline{\beta}} V_{\underline{\alpha}}(x)|_{x=0} &= \int_{\partial B(R)} \partial^{\underline{\beta}} E(u-x)|_{x=0} d\sigma_u V_{\underline{\alpha}}(u) \\ &= \int_{\partial B(R)} W_{\underline{\beta}}(u) d\sigma_u V_{\underline{\alpha}}(u) \\ &= \delta_{\underline{\beta},\underline{\alpha}}.\end{aligned}$$

(ii) By Cauchy's Theorem and the fact that $W_{\underline{\alpha}}$ is right monogenic and $W_{\underline{\beta}}$ is left monogenic in $\mathbf{R}_0^{m+1}$, the left hand side in (ii) is independent of $R > 0$. Hence

$$\begin{aligned}&\int_{\partial B(R)} W_{\underline{\alpha}}(u) d\sigma_u W_{\underline{\beta}}(u) \\ &= \lim_{R\to\infty} \frac{1}{R^{|\underline{\alpha}|+|\underline{\beta}|+m}} \int_{S^m} W_{\underline{\alpha}}(\omega) d\sigma_\omega W_{\underline{\beta}}(\omega) \\ &= 0. \blacksquare\end{aligned}$$

Remark 1.5.8. Taking into account Remark 1.5.2(2) and using either Cauchy's Formula or Cauchy's Theorem we also obtain the following orthogonality relations.

(1) Case of the Dirac operator
For all $\underline{\alpha}, \underline{\beta} \in \mathbf{N}^m$ and $R > 0$

$$\begin{aligned}\int_{\partial B(R)} \overline{V}_{\underline{\alpha}}(u) d\sigma_u W_{\underline{\beta}}(u) &= \delta_{\underline{\alpha},\underline{\beta}} ; \\ \int_{\partial B(R)} \overline{V}_{\underline{\alpha}}(u) d\sigma_u V_{\underline{\beta}}(u) &= 0.\end{aligned}$$

(2) Case of the Cauchy-Riemann operator
For all $\underline{\alpha}, \underline{\beta} \in \mathbf{N}^m$ and $R > 0$

$$\int_{\partial B(R)} V_{\underline{\alpha}}(u) d\sigma_u V_{\underline{\beta}}(u) = 0.$$

Theorem 1.5.9.

(i) The set $\{W_{\underline{\alpha}} : |\underline{\alpha}| = k\}$ *is a basis for* $M^-(k; C)$.

(ii) If $\{v_1, \ldots, v_d\}$ *is a basis for* S, *then the set* $\{W_{\underline{\alpha}} v_j : |\underline{\alpha}| = k; j = 1, \ldots, d\}$ *is a basis for* $M^-(k; S)$.

Proof.

(i) Let us first show that the set $\{W_{\underline{\alpha}} : |\underline{\alpha}| = k\}$ is linearly independent in the right $\mathcal{C}$-module $M^-(k;\mathcal{C})$. Suppose

$$\sum_{|\underline{\alpha}|=k} W_{\underline{\alpha}}(x) a_{\underline{\alpha}} = 0.$$

Then, for any $\underline{\beta} \in \mathbf{N}^m$ with $|\underline{\beta}| = k$, we obtain by 1.5.8 (1) that for all $R > 0$

$$\begin{aligned} 0 &= \sum_{|\underline{\alpha}|=k} \int_{\partial B(R)} \overline{V}_{\underline{\beta}}(x) d\sigma_x W_{\underline{\alpha}}(x) a_{\underline{\alpha}} \\ &= a_{\underline{\beta}}. \end{aligned}$$

As $\dim M^-(k;\mathcal{C}) = \frac{(k+m-1)!}{k!(m-1)!}$ is exactly the number of elements $W_{\underline{\alpha}}$, $|\underline{\alpha}| = k$, we indeed find that $\{W_{\underline{\alpha}} : |\underline{\alpha}| = k\}$ is a basis for $M^-(k;\mathcal{C})$.

(ii) Follows immediately from (i). ■

1.6 Laurent and Taylor series

In this subsection we study the local behaviour of monogenic functions. As in classical complex analysis, Cauchy's Integral Formula and series expansions of the Cauchy kernel play a preponderant role in this study.

In what follows all results are worked out explicitly for the case of the Dirac operator. When relevant, some results are added which are typical for the case of the Cauchy-Riemann operator.

Furthermore, free use will be made of results concerning Gegenbauer polynomials, for which we refer to [40], [28].

1.6.1 Expansions for the Cauchy-kernel

The Cauchy kernel for the operator ∂_x is given by (see 0.2.2)

$$\begin{aligned} E(y-x) &= \frac{1}{A_{m+1}} \frac{\bar{y}-\bar{x}}{|y-x|^{m+1}} \\ &= \frac{1}{A_{m+1}} \frac{1}{m-1} \bar{\partial}_x \frac{1}{|x-y|^{m-1}} \\ &= \frac{1}{A_{m+1}} \frac{1}{1-m} \bar{\partial}_y \frac{1}{|x-y|^{m-1}}. \end{aligned}$$

Furthermore, as is well-known, for $\nu \in \mathbf{C}$ with $Re\nu > -\frac{1}{2}$ we have that for all $x, t \in \mathbf{R}$

$$(1 - 2tx + x^2)^{-\nu} = \sum_{k=0}^{\infty} C_k^x(t)x^k$$

where C_k^ν is the Gegenbauer polynomial of degree k associated to ν.

Consequently, putting for $x, y \in \mathbf{R}^{m+1}$, $x = |x|\xi$, $y = |y|\omega$, $r = |x|/|y|$ and $t =< \xi, \omega >$, we have for

$$\frac{1}{|x-y|^{m-1}} = \frac{1}{|y|^{m-1}} \frac{1}{(1 - 2tr + r^2)^{\frac{m-1}{2}}}$$

that for $|x| < |y|$

$$\frac{1}{|x-y|^{m-1}} = \sum_{k=0}^{\infty} \frac{|x|^k}{|y|^{m+k-1}} C_k^{(m-1)/2}(t). \tag{1.3}$$

(1) A first Expansion

Let $y \in \mathbf{R}_0^{m+1}$ be fixed.
Then the series (1.4) converges normally in $\{x \in \mathbf{R}^{m+1} : |x| \leq (1-\varepsilon)|y|\}$, $0 < \varepsilon < 1$ being arbitrary, its sum is analytic in $|x| < |y|$ and for each $k \in \mathbf{N}$, the term

$$\frac{|x|^k}{|y|^{k+m-1}} C_k^{(m-1)/2}(t)$$

is a $\mathbf{C}$-valued homogeneous harmonic polynomial of degree k in x.

Now let us calculate

$$\bar{\partial}_x\left(C_{k+1}^{(m-1)/2}(t)|x|^{k+1}\right) \quad , \quad k \geq 0.$$

Passing to polar coordinates $x = r\xi$ and $y = \rho\omega$ we get

$$\begin{aligned} &\partial_x(C_{k+1}^{(m-1)/2}(t)|x|^{k+1}) \\ &= \xi((\partial_r + \frac{1}{r}\Gamma_\xi)(r^{k+1}C_{k+1}^{(m-1)/2}(t)) \\ &= r^k\xi[(k+1)C_{k+1}^{(m-1)/2}(t) + \Gamma_\xi(C_{k+1}^{(m-1)/2}(t))]. \end{aligned} \tag{1.4}$$

As $\frac{d}{dt}C_k^\nu(t) - 2\nu C_{k-1}^{\nu+1}(t)$, we obtain that

$$\Gamma_\xi[C_{k+1}^{(m-1)/2}(t)] = (m-1)C_k^{(m+1)/2}(t)\Gamma_\xi[t].$$

But, as

$$\Gamma_\xi = \bar{e_0}(x_0\partial_{\underline{x}} - \underline{x}\partial_{x_0}) - \underline{x} \wedge \partial_{\underline{x}}$$

and

$$\begin{aligned}\Gamma_\xi[t] &= \Gamma_\xi < \xi, \omega > \\ &= \frac{1}{r\rho}\Gamma_\xi < x, y >\end{aligned}$$

we obtain

$$\begin{aligned}\Gamma_\xi[t] &= \frac{1}{r\rho}(\bar{e_0}(x_0\partial_{\underline{x}} - \underline{x}\partial_{x_0}) + \sum_{i<j} e_{ij}(x_j\partial_{x_i} - x_i\partial_{x_j}))(x_0y_0 + < \underline{x}, \underline{y} >) \\ &= \frac{1}{r\rho}(\bar{e_0}(x_0\underline{y} - y_0\underline{x}) + \underline{y}\wedge\underline{x}) \\ &= \bar{e_0}(\xi_0\underline{\omega} - \omega_0\underline{\xi}) + \underline{\omega}\wedge\underline{\xi}. \qquad (1.5)\end{aligned}$$

Consequently (1.4) and (1.5) yield

$$\begin{aligned}&\partial_x(C_{k+1}^{(m-1)/2}(t)|x|^{k+1}) \\ &\quad = r^k\xi[(k+1)C_{k+1}^{(m-1)/2}(t) + (m-1)C_k^{(m+1)/2}(t)(\bar{e_0}(\xi_0\underline{\omega} - \omega_0\underline{\xi}) + \underline{\omega}\wedge\underline{\xi})]\end{aligned}$$

and so

$$\begin{aligned}&\bar{\partial}_x(C_{k+1}^{(m-1)/2}(t)|x|^{k+1}) \\ &\quad = |x|^k[(k+1)C_{k+1}^{(m-1)/2}(t) + (1-m)C_k^{(m+1)/2}(t)((\xi_0\underline{\omega} - \omega_0\underline{\xi})e_0 + \underline{\omega}\wedge\underline{\xi})]\bar{\xi}.\end{aligned}$$

Putting for each $k \in \mathbf{N}$ and $\xi, \omega \in S^m$

$$\begin{aligned}&C_{m+1,k}^-(\omega, \xi) \\ &\quad = \frac{1}{m-1}\Big[(k+1)C_{k+1}^{(m-1)/2}(t) \\ &\qquad\qquad + (1-m)C_k^{(m+1)/2}(t)((\xi_0\underline{\omega} - \omega_0\underline{\xi})e_0 + \underline{\omega}\wedge\underline{\xi})\Big] \qquad (1.6)\end{aligned}$$

we get, by letting act $\frac{1}{m-1}\bar{\partial}_x$ on both sides of (1.3)

$$\begin{aligned}\frac{\bar{y} - \bar{x}}{|y-x|^{m+1}} &= \frac{1}{m-1}\sum_{k=0}^{\infty}\frac{1}{|y|^{m+k}}\bar{\partial}_x(C_{k+1}^{(m-1)/2}(t)|x|^{k+1}) \\ &= \sum_{k=0}^{\infty}\frac{|x|^k}{|y|^{m+k}}C_{m+1,k}^-(\omega, \xi)\bar{\xi}. \qquad (1.7)\end{aligned}$$

Notice here that for each $k \in \mathbf{N}$ and $y \in \mathbf{R}_0^{m+1}$ fixed

$$\frac{|x|^k}{|y|^{m+k}}C_{m+1,k}^-(\omega, \xi)\bar{\xi} \qquad (1.8)$$

is the $\bar{\partial}_x$-derivative of a homogeneous **C**-valued harmonic polynomial of degree $k+1$, whence it is a homogeneous (left and right) monogenic polynomial of degree k in $x \in \mathbf{R}^{m+1}$ for $y \in \mathbf{R}_0^{m+1}$ fixed.

(2) A second expansion

Let $x \in \mathbf{R}^{m+1}$ be fixed.
Then the series (1.4), together with each of its derived series, converges normally on each closed annular domain of the form $R' \leq |y| \leq R''$ where $R' > 0$ is taken in such a way that $R' > |x|$.

Let us now calculate

$$\bar{\partial}_y\Big(C_k^{(m-1)/2}(t)\frac{1}{|y|^{m+k-1}}\Big).$$

Then, passing to polar coordinates $x = r\xi$ and $y = \rho\omega$, similar computations as effected in the preceding case yield

$$\partial_y\Big(C_k^{(m-1)/2}(t)\frac{1}{|y|^{m+k-1}}\Big)$$

$$= \frac{\omega}{\rho^{m+k}}[-(m+k-1)C_k^{(m-1)/2}(t)$$

$$+ (m-1)C_{k-1}^{(m+1)/2}(t)(\bar{e}_0(\omega_0\underline{\xi} - \xi_0\underline{\omega}) + \underline{\xi}\wedge\underline{\omega})]$$

whence

$$\bar{\partial}_y\Big(C_k^{(m-1)/2}(t)\frac{1}{|y|^{m+k-1}}\Big)$$

$$= \frac{1}{|y|^{m+k}}[-(m+k-1)C_k^{(m-1)/2}(t)$$

$$+ (1-m)C_{k-1}^{(m+1)/2}(t)((\omega_0\underline{\xi} - \xi_0\underline{\omega})e_0 + \underline{\xi}\wedge\underline{\omega})]\bar{\omega}.$$

Putting for each $k \in \mathbf{N}$ and $\xi, \omega \in S^m$

$$C_{m+1,k}^{+}(\xi,\omega)$$

$$= \frac{1}{1-m}[-(m+k-1)C_k^{(m-1)/2}(t)$$

$$+ (1-m)C_{k-1}^{(m+1)/2}(t)((\omega_0\underline{\xi} - \xi_0\underline{\omega})e_0 + \underline{\xi}\wedge\underline{\omega})]$$

we get, by letting act $\frac{1}{1-m}\bar{\partial}_y$ on both sides of (1.3)

$$\frac{\bar{y}-\bar{x}}{|y-x|^{m+1}} = \frac{1}{1-m}\sum_{k=0}^{\infty}|x|^k\bar{\partial}_y\Big(C_k^{(m-1)/2}(t)\frac{1}{|y|^{m+k-1}}\Big)$$

$$= \sum_{k=0}^{\infty} \frac{|x|^k}{|y|^{m+k}} C^+_{m+1,k}(\xi,\omega)\bar{\omega}. \tag{1.9}$$

Now we claim that for each $k \in \mathbf{N}$ and $y \in \mathbf{R}_0^{m+1}$ fixed,

$$\frac{|x|^k}{|y|^{m+k}} C^+_{m+1,k}(\xi,\omega)\bar{\omega} \tag{1.10}$$

is a homogeneous polynomial of degree k in x.

Indeed, as is well known,

$$C_k^\nu(t) = \sum_{j=0}^{[k/2]} \frac{(-1)^j(\nu)_{k-j}}{j!(k-2j)!}(2t)^{k-2j}.$$

Consequently

$$\begin{aligned} |x|^k C_k^{(m-1)/2}(t) &= |x|^k \sum_{j=0}^{[k/2]} \frac{(-1)^j(\frac{m-1}{2})_{k-j}}{j!(k-2j)!}(2 < \xi,\omega >)^{k-2j} \\ &= \sum_{j=0}^{[k/2]} \frac{(-1)^j(\frac{m-1}{2})_{k-j}}{j!(k-2j)!}|x|^{2j}(2 < x,\omega >)^{k-2j} \end{aligned}$$

where for each j, $|x|^{2j}(2 < x,\omega >)^{k-2j}$ is a homogeneous (scalar valued) polynomial of degree k in $\mathbf{R}^{m+1}$.

The same is true for

$$\begin{aligned} &|x|^k((\omega_0\underline{\xi} - \xi_0\underline{\omega})e_0 + \underline{\xi} \wedge \underline{\omega})C_{k-1}^{(m+1)/2}(t) \\ &= ((\omega_0\underline{x} - x_0\underline{\omega})e_0 + \underline{x} \wedge \underline{\omega})(|x|^{k-1}C_{k-1}^{(m+1)/2}(t)) \end{aligned}$$

since $(\omega_0\underline{x} - x_0\underline{\omega})e_0 + \underline{x} \wedge \underline{\omega}$ is homogeneous of degree 1 and $|x|^{k-1}C_{k-1}^{(m+1)/2}(t)$ is homogeneous of degree $k-1$.

Moreover, as for $y \in \mathbf{R}_0^{m+1}$ fixed, the series of homogeneous polynomials given by the right hand side of (1.7) converges normally in $\{x \in \mathbf{R}^{m+1} : |x| \leq (1-\epsilon)|y|\}$, its sum being the monogenic function

$$\frac{\bar{y} - \bar{x}}{|y-x|^{m+1}}$$

in $|x| < |y|$, for each $k \in \mathbf{N}$,

$$\frac{|x|^k}{|y|^{m+k}} C^+_{m+1,k}(\xi,\omega)\bar{\omega}$$

is a homogeneous (left and right) monogenic polynomial of degree k in $x \in \mathbf{R}^{m+1}$ for $y \in \mathbf{R}_0^{m+1}$ fixed.

From (1.7) and (1.9) it thus follows that for all $x \in \mathbf{R}^{m+1}$ and $y \in \mathbf{R}_0^{m+1}$,

$$\frac{|x|^k}{|y|^{m+k}} C^-_{m+1,k}(\omega,\xi)\bar{\xi} = \frac{|x|^k}{|y|^{m+k}} C^+_{m+1,k}(\xi,\omega)\bar{\omega} \tag{1.11}$$

i.e. for all $\xi, \omega \in S^m$,

$$C^-_{m+1,k}(\omega,\xi)\bar{\xi} = C^+_{m+1,k}(\xi,\omega)\bar{\omega}. \tag{1.12}$$

(3) A third expansion

As for $y \in \mathbf{R}_0^{m+1}$ fixed, the function

$$\frac{\bar{y} - \bar{x}}{|y - x|^{m+1}}$$

is analytic in $|x| < |y|$, we have that its Taylor series at the origin is given by

$$\frac{\bar{y} - \bar{x}}{|y - x|^{m+1}} = \sum_{k=0}^{\infty} \frac{(-1)^k}{k!} (< x, \partial_y >)^k \frac{\bar{y}}{|y|^{m+1}} \tag{1.13}$$

where $< x, \partial_y >= \sum_{i=o}^m x_i \partial_{y_i}$.

We thus obtain from (1.7) and (1.13) that for each $k \in \mathbf{N}$, $x \in \mathbf{R}^{m+1}$ and for each $y \in \mathbf{R}_0^{m+1}$,

$$\frac{|x|^k}{|y|^{m+k}} C^-_{m+1,k}(\omega,\xi)\bar{\xi} \frac{(-1)^k}{k!} (< x, \partial_y >)^k \frac{\bar{y}}{|y|^{m+1}}. \tag{1.14}$$

As for $x \in \mathbf{R}^{m+1}$ fixed, the right hand side of (1.14) is (left and right) monogenic in $\mathbf{R}_0^{m+1}$, the same is true for the left hand side and hence also, by means of (1.11) , for

$$\frac{|x|^k}{|y|^{m+k}} C^+_{m+1,k}(\xi,\omega)\bar{\omega}.$$

But now for all $x \in \mathbf{R}^{m+1}$ and $y \in \mathbf{R}_0^{m+1}$,

$$\begin{aligned} &\frac{1}{A_{m+1}} \frac{(-1)^k}{k!} (< x, \partial_y >)^k \frac{\bar{y}}{|y|^{m+1}} \\ &= \frac{1}{A_{m+1}} \frac{(-1)^k}{k!} \sum_{l_1=0}^{m} \cdots \sum_{l_k=0}^{m} x_{l_1} \ldots x_{l_k} \partial_{y_{l_1}} \ldots \partial_{y_{l_k}} \frac{\bar{y}}{|y|^{m+1}} \\ &= \frac{(-1)^k}{k!} \sum_{l_1=0}^{m} \cdots \sum_{l_k=0}^{m} x_{l_1} \ldots x_{l_k} \partial_{y_{l_1}} \ldots \partial_{y_{l_k}} E(y) \end{aligned}$$

so that the restriction to $\mathbf{R}^m$ (i.e. by putting $x_0 = 0$) becomes

$$\frac{1}{k!}\sum_{l_1=1}^{m}\cdots\sum_{l_k=1}^{m} x_{l_1}\ldots x_{l_k} W_{l_1\ldots l_k}(y)$$

which is equal to the restriction to $\mathbf{R}^m$ of respectively

$$\sum_{|\underline{\alpha}|=k} V_{\underline{\alpha}}(x)W_{\underline{\alpha}}(y)$$

and

$$\sum_{|\underline{\alpha}=k} W_{\underline{\alpha}}(y)\overline{V}_{\underline{\alpha}}(x).$$

Hence, by the Cauchy-Kovalevska extension theorem

$$\begin{aligned} \frac{1}{A_{m+1}}\frac{|x|^k}{|y|^{m+k}}C^-_{m+1,k}(\omega,\xi)\bar{\xi} &= \frac{1}{A_{m+1}}\frac{|x|^k}{|y|^{m+k}}C^+_{m+1,k}(\xi,\omega)\bar{\omega} \\ &= \sum_{|\underline{\alpha}=k} V_{\underline{\alpha}}(x)W_{\underline{\alpha}}(y). \\ &= \sum_{|\underline{\alpha}|=k} W_{\underline{\alpha}}(y)\overline{V}_{\underline{\alpha}}(x). \qquad (1.15) \end{aligned}$$

Consequently, in $|x|<|y|$ we have that

$$\begin{aligned} E(y-x) &= \sum_{k=0}^{\infty}\Big(\sum_{|\underline{\alpha}|=k} V_{\underline{\alpha}}(x)W_{\underline{\alpha}}(y)\Big) \\ &= \sum_{k=0}^{\infty}\Big(\sum_{\underline{\alpha}=k} W_{\underline{\alpha}}(y)\overline{V}_{\underline{\alpha}}(x)\Big) \qquad (1.16) \end{aligned}$$

Remark. In the case of the Cauchy-Riemann operator we find that in $|x|<|y|$ (see also 1.5.2 Remark (2)) :

$$\begin{aligned} E(y-x) &= \sum_{k=0}^{\infty}\Big(\sum_{|\underline{\alpha}|=k} V_{\underline{\alpha}}(x)W_{\underline{\alpha}}(y)\Big) \\ &= \sum_{k=0}^{\infty}\Big(\sum_{|\underline{\alpha}|=k} W_{\underline{\alpha}}(y)V_{\underline{\alpha}}(x)\Big) \end{aligned}$$

and this since then

$$\begin{aligned} \frac{1}{A_{m+1}}\frac{|x|^k}{|y|^{m+k}}C^-_{m+1,k}(\omega,\xi)\bar{\xi} &= \frac{1}{A_{m+1}}\frac{|x|^k}{|y|^{m+k}}C^+_{m+1,k}(\xi,\omega)\overline{\omega} \\ &= \sum_{|\alpha|=k} V_{\underline{\alpha}}(x)W_{\underline{\alpha}}(y) \\ &= \sum_{|\alpha|=k} W_{\underline{\alpha}}(y)V_{\underline{\alpha}}(x). \end{aligned}$$

We end this subsection with some relations concerning the Cauchy kernel

$$\frac{\bar{\omega}-r\bar{\xi}}{|\omega-r\xi|^{m+1}}$$

considered as a function in $L_2(S^m \times S^m)$, where either
$0<r<1$ or $r>1$.

Let $x=r\xi\in \mathring{B}(1)$ and $y=\omega\in S^m$. Then, referring to (1.7) we have

$$\frac{\bar{\omega}-r\bar{\xi}}{|\omega-r\xi|^{m+1}}=\sum_{k=0}^{\infty} r^k C^-_{m+1,k}(\omega,\xi)\bar{\xi}.$$

As for each $\xi\in S^m$ fixed, $C^-_{m+1,k}(\omega,\xi)\bar{\xi}\in \mathcal{M}^-(k;C)$ we find for $j\neq k$ that by virtue of Theorem 1.3.2

$$\begin{aligned}&\int_{S^m\times S^m}\overline{C^-_{m+1,k}(\omega,\xi)\bar{\xi}}C^-_{m+1,j}(\omega,\xi)\bar{\xi}d\Sigma(\omega)d\Sigma(\xi)\\ &= \int_{S^m}\xi\left[\int_{S^m}\overline{C^-_{m+1,k}(\omega,\xi)}C^-_{m+1,j}(\omega,\xi)d\Sigma(\omega)\right]\bar{\xi}d\Sigma(\xi)d\xi\\ &= 0.\end{aligned}$$

Consequently for each $0<r<1$

$$\left\|\frac{\bar{\omega}-r\xi}{|\omega-r\xi|^{m+1}}\right\|^2_{L_2}(S^m\times S^m)=\sum_{k=0}^{\infty} r^{2k}\|C^-_{m+1,k}(\omega,\xi)\bar{\xi})\|^2_{L_2(S^m\times S^m)}. \tag{1.17}$$

Analogously, as for each $\omega\in S^m$ fixed, $C^+_{m+1,k}(\xi,\omega)\bar{\omega}\in\mathcal{M}^+(k;C)$, we find that for $r>1$

$$\begin{aligned}&\left\|\frac{\bar{\omega}-r\xi}{|\omega-r\xi|^{m+1}}\right\|^2_{L_2(S^m\times S^m)}\\ &= \sum_{k=0}^{\infty}\frac{1}{r^{2k}}\|C^+_{m+1,k}(\xi,\omega)\bar{\omega}\|^2_{L_2(S^m\times S^m)}.\end{aligned} \tag{1.18}$$

1.6.2 The Cauchy Transform in $L_2(\partial B(R);V)$

Let $R>0$ be given, let $f\in L_2(\partial B(R);V)$ and let $\hat{f}$ be the function given by

$$\begin{aligned}\hat{f}(x) &= \frac{1}{A_{m+1}}\int_{\partial B(R)}\frac{\bar{y}-\bar{x}}{|y-x|^{m+1}}d\sigma_y f(y)\\ &= \frac{1}{A_{m+1}}\int_{S^m}\frac{\bar{\omega}-\bar{x}/R}{|\omega-x/R|^{m+1}}d\sigma_\omega f(R\omega).\end{aligned} \tag{1.19}$$

Then clearly $\hat{f}$ is left monogenic in $\mathbf{R}^{m+1}\backslash\partial B(R)$ and $\lim_{x\to\infty}\hat{f}(x)=0$; it is called the *Cauchy transform* of f.

Using the expansion (1.9) we obtain for all $x \in \mathring{B}(R)$:

$$\begin{aligned}\hat{f}(x) &= \frac{1}{A_{m+1}} \sum_{k=0}^{\infty} \frac{|x|^k}{R^k} \int_{S^m} C^+_{m+1,k}(\xi,\omega) f(R\omega) d\Sigma(\omega) \\ &= \sum_{k=0}^{\infty} \mathbf{P}_k f(x) \qquad (1.20)\end{aligned}$$

where for each $k \in \mathbf{N}$

$$\mathbf{P}_k f(x) = \frac{1}{A_{m+1}} \frac{|x|^k}{R^k} \int_{S^m} C^+_{k,m+1}(\xi,\omega) f(R\omega) d\Sigma(\omega) \qquad (1.21)$$

is an inner spherical monogenic of order k, i.e.

$$\mathbf{P}_k f \in M^+(k; V).$$

Moreover the convergence of the series (1.2.1) is normal on each $\bar{B}(r)$ with $0 < r < R$.

Using the expansion (1.7) where we interchange the role of x and y, we obtain for all $x \in \mathbf{R}^{m+1} \backslash \bar{B}(R)$.

$$\begin{aligned}\hat{f}(x) &= \frac{-1}{A_{m+1}} \sum_{k=0}^{\infty} \frac{R^{m+k}}{|x|^{m+k}} \int_{S^m} C^-_{m+1,k}(\xi,\omega) f(R\omega) d\Sigma(\omega) \\ &= -\sum_{k=0}^{\infty} \mathbf{Q}_k f(x) \qquad (1.22)\end{aligned}$$

where for each $k \in \mathbf{N}$

$$\mathbf{Q}_k f(x) = \frac{1}{A_{m+1}} \frac{R^{m+k}}{|x|^{m+k}} \int_{S^m} C^-_{m+1,k}(\xi,\omega) f(R\omega) d\Sigma(\omega) \qquad (1.23)$$

is an outer spherical monogenic of order k, i.e. $\mathbf{Q}_k f \in M^-(k; V)$.

Moreover the convergence of the series (1.2.3) is normal on each annular domain $\mathbf{R}^{m+1} \backslash \mathring{B}(R_1)$ with $R < R_1$.

Summarizing we thus have, using the notation $Y(t)$ for the Heaviside function, that in $\mathbf{R}^{m+1} \backslash \partial B(R)$

$$\hat{f}(x) = \sum_{k=0}^{\infty} Y(R - |x|) \mathbf{P}_k f(x) - \sum_{k=0}^{\infty} Y(|x| - R) \mathbf{Q}_k f(x). \qquad (1.24)$$

1.6.3 The Taylor expansion

Let $R > 0$ be given and let f be left monogenic in the open ball $\mathring{B}(R)$. Then for any $0 < r < R$ and any $x \in \mathring{B}(r)$, taking into account (1.7) and (1.12)

$$\begin{aligned}f(x) &= \frac{1}{A_{m+1}} \int_{\partial B(r)} \frac{\bar{y} - \bar{x}}{|y - x|^{m+1}} d\sigma_y f(y) \\ &= \sum_{k=0}^{\infty} \mathbf{P}_k f(x)\end{aligned}$$

with

$$\mathbf{P}_k f(x) = \frac{1}{A_{m+1}} \frac{|x|^k}{r^k} \int_{S^m} C^+_{m+1,k}(\xi, \omega) f(r\omega) d\Sigma(\omega).$$

Moreover the series converges normally on each $\bar{B}(r')$ with $0 < r' < r$.
As $0 < r < R$ may be chosen arbitrarily we obtain

Theorem 1. (Taylor expansion). *Let f be monogenic in $\mathring{B}(R)$. Then there exists a sequence $(\mathbf{P}_k f)_{k\in\mathbf{N}}$ of inner spherical monogenics such that in $\mathring{B}(R)$*

$$f(x) = \sum_{k=0}^{\infty} \mathbf{P}_k f(x) \tag{1.25}$$

where for each $k \in \mathbf{N}$

$$\mathbf{P}_k f(x) = \frac{1}{A_{m+1}} \frac{|x|^k}{r^k} \int_{S^m} C^+_{m+1,k}(\xi, \omega) f(r\omega) d\Sigma(\omega)$$

$0 < r < R$ being chosen arbitrarily. Moreover this series converges normally on each $\bar{B}((1-\varepsilon)R)$.

The preceding theorem enables us to prove immediately

Theorem 2. (Liouville's Theorem). *Let f be monogenic in $\mathbf{R}^{m+1}$.*

(i) If there exists a constant $C > 0$ and $k \in \mathbf{N}$ such that in $\mathbf{R}^{m+1}$

$$|f(x)| \leq C(1 + |x|)^k,$$

then f is a polynomial of degree k.

(ii) If there exists a constant $C > 0$ such that in $\mathbf{R}^{m+1}$ $|f(x)| \leq C$, then f is a constant function.

Remarks.

(1) By means of (1.13) and (1.15) we also have the following expression for the Taylor series of a monogenic function f in $\mathring{B}(R)$:

$$f(x) = \sum_{k=0}^{\infty} \mathbf{P}_k f(x)$$

where for $0 < r < R$ arbitrarily chosen

$$\mathbf{P}_k f(x) = \frac{1}{A_{m+1}} \frac{(-1)^k}{k!} \int_{\partial B(r)} (< x, \partial_y >)^k \frac{\bar{y}}{|y|^{m+1}}) d\sigma_y f(y).$$

In this way a direct generalization is obtained of the Taylor expansion for a function f which is holomorphic in the open disc $\mathring{B}(R)$ of the complex plane :

$$f(z)=\sum_{k=0}^{\infty} a_k z^k$$

where for $0<r<R$

$$a_k z^k = \frac{(-1)^k}{k!}\frac{1}{2\pi}\int_{\partial B(r')}\Big[(z\frac{d}{du})^k\frac{1}{u}\Big]\frac{du}{i}f(u).$$

(2) By virtue of (1.15) and (1.16) we also have that for $0<r<R$,

$$\mathbf{P}_k f(x)=\sum_{|\underline{\alpha}|=k} V_{\underline{\alpha}}(x)\int_{\partial B(r)} W_{\underline{\alpha}}(y)d\sigma_y f(y).$$

1.6.4 The Laurent expansion

Let $0<R_1<R_2$ be given and let f be left monogenic in the open annular domain $\mathring{B}(R_2)\backslash \bar{B}(R_1)$.

Then for any $0<R_1<r_1<r_2<R_2$ and any $x\in \mathring{B}(r_2)\backslash\bar{B}(r_1)$ we have

$$\begin{aligned} f(x) &= \frac{1}{A_{m+1}}\int_{\partial B(r_2)}\frac{\bar{y}-\bar{x}}{|y-x|^{m+1}}d\sigma_y f(y)\\ &= \frac{1}{A_{m+1}}\int_{\partial B(r_1)}\frac{\bar{y}-\bar{x}}{|y-x|^{m+1}}d\sigma_y f(y).\end{aligned}$$

Put

$$\tilde{f}_1(x)=\frac{1}{A_{m+1}}\int_{\partial B(r_2)}\frac{\bar{y}-\bar{x}}{|y-x|^{m+1}}d\sigma_y f(y)$$

and

$$\tilde{f}_2(x)=\frac{-1}{A_{m+1}}\int_{\partial B(r_1)}\frac{\bar{y}-\bar{x}}{|y-x|^{m+1}}d\sigma_y f(y).$$

Then $\tilde{f}_1$ is left monogenic in $\mathring{B}(r_2)$ and, as $0<R_1<r_2<R_2$ has been chosen arbitrarily, by means of Cauchy's Theorem, $\tilde{f}_1$ may be monogenically extended to $\mathring{B}(R_2)$; call this extension f_1. Taking account of (1.23), f_1 admits in $\mathring{B}(R_2)$ the expansion

$$f_1(x)=\sum_{k=0}^{\infty}\mathbf{P}_k f(x)$$

with

$$\mathbf{P}_k f(x) = \frac{1}{A_{m+1}} \frac{|x|^k}{r^k} \int_{S^m} C^+_{m+1,k}(\xi,\omega) f(r\omega) d\Sigma(\omega),$$

$0 < R_1 < r < R_2$ being taken arbitrarily. Moreover the series representing f_1 converges normally on each $\bar{B}(r_2)$ where $0 < r_2 < R_2$.

As to $\tilde{f}_2$, it is clearly a left monogenic function in the domain $\mathbf{R}^{m+1}\backslash\bar{B}(r_1)$ and, as $0 < R_1 < r_1 < R_2$ has been chosen arbitrarily, by means of Cauchy's Theorem, $\tilde{f}_2$ may be monogenically extended to $\mathbf{R}^{m+1}\backslash\mathring{B}(r_1)$; call this extension f_2.

Notice that $\lim_{x\to\infty} f_2(x) = 0$. Furthermore, taking account of (1.25) we have

$$f_2(x) = \sum_{k=0}^{\infty} \mathbf{Q}_k f(x)$$

with

$$\mathbf{Q}_k f(x) = \frac{1}{A_{m+1}} \frac{r^{m+k}}{|x|^{m+k}} \int_{S^m} C^-_{m+1,k}(\xi,\omega) f(r\omega) d\Sigma(\omega),$$

$0 < R_1 < r < R_2$ being taken arbitrarily.

Moreover the series representing f_2 converges normally on each $\bar{B}(r_1'')\backslash\mathring{B}(r_1')$ where $R_1 < r_1' < r_1''$.

Bringing together all these results we have

Theorem 1. (Laurent expansion). *Let f be left monogenic in the open annular domain $G = \mathring{B}(R_2)\backslash\bar{B}(R_1)$. Then*

(i) $f = f_1 + f_2$ in G with f_1 left monogenic in $\mathring{B}(R_2)$ and f_2 left monogenic in $\mathbf{R}^{m+1}\backslash\bar{B}(R_1)$. Moreover $\lim_{x\to\infty} f_2(x) = 0$.

(ii) There exist sequences $(\mathbf{P}_k f)_{k\in\mathbf{N}}$ and $(\mathbf{Q}_k f)_{k\in\mathbf{N}}$ of inner and outer spherical monogenics respectively such that in G

$$f(x) = \sum_{k=0}^{\infty} \mathbf{P}_k f(x) + \sum_{k=0}^{\infty} \mathbf{Q}_k f(x),$$

the convergence being normal on each closed annular domain $\overline{B}(r_2')\backslash\mathring{B}(r_1')$. Moreover

$$\mathbf{P}_k f(x) = \frac{1}{A_{m+1}} \frac{|x|^k}{r^k} \int_{S^m} C^+_{m+1,k}(\xi,\omega) f(r\omega) d\Sigma(\omega)$$

and

$$\mathbf{Q}_k f(x) = \frac{1}{A_{m+1}} \frac{r^{m+k}}{|x|^{m+k}} \int_{S^m} C^-_{m+1,k}(\xi,\omega) f(r\omega) d\Sigma(\omega),$$

$0 < R_1 < r < R_2$ being taken arbitrarily.

Theorem 2. (Laurent expansion outside a ball). *Let f be left monogenic in the domain $\mathbf{R}^{m+1}\backslash\bar{B}(R)$ such that $\lim_{x\to\infty} f(x) = 0$. Then there exists a sequence $(\mathbf{Q}_k f)_{k\in\mathbf{N}}$ of outer spherical monogenics such that*

$$f(x) = \sum_{k=0}^{\infty} \mathbf{Q}_k f(x)$$

with

$$\mathbf{Q}_k f(x) = \frac{1}{A_{m+1}} \frac{r^{m+k}}{|x|^{m+k}} \int_{S^m} C^-_{m+1,k}(\xi, \omega) f(r\omega) d\Sigma(\omega),$$

$r > R$ being taken arbitrarily.

Moreover the series above converges normally on each closed annular domain $\overline{B}(r_2)_V \mathbf{R}^m \backslash B(r_1)$ where $R < r_1 < r_2$.

Remarks.

(1) By means of (1.13) and (1.15) we also have the following expression for the Laurent series of a monogenic function f in the open annular domain $G = \mathring{B}(R_2) \backslash \bar{B}(R_1)$:

$$f(x) = \sum_{k=0}^{\infty} \mathbf{P}_k f(x) + \sum_{k=0}^{\infty} \mathbf{Q}_k f(x)$$

where for $R_1 < r < R_2$ arbitrarily chosen,

$$\mathbf{P}_k f(x) = \frac{1}{A_{m+1}} \frac{(-1)^k}{k!} \int_{\partial B(r)} (< x, \partial_y >)^k \frac{\bar{y}}{|y|^{m+1}}) d\sigma_y f(y)$$

and

$$\mathbf{Q}_k f(x) = \frac{1}{A_{m+1}} \frac{(-1)^k}{k!} \int_{\partial B(r)} \Big(< y, \partial_x >)^k \frac{\bar{x}}{|x|^{m+1}} \Big) d\sigma_y f(y).$$

In such a way an immediate analogy is obtained with the Laurent expansion for a function f which is holomorphic in an open annular domain $G = \mathring{B}(R_2) \backslash \bar{B}(R_1)$ of the complex plane :

$$f(z) = \sum_{k=0}^{+\infty} a_k z^k + \sum_{k=0}^{\infty} \frac{b_k}{z^{k+1}}$$

where for $R_1 < r < R_2$

$$a_k z^k = \frac{(-1)^k}{k!} \frac{1}{2\pi} \int_{\partial B(r)} \Big[\Big(z \frac{d}{du} \Big)^k \frac{1}{u} \Big] \frac{du}{i} f(u)$$

and

$$\frac{b_k}{z^{k+1}} = \frac{(-1)^k}{k!} \frac{1}{2\pi} \int_{\partial B(r)} \Big[(u \frac{d}{dz})^k \frac{1}{z} \Big] \frac{du}{i} f(u).$$

(2) As we already pointed out in 1.6.3 Remark 2, for $R_1 < r < R_2$,

$$\mathbf{P}_k f(x) = \sum_{|\underline{\alpha}|=k} V_{\underline{\alpha}}(x) \int_{\partial B(r)} W_{\underline{\alpha}}(y) d\sigma_y f(y).$$

By virtue of (1.16), where the roles of x and y are interchanged, we get that for $R_1 < r < R_2$

$$\mathbf{Q}_k f(x) = \sum_{|\underline{\alpha}|=k} W_{\underline{\alpha}}(x) \int_{\partial B(r)} \overline{V}_{\underline{\alpha}}(y) d\sigma_y f(y).$$

(3) In the case of the Cauchy-Riemann operator we have for $R_1 < r < R_2$ (see also 1.6.1 Remark)

$$\mathbf{Q}_k f(x) = \sum_{|\underline{\alpha}|=k} W_{\underline{\alpha}}(x) \int_{\partial B(r)} V_{\underline{\alpha}}(y) d\sigma_y f(y).$$

1.6.5 The inversion operator revisited

The Laurent expansion for a function f which is monogenic outside a closed ball $\bar{B}(R)$ in $\mathbf{R}^{m+1}$ and vanishes at infinity enables us to prove

Theorem 1. *Let $R > 0$ be given. Then the inversion* $\mathbf{I}$ *is an isomorphism of* $M(\mathring{B}(R); V)$ *onto* $M_0(\mathbf{R}^{m+1} \backslash \bar{B}(1/R); V)$.

Proof. Let $f \in M(\mathring{B}(R; V))$ and put $g = \mathbf{I}f$. Then from Theorem 0.2.14 it follows that $g \in M_0(\mathbf{R}^{m+1} \backslash \bar{B}(1/R))$.

Conversely, take $g \in M_0(\mathbf{R}^{m+1} \backslash \bar{B}(1/R); V)$. Then by means of the preceding Theorem 2 we have that

$$g(x) = \sum_{k=0}^{\infty} \mathbf{Q}_k g(x)$$

where the series is normally and hence uniformly convergent on each closed annular subset of $\mathbf{R}^{m+1} \backslash \bar{B}(1/R)$.

Consequently, using again Theorem 0.2.14 it follows that

$$\tilde{f}(x) = \sum_{k=0}^{\infty} \mathbf{I}\mathbf{Q}_k g(x)$$

converges uniformly on each compact subset of $\mathring{B}(R) \backslash \{0\}$ and hence it determines a monogenic function $\tilde{f}$ in $\mathring{B}(R) \backslash \{0\}$.

But, as it is a series of homogeneous monogenic polynomials, $\tilde{f}$ may be monogenically extended to the origin.

Because of $\mathbf{I}^2 g = g$ it follows that $\mathbf{I}\tilde{f} = g$. ∎

The foregoing theorem thus tells us that given $g \in M_0(\mathbf{R}^{m+1}\backslash\bar{B}(R);V)$, there exists a unique $f \in M(\mathring{B}(1/R);V)$ such that

$$\begin{aligned} g(x) &= \mathbf{I}f(x) \\ &= \frac{\bar{x}}{|x|^{m+1}} f\Big(\frac{\bar{x}}{|x|^2}\Big). \end{aligned}$$

This remark yields

Theorem 2. *Let $f \in M(\mathring{B}(R);V)$ and $g \in M_0(\mathbf{R}^{m+1}\backslash\bar{B}(R');V)$ where $R' < 1 < R$. Then*

$$\int_{S^m} \bar{g}(\omega) f(\omega) d\Sigma(\omega) = 0,$$

i.e.

$$(g|S^m, f|S^m) = 0.$$

Proof. There exists $h \in M(\mathring{B}(1/R');V)$ such that

$$g(x) = \mathbf{I}h(x) = \frac{\bar{x}}{|x|^{m+1}} h\Big(\frac{\bar{x}}{|x|^2}\Big).$$

Hence on S^m

$$\bar{g}(\omega) = \bar{h}(\bar{\omega})\omega.$$

But now, $\bar{h}(\bar{x})$ being right monogenic in $\mathring{B}(1/R')$ where $1/R' > 1$, we have by Cauchy's Theorem that

$$\begin{aligned} 0 &= \int_{S^m} \bar{h}(\bar{\omega}) d\sigma_\omega f(\omega) \\ &= \int_{S^m} \bar{h}(\bar{\omega}) \omega f(\omega) d\Sigma(\omega) \\ &= \int_{S^m} \bar{g}(\omega) f(\omega) d\Sigma(\omega) \\ &= (g|S^m, f|S^m). \quad \blacksquare \end{aligned}$$

1.7 Integral representations for the projection operators $\mathbf{P}(k)$ *and* $\mathbf{Q}(k)$.

In 1.3 we introduced the orthogonal projections $\mathbf{P}(k) : L_2(S^m;V) \to \mathcal{M}^+(k,V)$ and $\mathbf{Q}(k) : L_2(S^m;V) \to \mathcal{M}^-(k,V)$.

Now we prove

Theorem 1.7.1. *For each $f \in L_2(S^m;V)$ and $k \in \mathbf{N}$*

$$\mathbf{P}(k)f(\xi) = \frac{1}{A_{m+1}} \int_{S^m} C^+_{m+1,k}(\xi,\omega) f(\omega) d\Sigma(\omega)$$

and

$$\mathbf{Q}_k f(\xi) = \frac{1}{A_{m+1}} \int_{S^m} C^-_{m+1,k}(\xi,\omega) f(\omega) d\Sigma(\omega).$$

Proof. Take $V = \mathcal{C}$, the case $V = \mathcal{S}$ being similar. Let us first recall that for each $k \in \mathbf{N}$ and all $\xi, \omega \in S^m$ (see (1.15)(1.16))

$$\frac{1}{A_{m+1}} C^+_{m+1,k}(\xi,\omega)\bar{\omega} = \sum_{|\underline{\alpha}|=k} V_{\underline{\alpha}}(\xi) W_{\underline{\alpha}}(\omega) \tag{1.26}$$

and

$$\frac{1}{A_{m+1}} C^-_{m+1,k}(\omega,\xi)\bar{\xi} = \sum_{|\underline{\alpha}|=k} W_{\underline{\alpha}}(\xi) \overline{V}_{\underline{\alpha}}(\omega). \tag{1.27}$$

As for $f \in L_2(S^m; \mathcal{C})$, $\mathbf{P}(k)f \in \mathcal{M}^+(k; \mathcal{C})$, there ought to exist $\lambda_{\underline{\beta}} \in V$ such that

$$\mathbf{P}(k)f(\omega) = \sum_{|\underline{\beta}|=k} V_{\underline{\beta}}(\omega)\lambda_{\underline{\beta}}.$$

So we have by means of (1.27) and Lemma 1.5.7 that

$$\begin{aligned}
&\frac{1}{A_{m+1}} \int_{S^m} C^+_{m+1,k}(\xi,\omega)\overline{\omega}\mathbf{P}(k)f(\omega)d\Sigma(\omega) \\
&= \sum_{\substack{|\underline{\alpha}|=k \\ |\underline{\beta}|=k}} V_{\underline{\alpha}}(\xi)\Big(\int_{S^m} W_{\underline{\alpha}}(\omega) d\sigma_\omega V_{\underline{\beta}}(\omega)\Big)\lambda_{\underline{\beta}} \\
&= \sum_{|\underline{\beta}|=k} V_{\underline{\beta}}(\xi)\lambda_{\underline{\beta}} \\
&= \mathbf{P}(k)f(\xi).
\end{aligned}$$

As for $f \in L_2(S^m; \mathcal{C})$, $\mathbf{Q}(k)f \in \mathcal{M}^-(k; \mathcal{C})$, there ought to exist $\mu_{s_1 \ldots s_k} \in V$ such that

$$\mathbf{Q}(k)f(\omega) = \sum_{|\underline{\beta}|=k} W_{\underline{\beta}}(\omega)\mu_{\underline{\beta}}.$$

So we have that, by virtue of (1.27) and 1.5.8. Remark (1),

$$\frac{1}{A_{m+1}} \int_{S^m} C^-_{m+1,k}(\xi,\omega)\mathbf{Q}(k)f(\omega)d\Sigma(\omega)$$

$$= \sum_{\substack{|\underline{\alpha}|=k \\ |\underline{\beta}|=k}} W_{\underline{\alpha}}(\xi)\Big(\int_{S^m} \overline{V}_{\underline{\alpha}}(\omega)d\sigma_{(\omega)}W_{\underline{\beta}}(\omega)\Big)\mu_{\underline{\beta}}$$

$$= \sum_{|\underline{\beta}|=k} W_{\underline{\beta}}(\xi)\mu_{\underline{\beta}}$$

$$= \mathbf{Q}(k)f(\omega).$$

By the same reason we obtain that for all $f \in L_2(S^m;C)$, $k \in \mathbf{N}$ fixed and $j \in \mathbf{N}$, $j \neq k$,

$$\frac{1}{A_{m+1}}\int_{S^m} C^+_{m+1,k}(\xi,\omega)\mathbf{P}(j)f(\omega)d\Sigma(\omega)$$

$$= \frac{1}{A_{m+1}}\int_{S^m} C^-_{m+1,k}(\xi,\omega)\mathbf{Q}(j)f(\omega)d\Sigma(\omega)$$

$$= 0$$

while for all $j \in \mathbf{N}$

$$\frac{1}{A_{m+1}}\int_{S^m} C^+_{m+1,k}(\xi,\omega)\mathbf{Q}(j)f(\omega)d\Sigma(\omega)$$

$$= \frac{1}{A_{m+1}}\int_{S^m} C^-_{m+1,k}(\xi,\omega)\mathbf{P}(j)f(\omega)d\Sigma(\omega)$$

$$= 0.$$

Consequently, if $f \in L_2(S^m;C)$ has the decomposition (see also Theorem 1.3.1)

$$f(\omega) = \sum_{j=0}^{\infty}\mathbf{P}(j)f(\omega) + \sum_{j=0}^{\infty}\mathbf{Q}(j)f(\omega)$$

we find that for $k \in \mathbf{N}$ fixed

$$\frac{1}{A_{m+1}}\int_{S^m} C^+_{m+1,k}(\xi,\omega)\overline{\omega}f(\omega)d\Sigma(\omega)$$

$$= \frac{1}{A_{m+1}}\int_{S^m} C^+_{m+1,k}(\xi,\omega)\overline{\omega}\mathbf{P}(k)f(\omega)d\Sigma(\omega)$$

$$= \mathbf{P}(k)f(\xi)$$

and

$$\frac{1}{A_{m+1}}\int_{S^m} C^-_{m+1,k}(\xi,\omega)f(\omega)d\Sigma(\omega)$$

$$= \frac{1}{A_{m+1}} \int_{S^m} C^-_{m+1,k}(\xi,\omega)\mathbf{Q}(k)f(\omega)d\Sigma(\omega)$$

$$= \mathbf{Q}(k)f(\xi). \quad \blacksquare$$

Remarks.

(1) In the case of the Dirac operator we also have

Theorem 1.7.2. *For each $f \in L_2(S^m; V)$ and $k \in \mathbf{N}$ we have that*

$$\mathbf{P}(k)f(\xi) = \frac{1}{A_{m+1}} \int_{S^m} \left[C_k^{\frac{m+1}{2}}(t) + \xi\omega C_{k-1}^{\frac{m+1}{2}}(t)\right] f(\omega)d\Sigma(\omega)$$

and

$$\mathbf{Q}(k)f(\xi) = \frac{-1}{A_{m+1}} \int_{S^m} \left[C_{k-1}^{\frac{m+1}{2}}(t) + \xi\omega C_k^{\frac{m+1}{2}}(t)\right] f(\omega)d\Sigma(\omega).$$

Proof. On the one hand

$$\xi\omega C_{k-1}^{\frac{m+1}{2}}(t) = \xi \wedge \omega C_{k-1}^{\frac{m+1}{2}}(t) - tC_{k-1}^{\frac{m+1}{2}}(t)$$

while on the other hand, using the recurrence relation for Gegenbauer polynomials

$$(2\nu + k)C_k^\nu(t) = 2t(C_k^{\nu+1}(t) - tC_{k-1}^{\nu+1}(t))$$

we obtain for $\nu = \frac{m-1}{2}$ that

$$(k + m - 1)C_k^{\frac{m-1}{2}} t) = (m-1)(C_k^{\frac{m+1}{2}}(t) - tC_{k-1}^{\frac{m+1}{2}}(t)). \quad \blacksquare$$

(2) Notice that the inner and outer spherical monogenics determining the spherical monogenics in Theorem 1.7.1 are given by

$$\mathbf{P}(k)f(x) = \frac{|x|^k}{A_{m+1}} \int_{S^m} C^+_{m+1,k}(\xi,\omega)f(\omega)d\Sigma(\omega)$$

and

$$\mathbf{Q}(k)f(x) = \frac{1}{A_{m+1}} \frac{1}{|x|^{m+k}} \int_{S^m} C^-_{m+1,k}(\xi,\omega)f(\omega)d\Sigma(\omega). \tag{1.28}$$

(3) The proof of Theorem 1.7.1 has been given for the Dirac case. Using 1.6.1 Remark, a similar proof may be worked out for the case of the Cauchy-Riemann operator.

1.8 L_2-boundary values of monogenic functions on S^m

In this section we prove that each $f \in L_2(S^m; V)$ may be written as a sum of boundary values of functions, respectively monogenic inside and outside the unit ball.

Moreover the space $\mathcal{M}_2(S^m; V)$ is introduced which generalizes in a natural way the classical Hardy space $\mathfrak{H}_2$.

Finally the Hilbert-Riesz transform on $L_2(S^m; V)$ is defined.

1.8.1 The Cauchy transform revisited

Let $f \in L_2(S^m; V)$. Then we saw in 1.6.2 that for the Cauchy transform $\hat{f}$ of f

$$\begin{aligned}\hat{f}(x) &= \frac{1}{A_{m+1}} \int_{S^m} \frac{\bar{\omega} - \bar{x}}{|\omega - x|^{m+1}} d\sigma_\omega f(\omega) \\ &= \sum_{k=0}^{\infty} Y(1-|x|)\mathbf{P}_k f(x) - \sum_{k=0}^{\infty} Y(|x|-1)\mathbf{Q}_k f(x) \\ &= f^+(x) + f^-(x)\end{aligned}$$

where $f^+ \in M(\mathring{B}(1); V)$, $f^- \in M(\mathbf{R}^{m+1} \backslash \bar{B}(1); V)$,

$$\mathbf{P}_k f(x) = \frac{1}{A_{m+1}} |x|^k \int_{S^m} C^+_{m+1,k}(\xi, \omega) f(\omega) d\Sigma(\omega)$$

and

$$\mathbf{Q}_k f(x) = \frac{1}{A_{m+1}} \frac{1}{|x|^{m+k}} \int_{S^m} C^-_{m+1,k}(\xi, \omega) f(\omega) d\Sigma(\omega).$$

It thus follows that for any $0 < r < 1$ fixed

$$f^+(r\xi) \in L_2(S^m; V)$$

while for any $r > 1$ fixed

$$f^-(r\xi) \in L_2(S^m; V).$$

Furthermore by virtue of (1.28) we have that on S^m

$$\mathbf{P}_k f(\xi) = \mathbf{P}(k) f(\xi)$$

and

$$\mathbf{Q}_k f(\xi) = \mathbf{Q}(k) f(\xi). \qquad (1.29)$$

Putting

$$f^+(\xi) = \sum_{k=0}^{\infty} \mathbf{P}(k) f(\xi)$$

we have that $f^+ \in L_2(S^m; V)$ and

$$\|f^+(\xi) - f^+(r\xi)\|^2 = \sum_{k=0}^{\infty}(1 - r^k)^2\|\mathbf{P}(k)f\|^2$$

whence

$$\lim_{\substack{r \to 1 \\ <}} f^+(r\xi) = f^+(\xi).$$

We call $f^+ = BV_+\hat{f}$ the *inner boundary value* of $\hat{f}$.

Analogously, putting

$$f^-(\xi) = -\sum_{k=0}^{\infty} \mathbf{Q}(k)f(\xi)$$

we find that

$$\lim_{\substack{r \to 1 \\ >}} f^-(r\xi) = f^-(\xi).$$

We call $f^- = BV_-\hat{f}$ the *outer boundary value* of $\hat{f}$.

We thus have

Theorem 1. *Let $f \in L_2(S^m; V)$. Then $f = BV_+\hat{f} - BV_-\hat{f}$, $\hat{f}$ being the Cauchy transform of f.*

1.8.2 The spaces $\mathcal{M}_2(S^m; V)$ and $\mathcal{M}_{0,2}(S^m; V)$

Call $\mathcal{M}_2(S^m; V)$ the space consisting of those elements f in $L_2(S^m; V)$ such that their decomposition into spherical monogenics is of the form

$$f(\omega) = \sum_{k=0}^{\infty} \mathbf{P}(k)f(\omega),$$

i.e. $\mathbf{Q}(k)f(\omega) = 0$ for all $k \in \mathbf{N}$.

It is clear that $\mathcal{M}_2(S^m; V)$ is a closed subspace of $L_2(S^m; V)$.

A characterization of this space in terms of monogenic functions in the unit ball is now given in

Theorem 2. *A function f belongs to $\mathcal{M}_2(S^m; V)$ if and only if there exists a function g in $M(\mathring{B}(1); V)$ such that*

$$f(\xi) = \lim_{\substack{r \to 1 \\ <}} g(r\xi),$$

the convergence being taken in $L_2(S^m; V)$.

Proof. Let $f \in \mathcal{M}_2(S^m; V)$. Then using the results of the preceding subsection, its Cauchy transform $\hat{f}$ reduces to f^+, where $f^+ \in M(\mathring{B}(1); V)$ and $\lim_{r \underset{<}{\to} 1} f^+(r\xi) = f(\xi)$.

Conversely, assume that $g \in M(\mathring{B}(1); V)$ is such that $\lim_{r \underset{<}{\to} 1} g(r\xi) = f(\xi)$ belongs to $L_2(S^m; V)$.

Considering the Taylor series of g we have

$$g(x) = \sum_{k=0}^{\infty} \mathbf{P}_k g(x)$$

whence for all $0 < r < 1$

$$g(r\xi) = \sum_{k=0}^{\infty} r^k \mathbf{P}(k) g(\xi)$$

and so $g(r\xi) \in \mathcal{M}_2(S^m; V)$.

Consequently

$$\lim_{\substack{r \to 1 \\ <}} g(r\xi) = f(\xi) \in \mathcal{M}_2(S^m; V). \blacksquare$$

The foregoing theorem thus tells us that $\mathcal{M}_2(S^m; V)$ may be identified with the subspace of $M(\mathring{B}(1); V)$ consisting of those functions in $M(\mathring{B}(1); V)$ having boundary values in the $L_2(S^m; V)$-sense.

In such a way it is a generalization of the Hardy space $\mathfrak{H}_2$.

Calling $\mathcal{M}_{0,2}(S^m; V)$ the subspace of $L_2(S^m; V)$ consisting of those elements $f \in L_2(S^m; V)$ for which the decomposition into spherical monogenics is given by

$$f(\omega) = \sum_{k=0}^{\infty} \mathbf{Q}(k) f(\omega),$$

similar reasonings as made before lead to the following

Theorem 3. *A function f belongs to $\mathcal{M}_{0,2}(S^m; V)$ if and only if there exists a function $g \in M_0(\mathbf{R}^{m+1} \backslash \bar{B}(1); V)$ such that*

$$\lim_{\substack{r \to 1 \\ >}} g(r\xi) = f(\xi),$$

the convergence being taken in $L_2(S^m; V)$.

Remark. In [10] we denoted the space $\mathcal{M}_2(S^m; \mathcal{C})$ by $ML_{2(r)}(S^m; \mathcal{A})$ and we proved that it is a right Hilbert $\mathcal{C}$-module with reproducing kernel $S(\omega, t)$, $\omega \in S^m$, $t \in \mathring{B}(1)$ given by

$$S(\omega, t) = \frac{1 - \omega \bar{t}}{|1 - \bar{\omega} t|^{m+1}}.$$

This kernel generalizes the classical *Szegökernel* for $\mathfrak{H}_2$.

1.8.3 The projection operators $\mathbf{\Pi}_+$ and $\mathbf{\Pi}_-$; the Hilbert-Riesz transform $\mathbf{H}_\xi$

By means of Theorem 2 and Theorem 3 we have that

$$L_2(S^m; V) = \mathcal{M}_2(S^m; V) \oplus \mathcal{M}_{0,2}(S^m; V)$$

the direct sum being orthogonal.

Hence we may define orthogonal projections $\mathbf{\Pi}_+$ and $\mathbf{\Pi}_-$ on $L_2(S^m; V)$ where, if $f \in L_2(S^m; V)$ has the decomposition

$$f = \sum_{k=0}^{\infty} \mathbf{P}(k)f + \sum_{k=0}^{\infty} \mathbf{Q}(k)f,$$

$$\mathbf{\Pi}_+ f = \sum_{k=0}^{\infty} \mathbf{P}(k)f = BV_+\hat{f} \in \mathcal{M}_2(S^m; V)$$

and

$$\mathbf{\Pi}_- f = \sum_{k=0}^{\infty} \mathbf{Q}(k)f = -BV_-\hat{f} \in \mathcal{M}_{0,2}(S^m, V)$$

We thus have in $L_2(S^m; V)$ that

$$\begin{aligned} \mathbf{1} &= \mathbf{\Pi}_+ + \mathbf{\Pi}_- \\ \mathbf{0} &= \mathbf{\Pi}_+\mathbf{\Pi}_- = \mathbf{\Pi}_-\mathbf{\Pi}_+ \end{aligned}$$

By means of the projection operators $\mathbf{\Pi}_+$ and $\mathbf{\Pi}_-$ we introduce the *Hilbert-Riesz transform* $\mathbf{H}_\xi$ on $L_2(S^m; V)$ with

$$\mathbf{H}_\xi = \mathbf{\Pi}_- - \mathbf{\Pi}_+.$$

It is clear that

$$\mathbf{H}_\xi^2 = \mathbf{1}.$$

1.8.4 The inversion operator $\mathbf{I}$ and the antipodal map $\mathbf{A}$

In 1.6.5. Theorem 1 we saw that the inversion operator $\mathbf{I}$ is an isomorphism of $M(\mathring{B}(1); V)$ onto $M_0((\mathbf{R}^{m+1}\backslash\bar{B}(1); V)$.

Restricting this operator to the subspaces of $M(\mathring{B}(1); V)$ and $M_0(\mathbf{R}^{m+1}\backslash\bar{B}(1); V)$ which may be identified with $\mathcal{M}_2(S^m; V)$ and $\mathcal{M}_{0,2}(S^m; V)$ respectively, we may extend this operator to $L_2(S^m; V)$ by defining for $f \in L_2(S^m; V)$

$$\mathbf{I}f(\omega) = \bar{\omega}f(\bar{\omega}). \tag{1.30}$$

In this way $\mathbf{I}$ is a unitary operator on $L_2(S^m; V)$ such that

$$\mathbf{I}(\mathcal{M}_2(S^m; V)) = \mathcal{M}_{0,2}(S^m; V)$$

and

$$\mathbf{I}(\mathcal{M}_{0,2}(S^m;V)) = \mathcal{M}_2(S^m;V).$$

Notice in particular that for $P_k \in \mathcal{M}^+(k,V)$

$$\mathbf{I}P_k(\omega) = \bar{\omega}P_k(\bar{\omega}) \in \mathcal{M}^-(k,V)$$

while for $Q_k \in \mathcal{M}^-(k,V)$,

$$\mathbf{I}Q_k(\omega) = \bar{\omega}Q_k(\bar{\omega}) \in \mathcal{M}^+(k,V).$$

We thus have that

$$\mathbf{IQ}(k) = \mathbf{P}(k)\mathbf{I}$$

or, since $\mathbf{I}^2 = \mathbf{1}$, that

$$\mathbf{Q}(k) = \mathbf{IP}(k)\mathbf{I}$$

and

$$\mathbf{P}(k) = \mathbf{IQ}(k)\mathbf{I}.$$

The antipodal map $\mathbf{A}$ on $L_2(S^m;V)$ is given by

$$\mathbf{A}f(\omega) = f(-\omega).$$

Notice that $\mathbf{A}^2 = \mathbf{1}$.

1.9 Conformal invariance - The inversion $\mathbf{I}^*$

As is well known the conformal group in $\mathbf{R}^{m+1}$, $m > 1$, is generated by four elementary transformations, namely the rotation, the homothetic transformation, the translation and the inversion. Another way of looking at the conformal group is by saying that it is generated by the reflections with respect to spheres having positive radius, including hyperplanes (which have infinite radius).

Hence to show the invariance of the theory of monogenic functions under conformal mappings, it suffices to show how monogenic functions transform under spherical reflections. Here monogenicity is meant with respect to the **Dirac** operator.

1.9.1 Reflections with respect to hyperplanes

Let H be a hyperplane through a point y and let w be the unit normal to H. Then the reflection with respect to H is given by the map

$$x \to y + w(x-y)w.$$

It is easily seen that if a function f is monogenic then so is the reflected function $R_H f$ with

$$\mathbf{R}_H f(x) = wf(y + w(x-y)w).$$

Note that the orientation on H determined by w is essential. Changing the orientation on H is like changing $w \to -w$. We therefore say that monogenic functions transform under reflections with respect to oriented hyperplanes.

In particular we know that the elements s belonging to Spin$(m+1)$ are products of the form (see I. 1.10).

$$s = w_1 \ldots w_{2\ell} \quad , \quad w_j \in S^m.$$

From the above definition of the reflection $R_H(f)$ of a function f we may derive that a monogenic function f transforms under the group Spin$(m+1)$ like

$$f \to L(s)f$$

where

$$L(s)f(x) = sf(\bar{s}xs).$$

1.9.2 Reflections with respect to spheres

Let S be a sphere with center u and radius R. Then the reflection with respect to S is given by

$$x \to u + R^2 \frac{x-u}{|x-u|^2}.$$

Now let f be (left) monogenic ; then we define the reflected function $R_S f$ by

$$R_S f(x) = R^m \frac{x-u}{|x-u|^{m+1}} f(u + R^2 \frac{x-u}{|x-u|^2}).$$

Putting $R=1$ and $u=0$, we obtain the inversion $\mathbf{I}^\star$ given by

$$\mathbf{I}^\star f(x) = \frac{x}{|x|^{m+1}} f(\frac{x}{|x|^2}). \tag{1.31}$$

Clearly $\mathbf{I}^{\star 2} = -\mathbf{1}$.

Of course, just as in Theorem 0.2.14, we have that $\mathbf{I}^* f$ is monogenic if f is monogenic and so, using elementary arguments, we have that $R_S f$ is monogenic whenever f itself is monogenic.

Also here the definition of reflection involves the choice of an orientation on S by the unit normal. Indeed, if we change this orientation, i.e. consider - S instead of S, we have to define $R_{-S} f = -R_S f$ in order to have a fully consistent picture.

Note also that, in order to illustrate the consistence of the notions introduced, we in fact have to show that the reflection of f with respect to an oriented plane H is indeed a limit case of reflections with respect to spheres. Therefore, let H be a plane through the origin, say with unit normal w. Then H can be considered as the limit of spheres with radius R and center $-Rw$, where we assume the outer unit normal orientation. The reflection with respect to S may then be rewritten as

$$R_S f(x) = \frac{\frac{x}{R} + w}{|\frac{x}{R} + w|^{m+1}} f(Tx)$$

where

$$Tx = \frac{x + 2w(x \cdot w) - \frac{w}{R}|x|^2}{\frac{|x|^2}{R^2} - \frac{2}{R} x \cdot w + 1}.$$

We thus have that

$$\lim_{R \to \infty} R_S f(x) = w f(x + 2w(x \cdot w)).$$

But, as

$$wxw = x + 2w(x \cdot w)$$

we finally get that

$$\begin{aligned} \lim_{R \to \infty} R_S f(x) &= w f(wxw) \\ &= R_H f(x). \end{aligned}$$

Note that in the above expression for $R_S f$, the transformation $R \to -R, w \to -w$ is equivalent to the change of orientation $S \to -S$. Moreover, in the limit for R going to infinity, we still obtain the reflection of f with respect to the oriented hyperplane H (as one should expect). This makes the theory fully consistent.

Finally notice that if we parametrize the oriented sphere S or the hyperplane H by the pair $(p, w) = (\frac{1}{R}, w) \in \mathbf{R}^+ \times S^m$, we can use one single formula for the general spherical reflection of f :

$$R_{(p,w)} f(x) = \frac{px + w}{|px + w|^{m+1}} f\left(\frac{wxw + |x|^2 pw}{|w + px|^2}\right).$$

1.9.3 The inversion $\mathbf{I}^*$

By means of the inversion $\mathbf{I}^*$ a function $f \in M(\mathring{B}(R), V)$ is transformed into $\mathbf{I}^* f \in M_0(\mathbf{R}^{m+1}\backslash\overline{B}(1/R); V)$.

Notice that by formally substituting e_0 by 1 and e_i by ε_i, $i = 1, \ldots, m$ in (1.31) we do not longer obtain results that remain valid for nullsolutions of the Weyl equation.

However, in the case of the Dirac operator we have that for all $P_k \in M^+(k; V)$

$$\begin{aligned} \mathbf{I}^* P_k(x) &= \frac{x}{|x|^{m+1}} P_k\left(\frac{x}{|x|^2}\right) \\ &= (-1)^{k+1} \mathbf{I} P_k(x) \end{aligned} \tag{1.32}$$

whence $\mathbf{I}^*$ is again an isomorphism from $M^+(k; V)$ onto $M^-(k; V)$.

Again identifying $\mathcal{M}_2(S^m V)$ and $\mathcal{M}_{0,2}(S^m; V)$ with appropriate subspaces of $M(\mathring{B}(1); V)$ and $M_0(\mathbf{R}^{m+1}\backslash\overline{B}(1); V)$ respectively, we may extend the operator $\mathbf{I}^*$ to $L_2(S^m; V)$ by putting for $f \in L_2(S^m; V)$

$$\mathbf{I}^* f(\omega) = \omega f(\omega).$$

By restricting $\mathbf{I}^*$ to $\mathcal{M}^+(k; V)$ we get that

$$\mathbf{I}^* P_k(\xi) = \xi P_k(\xi) \tag{1.33}$$

so that on $L_2(S^m; V)$

$$\mathbf{I}^* \mathbf{Q}(k) = \mathbf{P}(k) \mathbf{I}^*$$

or

$$\mathbf{Q}(k) = -\mathbf{I}^* \mathbf{P}(k) \mathbf{I}^*$$

and

$$\mathbf{P}(k) = -\mathbf{I}^* \mathbf{Q}(k) \mathbf{I}^*.$$

If $V = S$ and so $L_2(S^m; S)$ is regarded as a complex inner product space, we obtain that $\mathbf{I}^*$ is a unitary operator in $L_2(S^m; S)$ having eigenvalues i and $-i$.

The projections on the corresponding eigenspaces are then given by

$$\frac{1}{2}(1 \pm i\mathbf{I}^*) f(\xi) = \frac{1}{2}(1 \pm i\xi) f(\xi)$$

where, putting

$$\alpha = \frac{1}{2}(1 + i\xi)$$

and

$$\beta = \frac{1}{2}(1 - i\xi),$$

α and β satisfy the relations

$$\begin{aligned} \alpha^2 = \alpha \quad &; \quad \beta^2 = \beta \\ \alpha\beta \quad &= \quad \beta\alpha = 0 \\ \alpha + \beta \quad &= \quad 1. \end{aligned}$$

Finally notice that since $\mathbf{I}^\star$ and $\mathbf{A}$ are anticommuting on $L_2(S^m; S)$, we have that $\mathbf{I}^\star \circ \mathbf{A}^2 = \mathbf{1}$.

The operators $\mathbf{I}^\star$ and $i\mathbf{A}$, both being unitary operators on $L_2(S^m; S)$, thus generate a real algebra which is isomorphic to the algebra $\mathbf{H}$ of real quaternions, and this by identifying $\mathbf{i}$ and $\mathbf{j}$ with $\mathbf{I}^\star$ and $i\mathbf{A}$ respectively.

As $\text{Spin}(3) \cong S^3 \cong \text{Sp}(1)$ (see also 0. 4.3.4), by the identification of an element $\omega = (\omega_0, \omega_1, \omega_2, \omega_3) \in S^3$ with $\omega_0 + \omega_1 \mathbf{I}^\star + \omega_2(i\mathbf{A}) + \omega_3(i\mathbf{I}^\star \circ \mathbf{A})$, we thus obtain a representation of Spin(3) into the algebra of unitary operators on $L_2(S^m; S)$.

1.10 The Fischer decomposition into spherical monogenics

As is well known the Fischer decomposition of homogeneous polynomials in terms of solid spherical harmonics is a basic tool in quite a lot of applications.

In the case of the Dirac operator this decomposition can be substantially refined and this by using inner spherical monogenics.

Theorem 1.10.1. *Let $k \in \mathbf{N}$. Then*

$$\mathcal{P}(k; V) = \sum_{s=0}^{k} \oplus x^s M^+(k - s; V),$$

i.e. for each $R_k \in \mathcal{P}(k; V)$, there exist unique $P_{k-s} \in M^+(k - s; V)$ such that

$$R_k(x) = \sum_{s=0}^{k} x^s P_{k-s}(x).$$

Proof. Let us start from the Fischer decomposition of R_k into solid spherical harmonics (see [29]) :

$$R_k(x) = \sum_{s=0}^{[k/2]} r^{2s} S_{k-2s}(x)$$

with S_{k-2s} a V-valued solid harmonic of degree $k - 2s$.

As for each $s = 0, \ldots, [k/2]$,

$$r^{2s} = (-1)^s x^{2s}$$

and

$$S_{k-2s}(x) = |x|^{k-2s} S_{k-2s}(\xi)$$

and moreover there exist unique element $P_{k-2s} \in \mathcal{M}^+(k-2s; V)$ and unique element $P_{k-2s-1} \in \mathcal{M}^-(k-2s-1; V)$ such that (see Theorem 1.3.2 (4), (1.3.5) and (1.3.6))

$$(-1)^s S_{k-2s})(\xi) = P_{k-2s}(\xi) + \xi P_{k-2s-1}(\xi)$$

we obtain

$$\begin{aligned} R_k(x) &= \sum_{s=0}^{[k/2]} x^{2s}|x|^{k-2s}(P_{k-2s}(\xi) + \xi P_{k-2s-1}(\xi)) \\ &= \sum_{s=0}^{[k/2]} x^{2s}(P_{k-2s}(x) + |x|\xi P_{k-2s-1}(x)) \\ &= \sum_{s=0}^{[k/2]} (x^{2s} P_{k-2s}(x) + x^{2s+1} P_{k-(2s+1)}(x)) \\ &= \sum_{j=0}^{k} x^j P_{k-j}(x). \blacksquare \end{aligned}$$

It should be noted that we may obtain the Fischer decomposition of homogeneous polynomials into spherical monogenics directly and this by using a classical technique (see e.g. [65]).

To this end, let us first recall some standard notations. For any multi-index $\alpha = (\alpha_0, \alpha_1, \ldots, \alpha_m) \in \mathbf{N}^{m+1}$, we put

$$\begin{aligned} x^\alpha &= x_0^{\alpha_0} x_1^{\alpha_1} \ldots x_m^{\alpha_m}; \\ \alpha! &= \alpha_0!\alpha_1! \ldots \alpha_m!; \\ |\alpha| &= \sum_{j=0}^{m} \alpha_j; \\ \partial^\alpha &= \frac{\partial^{|\alpha|}}{\partial_{x_0}^{\alpha_0} \partial_{x_1}^{\alpha_1} \ldots \partial_{x_m}^{\alpha_m}}. \end{aligned}$$

For convenience, let us only consider the case where $V = \mathcal{C}$, the case $V = \mathcal{S}$ being similar.

Let $\{x^\alpha : |\alpha| = k\}$ be the standard basis of $\mathcal{P}(k; \mathcal{C})$. Then on the complex vector space $\mathcal{P}(k; \mathcal{C})$ an inner product $<,>_k$ is defined as follows : for all $R_k^1, R_k^2 \in \mathcal{P}(k; \mathcal{C})$ with

$$R_k^i(x) = \sum_{|\alpha|=k} x^\alpha a_\alpha^i \quad , \quad i = 1, 2,$$

where $a_\alpha^i \in \mathcal{C}$, we put

$$< R_k^1, R_k^2 >_k = \sum_{|\alpha|=k} \alpha! [\overline{a}_\alpha^1 a_\alpha^2]_0$$

If for $R_k \in \mathcal{P}(k;\mathcal{C})$ we call $R_k(\frac{\partial}{\partial x})$ the differential operator obtained from R_k by replacing in it x_i by $\frac{\partial}{\partial x_i}$, $i = 0, 1, \ldots, m$, we then have

Lemma 1.10.2. *For all $R_k^1, R_k^2 \in \mathcal{P}(k;\mathcal{C})$,*

$$< R_k^1, R_k^2 >_k = [\overline{R_k^1}(\frac{\partial}{\partial x}) R_k^2]_0$$

Proof. Use the fact that

$$\partial^\alpha x^\beta = \begin{cases} \alpha! & \text{if} \quad \alpha = \beta \\ 0 & \text{if} \quad \alpha \neq \beta. \blacksquare \end{cases}$$

Corollary 1.10.3. *For all $R_{k-1} \in \mathcal{P}(k-1;\mathcal{C})$ and $R_k \in \mathcal{P}(k;\mathcal{C})$*

$$< x R_{k-1}, R_k >_k = - < R_{k-1}, \partial_x R_k >_{k-1} .$$

Theorem 1.10.4. *For each $k \in \mathbf{N}$,*

$$\mathcal{P}(k;\mathcal{C}) = M^+(k;\mathcal{C}) \oplus x\mathcal{P}(k-1;\mathcal{C}).$$

Moreover, the subspaces $M^+(k;\mathcal{C})$ and $x\mathcal{P}(k-1;\mathcal{C})$ of $\mathcal{P}(k;\mathcal{C})$ are orthogonal w.r.t. $<,>_k$.

Proof. As $\mathcal{P}(k;\mathcal{C}) = x\mathcal{P}(k-1;\mathcal{C}) \oplus (x\mathcal{P}(k-1;\mathcal{C})^\perp$ it suffices to prove that $M^+(k;\mathcal{C}) = (x\mathcal{P}(k-1;\mathcal{C}))^\perp$.

Take $R_{k-1} \in \mathcal{P}(k-1;\mathcal{C})$ arbitrarily and assume that for some $R_k \in \mathcal{P}(k;\mathcal{C})$,

$$< x R_{k-1}, R_k >_k = 0.$$

Then $< R_{k-1}, \partial_x R_k >_{k-1} = 0$ for all $R_{k-1} \in \mathcal{P}(k-1;\mathcal{C})$ so that, by putting $R_{k-1} = \partial_x R_k$, we get that $\partial_x R_k = 0$ or $R_k \in M^+(k;\mathcal{C})$.

Consequently $(x\mathcal{P}(k-1;\mathcal{C}))^\perp \subset M^+(k;\mathcal{C})$.

Conversely, let $P_k \in M^+(k;\mathcal{C})$. Then for each $R_{k-1} \in \mathcal{P}(k-1;\mathcal{C})$,

$$\begin{aligned} < x R_{k-1}, P_k >_k &= - < R_{k-1}, \partial_x P_k >_{k-1} \\ &= 0 \end{aligned}$$

whence $P_k \in (x\mathcal{P}(k-1;\mathcal{P}))^\perp$. $\blacksquare$

Corollary 1.10.5. (Fischer decomposition)

$$\mathcal{P}(k;\mathcal{C}) = \sum_{s=0}^{k} \oplus x^s M^+(k-s,\mathcal{C}).$$

Remark. The Fischer decomposition leads to the following useful decomposition of V-valued (real) analytic functions.

Let Ω be an open subset in $\mathbf{R}^{m+1}$ containing the origin and let $f \in \mathcal{A}(\Omega;V)$. Then in a suitable open SO$(m+1)$-invariant neighbourhood Λ of the origin, f admits the multiple series expansion

$$f(x) = \sum_{k=0}^{\infty} \frac{1}{k!} \sum_{\ell_1=0}^{m} \cdots \sum_{\ell_k=0}^{m} x_{\ell_1} \ldots x_{\ell_k} \partial_{x_{\ell_1}} \ldots \partial_{x_{\ell_k}} f(0)$$

which converges normally in Λ.

Bracketing terms together we arrive at

$$f(x) = \sum_{k=0}^{\infty} R_k(x)$$

where $R_k \in \mathcal{P}(k;V)$ for each $k \in \mathbf{N}$.

Using the Fischer decomposition into spherical monogenics, we obtain that

$$f(x) = \sum_{k=0}^{\infty} (\sum_{s=0}^{k} x^s P_{k,s}(x))$$

where for all $k, s \in \mathbf{N}$, $P_{k,s} \in M^+(k-s;V)$.

Passing to polar coordinates $x = \rho\omega$ and observing that $x^s P_{k,s}(x)$ admits the form $A_{k,\omega}^{(s)}(\rho) + \omega B_{k,\omega}^{(s)}(\rho)$ where for ρ fixed, $A_{k,\omega}^{(s)}(\rho)$ and $B_{k,\omega}^{(s)}(\rho)$ belong to $\mathcal{M}^+(k-s;V)$, we have that by reordering terms, f admits in Λ the decomposition

$$f(x) = \sum_{k=0}^{\infty} (A_{k,\omega}(\rho) + \omega B_{k,\omega}(\rho)),$$

where for ρ fixed, $A_{k,\omega}(\rho)$ and $B_{k,\omega}(\rho)$ belong to $\mathcal{M}^+(k;V)$.

This decomposition will be frequently used in Chapter III but then in the case where real analytic functions in $\mathbf{R}^m$ are considered.

1.11 Spherical monogenics and Casimir operators on S^m

Let h (resp. ℓ) stand for the vectorial representation of Spin$(m+1)$ on $\mathbf{R}^{m+1}$ (resp. for the representation of Spin$(m+1)$ on $\mathcal{C}$ obtained by left multiplication).

Then these representations give rise to representations H and L of Spin$(m+1)$ on $L_2(\mathcal{R};V)$ where $\mathcal{R}$ is any domain in $\mathbf{R}^{m+1}$ invariant under SO$(m+1)$ and V is, as usual, either S or $\mathcal{C}$. The corresponding infinitesimal representations dH and dL then lead to representations of the Lie algebra $\mathbf{R}_{0,m+1}^2$ of Spin$(m+1)$. Since $\mathbf{R}_{0,m+1}^2$

carries a bilinear form B, the associated Casimir operators $C_B(H)$ and $C_B(L)$ may be constructed. It then appears that the spaces $\mathcal{M}^+(k;V)$ and $\mathcal{M}^-(k;V)$ of spherical monogenics are precisely the simultaneous eigenspaces of the Casimir operators $C_B(H) = \Delta_\xi$ and $C_B(L) = \Delta_\xi + \Gamma_\xi - \frac{1}{4}\binom{m+1}{2}\mathbf{1}$.

It should be noted that the results thus obtained are only valid in the case of the **Dirac** operator.

1.11.1 Preliminaries

Let us recall that the spin group Spin$(m+1)$ is given by

$$\mathrm{Spin}(m+1) = \{\Pi_{i=1}^{2k} s_i : k \in \mathbf{N}; s_i \in S^m\}$$

and that its Lie algebra is isomorphic to the space $\mathbf{R}^2_{0,m+1}$ of bi-vectors in $\mathbf{R}_{0,m+1}$ (see also I. 1.10.4).

Denoting henceforth the rectorial representation of Spin$(m+1)$ on $\mathbf{R}^{m+1}$ by h, we thus have for all $s \in \mathrm{Spin}(m+1)$ that (see also I. 1.10.2)

$$h(s)x = sx\bar{s} \quad , \quad x \in \mathbf{R}^{m+1}.$$

A representation ℓ of Spin$(m+1)$ on V is obtained by left multiplication in $\mathcal{C}$, i.e. for all $s \in \mathrm{Spin}(m+1)$,

$$\ell(s)a = sa \quad , \quad a \in V.$$

Now let $\mathcal{R}$ be any domain in $\mathbf{R}^{m+1}$ which is invariant under SO$(m+1)$. Then h induces a representation - still denoted by h - on $L_2(\mathcal{R})$, namely

$$h(s)f(x) = f(\bar{s}xs), f \in L_2(\mathcal{R}).$$

Using the trivial representation $\mathbf{1}$ of Spin$(m+1)$ on V we may define the representation H of Spin$(m+1)$ on $L_2(\mathcal{R};V) \cong L_2(\mathcal{R}) \otimes_{\mathbf{C}} V$ by

$$H = h \otimes \mathbf{1}.$$

We thus have that for each $s \in \mathrm{Spin}(m+1)$,

$$H(s)f(x) = f(\bar{s}xs) \quad , \quad f \in L_2(\mathcal{R};V).$$

The tensor product of the representations H and ℓ is denoted by L, i.e.

$$L = H \otimes \ell.$$

So for all $s \in \mathrm{Spin}(m+1)$,

$$L(s)f(x) = sf(\bar{s}xs) \quad , \quad f \in L_2(\mathcal{R};V).$$

Notice that H and L are unitary representations on, respectively, the right Hilbert $\mathcal{C}$-module $L_2(S^m;\mathcal{C})$ if $V=\mathcal{C}$ and the complex Hilbert space $L_2(S^m;S)$ if $V=S$.

Indeed, in the case $V=\mathcal{C}$ for instance, we obtain that

$$\begin{aligned}(H(s)f,H(s)g) &= \int \overline{f(\bar{s}xs)}g(\bar{s}xs)dx \\ &= \int_{\mathcal{R}} \overline{f(x)}g(x)dx \\ &= (f,g)\end{aligned}$$

and

$$\begin{aligned}(L(s)f,L(s)g) &= \int_{\mathcal{R}} \overline{sf(\bar{s}xs)}sg(\bar{s}xs)dx \\ &= \int_{\mathcal{C}} \overline{f(\bar{s}xs)}\bar{s}sg(\bar{s}xs)dx \\ &= \int_{\mathcal{R}} \overline{f(\bar{s}xs)}g(\bar{s}xs)dx \\ &= (f,g).\end{aligned}$$

Now define the bilinear form B on $\mathbf{R}^2_{0,m+1}$ - the Lie algebra of Spin$(m+1)$ - by

$$B(w,w') = 4[\overline{w}w']_0.$$

Then clearly B is non-degenerate, symmetric and associative and the set $\{\frac{1}{2}e_{ij} : i<j; i,j=0,1,\ldots,m\}$ is an orthonormal basis for $(\mathbf{R}^2_{0,m+1};B)$.

If R stands for any of the representations H or L and dR is the associated infinitesimal representation of $\mathbf{R}^2_{0,m+1}$, i.e. for each $w\in\mathbf{R}^2_{0,m+1}$ and $f\in\mathcal{E}(R;V)$,

$$dR(w)f(x) = \lim_{\varepsilon\to 0}\Big(\Big(\frac{R(e^{\varepsilon w})-1}{\varepsilon}\Big)f(x)\Big), \tag{1.34}$$

then the corresponding *Casimir operator* $C_B(R)$ is defined by (see also [43])

$$C_B(R) = \frac{1}{4}\sum_{i<j}(dR(e_{ij}))^2. \tag{1.35}$$

As is well known the Casimir operator $C_B(R)$ commutes with each $R(s)$, $s\in$ Spin$(m+1)$ and hence also with each $dR(w)$, $w\in\mathbf{R}^2_{m+1}$ (see e.g. [11]).

1.11.2 The Casimir operators $C_B(H)$ and $C_B(L)$

Approximating $e^{\varepsilon w}$ by $1+\varepsilon w$ in (1.34) and taking $w = e_{ij}$, we get for any $f \in \mathcal{E}(R; V)$ that

$$\begin{aligned}
dH(e_{ij})f(x) &= \lim_{\varepsilon\to 0} \frac{f(\overline{(1+\varepsilon e_{ij})}x(1+\varepsilon e_{ij})) - f(x)}{\varepsilon} \\
&= \lim_{\varepsilon\to 0} \frac{((1-\varepsilon e_{ij})x(1+\varepsilon e_{ij})) - f(x)}{\varepsilon} \\
&= \lim_{\varepsilon\to 0} \frac{1}{\varepsilon}[f(x + 2\varepsilon(e_i x_j - e_j x_i)) - f(x)] \\
&= \lim_{\varepsilon\to 0} \frac{1}{\varepsilon}[f(x + 2\varepsilon(e_i x_j - e_j x_i)) - f(x - 2\varepsilon e_j x_i) \\
&\qquad + f(x - 2\varepsilon e_j x_i) - f(x)] \\
&= 2x_j \frac{\partial f}{\partial x_i} - 2x_i \frac{\partial f}{\partial x_j} \\
&= -2L_{ij}
\end{aligned}$$

where

$$L_{ij} = x_i \partial_{x_j} - x_j \partial_{x_i}, \quad i < j. \tag{1.36}$$

Consequently

$$C_B(H) = \sum_{i<j} L_{ij}^2. \tag{1.37}$$

Again approximating $e^{\varepsilon w}$ by $1 + \varepsilon w$ in (1.34) and taking $w = e_{ij}$, we get for any $f \in \mathcal{E}(R; V)$ that

$$\begin{aligned}
dL(e_{ij}f(x) &= \lim_{\varepsilon\to 0} \frac{(1+\varepsilon e_{ij})f((1-\varepsilon e_{ij})x(1+\varepsilon e_{ij})) - f(x)}{\varepsilon} \\
&= \lim_{\varepsilon\to 0} \frac{1}{\varepsilon}[f((1-\varepsilon e_{ij})x(1+\varepsilon e_{ij})) - f(x)] \\
&\quad + \lim_{\varepsilon\to 0} \frac{1}{\varepsilon}[\varepsilon e_{ij} f((1-\varepsilon e_{ij})x(1+\varepsilon e_{ij}))] \\
&= dH(e_{ij})f(x) + e_{ij}f(x) \\
&= (dH(e_{ij}) + e_{ij}\mathbf{1})f(x)
\end{aligned}$$

whence

$$dL(e_{ij}) = dH(e_{ij}) + e_{ij}\mathbf{1}.$$

Consequently

$$\begin{aligned} C_B(L) &= \frac{1}{4}\sum_{i<j}(dL(e_{ij}))^2 \\ &= \frac{1}{4}\sum_{i<j}(dH(e_{ij}) + e_{ij}\mathbf{1})^2 \\ &= C_B(H) + \frac{1}{2}\sum_{i<j} e_{ij}dH(e_{ij}) - \frac{1}{4}\binom{m+1}{2}\mathbf{1} \end{aligned} \tag{1.38}$$

1.11.3 The case $L_2(S^m; V)$

When $R = S^m$ we have

Theorem 1. *On $\mathcal{E}(S^m; V)$ the Casimir operators $C_B(H)$ and $C_B(L)$ are given by*

$$(i) C_B(H) = \Delta$$

$$\begin{aligned} (ii) C_B(L) &= \Delta_\xi + \Gamma_\xi - \tfrac{1}{4}\tbinom{m+1}{2}\mathbf{1} \\ &= (m\mathbf{1} - \Gamma_\xi)\Gamma_\xi - \tfrac{1}{4}\tbinom{m+1}{2}\mathbf{1}. \end{aligned}$$

Proof.

(i) As we know (see also (0.16) and (0.17))

$$\begin{aligned} \Delta_\xi &= ((m-1)\mathbf{1} - \Gamma_\xi)\Gamma_\xi \\ &= [-\Gamma_\xi^2]_0 \end{aligned}$$

with

$$\Gamma_\xi = -\sum_{i<j} e_{ij}L_{ij}.$$

But, as

$$[-\Gamma_\xi^2]_0 = \sum_{i<j} L_{ij}^2$$

we indeed obtain that

$C_B(H) = \Delta_\xi.$

(ii) By virtue of Property (i) and (1.40) and the fact that

$$\begin{aligned}\frac{1}{2}\sum_{i<j} e_{ij} dH(e_{ij}) &= -\sum_{i<j} e_{ij}L_{ij} \\ &= \Gamma_\xi,\end{aligned}$$

we have that

$$\begin{aligned}C_B(L) &= \Delta_\xi + \Gamma_\xi - \frac{1}{4}\binom{m+1}{2}\mathbf{1} \\ &= (m\mathbf{1} - \Gamma_\xi)\Gamma_\xi - \frac{1}{4}\binom{m+1}{2}\mathbf{1}. \blacksquare\end{aligned}$$

Since the operators $C_B(H)$ and $C_B(L)$ are polynomials in Γ_ξ, Theorem 1.3.7 tells us that both admit a selfadjoint extension to $L_2(S^m; V)$. Moreover, maintaining the notation for these extensions, we have that $C_B(H)$ and $C_B(L)$ are selfadjoint operators with pure pointspectrum such that dom $C_B(H)$ = dom $C_B(L)$ = dom Δ_ξ.

As an easy computation shows, the spectrum $\sigma_p(C_B(L))$ consists of the eigenvalues $(-k)(k+m) - \frac{1}{4}\binom{m+1}{2}$, $k \in \mathbf{N}$, the corresponding eigenspace being given by the space of spherical monogenics $\mathcal{M}(k, V) = \mathcal{M}^+(k; V) \oplus \mathcal{M}^-(k; V)$.

Of course, the spectrum of $C_B(H) = C_B(\Delta_\xi)$ consists of the eigenvalues $(-k)(k+m-1)$, $k \in \mathbf{N}$, the corresponding eigenspaces being the space of spherical harmonics $\mathcal{H}(k; V) = \mathcal{M}^+(k; V) \oplus \mathcal{M}^-(k-1, V)$.

Summarizing we thus obtain

Theorem 2. *The Casimir operators $C_B(H) = \Delta_\xi$ and $C_B(L) = \Delta_\xi + \Gamma_\xi - \frac{1}{4}\binom{m+1}{2}\mathbf{1}$ are selfadjoint operators with pure point spectrum on $L_2(S^m; V)$. Moreover, for each $k \in \mathbf{N}$, we have*

(i) $\mathcal{M}(k; V) = \mathcal{M}^+(k; V) \oplus \mathcal{M}^-(k; V)$ is an eigenspace of $C_B(L)$ corresponding to the eigenvalue $(-k)(k+m) - \frac{1}{4}\binom{m+1}{2}$.

(ii) $\mathcal{H}(k; V) = \mathcal{M}^+(k; V) \oplus \mathcal{M}^-(k-1; V)$ is an eigenspace of $C_B(H)$ corresponding to the eigenvalue $(-k)(k+m-1)$.

*(iii) $\mathcal{M}^+(k; V)$ and $\mathcal{M}^-(k; V)$ are simultaneous eigenspaces of the operators $C_B(H)$
and $C_B(L)$.*

Corollary. *For each $k \in \mathbf{N}$, we have that*

(i) $[\mathbf{\Pi}(k), C_B(L)] = 0$

(ii) $[\mathbf{P}(k), C_B(L)] = 0$

(iii) $[\mathbf{Q}(k), C_B(L)] = 0$,

i.e. the projection operators $\mathbf{\Pi}(k), \mathbf{P}(k)$ *and* $\mathbf{Q}(k)$ *commute with* $C_B(L)$.

Remark. Theorem 2 tells us that the spaces $\mathcal{M}^+(k; V)$ and $\mathcal{M}^-(k; V)$ may also be interpreted as follows : they are the only simultaneous eigenspaces of the Casimir operators $C_B(H)$ and $C_B(L)$ acting on $L_2(S^m; V)$.

1.11.4 The Hilbert-Riesz transform revisited

In 1.8.3 we introduced the Hilbert-Riesz transform $\mathbf{H}_\xi$ on $L_2(S^m; V)$. Defining on $L_2(S^m; V)$ the operator

$$\begin{aligned} \Omega_\xi &= C_B(L) + \frac{1}{4}\binom{m+1}{2}\mathbf{1} \\ &= \Delta_\xi + \Gamma_\xi, \end{aligned}$$

the importance of the operator $\mathbf{H}_\xi$ becomes apparent from the following

Theorem. *On* $L_2(S^m; V)$*, we have that*

(i) $\sqrt{(\frac{m-1}{2})^2\mathbf{1} - \Delta_\xi} = \mathbf{H}_\xi(\Gamma_\xi - \frac{m-1}{2}\mathbf{1})$.

(ii) $\sqrt{(\frac{m}{2})^2\mathbf{1} - \Omega_\xi} = \mathbf{H}_\xi(\Gamma_\xi - \frac{m}{2}\mathbf{1})$.

Proof. For the sake of simplicity, we take $V = C$, the case $V = S$ being similar.

First of all notice that the operators $\mathbf{H}_\xi$ and Γ_ξ are commuting (on dom Γ_ξ).

Secondly, the operators $(\frac{m-1}{2})^2\mathbf{1} - \Delta_\xi$ and $(\frac{m^2}{2})\mathbf{1} - \Omega_\xi$ are positive definite and have pure point spectrum. Indeed, both operators are polynomials in Γ_ξ and are thus selfadjoint operators with pure point spectrum (see Theorem 1.3.7). Moreover, their domain is given by dom Δ_ξ.

Now taking for any $k \in \mathbf{N}$, orthonormal bases $\{P_k^{(i)} : i = 1, \ldots, K(m,k)\}$ and $\{Q_{k-1}^{(i)} : j = 1, \ldots, K(m; k-1)\}$ in $\mathcal{M}^+(k; C)$ and $\mathcal{M}^-(k-1; C)$ respectively, we have that

$$((\frac{m-1}{2})^2\mathbf{1} - \Delta_\xi)\left\{\begin{array}{c} P_k^{(i)} \\ Q_{k-1}^{(j)} \end{array}\right\} = (k + \frac{m-1}{2})^2\left\{\begin{array}{c} P_k^{(i)} \\ Q_{k-1}^{(j)} \end{array}\right\},$$

$$((\frac{m}{2})^2 - \Omega_\xi)P_k^{(i)} = (k + \frac{m}{2})^2 P_k^{(i)},$$

$$\left((\frac{m}{2})^2 - \boldsymbol{\Omega}_\xi\right)Q_{k-1}^{(j)} = (\frac{m}{2} + k - 1)^2 Q_{k-1}^{(j)}.$$

It thus follows that $\sqrt{(\frac{m-1}{2})^2\mathbf{1} - \Delta_\xi}$ and $\sqrt{(\frac{m}{2})^2\mathbf{1} - \boldsymbol{\Omega}_\xi}$ are selfadjoint operators having as domain the energy space H_{B_ξ} of $B_\xi = -\Delta_\xi$.

Here (see [84])

$$\begin{aligned} H_{B_\xi} &= \{f \in L_2(S^m; \mathcal{C}) : \textstyle\sum_{k=0}^{\infty} k(k+m-1)(\sum_{i=1}^{K(m;k)} |(f, P_k^{(i)})|^2 \\ &\qquad\qquad + \textstyle\sum_{j=1}^{K(m;k-1)} |(f, Q_{k-1}^{(j)})|^2) < +\infty\} \\ &= \{f \in L_2(S^m; \mathcal{C}) : \\ &\qquad \textstyle\sum_{k=0}^{\infty} k(k+m-1)(\|\mathbf{P}(k)f\|^2 + \|Q_{k-1}f\|^2) < +\infty\}. \end{aligned}$$

But, as

$$\operatorname{dom}\Gamma_\xi = \{f \in L_2(S^m; \mathcal{C}) : \sum_{k=0}^{\infty} k^2\|\mathbf{P}(k)f\|^2 + (k+m-1)^2\|Q_{k-1}f\|^2 < +\infty\}$$

we obtain that

$$\begin{aligned} H_{B_\xi} &= \operatorname{dom}\Gamma_\xi \\ &= \operatorname{dom}\mathbf{H}_\xi(\Gamma_\xi - \frac{m-1}{2}\mathbf{1}) \\ &= \operatorname{dom}\mathbf{H}_\xi(\Gamma_\xi - \frac{m}{2}\mathbf{1}). \end{aligned}$$

Finally

$$\begin{aligned} \mathbf{H}_\xi(\Gamma_\xi - \frac{m-1}{2}\mathbf{1})P_k^{(i)} &= (k + \frac{m-1}{2})P_k^{(i)}; \\ \mathbf{H}_\xi(\Gamma_\xi - \frac{m-1}{2}\mathbf{1})Q_{k-1}^{(j)} &= (k + \frac{m-1}{2})Q_{k-1}^{(j)}; \\ \mathbf{H}_\xi(\Gamma_\xi - \frac{m}{2}\mathbf{1})P_k^{(i)} &= (k + \frac{m}{2})P_k^{(i)} \end{aligned}$$

and

$$\mathbf{H}_\xi(\Gamma_\xi - \frac{m}{2}\mathbf{1})Q_{k-1}^{(j)} = (\frac{m}{2} + k - 1)Q_{k-1}^{(j)}. \blacksquare$$

1.12 Some operator equalities and commutation relations

In this subsection we gather some computational rules connected with the **Dirac** operator.

1.12.1 Some operator equalities

Let ∂_x be the Dirac operator in $\mathbf{R}^{m+1}$. Then passing to polar coordinates $x = r\xi$ we have the well-known operator equalities (see also §0)

$$
\begin{aligned}
&\text{(i) } \partial_x &&= \xi(\partial_r + \tfrac{1}{r}\Gamma_\xi) \; ; \\
&\text{(ii) } \Gamma_\xi &&= -x \wedge \partial_x \\
& &&= -\textstyle\sum_{i<j} e_{ij}(x_i\partial_{x_j} - x_j\partial_{x_i}) \\
& &&= -\textstyle\sum_{i<j} e_{ij}L_{ij} \; ; \\
&\text{(iii) } \xi\Gamma_\xi\xi = (\Gamma_\xi - m\mathbf{1}) \; ; && \\
&\text{(iv) } x\partial_x &&= x \cdot \partial_x + x \wedge \partial_x \\
& &&= -< x, \partial_x > + x \wedge \partial_x \\
& &&= -\textstyle\sum_{j=0}^{m} x_j\partial_{x_j} + \sum_{i<j} e_{ij}(x_i\partial_{x_j} - x_j\partial_{x_i}) \\
& &&= -r\partial_r - \Gamma_\xi.
\end{aligned}
\tag{1.39}
$$

These equalities allow us to obtain the following propositions.

Proposition 1. *Let f be a smooth function in $\mathbf{R}^{m+1}$. Then*

(i) $\partial_x(rf) = \xi f + r\partial_x f$.

(ii) $\partial_x(\xi f) = -\frac{m}{r} f + \xi\partial_x f + 2\frac{\Gamma_\xi}{r} f$.

(iii) $\partial_x(xf) = -(m+1)f + x\partial_x f + 2\Gamma_\xi f$.

Proof.

$$
\begin{aligned}
\text{(i)}\ \partial_x(rf) &= (\xi(\partial_r + \tfrac{1}{r}\Gamma_\xi))(rf) \\
&= \xi f + r\xi(\partial_r + \tfrac{1}{r}\Gamma_\xi)f \\
&= \xi f + r\partial_x f.
\end{aligned}
$$

$$
\begin{aligned}
\text{(ii)}\ \partial_x(\xi f) &= \xi(\partial_r + \tfrac{1}{r}\Gamma_\xi)(\xi f) \\
&= \xi\partial_r\xi f + \tfrac{1}{r}\xi\Gamma_\xi\xi f \\
&= -\partial_r f + \tfrac{1}{r}\Gamma_\xi f - \tfrac{m}{r} f \\
&= -\tfrac{m}{r} f + \xi\partial_x f + 2\tfrac{\Gamma_\xi}{r} f.
\end{aligned}
$$

(iii) By means of (i) and (ii) we find :

$$
\begin{aligned}
\partial_x(xf) &= \partial_x(r(\xi f)) \\
&= \xi(\xi f) + r\partial_x(\xi f) \\
&= -f - mf + r\xi\partial_x f + 2\Gamma_\xi f \\
&= -(m+1)f + x\partial_x f + 2\Gamma_\xi f. \blacksquare
\end{aligned}
$$

Corollary. *We have :*

$$\begin{aligned} x\partial_x - \partial_x x &= (m+1)1 - 2\Gamma_\xi \\ &= 1 - \Gamma_\xi - \xi\Gamma_\xi\xi. \end{aligned}$$

Proposition 2. *For all $P_k \in M^+(k;V)$ and smooth* C*-valued functions f in $]0,+\infty[$ we have :*

(i) $\partial_x(f(r)P_k(x)) = \xi f'(r)P_k(x)$

(ii) $\partial_x(\xi f(r)P_k(x)) = -(f'(r) + \frac{m+2k}{r}f(r))P_k(x)$.

Proof.

$$\begin{aligned} \text{(i) } \partial_x(f(r)P_k(x)) &= (\partial_x(f(r))P_k(x) + f(r)\partial_x P_k(x) \\ &= (\partial_x(f(r)))P_k(x) \\ &= [(\xi(\partial_r + \tfrac{1}{r}\Gamma_\xi))(f(r))]P_k(x) \\ &= \xi f'(r)P_k(x) \end{aligned}$$

and this since Γ_ξ is an angular operator.

(ii) By means of (i) and Proposition 1 (ii) we get :

$$\begin{aligned} \partial_x(\xi(f(r)P_k(x)) &= -\frac{m}{r}f(r)P_k(x) + \xi\partial_x(f(r)P_k(x)) + 2\frac{\Gamma_\xi}{r}(f(r)P_k(x)) \\ &= -\frac{m}{r}f(r)P_k(x) + \xi\xi f'(r)P_k(x) + \frac{2}{r}f(r)\Gamma_\xi(P_k(x)) \\ &= -(\frac{m}{r}f(r) + f'(r) + \frac{2k}{r}f(r))P_k(x) \\ &= -(\frac{m+2k}{r}f(r) + f'(r))P_k(x). \blacksquare \end{aligned}$$

Remark. Functions of the form $f(r)P_k(x)$ and $\xi f(r)P_k(x)$ are met explicitly when one deals with expressions like

$$(\sum_{j=0}^{n} a_j x^j)P_k(x), \quad a_j \in \mathbf{C}.$$

Indeed, passing to polar coordinates $x = r\xi$, we find, since

$$\begin{aligned}\sum_{j=0}^{n} a_j x^j &= \sum_{j \text{ even}} a_j x^j + \sum_{j \text{ odd}} a_j x^j \\ &= \sum_{j=2\ell} a_{2\ell}(-1)^\ell r^{2\ell} + \sum_{j=2\ell+1} a_{2\ell+1}(-1)^\ell r^{2\ell+1}\xi\end{aligned}$$

that

$$\sum_{j=0}^{n} a_j x^j = f(r) + \xi g(r).$$

By means of Proposition 2 we have at our disposal some rules for calculating $\partial_x((\sum_{j=0}^n a_j x^j)P_k(x))$.

Because of their importance we now give explicit formulas for the particular case $x^s P_k(x)$, $s \in \mathbf{N}$.

Proposition 3. *Let $s \in \mathbf{N}$ and $P_k \in M^+(k;V)$. Then*

(i) $$\Gamma_\xi(x^s P_k(x)) = \begin{cases} (-k)x^s P_k(x) & \text{for } s \text{ even} \\ (k+m)x^s P_k(x) & \text{for } x \text{ odd} \end{cases}$$

(ii) $$\partial_x(x^s P_k(x)) = \begin{cases} -sx^{s-1}P_k(x) & \text{for } s \text{ even} \\ -(s+2k+m)x^{s-1}P_k(x) & \text{for } s \text{ odd} \end{cases}$$

Proof.

(i) Let $s = 2\ell$ be even. Then, as $x^{2\ell}P_k(x) = (-1)^\ell r^{2\ell+k}P_k(\xi)$ we get, Γ_ξ being an angular operator, that

$$\begin{aligned}\Gamma_\xi(x^{2\ell}P_k(x)) &= (-1)^\ell r^{2\ell+k}\Gamma_\xi(P_k(\xi)) \\ &= (-1)^\ell r^{2\ell+k}(-k)P_k(\xi) \\ &= (-k)x^{2\ell}P_k(x)\end{aligned}$$

If $s = 2\ell+1$ is odd, then, as $x^{2\ell+1}P_k(x) = (-1)^\ell r^{2\ell+1+k}\xi P_k(\xi)$, we obtain :

$$\begin{aligned}\Gamma_\xi(x^{2\ell+1}P_k(x)) &= (-1)^\ell r^{2\ell+1+k}\Gamma_\xi(\xi P_k(\xi)) \\ &= (-1)^\ell r^{2\ell+1+k}(k+m)(\xi P_k(\xi)) \\ &= (k+m)x^{2\ell+1}P_k(x).\end{aligned}$$

(ii) If $s = 2\ell$ is even we have, by virtue of Proposition 2 (i), that

$$\begin{aligned}
\partial_x(x^{2\ell}P_k(x)) &= \partial_x((-1)^\ell r^{2\ell}P_k(x)) \\
&= (-1)^\ell \xi(2\ell)r^{2\ell-1}P_k(x) \\
&= (-1)^\ell(2\ell)r^{2\ell-2}(r\xi)P_k(x) \\
&= (-1)(2\ell)x^{2\ell-2}xP_k(x) \\
&= -sx^{2\ell-1}P_k(x).
\end{aligned}$$

If $s = 2\ell + 1$ is odd then, by means of Proposition 1 (iii) we get :

$$\begin{aligned}
&\partial_x(x^{2\ell+1}P_k(x)) \\
&= \partial_x(x(x^{2\ell}P_k(x)) \\
&= -(m+1)x^{2\ell}P_k(x) + x\partial_x(x^{2\ell}P_k(x)) + 2\Gamma_\xi(x^{2\ell}P_k(x)) \\
&= -(m+1)x^{2\ell}P_k(x) - (s-1)x^{2\ell}P_k(x) - (2k)x^{2\ell}P_k(x) \\
&= -(s+2k+m)x^{s-1}P_k(x). \blacksquare
\end{aligned}$$

Corollary 1.

(i) For each $P_k \in M^+(k;V)$,

$$\partial_x(xP_k(x)) = -(2k+m+1)P_k(x)$$

(ii) For each $s \in \mathbf{N}$,

$$\partial_x(x^s) = \begin{cases} -sx^{s-1} & \textit{for} \;\; s \textit{ even} \\ -(s+m)x^{s-1} & \textit{for} \;\; s \textit{ odd} \end{cases}$$

To conclude this subsection we give a characterization of polynomials of the form $x^sP_k(x)$.

As we know,

$$\begin{aligned}
x\partial_x &= x.\partial_x + x \wedge \partial_x \\
&= -r\partial_r - \Gamma_\xi.
\end{aligned}$$

As $x^s P_k(x)$ is a homogeneous polynomial of degree $(k+s)$ in $(x_0, x_1, \ldots, x_m)$, we have by Euler's formula that

$$\begin{aligned}(r\partial_r)(x^s P_k(x)) &= (\sum_{j=0}^{m} x_j \partial_{x_j})(x^s P_k(x)) \\ &= (k+s)(x^s P_k(x)).\end{aligned}$$

Furthermore (see Proposition 3 (i))

$$\Gamma_\xi(x^s P_k(x)) = \mu x^s P_k(x)$$

with $\mu = (-k)$ for s even and $\mu = (k+m)$ for s odd.

Now putting $E = x.\partial_x$, we prove

Proposition 4. *Let $P(x)$ be a V-valued polynomial of degree ℓ in $(x_0, x_1, \ldots, x_m)$ such that for some $\lambda, \mu \in \mathbf{R}$,*

$$E(P(x)) = \lambda P(x)$$

and

$$\Gamma_\xi(P(x)) = \mu P(x).$$

Then there exists a unique $s \leq \ell$ such that

$$P(x) = x^s P_{\ell-s}(x).$$

Proof. We may write out $P(x)$ as

$$P(x) = \sum_{k=0}^{\ell} R_k(x)$$

where each R_k is a homogeneous polynomial of degree k in $(x_0, x_1, \ldots, x_m)$.

We thus obtain that

$$\begin{aligned}E(P(x)) &= \sum_{k=0}^{\ell} k R_k(x) \\ &= \sum_{k=0}^{\ell} \lambda R_k(x)\end{aligned}$$

whence $\lambda = \ell$ or $P(x)$ is a homogeneous V-valued polynomial of degree ℓ.

By the Fischer decomposition (see Theorem 1.10.1) we have that

$$\begin{aligned} P(x) &= \sum_{s+0}^{\ell} x^s P_{\ell-s}(x) \\ &= \sum_{s \text{ even}} x^s P_{\ell-s}(x) + \sum_{s \text{ odd}} x^s P_{\ell-s}(x). \end{aligned}$$

Consequently, by means of Proposition 3 (i) and the assumption made,

$$\begin{aligned} \Gamma_\xi(P(x)) &= \sum_{s \text{ even}} -(\ell - s)x^s P_{\ell-s}(x) + \sum_{s \text{ odd}} (\ell - s + m)x^s P_{\ell-s}(x) \\ &= \sum_{s \text{ even}} \mu x^s P_{\ell-s}(x) + \sum_{s \text{ odd}} \mu x^s P_{\ell-s}(x), \end{aligned}$$

whence there ought to exist $s \leq \ell$ such that either $\mu = -(\ell - s)$ or $\mu = (\ell - s + m)$.
In both cases, $P(x) = x^s P_{\ell-s}(x)$.■

1.12.2 Some commutation relations

In this subsection we investigate commutation relations among operators acting on a suitable space of functions, all of them being connected in some sense or another with the Dirac operator ∂_x.

First notice that by the Corollary to Proposition 1 in 1.12.1,

$$2\Gamma_\xi = (m+1)\mathbf{1} - (x\partial_x - \partial_x x)$$

Consequently Γ_ξ belongs to the (complex) algebra of operators generated by the operators x and ∂_x.

Proposition 1. *For all $s \in Spin(m+1)$, the operator $L(s)$ commutes with the operators x and ∂_x, i.e.*

(i) $[x, L(s)] = 0$

(ii) $[\partial_x, L(s)] = 0$.

Proof.

$$\begin{aligned} \text{(i) } L(s)(xf(x)) &= s(\bar{s}xs)f(\bar{s}xs) \\ &= xsf(\bar{s}xs) \\ &= xL(s)f(x). \end{aligned}$$

(ii) Putting $\partial_x f(x) = g(x)$ and $y = \bar{s}xs$ we have :

$$\begin{aligned} L(s)g(x) &= sg(\bar{s}xs) \\ &= sg(y) \\ &= s\partial_y f(y) \\ &= s\partial_{\bar{s}xs} f(\bar{s}xs). \end{aligned}$$

Now $\partial_{\bar{s}xs} = \bar{s}\partial_x s$.

Indeed, as for $s \in \text{Spin}(m+1)$, $x \to \bar{s}xs$ is a rotation, there exists $Q \in \text{SO}(m+1)$ such that with respect to the standard orthonormal basis of $\mathbf{R}^{m+1}$, $y = Qx$ whence

$$\begin{pmatrix} \partial_{y_0} \\ \partial_{y_1} \\ \vdots \\ \partial_{y_m} \end{pmatrix} = Q \begin{pmatrix} \partial_{x_0} \\ \partial_{x_1} \\ \vdots \\ \partial_{x_m} \end{pmatrix}.$$

Consequently

$$\begin{aligned} \partial_y &= \sum_{i=0}^{m} e_i \partial_{y_i} \\ &= \sum_{i=0}^{m} e_i (\sum_{k=0}^{m} Q_{ik} \partial_{x_k}) \\ &= \sum_{k=0}^{m} (\sum_{i=0}^{m} e_i Q_{ik}) \partial_{x_k}. \end{aligned}$$

But for each $k = 0, 1, \ldots, m$,

$$\sum_{i=0}^{m} e_i Q_{i_k} = e'_k = \bar{s} e_k s$$

so that

$$\partial_y = \sum_{k=0}^{m} \bar{s} e_k s \partial_{x_k} = \bar{s}\partial_x s.$$

We thus obtain that

$$\begin{aligned} L(s)(\partial_x f(x)) &= s\bar{s}\partial_x s f(\bar{s}xs) \\ &= \partial_x s f(\bar{s}xs) \\ &= \partial_x (L(s) f(x)). \blacksquare \end{aligned}$$

Corollary 1. *For all $i, j = 0, 1, \ldots, m$, $i \neq j$, the operator $dL(e_{ij})$ commutes with the operators x and ∂_x, i.e.*

(i) $[x, dL(e_{ij})] = 0$

(ii) $[\partial_x, dL(e_{ij})] = 0$.

Proof. As by definition

$$dL(e_{ij})f = \lim_{\varepsilon \to 0} \frac{1}{\varepsilon}[L(1 + \varepsilon e_{ij}) - L(1)]f$$

we obtain the desired results from the corresponding ones concerning $L(s)$ by taking limits. ■

Corollary 2.

(i) $[\partial_x, C_B(L)] = 0$

(ii) $[\partial_x, \Pi(k)] = 0$ *for all* $k \in \mathbf{N}$.

Proof.

(i) As $C_B(L) = \frac{1}{4}\sum_{i<j}(dL(e_{ij}))^2$ and $[\partial_x, dL(e_{ij})] = 0$ for all $i < j$, we obtain the desired result.

(ii) $\Pi(k)$ is the projection operator on the eigenspace

$$\mathcal{M}(k; V) = \mathcal{M}^+(k; V) \oplus \mathcal{M}^-(k; V)$$

of $C_B(L)$. ■

Remark. Once we have the relation $[x, dL(e_{ij})] = 0$, we may derive the relation $[\partial_x, dL(e_{ij}] = 0$ from it straightforwardly by using the Fourier transform $\mathcal{F}^+$.
Indeed, as

$$dL(e_{ij}) = x_i\partial_{x_j} - x_j\partial_{x_i} + e_{ij}\mathbf{1}$$

we have that

$$\mathcal{F}^+(dL(e_{ij})f(x))(y) = dL(e_{ij})((\mathcal{F}^+ f)(y)).$$

Hence $[x, dL(e_{ij})]f(x) = 0$ implies that

$$[-i\partial_y, dL(e_{ij})]\mathcal{F}^+ f(y) = 0$$

or

$$[\partial_y, dL(e_{ij})] = 0.$$

Proposition 2. *On* $L_2(S^m; V)$ *we have that*

(i) $\Gamma_\xi(\xi f(\xi)) = (\xi(m\mathbf{1} - \Gamma_\xi))f(\xi)$ or $\Gamma_\xi I^\star = I^\star(m\mathbf{1} - \Gamma_\xi)$

(ii) $[C_B(L), I^\star] = 0.$

Proof.

(i) From (1.41) (iii) we obtain that

$$\begin{aligned}\Gamma_\xi(\xi f) &= \overline{\xi}(\Gamma_\xi - m\mathbf{1})f \\ &= \xi(m\mathbf{1} - \Gamma_\xi)f\end{aligned}$$

whence

$$\Gamma_\xi I^\star = I^\star(m\mathbf{1} - \Gamma_\xi).$$

(ii) As $C_B(L) = (m\mathbf{1} - \Gamma_\xi)\Gamma_\xi - \frac{1}{4}\binom{m+1}{2}\mathbf{1}$
and $((m\mathbf{1} - \Gamma_\xi)\Gamma_\xi)I^\star = I^\star(\Gamma_\xi(m\mathbf{1} - \Gamma_\xi))$, the result follows immediately. ■

We end up this subsection by enumerating some obvious commutation relations.

Proposition 3.

(i) $[C_B(L), \partial_r] = 0$

(ii) $[C_B(H), \partial_r] = 0$

(iii) $[C_B(L), \Gamma_\xi] = 0$

(iv) $[C_B(H), \Gamma_\xi] = 0.$

1.13 From Clifford analysis in $\mathbf{R}^2$ *to complex analysis in the plane*

Originally quaternionic analysis and Clifford analysis started, respectively, as generalizations of complex analysis in the plane to four and $(m+1)$-dimensional Euclidean space. For a historical account we refer to [10].

In this section we want to show how complex analysis in the plane completely fits into the general scheme of Clifford analysis and this by considering Weyl's equation in $\mathbf{R}^2$.

It is thus nice to see how formulae familiar in complex analysis reappear in studying solutions to Weyl's equation in $\mathbf{R}^2$.

We shall indicate this explicitly for some basic formulae established in the foregoing sections.

Let us first consider Dirac's and Weyl's equation for the case $m = 1$.

The Clifford algebra $\mathbf{C}_2 = \mathbf{R}_{0,2} \otimes_{\mathbf{R}} \mathbf{C}$ is four dimensional and has $\{1, e_0, e_1, e_0 e_1\}$ as basis . In the case of the Dirac equation, $\mathbf{R}^2$ is identified with the real subspace $\text{span}_{\mathbf{R}}\{e_0, e_1\}$ of $\mathbf{C}_2$, i.e. $\mathbf{R}^2 \cong \{x = x_0 e_0 + x_1 e_1 : x_0, x_1 \in \mathbf{R}\}$ and for an open subset $\Omega \subset \mathbf{R}^2$ we are looking at null-solutions of $f \in C_1(\Omega; V)$ to the equation

$$\partial_x f = 0$$

where $\partial_x = e_0 \partial_{x_1} + e_1 \partial_{x_2}$ and $V = \mathbf{C}_2$ or S_2.

In the case of the Weyl equation, we put $\varepsilon_1 = \bar{e}_0 e_1$ and we identify $\mathbf{R}^2$ with $\text{span}_{\mathbf{R}}\{1, \varepsilon_1\}$, i.e. $\mathbf{R}^2 = \{x = x_0 + x_1 \varepsilon_1 : x_0, x_1 \in \mathbf{R}\}$. For an open subset $\Omega \subset \mathbf{R}^2$ we are studying solutions $f \in C_1(\Omega; V)$ to the equation

$$D_x f = 0$$

where $D_x = \partial_0 + \varepsilon_1 \partial_1$ and $V = \mathbf{C}_2$ or S_2^+.

Let us take the case $V = S_2^+ = \mathbf{C}_2^+ I$ where $\mathbf{C}_2^+ = \text{span}_{\mathbf{C}}\{1, \varepsilon_1\}$ and $I = \frac{1}{2}(1 + i\varepsilon_1)$. Then by straightforward calculations we find that

$$S_2^+ = \mathbf{C}_2^+ I = (\mathbf{R} \oplus \mathbf{R}\varepsilon_1) I \cong \mathbf{R} \oplus \mathbf{R}\varepsilon_1 \cong \mathbf{C}.$$

Consequently S_2^+ is a field isomorphic to $\mathbf{R} \oplus \mathbf{R}\varepsilon_1$ and S_2^+-valued maps may thus be identified with $\mathbf{R} \oplus \mathbf{R}\varepsilon_1$-valued ones. Henceforth we put $K = \mathbf{R} \oplus \mathbf{R}\varepsilon_1$ and, as we saw, $K \cong \mathbf{C}$.

In the following table we reinterpret some basic function theoretic results obtained in the foregoing sections into the language of complex analysis. For the space of functions holomorphic on some open subset $\Omega \subset \mathbf{C}$ we here use the standard notation $\mathcal{O}(\Omega)$.

	Clifford analysis in $\mathbf{R}^2$	Complex analysis in $\mathbf{C}$
variable	$x = x_0 + x_1\varepsilon_1$	$z = x + iy$
functions	$f : \mathbf{R}^2 \to K$	$f : \mathbf{C} \to \mathbf{C}$
equation	Weyl : $(\partial_{x_0} + \varepsilon_1\partial_{x_1})f = 0$	Cauchy-Riemann : $\partial_{\bar{z}} f = 0$
Cauchy kernel	$E(x) = \frac{1}{2\pi}\frac{1}{x}$	$E(z) = \frac{1}{2\pi i}\frac{1}{z}$
differential form	$d\sigma = \frac{dx_0 + \varepsilon_1 dx_1}{\varepsilon_1}$	$\frac{dz}{i}$
Cauchy's Theorem	$\int_{\partial\Sigma} d\sigma f = 0$	$\int_{\partial\Sigma} \frac{dz}{i} f(z) = 0$
Cauchy's Formula	$f(x) = \frac{1}{A_1}\int_{\partial\Sigma} \frac{\bar{x}-\bar{y}}{\lvert x-y\rvert^2} d\sigma_y f(y)$	$f(z) = \frac{1}{2\pi}\int_{\partial\Sigma} \frac{1}{u-z}\frac{du}{i} f(u)$ $= \frac{1}{2\pi i}\int_{\partial\Sigma} \frac{f(u)du}{u-z}$

Inversion	$\mathbf{I}f(x) = \frac{\bar{x}}{\lvert x\rvert^2} f(\frac{\bar{x}}{\lvert x\rvert^2})$	$\mathbf{I}f(z) = \frac{1}{z} f(\frac{1}{z})$
Space of homogeneous monogenic (resp. holomorphic) polynomials of degree k Basic elements	$M^+(k;K)$ $k=0: V_0(x) = 1$ $k=1: V_1(x) = z_1 = x_1 - x_0\varepsilon_1$ $k \quad : V_k(x) = V_{\underbrace{1\ldots1}_{k}}(x) = \frac{1}{k!}z_1^k$	$\mathcal{O}_k^+$ $z^0 = 1$ $\frac{z}{i}$ $\frac{1}{k!}(\frac{z}{i})^k$
$\mathbf{I}M^+(k;K)(\text{resp.} \mathbf{I}\mathcal{O}_k^+)$ Basic elements	$M^-(k;K)$ $k=0: W_0(x) = \frac{1}{2\pi}\frac{1}{x}$ $k: W_{\underbrace{1\ldots1}_{k}}(x) = (-1)^k \partial_{x_1}^k E(x)$ $= \frac{k!}{2\pi}(\varepsilon_1)^k \frac{1}{x^{k+1}}$	$\mathcal{O}_k^-$ $\frac{1}{2\pi}\frac{1}{z}$ $\frac{k!}{2\pi}(i)^k \frac{1}{z^{k+1}}$
Space of spherical harmonics of degree k	$\mathcal{H}(k;K) = \mathcal{M}^+(k) \oplus \mathcal{M}^-(k-1)$ $= \text{span}_K\{e^{\varepsilon_1 k\theta}, e^{-\varepsilon_1 k\theta} : k \in \mathbf{N}\}$	$\mathcal{H}_k = \text{span}_{\mathbf{C}}\{e^{ik\theta}, e^{-ik\theta} : k \in \mathbf{N}\}$

Taylor series in $\mathring{B}(R)$	$f(x) = \sum_{k=0}^{\infty} P_k f(x)$ $= \sum_{k=0}^{\infty} V_k(x)\partial_{x_1}^k f(0)$ $= \sum_{k=0}^{\infty} \frac{1}{k!} z_1^k \partial_{x_1}^k f(0)$	$f(z) = \sum_{k=0}^{\infty} \frac{1}{k!}(\frac{z}{i})^k (iD_z)^k f(0)$ $= \sum_{k=0}^{\infty} \frac{z^k}{k!} D_z^k f(0)$
Laurent series in $B'(R) = \mathring{B}(R)\setminus\{0\}$	$f(x) = \sum_{k=0}^{\infty}(P_k f(x) + Q_k f(x))$ $= \sum_{k=0}^{\infty}(V_k(x)\lambda_k + W_k(x)\mu_k)$ with $\lambda_k = \int_{\partial B(r)} W_k(y) d\sigma_y f(y)$ and $\mu_k = \int_{\partial B(r)} V_k(y) d\sigma_y f(y)$ or $f(x) = \sum_{k=0}^{\infty} z_1^k \frac{(\varepsilon_1)^k}{2\pi} \int_{\partial B(r)} \frac{1}{y^{k+1}} d\sigma_y f(y)$ $+ \sum_{k=0}^{\infty} \frac{k!}{2\pi}(\varepsilon_1)^k \frac{1}{x^{k+1}} \int_{\partial B(r)} w_1^k d\sigma_y f(y)$ with $w_1 = y_1 - y_0\varepsilon_1$	$f(z) = \sum_{k=0}^{\infty}(\frac{z}{i})^k \frac{i^k}{2\pi} \int_{\partial B(r)} \frac{1}{u^{k+1}} \frac{du}{i} f(u)$ $+ \sum_{k=0}^{\infty} \frac{k!}{2\pi} i^k \frac{1}{z^{k+1}} \int_{\partial B(r)} \frac{1}{k!}(\frac{u}{i})^k \frac{du}{i} f(u)$ $= \sum_{k=0}^{\infty} z^k \frac{1}{2\pi i} \int_{\partial B(r)} \frac{f(u)du}{u^{k+1}}$ $+ \sum_{k=0}^{\infty} \frac{1}{z^{k+1}} \frac{1}{2\pi i} \int_{\partial B(r)} f(u) u^k du$

2 Polyaxially monogenic functions

In this section we study nullsolutions of the Dirac operator in an open subset $\underline{\Omega}$ of $\mathbf{R}^m$ where it is assumed that, when $\mathbf{R}^m$ is split into $\mathbf{R}^m = \mathbf{R}^{m_1} \times \mathbf{R}^{m_2} \times \ldots \times \mathbf{R}^{m_n} (n \geq 2)$, $\underline{\Omega}$ is invariant under the compact subgroup $\text{Spin}(m_1) \times \text{Spin}(m_2) \times \ldots \times \text{Spin}(m_n)$ of $\text{Spin}(m)$.

For such functions, called polyaxially monogenic functions, a Laurent expansion is obtained into special functions which are solutions of first order systems of partial differential equations, called Laurent equations. These equations are written out more explicitly in some special cases.

2.1 Notations and definitions

Let $m \in \mathbf{N}$ and let $m_1, \ldots, m_n \in \mathbf{N}$ be such that $m_j \geq 1$ and $\sum_{j=1}^n m_j = m\,(n \geq 2)$. Furthermore let S^{m_j-1} be the unit sphere in $\mathbf{R}^{m_j}$ and consider the set $\mathcal{S} = \prod_{j=1}^n S^{m_j-1}$ in $\mathbf{R}^m = \prod_{j=1}^n \mathbf{R}^{m_j}$.

In what follows arbitrary elements $\underline{x} \in \mathbf{R}^m$ and $\underline{\omega} \in \mathcal{S}$ will be written as

$$\underline{x} = (\underline{x}_1, \ldots, \underline{x}_n), \quad \underline{x}_j \in \mathbf{R}^{m_j}$$

and

$$\underline{\omega} = (\underline{\omega}_1, \ldots, \underline{\omega}_n), \quad \underline{\omega}_j \in S^{m_j-1}.$$

Moreover $d\Sigma(\underline{\omega}) = \prod_{j=1}^n d\Sigma(\underline{\omega}_j)$ and $d\underline{x} = \prod_{j=1}^n d\underline{x}_j$ stand for, respectively, the elementary surface and volume elements on $\mathcal{S}$ and $\mathbf{R}^m$.

Now let $e = \{e_1, \ldots, e_m\}$ be an orthonormal basis for $\mathbf{R}^{0,m}$ and consider the Clifford algebra $\mathcal{C} = \mathbf{R}_{0,m} \otimes_{\mathbf{R}} \mathbf{C}$.

Putting for $j = 1, \ldots, n$,

$$M_j = \{m_1 + \ldots + m_{j-1} + 1, \ldots, m_1 + \ldots + m_j\}$$

we have that $\text{Pin}(m_j)$, the group generated by $\{e_k : k \in M_j\}$ is then a subgroup of $\text{Pin}(m)$, whence any representation space for $\text{Pin}(m)$ automatically becomes a representation space for $\text{Pin}(m_j)$ too.

We also write for $\underline{x}_j \in \mathbf{R}^{m_j}$ and $\underline{\omega}_j \in S^{m_j-1}$,

$$\underline{x}_j = \sum_{k \in M_j} e_k x_k$$

$$\underline{\omega}_j = \sum_{k \in M_j} e_k \omega_k,$$

$$\underline{x}_j = r_j \underline{\omega}_j \quad , \quad r_j = |\underline{x}_j|.$$

Introducing the Dirac operators $\partial_{\underline{x}}$ and $\partial_{\underline{x}_j}$ on $\mathbf{R}^m$ and $\mathbf{R}^{m_j}$ respectively, we thus have that

$$\partial_{\underline{x}_j} = \sum_{k \in M_j} e_k \partial_{x_k} = \underline{\omega}_j(\partial_{r_j} + \frac{1}{r_j}\Gamma_{\underline{\omega}_j})$$

and

$$\partial_{\underline{x}} = \sum_{j=1}^{n} \partial_{\underline{x}_j}.$$

Here

$$\begin{aligned} \Gamma_{\underline{\omega}_j} &= -\underline{x}_j \wedge \partial_{\underline{x}_j} \\ &= -\sum_{\substack{k,l \in M_j \\ k<l}} e_{kl}(x_k \partial_{x_l} - x_l \partial_{x_k}). \end{aligned}$$

is the spherical Dirac operator on S^{m_j-1}.

Now consider the space $L_2(\mathcal{S}; V)$ of V-valued square integrable functions on $\mathcal{S}, V$ being a basic representation space S of $\mathrm{Pin}(m)$ or a direct sum of such representation spaces.

If we provide $L_2(\mathcal{S}; V)$ with the inner product

$$(f,g) = \int_{\mathcal{S}} \bar{f}(\underline{\omega}) g(\underline{\omega}) d\Sigma(\underline{\omega}),$$

then for each $j = 1, \ldots, n$, the following representations H_j and L_j of the compact subgroup $\mathrm{Spin}(m_1) \times \mathrm{Spin}(m_2) \times \ldots \times \mathrm{Spin}(m_n)$ of $\mathrm{Spin}(m)$ on $L_2(\mathcal{S}; V)$ may be introduced.

Take $s = (s_1, \ldots, s_n) \in \mathrm{Spin}(m_1) \times \ldots \times \mathrm{Spin}(m_n)$ and $f(\underline{\omega}) \in L_2(\mathcal{S}; V)$, then

$$H_j(s) f(\underline{\omega}) = f(\bar{s}_j \underline{\omega} s_j)$$

and

$$L_j(s) f(\underline{\omega}) = s_j f(\bar{s}_j \underline{\omega} s_j).$$

Clearly H_j and L_j are unitary representations admitting, with respect to the bilinear form $B(w, w') = \frac{1}{2}[\overline{w} w']_0$ on $\mathrm{Lie}(\mathrm{Spin}(m)) = \mathbf{R}^2_{0,m}$, the Casimir operators

$$C(H_j) = \Delta_{\underline{\omega}_j} = \Gamma_{\underline{\omega}_j}(m_j - 2 - \Gamma_{\underline{\omega}_j})$$

and

$$\begin{aligned} C(L_j) &= \Delta_{\underline{\omega}_j} + \Gamma_{\underline{\omega}_j} - \frac{1}{4}\binom{m_j}{2}\mathbf{1} \\ &= \Gamma_{\underline{\omega}_j}(m_j - 1 - \Gamma_{\underline{\omega}_j}) - \frac{1}{4}\binom{m_j}{2}\mathbf{1}. \end{aligned}$$

Here $\Delta_{\underline{\omega}_j}$ is the Laplace-Beltrami operator on S^{m_j-1}.

Notice that for $i \neq j$

$$\begin{aligned} \underline{x}_i\underline{x}_j + \underline{x}_j\underline{x}_i &= 0; \\ \partial_{\underline{x}_i}\partial_{\underline{x}_j} + \partial_{\underline{x}_j}\partial_{\underline{x}_i} &= 0; \\ \Gamma_{\underline{\omega}_i}\Gamma_{\underline{\omega}_j} &= \Gamma_{\underline{\omega}_j}\Gamma_{\underline{\omega}_i}; \\ \Gamma_{\underline{\omega}_i}\underline{x}_j^s &= \underline{x}_j^s\Gamma_{\underline{\omega}_i}; \\ \Gamma_{\underline{\omega}_i}C(L_j) &= C(L_j)\Gamma_{\underline{\omega}_i}. \end{aligned} \tag{2.1}$$

These commutation or anti-commutation relations surely hold on $L_2(S;V) \cap \mathcal{E}(S;V)$ but, in analogy with the discussion made in section 1.3, they remain valid on appropriate domains of definition of the operators considered.

2.2 Separately spherical monogenics

Let again V be a basic representation space S of $\mathrm{Pin}(m)$ or a direct sum of such representation spaces and let $\mathbf{R}^m$ admit the splitting $\mathbf{R}^m = \mathbf{R}^{m_1} \times \mathbf{R}^{m_2} \times \ldots \times \mathbf{R}^{m_n}$.

Furthermore let $(k) = (k_1, \ldots, k_n) \in \mathbf{N}^n$ and let $P_{(k)}(\underline{x})$ denote a V-valued polynomial in $\underline{x}$ which is homogeneous of degree k_j in $\underline{x}_j, j = 1, \ldots, n$. Then we also say that $P_{(k)}$ is a homogeneous V-valued polynomial of degree (k) in $\underline{x}$.

If moreover

$$\partial_{\underline{x}_j} P_{(k)} = 0, \quad j = 1, \ldots, n,$$

then $P_{(k)}$ is called an *inner separately spherical monogenic polynomial of degree* (k). The space of such polynomials is denoted by $M^+_{(k)}$ while $\mathcal{M}^0_{(k)}$ stands for the space consisting of the restrictions of elements in $M^+_{(k)}$ to S.

Putting for each $\sigma \in \{0,1\}^n$,

$$\underline{\omega}^\sigma = \underline{\omega}_1^{\sigma_1}\underline{\omega}_2^{\sigma_2} \ldots \underline{\omega}_n^{\sigma_n},$$

and

$$\mathcal{M}^\sigma_{(k)} = \{\underline{\omega}^\sigma P_{(k)}(\underline{\omega}) : P_{(k)} \in \mathcal{M}^0_{(k)}\},$$

we call

$$\mathcal{M}_{(k)} = \sum_{\sigma \in \{0,1\}^n} \oplus \mathcal{M}^\sigma_{(k)}$$

the space of *separately spherical monogenics of degree* (k).

Taking account of the relations (2.1) and of Theorem 1.3.6, we easily obtain that for each $P \in \mathcal{M}^\sigma_{(k)}$ and $j = 1, \ldots, n$,

$$\Gamma_{\underline{\omega}_j} P(\underline{\omega}) = ((-1)^{\sigma_j+1}k_j + \sigma_j(m_j-1))P(\underline{\omega}).$$

Moreover it is clear that the spaces $\mathcal{M}_{(k)}$ are simultaneous eigenspaces of the operators $C(L_j), j=1,\ldots,n$.

Now let $f(\underline{\omega}) = f(\underline{\omega}_1,\ldots,\underline{\omega}_n) \in L_2(\mathcal{S};V)$ and let $j \in \{1,\ldots,n\}$ be fixed. Then for a.e. $(\underline{\omega}_1,\ldots,\underline{\omega}_{j-1},\underline{\omega}_{j+1},\ldots,\omega_n) \in \prod_{i\neq j} S^{m_i-1}$ fixed, $f(\underline{\omega}_1,\ldots,\underline{\omega}_n) \in L_2(S^{m_j-1};V)$ and hence it admits the decomposition (see Theorem 1.3.4)

$$f(\underline{\omega}_1,\ldots,\underline{\omega}_j,\ldots,\omega_n) = \sum_{k_j=0}^{\infty} \Pi(k_j)f(\underline{\omega}_1,\ldots,\underline{\omega}_j,\ldots,\underline{\omega}_n)$$

where

$$\Pi(k_j) = \mathbf{P}(k_j) + \mathbf{Q}(k_j).$$

Moreover $\mathbf{P}(k_j)$ admits the integral representation (see Theorem 1.7.1)

$$\mathbf{P}(k_j)f(\underline{\omega}_1,\ldots,\underline{\omega}_j,\ldots,\underline{\omega}_n) = \frac{1}{A_{m_j}}\int_{S^{m_j-1}} C^+_{m_j,k_j}(\underline{\omega},\underline{\xi}_j)f(\underline{\omega}_1,\ldots,\underline{\xi}_j,\ldots,\underline{\omega}_n)d\Sigma(\underline{\xi}_j)$$

with

$$C^+_{m_j,k_j}(\underline{\omega}_j,\underline{\xi}_j) = \frac{1}{m_j-2}((k_j+m_j-2)C_{k_j}^{\frac{m_j}{2}-1}(\theta_j) + (m_j-2)\underline{\omega}_j \wedge \underline{\xi}_j C_{k_j-1}^{\frac{m_j}{2}}(\theta_j))$$

and

$$\theta_j = <\underline{\omega}_j, \underline{\xi}_j>.$$

Furthermore, for $f \in L_2(\mathcal{S};V)$ we put

$$\mathbf{P}^0_{(k)}f(\underline{\omega}) = \frac{1}{\prod_{j=1}^n A_{m_j}}\int_{\mathcal{S}}(\prod_{j=1}^n C^+_{m_j,k_j}(\underline{\omega}_j,\underline{\xi}_j))f(\underline{\xi})d\Sigma(\underline{\xi}),$$

while for $\sigma \in \{0,1\}^n$,

$$\mathbf{P}^\sigma_{(k)}f(\underline{\omega}) = \underline{\omega}^\sigma \mathbf{P}^0_{(k)}(\overline{\underline{\omega}}^\sigma f)(\underline{\omega}). \tag{2.2}$$

Finally we call

$$\Pi_{(k)} = \sum_{\sigma\in\{0,1\}^n} \mathbf{P}^\sigma_{(k)}. \tag{2.3}$$

We are now ready to prove

Theorem 1.2.1. *Let $f \in L_2(\mathcal{S},V)$. Then f admits the orthogonal decomposition*

$$f(\underline{\omega}) = \sum_{(k)} \Pi_k f(\underline{\omega}).$$

Proof. The case $n = 1$ was met in Theorem 1.3.4.

As to the case $n = 2$ we proceed as follows. Let $f(\underline{\omega}_1, \underline{\omega}_2) \in L_2(S^{m_1-1} \times S^{m_2-1}; V)$. Then for a.e. $\underline{\omega}_2 \in S^{m_2-1}$ fixed, $f(.,\underline{\omega}_2) \in L_2(S^{m_1-1}; V)$ whence

$$f(\underline{\omega}_1, \underline{\omega}_2) = \sum_{k_1=0}^{\infty} \Pi(k_1) f(\underline{\omega}_1, \underline{\omega}_2)$$

where

$$\begin{aligned} \Pi(k_1)f &= \mathbf{P}(k_1)f + \underline{\omega}_1 \mathbf{P}(k_1)(-\underline{\omega}_1 f) \\ &= \sum_{\sigma_1 \in \{0,1\}} \underline{\omega}_1^{\sigma_1} \mathbf{P}(k_1)(\overline{\underline{\omega}}_1^{\sigma_1} f). \end{aligned}$$

Putting for $\sigma_1 \in \{0, 1\}$,

$$\mathbf{P}_{k_1}^{\sigma_1} f = \mathbf{P}(k_1)(\overline{\underline{\omega}}_1^{\sigma_1} f)$$

we thus have that

$$\Pi(k_1)f = \sum_{\sigma_1 \in \{0,1\}} \underline{\omega}_1^{\sigma_1} \mathbf{P}_{k_1}^{\sigma_1} f.$$

Taking account of the integral representation of $\mathbf{P}(k_1)$, we have that, as $C^+_{m_1,k_1}(\underline{\omega}_1, \underline{\xi}_1) \in L_2(S^{m_1-1} \times S^{m_1-1}; V)$ and $\overline{\underline{\xi}}_1^{\sigma_1} f(\underline{\xi}_1, \underline{\omega}_2) \in L_2(S^{m_1-1} \times S^{m_2-1}; V)$,

$$P_{k_1}^{\sigma_1}(f)(\underline{\omega}_1, \underline{\omega}_2) \in L_2(S^{m_1-1} \times S^{m_2-1}; V).$$

Hence, for a.e. $\underline{\omega}_1 \in S^{m_1-1}$ fixed,

$$\Pi(k_1)f \in L_2(S^{m_1-1} \times S^{m_2-1}; V).$$

Consequently for a.e. $\underline{\omega}_1 \in S^{m_1-1}$ fixed,

$$\Pi(k_1)f(\underline{\omega}_1, \underline{\omega}_2) = \sum_{k_2=0}^{\infty} \Pi(k_2)(\Pi(k_1)f)(\underline{\omega}_1, \underline{\omega}_2)$$

with

$$\begin{aligned} &\Pi(k_2)(\Pi(k_1)f) \\ &= \sum_{\sigma_1 \in \{0,1\}} \Pi(k_2)(\underline{\omega}_1^{\sigma_1} \mathbf{P}_{k_1}^{\sigma_1} f). \\ &= \sum_{\sigma_1 \in \{0,1\}} (\mathbf{P}(k_2)(\underline{\omega}_1^{\sigma_1} \mathbf{P}_{k_1}^{\sigma_1} f) + \underline{\omega}_2 \mathbf{P}(k_2)(\underline{\omega}_2 \underline{\omega}_1^{\sigma_1} \mathbf{P}_{k_1}^{\sigma_1} f)). \\ &= \sum_{\sigma_1, \sigma_2} \underline{\omega}_2^{\sigma_2} \mathbf{P}(k_2)(\overline{\underline{\omega}}_2^{\sigma_2} \underline{\omega}_1^{\sigma_1} \mathbf{P}_{k_1}^{\sigma_1} f). \end{aligned}$$

But, looking at the kernel $C^+_{m_2,k_2}$ appearing in the integral representation of $\mathbf{P}(k_2)$, we have that it is $\mathbf{R} \oplus \mathbf{R}^2_{0,m_2}$-valued where

$$\mathbf{R}^2_{0,m_2} = \operatorname{span}_{\mathbf{C}}\{e_{st} : s, t \in M_2\} = \operatorname{Lie}(\operatorname{Spin}(m_2)).$$

Hence it commutes with $\underline{\omega}_1^{\sigma_1}$ so that

$$\begin{aligned} &\underline{\omega}_2^{\sigma_2}\mathbf{P}(k_2)(\overline{\underline{\omega}}_2^{\sigma_2}\underline{\omega}_1^{\sigma_1}\mathbf{P}^{\sigma_1}_{k_1}f) \\ &= \underline{\omega}_1^{\sigma_1}\underline{\omega}_2^{\sigma_2}\mathbf{P}(k_2)(\overline{\underline{\omega}}_2^{\sigma_2}\mathbf{P}^{\sigma_1}_{k_1}f). \end{aligned}$$

Using again the integral representations of $\mathbf{P}(k_1)$ and $\mathbf{P}(k_2)$, we find that

$$\mathbf{P}(k_2)(\overline{\underline{\omega}}_2^{\sigma_2}\mathbf{P}^{\sigma_1}_{k_1}f)(\underline{\omega}_1, \underline{\omega}_2) = \frac{1}{A_{m_1}A_{m_2}}\int_{S^{m_2-1}} C^+_{m_2,k_2}(\underline{\omega}_2, \underline{\xi}_2)\overline{\underline{\xi}}_2^{\sigma_2}\cdot$$

$$\cdot[\int_{S^{m_1-1}} C^+_{m_1,k_1}(\underline{\omega}_1, \underline{\xi}_1)\overline{\underline{\xi}}_1^{\sigma_1} f(\underline{\xi}_1, \underline{\xi}_2)d\Sigma(\underline{\xi}_1)]d\Sigma(\underline{\xi}_2).$$

But, as $C^+_{m_1,k_1}$ and $C^+_{m_2,k_2}$ commute and $\underline{\xi}_2^{\sigma_2}$ commutes with $C^+_{m_1,k_1}$, we get that

$$\begin{aligned} &\mathbf{P}(k_2)(\overline{\underline{\omega}}_2^{\sigma_2}\mathbf{P}^{\sigma_1}_{k_1}f)(\underline{\omega}_1\underline{\omega}_2) \\ &= \frac{1}{\prod_{j=1}^2 A_{m_j}}\int_{\prod_{j=1}^2 S^{m_j-1}}(\prod_{j=1}^2 C^+_{m_j,k_j}(\underline{\omega}_j, \underline{\xi}_j))\overline{\underline{\xi}}_2^{\sigma_2}\overline{\underline{\xi}}_1^{\sigma_1} f(\underline{\xi}_1, \underline{\xi}_2)d\Sigma(\underline{\xi}_1)d\Sigma(\underline{\xi}_2) \\ &= \mathbf{P}^0_{k_1,k_2}(\overline{\underline{\omega}}_2^{\sigma_2}\overline{\underline{\omega}}_1^{\sigma_1}f). \end{aligned}$$

We thus find that

$$\begin{aligned} &\mathbf{\Pi}(k_2)(\mathbf{\Pi}(k_1)f)(\underline{\omega}_1, \underline{\omega}_2) \\ &= \sum_{\sigma_1,\sigma_2}\underline{\omega}_1^{\sigma_1}\underline{\omega}_2^{\sigma_2}\mathbf{P}^0_{(k_1,k_2)}(\overline{\underline{\omega}}_2^{\sigma_2}\overline{\underline{\omega}}_1^{\sigma_1}f)(\underline{\omega}_1, \underline{\omega}_2) \\ &= \sum_{\sigma\in\{0,1\}^2}\mathbf{P}^\sigma_{(k_1,k_2)}f(\underline{\omega}_1, \underline{\omega}_2) \\ &= \mathbf{\Pi}_{(k_1,k_2)}f(\underline{\omega}_1, \underline{\omega}_2). \end{aligned}$$

By induction onto n ($n \geq 2$) the result may be proved in its full generality using techniques which are completely similar to those in the case $n = 2$. ■

Remarks 1.2.2.

(1) It is clear that the spaces $\mathcal{M}^\sigma_{(k)}$ are mutually orthogonal in $L_2(\mathcal{S}; V)$ and that the operators $\mathbf{P}^\sigma_{(k)}$ are the corresponding orthogonal projection operators. Moreover the operator $\mathbf{\Pi}_{(k)}$ is the orthogonal projection operator mapping $L_2(\mathcal{S}; V)$ onto

$$\sum_{\sigma\in\{0,1\}^n} \oplus \mathcal{M}^{\sigma}_{(k)} = \mathcal{M}_{(k)}.$$

(2) From the proof of the foregoing theorem it follows that for each $(k) = (k_1, \ldots, k_n) \in \mathbf{N}^n$,

$$\mathbf{\Pi}_{(k)} = \mathbf{\Pi}(k_n) \ldots \mathbf{\Pi}(k_1).$$

Of course, for $(k) \neq (\ell), \mathbf{\Pi}_{(k)}$ and $\mathbf{\Pi}_{(\ell)}$ are mutually orthogonal.

2.3 Polyaxially monogenic functions

Let $\underline{\Omega} \subset \mathbf{R}^m = \prod_{j=1}^n \mathbf{R}^{m_j}$ be open and invariant under $\text{Spin}(m_1) \times \ldots \times \text{Spin}\,(m_n)$, where for simplicity it is assumed that $\underline{\Omega} \cap \mathbf{R}^{m_j} = \emptyset$, $j = 1, \ldots, n$.

2.3.1 A decomposition of $L_2(\underline{\Omega}; V)$

Consider the space $L_2(\underline{\Omega}; V)$ of square integrable V-valued functions in $\underline{\Omega}$.

Putting

$$\tilde{\underline{\Omega}} = \{(r_1, \ldots, r_n) : r_j > 0,\ j = 1, \ldots, n \quad \text{and} \quad \sum_{j=1}^{n} r_j \underline{\omega}_j \in \underline{\Omega}\},$$

then clearly $\tilde{\underline{\Omega}} \times S$ is determining a coordinate system in $\underline{\Omega}$ in such a way that on $L_2(\underline{\Omega}; V)$ an inner product may be defined by

$$\begin{aligned}(f, g) &= \int_{\Omega} \overline{f(\underline{x})} g(\underline{x}) d\underline{x} \\ &= \int_{\tilde{\Omega}} (f, g)_S \Pi_{j=1}^n r_j^{m_j - 1} \Pi_{j=1}^n dr_j\end{aligned}$$

with

$$(f, g)_S = \int_S \overline{f}(r_1\underline{\omega}_1, \ldots, r_n\underline{\omega}_n) g(r_1\underline{\omega}_1, \ldots, r_n\underline{\omega}_n) d\Sigma(\underline{\omega}).$$

Now take $f \in L_2(\underline{\Omega}; V)$; then for a.e. $(r_1, \ldots, r_n) \in \tilde{\underline{\Omega}}$, $f(r_1\underline{\omega}_1, \ldots, r_n\underline{\omega}_n) \in L_2(S; V)$, whence for all $(k) \in \mathbf{N}^n$, $\mathbf{\Pi}_{(k)} f \in L_2(\underline{\Omega}; V)$ may be defined.

Indeed, using the definition of $\mathbf{\Pi}_{(k)}$ in $L_2(S; V)$ given in (2.2) and the integral representation of the projection operators involved, we obtain that $\mathbf{\Pi}_{(k)} f$ is measurable in $\tilde{\underline{\Omega}} \times S$, its square integrability thus following immediately. But because of $\mathbf{\Pi}_{(k)} \perp \mathbf{\Pi}_{(\ell)}$ for $(k) \neq (\ell)$ in $L_2(S; V)$, we finally have that $\mathbf{\Pi}_{(k)} \perp \mathbf{\Pi}_{(\ell)}$ for $(k) \neq (\ell)$ in $L_2(\underline{\Omega}; V)$.

Hence we obtain

Lemma. *Each $f \in L_2(\underline{\Omega}; V)$ admits the expression*

$$f(\underline{x}) = \sum_{(k)=0}^{\infty} \mathbf{\Pi}_{(k)} f(\underline{x}).$$

2.3.2 Some commutation relations

In this subsection we assume that the functions f under consideration belong to $L_2(\underline{\Omega}; V) \cap \mathcal{E}(\underline{\Omega}; V)$

Following Lemma 2.3.1 we have for such a function f that

$$f(\underline{x}) = \sum_{(k)=0}^{\infty} \Pi_{(k)} f(\underline{x})$$

where for each $(k) \in \mathbf{N}^n$,

$$\Pi_{(k)} f(\underline{x}) = \sum_{\sigma \in \{0,1\}^n} \mathbf{P}^{\sigma}_{(k)} f(\underline{x}). \tag{2.4}$$

Notice that for any $(r_1, \ldots, r_n) \in \tilde{\Omega}$ fixed

$$\mathbf{P}^{\sigma}_{(k)} f(r_1\underline{\omega}_1, \ldots, r_n\underline{\omega}_n) = \underline{\omega}^{\sigma} \mathbf{P}^{0}_{(k)}(\overline{\underline{\omega}}^{\sigma} f)(r_1\underline{\omega}_1, \ldots, r_n\underline{\omega}_n) \tag{2.5}$$

where for $g \in L_2(\underline{\Omega}; V) \cap \mathcal{E}(\underline{\Omega}; V)$,

$$\begin{aligned} &\mathbf{P}^{(0)}_{(k)} g(r_1\underline{\omega}_1, \ldots, r_n\underline{\omega}_n) \\ &= \frac{1}{\prod_{j=1}^n A_{m_j}} \int_S (\Pi_{j=1}^n C^{+}_{m_j.k_j}(\underline{\omega}_j, \underline{\xi}_j)) g(r_1\underline{\xi}_1, \ldots, r_n\underline{\xi}_n) d\Sigma(\underline{\xi}) \\ &= \left(\Pi_{j=1}^n \mathrm{P}(k_j)\right) g(r_1\underline{\omega}_1, \ldots, r_n\underline{\omega}_n). \end{aligned}$$

Now consider for $j = 1, \ldots, n$ the following maps from $\mathcal{E}(\underline{\Omega}; V)$ into $\mathcal{E}(\underline{\Omega}; V)$:

$$\begin{aligned} \partial_{r_j} &: f \longrightarrow \partial_{r_j} f; \\ R_j &: f \longrightarrow r_j f; \\ \Gamma_{\underline{\omega}_j} &: f \longrightarrow \Gamma_{\underline{\omega}_j} f; \\ \Omega_j &: f \longrightarrow \underline{\omega}_j f; \end{aligned}$$

Then we claim

Lemma. *For all $(k) \in \mathbf{N}^n$ and $j = 1, \ldots, n$, we have that on $L_2(\underline{\Omega}; V) \cap \mathcal{E}(\underline{\Omega}; V)$,*

(i) $[\Pi_{(k)}, \partial_{r_j}] = 0$

(ii) $[\Pi_{(k)}, R_j] = 0$

(iii) $[\Pi_{(k)}, \Gamma_{\underline{\omega}_j}] = 0$

(iv) $[\mathbf{\Pi}_{(k)}, \Omega_j] = 0.$

Proof. As $\mathbf{\Pi}_{(k)}$ is an angular and ∂_{r_j} and R_j are radial operators, (i) and (ii) follow immediately from the integral representation (2.5) combined with (2.4) and (2.3).

(iii) We already know that $\mathbf{\Pi}_{(k)}$ is the orthogonal projection of $L_2(S;V)$ onto $\mathcal{M}_{(k)}$ where the spaces $\mathcal{M}_{(k)}$ are the only simultaneous eigenspaces of the operators $C(L_j)$,
$j = 1, \ldots, n$, with

$$C(L_j) = \Gamma_{\underline{\omega}_j}(m_j - 1 - \Gamma_{\underline{\omega}_j}) - \frac{1}{4}\binom{m_j}{2}\mathbf{1}.$$

As for any $j, j' \in \{1, \ldots, n\}$, $[\Gamma_{\underline{\omega}_j}, \Gamma_{\underline{\omega}'_j}] = 0$, we have that $[C(L_j), \Gamma_{\underline{\omega}_{j'}}] = 0$ whence also $[\mathbf{\Pi}_{(k)}, \Gamma_{\underline{\omega}_{j'}}] = 0$.

(iv) As Ω_j leaves $\mathcal{M}_{(k)}$ invariant we immediately have that $[\mathbf{\Pi}_{(k)}, \Omega_j] = 0$. ∎

2.3.3 The Laurent expansion for polyaxially monogenic functions

Let again $\underline{\Omega} \subset \mathbf{R}^m$ be invariant under Spin$(m_1)\times$ Spin$(m_2) \times \ldots \times$ Spin(m_n). A V-valued function f which is monogenic in $\underline{\Omega}$ is called *polyaxially monogenic* in $\underline{\Omega}$.

Theorem. *Let f be monogenic in $\underline{\Omega}$. Then*

$$f(\underline{x}) = \sum_{(k)=0}^{\infty} \mathbf{\Pi}_{(k)} f(\underline{x})$$

where the convergence is uniform on all compact subsets of $\underline{\Omega}$.
Moreover for each $(k) \in \mathbf{N}^n$,

$$\partial_{\underline{x}} \mathbf{\Pi}_{(k)} f(\underline{x}) = 0 \quad \text{in} \quad \underline{\Omega}.$$

Proof. Let $\underline{K} \subset \underline{\Omega}$ be a regular compact set which is invariant under the group Spin$(m_1) \times \ldots \times$ Spin(m_n).

Then, as $f \in L_2(\mathring{\underline{K}}; V) \cap \mathcal{E}(\mathring{\underline{K}}; V)$, we obtain by Lemma 2.3.1 that in $L_2(\mathring{\underline{K}}; V)$

$$f(\underline{x}) = \sum_{(k)=0}^{\infty} \mathbf{\Pi}_{(k)} f(\underline{x}),$$

where, in virtue of (2.5), (2.4) and (2.3), $\mathbf{\Pi}_{(k)} f \in \mathcal{E}(\mathring{\underline{K}}; V)$ and this for all $(k) \in \mathbf{N}^n$.

But, as Lemma 2.3.2. is telling us that $\mathbf{\Pi}_{(k)}$ is commuting with each of the operators ∂_{r_j}, R_j, $\Gamma_{\underline{\omega}_j}$ and Ω_j and as moreover for each $j = 1, \ldots, n$,

$$\partial_{\underline{x}_j} = \underline{\omega}_j(\partial_{r_j} + \frac{1}{r_j}\Gamma_{\underline{\omega}_j})$$

and

$$\partial_{\underline{x}} = \sum_{j=1}^{n} \partial_{\underline{x}_j},$$

we find that $\partial_{\underline{x}}\Pi_{(k)}f = \Pi_{(k)}\partial_{\underline{x}}f = 0$ in $\underline{\mathring{K}}$, whence $\Pi_{(k)}f$ is monogenic in $\underline{\mathring{K}}$.

Consequently $\Pi_{(k)}f$ is monogenic in $\underline{\Omega}$ for all $(k) \in \mathbf{N}^n$ and so

$$f(\underline{x}) = \sum_{(k)=0}^{\infty} \Pi_{(k)}f(\underline{x})$$

is a series of monogenic functions which is locally L_2-convergent.

By virtue of Theorem 0.2.10 we obtain that this convergence is uniform on each compact subset of $\underline{\Omega}$. ■

2.3.4 The Laurent equations for polyaxially monogenic functions

Suppose again that f is monogenic in $\underline{\Omega}$. Then we proved in the foregoing subsection that for each $(k) \in \mathbf{N}^n$,

$$\partial_{\underline{x}}\Pi_{(k)}f(\underline{x}) = 0 \quad \text{in} \quad \underline{\Omega}.$$

Let us now write out this expression explicitly.

Putting for each $\sigma \in \{0,1\}^n$, $A^{\sigma}_{(k)} = A^{(\sigma_1,\ldots,\sigma_n)}_{(k_1,\ldots,k_n)}$ with

$$A^{\sigma}_{(k)}(r_1\underline{\omega}_1,\ldots,r_n\underline{\omega}_n) = \mathbf{P}^0_{(k)}(\overline{\underline{\omega}}^{\sigma}f)(r_1\underline{\omega}_1,\ldots,r_n\underline{\omega}_n) \tag{2.6}$$

we have that for $(r_1,\ldots,r_n) \in \tilde{\underline{\Omega}}$ fixed, $A^{\sigma}_{(k)} \in \mathcal{M}^{o}_{(k)}$. Moreover, in view of (2.2)

$$\Pi_{(k)}f(\underline{x}) = \sum_{\sigma\in\{0,1\}^n} \underline{\omega}^{\sigma} A^{\sigma}_{(k)}(\underline{x}).$$

Now we calculate for $j = 1,\ldots,n$,

$$\Gamma_{\underline{\omega}_j}(\underline{\omega}^{\sigma} A^{\sigma}_{(k)}) \quad \text{and} \quad \underline{\omega}_j(\underline{\omega}^{\sigma} A^{\sigma}_{(k)}).$$

First, observing that $\Gamma_{\underline{\omega}_j}$ is an angular operator and that each element of $\underline{\omega}^{\sigma}\mathcal{M}^{o}_{(k)}$ is an eigen function of $\Gamma_{\underline{\omega}_j}$ corresponding to either $(-k_j)$ if $\sigma_j = 0$ or $(k_j + m_j - 1)$ if $\sigma_j = 1$, we obtain that

$$\Gamma_{\underline{\omega}_j}(\underline{\omega}^{\sigma} A^{\sigma}_{(k)}) = ((-1)^{\sigma_j+1}k_j + \sigma_j(m_j-1))\underline{\omega}^{\sigma} A^{\sigma}_{(k)} \tag{2.7}$$

Next,

$$\begin{aligned}\underline{\omega}_j\underline{\omega}^{\sigma} &= \underline{\omega}_j\underline{\omega}_1^{\sigma_1}\ldots\underline{\omega}_n^{\sigma_n} \\ &= (-1)^{\sigma_1+\sigma_2+\ldots+\sigma_{j-1}}\underline{\omega}_1^{\sigma_1}\ldots\underline{\omega}_j^{\sigma_j+1}\ldots\underline{\omega}_n^{\sigma_n} \end{aligned} \tag{2.8}$$

or, putting $\tau_j = 1 - \sigma_j$ and taking account of

$$\begin{aligned}\underline{\omega}_j^{\sigma_j+1} &= (-1)\underline{\omega}_j^{-\tau_j}\\ &= (-1)^{\sigma_j}\underline{\omega}_j^{\tau_j},\end{aligned}$$

$$\underline{\omega}_j\underline{\omega}^\sigma = (-1)^{\sigma_1+\sigma_2+\ldots+\sigma_j}\underline{\omega}_1^{\sigma_1}\ldots\underline{\omega}_j^{\tau_j}\ldots\underline{\omega}_n^{\sigma_n}.$$

Hence, as $\partial_{\underline{x}} = \sum_{j=1}^n \underline{\omega}_j(\partial_{r_j} + \frac{1}{r_j}\Gamma_{\underline{\omega}_j})$,

$$\begin{aligned}\partial_{\underline{x}}\Pi_{(k)}f &= \sum_{j=1}^n \sum_{\sigma\in\{0,1\}^n} (-1)^{\sigma_1+\ldots+\sigma_j}\underline{\omega}_1^{\sigma_1}\ldots\underline{\omega}_j^{\tau_j}\ldots\underline{\omega}_n^{\sigma_n}\\ &\quad (\partial_{r_j} + \frac{1}{r_j}((-1)^{\sigma_j+1}k_j + \sigma_j(m_j-1))A_{(k)}^\sigma)\\ &= \sum_{j=1}^n \sum_{\sigma\in\{0,1\}^n} (-1)^{\sigma_1+\ldots+\sigma_j+1}\underline{\omega}_1^{\sigma_1}\ldots\underline{\omega}_j^{\sigma_j}\ldots\underline{\omega}_n^{\sigma_n}\\ &\quad (\partial_{r_j} + \frac{1}{r_j}((-1)^{\sigma_j}k_j + (1-\sigma_j)(m_j-1))A_{(k)}^{(\sigma_1\ldots,1-\sigma_j,\ldots,\sigma_r)})\\ &= \sum_{\sigma\in\{0,1\}^n} \underline{\omega}^\sigma[\sum_{j=1}^n(-1)^{\sigma_1+\ldots+\sigma_j+1}\\ &\quad (\partial_{r_j} + \frac{1}{r_j}((-1)^{\sigma_j}k_j + (1-\sigma_j)(m_j-1))A_{(k)}^{(\sigma_1,\ldots,1-\sigma_j,\ldots,\sigma_n)})]\\ &= \sum_{\sigma\in\{0,1\}^n} \underline{\omega}^\sigma B_\sigma\\ &= 0\end{aligned}$$

where each factor $B_\sigma \in \mathcal{M}_{(k)}^0$ for any $(r_1,\ldots,r_n) \in \underline{\tilde{\Omega}}$ and so $\underline{\omega}^\sigma B_\sigma \in \mathcal{M}_{(k)}^\sigma$ for any $(r_1,\ldots,r_n) \in \tilde{\Omega}$ fixed.

Consequently, as we are taking the summation over all possible $\sigma \in \{0,1\}^n$ and as for any $\sigma,\tau \in \{0,1\}^n$ with $\sigma \neq \tau$, $\mathcal{M}_{(k)}^\sigma \cap \mathcal{M}_{(k)}^\tau = \{0\}$, we find that for all $\sigma \in \{0,1\}^n$

$$\sum_{j=1}^n (-1)^{\sigma_1+\ldots+\sigma_j}(\partial_{r_j} + \frac{1}{r_j}((-1))^{\sigma_j}k_j + (1-\sigma_j)(m_j-1))A_{(k)}^{(\sigma_1,\ldots,1-\sigma_j,\ldots,\sigma_n)} = 0. \quad (2.9)$$

We thus obtain that the functions $A_{(k)}^{(\sigma_1,\ldots,1-\sigma_j,\ldots,\sigma_n)}$ satisfy 2^n first order partial differential equations, called the *Laurent equations of order* (k).

Note that the functions $A_{(k)}^{(\sigma_1,\ldots,1-\sigma_j,\ldots,\sigma_n)}$, $(k) \in \mathbf{N}^n$, $j = 1,\ldots,n$, are the building blocks of a polyaxially monogenic function.

2.4 Some special cases

Let again $\underline{\Omega} \subset \mathbf{R}^m = \mathbf{R}^{m_1} \times \ldots \times \mathbf{R}^{m_n}$ be open and invariant under the group $\text{Spin}(m_1) \times \ldots \times \text{Spin}(m_n)$ and let f be monogenic in $\underline{\Omega}$.

If $n = 2$, then f is said to be *biaxially monogenic* ; if $m_1 = 1$, we say that f is *axially monogenic.*

Finally, if $m_1 = m_2 = \ldots = m_n = 2$, we call f *toroidally monogenic.*

In each of these cases we shall now write out the Laurent equations explicitly.

2.4.1 The case $G = \text{Spin}(m_1) \times \text{Spin}(m_2)$: Biaxially monogenic functions

Take $n = 2$ and let thus $\underline{\Omega} \subset \mathbf{R}^m - \mathbf{R}^{m_1} \times \mathbf{R}^{m_2}$ be invariant under the group $\text{Spin}(m_1) \times \text{Spin}(m_2)$.

Putting $(k) = (k, l) \in \mathbf{N}^2$ and $\sigma = (\sigma_1, \sigma_2) \in \{0,1\}^2$ we obtain the following Laurent equations to be satisfied by the building blocks A_{kl}^{00}, A_{kl}^{11}, A_{kl}^{10}, A_{kl}^{01} of a biaxially monogenic function :

$$\begin{cases} (\partial_{r_1} + \frac{1}{r_1}(k + m_1 - 1))A_{kl}^{10} + (\partial_{r_2} + \frac{1}{r_2}(l + m_2 - 1))A_{kl}^{01} &= 0 \\ -(\partial_{r_1} - \frac{k}{r_1})A_{kl}^{01} + (\partial_{r_2} - \frac{l}{r_2})A_{kl}^{10} &= 0 \end{cases}$$

$$\begin{cases} (\partial_{r_1} + \frac{1}{r_1}(k + m_1 - 1))A_{kl}^{11} - (\partial_{r_2} - \frac{l}{r_2})A_{k,l}^{0,0} &= 0 \\ -(\partial_{r_1} - \frac{k}{r_1})A_{kl}^{00} - (\partial_{r_2} + \frac{1}{r_2}(l + m_2 - 1))A_{kl}^{11} &= 0 \end{cases}$$

2.4.2 The case $G = \text{Spin}(1) \times \text{Spin}(m_2)$: Axially monogenic functions

Take $n = 2$ and assume now that $m_1 = 1$. Putting again $(k) = (k, l) \in \mathbf{N}^2$ and $\sigma = (\sigma_1, \sigma_2) \in \{0,1\}^2$, we obtain that the building blocks A_{kl}^{00}, A_{kl}^{11}, A_{kl}^{10} and A_{kl}^{01} satisfy the Laurent equations :

$$\begin{cases} (\partial_{r_1} + \frac{k}{r_1})A_{kl}^{10} + (\partial_{r_2} + \frac{1}{r_2}(l + m_2 - 1))A_{kl}^{01} &= 0 \\ -(\partial_{r_j} - \frac{k}{r_1})A_{kl}^{01} + (\partial_{r_2} - \frac{l}{r_2})A_{kl}^{10} &= 0 \end{cases}$$

$$\begin{cases} (\partial_{r_1} + \frac{k}{r_1})A_{kl}^{11} - (\partial_{r_2} - \frac{l}{r_2})A_{kl}^{00} &= 0 \\ -(\partial_{r_1} - \frac{k}{r_1})A_{kl}^{00} - (\partial_{r_2} + \frac{1}{r_2}(l + m_2 - 1))A_{kl}^{11} &= 0 \end{cases}$$

Now we claim that $A_{kl}^{\sigma} = 0$ for $k > 0$.

To this end let us recall that for $(r_1, r_2) \in \tilde{\underline{\Omega}}$ fixed, $A_{kl}^{\sigma} \in \mathcal{M}_{(k,l)}$ for all $\sigma \in \{0,1\}^2$ with

$$A^{\sigma}_{kl} = \mathbf{P}^0_{(k,l)}(\underline{\omega}^{\sigma} f)$$

and (2.10)

$$\mathbf{P}^0_{(k,l)} = \mathbf{P}(k)\mathbf{P}(l).$$

Here $\mathbf{P}(k)$, $k \in \mathbf{N}$, is the projection operator of $L_2(S^0; V)$ onto $\mathcal{M}^+_k$.

As $S^0 = \{e_1, -e_1\}$, we have that

$$L_2(S^0; V) = \{f = (a, b) : a, b \in V\} = \sum_{k=0}^{\infty} \oplus \mathcal{M}_k$$

with

$$\mathcal{M}_k = \mathcal{M}^+_k \oplus \mathcal{M}^-_k$$

For $k = 0$ we get

$$\mathcal{M}^+_0 = \{f = (a, a) : a \in V\}$$

and

$$\mathcal{M}^-_0 = \{g = (b, -b) : b \in V\}$$

whence

$$L_2(S^0; V) = \mathcal{M}^+_0 \oplus \mathcal{M}^-_0. \tag{2.11}$$

Consequently, for $k > 0$, $\mathcal{M}_k = \{0\}$ and a fortiori $\mathcal{M}^+_k = \{0\}$ which means that $\mathbf{P}(k) = \mathbf{O}$ for $k > 0$. Taking account of (2.10) and (2.11) we thus obtain that $A^{\sigma}_{kl} = 0$ for all $k > 0$.

Hence we only have to consider the Laurent equations corresponding to $(0, l) = (k)$, namely :

$$\begin{cases} \partial_{r_1} A^{10}_{0\ell} + (\partial_{r_2} + \frac{1}{r_2}(\ell + m_2 - 1))A^{01}_{0\ell} = 0 \\ -\partial_{r_1} A^{01}_{0\ell} + (\partial_{r_2} - \frac{\ell}{r_2})A^{10}_{0\ell} = 0 \end{cases} \tag{2.12}$$

$$\begin{cases} \partial_{r_1} A^{11}_{0\ell} - (\partial_{r_2} - \frac{\ell}{r_2})A^{00}_{0\ell} = 0 \\ -\partial_{r_1} A^{00}_{0\ell} - (\partial_{r_2} + \frac{1}{r_2}(l + m_2 - 1))A^{11}_{0\ell} = 0 \end{cases} \tag{2.13}$$

Obviously (2.12) and (2.13) determine the same system of first order partial differential equations.

Solutions to either of the systems (2.12) or (2.13) are called *axially monogenic of degree k*.

2.4.3 The case G =Spin(2)×Spin(2) × ... × Spin(2) : Toroidally monogenic functions

For reasons which will become clear in the subsection 3.4.2, we first introduce a new notation for the unknown functions $A_{(k)}^{(\sigma_1,\ldots,1-\sigma_j,\ldots,\sigma_n)}$ appearing in the Laurent equations (2.9)

Defining for each $j = 1, \ldots, n$, $\gamma_j \in \mathbf{Z}$ by

$$\gamma_j = (-1)^{\sigma_j} k_j - \sigma_j,$$

we are thus able to rewrite the Laurent equations (2.9) into the form

$$\sum_{j=1}^{n} (\Pi_{\ell=1}^{j} \operatorname{sgn} \gamma_\ell)(\partial_{r_j} + \tfrac{\gamma_j+1}{r_j}) A_{(\gamma_1,\ldots,-\gamma_j-1,\ldots,\gamma_n)} = 0 \tag{2.14}$$

3 Separately monogenic functions in $\mathbf{R}^m$

In section 2.2 we already touched upon the notion of a separately monogenic polynomial. In this section we present an introduction to the theory of separately monogenic functions. It will be seen how this theory, although being subordinate to the one of monogenic functions, may nevertheless be regarded as being one of the cornerstones in understanding the nature of monogenic functions. Moreover it will be shown how complex analysis in several variables may be embedded into the theory of separately monogenic functions.

For more details concerning separately monogenic functions, the reader is referred to [20].

3.1 Preliminaries

Let $\mathbf{R}^m$ be split into $\mathbf{R}^m = \mathbf{R}^{m_1} \times \ldots \times \mathbf{R}^{m_n}$ $(n \geq 2)$. Then in the sequel we shall use the notations :

(i) $\underline{x} = (\underline{x}_1, \ldots, \underline{x}_n)$ for an arbitrary element of $\mathbf{R}^m$ with

$$\underline{x}_j = (\underline{x}_j^{(1)}, \ldots, \underline{x}_j^{(m_j)}) \in \mathbf{R}^{m_j};$$

(ii) $e = (e_1^{(1)}, \ldots, e_1^{(m_1)}, \ldots, e_n^{(1)}, \ldots, e_n^{(m_n)})$ for an orthonormal basis of $\mathbf{R}^{0,m}$, $e_j = (e_j^{(1)}, \ldots, e_j^{(m_j)})$ being an orthonormal basis for $\mathbf{R}^{0,m_j}$;

(iii) $\mathcal{C} = \mathbf{R}_{0,m} \otimes {}_{\mathbf{R}}\mathbf{C}$ for the complex Clifford algebra

(iv) $\partial_{\underline{x}_j} = \sum_{k=1}^{m_j} e_j^{(k)} \partial_{\underline{x}_j^{(k)}}$ for the Dirac operator in $\mathbf{R}^{m_j}$;

(v) V for a basic representation S of Pin(m) or a direct sum of such representation spaces.

Definition 3.1.1. *Let $\underline{\Omega}$ be open subset of $\mathbf{R}^m$ and let f be a V-valued function defined in $\underline{\Omega}$. Then f is said to be separately monogenic in $\underline{\Omega}$ if and only if for $j = 1, \ldots, n$ and for all $\underline{y} \in \underline{\Omega}$, f is monogenic in $\underline{x}_j$ on the slice $\underline{\Omega} \cap (\underline{y} +$ span$_{\mathbf{R}}(e_j^{(1)}, \ldots, e_j^{(m_j)}))$, i.e.*

$$\partial_{\underline{x}_j} f = 0 \quad \text{for} \quad j = 1, \ldots, n \text{ on } \underline{\Omega} \cap (\underline{y} + \text{ span}_{\mathbf{R}}(e_j^{(1)}, \ldots, e_j^{(m_j)})).$$

3.2 Some elementary properties

Notice that in the previous definition of separate monogenicity, we do not require the function f to belong to $C_1(\underline{\Omega}; V)$.

As will be established now this regularity is a consequence of separate monogenicity.

Theorem 3.2.1. *Let us suppose that f is separately monogenic in $\underline{\Omega}$ for the decomposition $\mathbf{R}^m = \mathbf{R}^{m_1} \times \ldots \times \mathbf{R}^{m_n}$. Then f is also separately monogenic in $\underline{\Omega}$ for the decomposition $\mathbf{R}^m = \mathbf{R}^{m_1+m_2} \times \mathbf{R}^{m_3} \times \ldots \times \mathbf{R}^{m_n}$.*

Proof. We only have to show that f is continuously differentiable in the variables $x_1^{(1)}, \ldots, x_1^{(m_1)}, x_2^{(1)}, \ldots, x_2^{(m_2)}$ jointly, the $x_j^{(k)}$ for $k = 3, \ldots, n$ being held constant. So we may as well assume $n = 2$.

Let $\underline{a} = (\underline{a}_1, \underline{a}_2) \in \underline{\Omega}$ be arbitrary and let $R > 0$ be such that

$$\mathring{B}_{m_1}(\underline{a}_1, \sqrt{2}R) \times \mathring{B}_{m_2}(\underline{a}_2, \sqrt{2}R) \subset \underline{\Omega},$$

denoting by $\mathring{B}_{m_j}(\underline{a}_j, r), j = 1, 2$, the open ball of radius r and center $\underline{a}_j$ in $\mathbf{R}^{m_j}$.

Choose $R' > 0$ such that

$$J_1 = \Pi_{k=1}^{m_1}]\underline{a}_1^{(k)} - R', \underline{a}_1^{(k)} + R'[\subset \mathring{B}_{m_1}(\underline{a}_1, R)$$

$$J_2 = \Pi_{k=1}^{m_2}]\underline{a}_2^{(k)} - R', \underline{a}_2^{(k)} + R'[\subset \mathring{B}_{m_2}(\underline{a}_2, R).$$

Then for fixed $\underline{x}_2 \in J_2$, f has a multiple Taylor series in $\underline{x}_1^{(1)}, \ldots, \underline{x}_1^{(m_1)}$ converging in J_1 (see also 0.2). The same holds with 1 and 2 interchanged, whence by a result of Siciak (see [71]), f is then real-analytic in the neighbourhood $J_1 \times J_2$ of $(\underline{a}_1, \underline{a}_2)$. ∎

By repeated application of this theorem, separate monogenicity of f for the trivial decomposition $\mathbf{R}^m = \mathbf{R}^m$ is obtained, which means that f is monogenic in $\underline{x}$.

Now let Σ_j be an m_j-dimensional compact and oriented manifold-with-boundary in $\mathbf{R}^{m_j}$, $j = 1, \ldots, n$, such that $\Sigma_1 \times \Sigma_2 \times \ldots \times \Sigma_n \subset \underline{\Omega}$ and let, as in section 0, the $\mathbf{C}_m$-valued surface elements $d\sigma_j$ on $\partial\Sigma_j$ be defined by

$$d\sigma_j = \sum_{k=1}^{m_j} (-1)^{k-1} e_j^{(k)} d\hat{\underline{x}}_j^{(k)}$$

where

$$d\underline{x}_j^k = d\underline{x}_j^{(1)} \wedge \ldots \wedge [d\underline{x}_j^{(k)}] \wedge \ldots \wedge d\underline{x}_j^{(m_j)}.$$

Furthermore, considering the fundamental solution E_j of $\partial_{\underline{x}_j}$, i.e.

$$E_j(\underline{x}_j) = \frac{1}{A_{m_j}} \frac{\overline{\underline{x}}_j}{|\underline{x}_j|^{m_j}},$$

by applying Cauchy's Integral Formula repeatedly, we obtain

Theorem 3.2.2. (Cauchy's Integral Formula on product domains)

Let f be separately monogenic in $\underline{\Omega} \subset \mathbf{R}^m$. Then for each $\underline{x} = (\underline{x}_1, \ldots, \underline{x}_n)$ with $\underline{x}_j \in \overset{0}{\Sigma}_j$, we have that

$$f(\underline{x}) = \int_{\prod_{j=1}^n \partial\Sigma_j} (\Pi_{j=1}^n E_j(\underline{y}_j - \underline{x}_j) d\sigma_j(\underline{y})) f(\underline{y}).$$

3.3 Inner separately spherical monogenics

In 2.2 we already introduced the notion of an inner separately spherical monogenic polynomial of degree (k) in $\mathbf{R}^m$.

Let us now construct a basis for the space of such polynomials.

Let the symbol $(\ell)_{j,r}$ denote any combination $\ell_{j,1}, \ldots, \ell_{j,r}$ of the numbers $2, \ldots, m_j$ in groups of r, repetition being allowed.

Defining the hypercomplex variables

$$z_j^{(p)} = \underline{x}_j^{(p)} - e_j^{(p)} e_j^{(1)} \underline{x}_j^{(1)} \quad \text{for} \quad p = 2, \ldots, m_j$$

and the Fueter polynomials (see also 1.5.4. Remark (2))

$$V_j^{(\ell)_{j,r}}(\underline{x}_j) = \frac{1}{r!} \sum_{\pi((\ell)_{j,r})} z_j^{(\ell_{j,1})} \ldots z_j^{(\ell_{j,r})}$$

where $\pi((\ell)_{j,r})$ stands for all permutations of a given $(\ell)_{j,r}$, a direct application of the results obtained in Theorem 1.5.3. and 1.5.4. Remark (3) yields.

Theorem 3.3.1. *Let* $(k) = (k_1, \ldots, k_n) \in \mathbf{N}^n$ *be fixed. Then*

(i) $V_{(m)_{1,k_1}, \ldots, (m)_{n,k_n}}(\underline{x}) = \prod_{j=1}^m V_j^{(m)_{j,k_j}}(\underline{x}_j)$ *is a C-valued inner separately spherical monogenic polynomial of degree* (k) *in* $\mathbf{R}^m$.

(ii) Each inner separately monogenic C-valued polynomial $P_{(k)}(\underline{x})$ *of degree* (k) *in* $\mathbf{R}^m$ *admits the unique decomposition*

$$P_{(k)}(\underline{x}) = \sum_{(m)_{1,k_1}, \ldots, (m)_{n,k_n}} V_{(m)_{1,k_1}, \ldots, (m)_{n,k_n}}(\underline{x}) a_{(m)_{1,k_1}, \ldots, (m)_{n,k_n}}.$$

Here $a_{(m)_{1,k_1}, \ldots, (m)_{n,k_n}} \in V$ *with*

$$a_{(m)_{1,k_1}, \ldots, (m)_{n,k_n}} = \left(\prod_{j=1}^n \frac{\partial}{\partial_{\underline{x}_j}^{(m)_{j,k_j}}} \right) P_{(k)}$$

with

$$\frac{\partial}{\partial_{\underline{x}_j}^{(m)_{j,k_j}}} = \frac{\partial}{\partial_{\underline{x}_j}^{(m_{j,1})}} \cdots \frac{\partial}{\partial_{\underline{x}_j}^{(m_{j,k_j})}}.$$

Obviously, if V is a representation space of $\mathrm{Pin}(m)$, a basis for the space of V valued inner separately spherical monogenic polynomials of degree (k) may be obtained by multiplying the elements $V_{(m)_{1,k_1},\ldots,(m)_{n,k_n}}$ on the right by the elements v_i, $i = 1, \ldots, d$, belonging to a basis of V (see also Theorem 1.5.3).

Now using the expansions (see also (1.17))

$$E_j(\underline{y}_1 - \underline{x}_j) = \sum_{k_j=0}^{\infty} \Big(\sum_{(m)_{j,k_j}} V_j^{(m)_{j,k_j}}(\underline{x}_j) W_j^{(m)_{j,k_j}}(\underline{y}_j) \Big)$$

valid uniformly in $|\underline{x}_j| < |\underline{y}_j|$, we obtain

Theorem 3.3.2 (Taylor series expansion) *Let f be separately monogenic in the product domain $\mathring{B}_{m_1}(R_1) \times \ldots \times \mathring{B}_{m_n}(R_n)$. Then there exists a unique sequence $(P_{(k)}f(\underline{x}))_{(k)\in\mathbf{N}^n}$ of inner separately spherical monogenic polynomials $P_{(k)}f(\underline{x})$ of degree (k) such that*

$$f(\underline{x}) = \sum_{(k)} P_{(k)}f(\underline{x}),$$

the convergence being uniformly on the compact subsets of $\mathring{B}_{m_1}(R_1) \times \ldots \times \mathring{B}_{m_n}(R_n)$.

Moreover, for each $(k) \in \mathbf{N}^n$,

$$P_{(k)}f(\underline{x}) = \sum_{(m)_{1,k_1},\ldots,(m)_{n,k_n}} V_{(m)_{1,k_1},\ldots,(m)_{n,k_n}}(\underline{x})\, a_{(m)_{1,k_1},\ldots,(m)_{n,k_n}}$$

with

$$a_{(m)_{1,k_1},\ldots,(m)_{n,k_n}} = \left(\prod_{j=1}^{n} \frac{\partial}{\partial_{\underline{x}_j}^{(m)_{j,k_j}}} \right) f(0).$$

As $\mathring{B}_{m_1}(R_1) \times \ldots \times \mathring{B}_{m_n}(R_n)$ is $\mathrm{Spin}(m_1) \times \ldots \times \mathrm{Spin}(m_n)$ invariant and a separately monogenic function is monogenic, the foregoing theorem should be compared with Theorem 2.3.3, of which it is of course a special case.

As will be seen in section 4, of particular importance is the decomposition of a homogeneous polynomial in $(\underline{x}_1, \underline{x}_2) \in \mathbf{R}^{m_1} \times \mathbf{R}^{m_2}$ into inner separately monogenic polynomials.

We have

Theorem 3.3.3. (Biaxial decomposition of homogeneous polynomials) *Let $\mathbf{R}^m$ be split into $\mathbf{R}^m = \mathbf{R}^{m_1} \times \mathbf{R}^{m_2}$ and let $R(\underline{x}_1, \underline{x}_2)$ be a V-valued polynomial homogeneous of degree (k, ℓ) in $(\underline{x}_1, \underline{x}_2)$. Then R admits a canonical decomposition of the form*

$$R(\underline{x}_1, \underline{x}_2) = \sum_{i=0}^{k} \sum_{j=0}^{\ell} \underline{x}_1^i \underline{x}_2^j P_{k-i,\ell-j}(\underline{x}_1, \underline{x}_2),$$

where $P_{k-i,\ell-j}$ are inner separately spherical monogenic V-valued polynomials of degree $(k-i, \ell-j)$, i.e. $P_{k-i,\ell-j} \in M^+_{(k-i,\ell-j)}$.

Proof. As $R(\underline{x}_1, \underline{x}_2)$ is a homogeneous polynomial of degree k in $\underline{x}_1$, we obtain by the Fischer decomposition that (see also 1.10)

$$R(\underline{x}_1, \underline{x}_2) = \sum_{i=0}^{k} \underline{x}_1^i R_{k-i}(\underline{x}_1, \underline{x}_2)$$

where $R_{k-i}(\underline{x}_1, \underline{x}_2)$ are V-valued polynomials homogeneous of degree $(k-i, \ell)$ which moreover satisfy

$$\partial_{\underline{x}_1} R_{k-i}(\underline{x}_1, \underline{x}_2) = 0$$

whence

$$\Gamma_{\underline{\omega}_1} R_{k-i}(\underline{x}_1, \underline{x}_2) = -(k-i) R_{k-i}(\underline{x}_1, \underline{x}_2). \tag{3.1}$$

As each $R_{k-i}(\underline{x}_1, \underline{\omega}_2)$ is a homogeneous polynomial of degree ℓ in $\underline{x}_2$, we get by the same Fischer decomposition that

$$R_{k-i}(\underline{x}_1, \underline{x}_2) = \sum_{j=0}^{\ell} \underline{x}_2^j P_{k-i,\ell-j}(\underline{x}_1, \underline{x}_2). \tag{3.2}$$

Here $P_{k-i,\ell-j}(\underline{x}_1, \underline{x}_2)$ are V-valued polynomials homogeneous of degree $(k-i, \ell-j)$ which moreover satisfy

$$\partial_{\underline{x}_2} P_{k-i,\ell-j}(\underline{x}_1, \underline{x}_2) = 0$$

whence

$$\Gamma_{\underline{\omega}_2} P_{k-i,\ell-j}(\underline{x}_1, \underline{x}_2) = -(\ell-j) P_{k-i,\ell-j}(\underline{x}_1, \underline{x}_2).$$

From (3.2) and the fact that $\underline{x}_2^j \Gamma_{\underline{\omega}_1} = \Gamma_{\underline{\omega}_1} \underline{x}_2^j$, we obtain :

$$\Gamma_{\underline{\omega}_1} R_{k-i}(\underline{x}_1, \underline{x}_2) = \sum_{j=0}^{\ell} \underline{x}_2^j \Gamma_{\underline{\omega}_1} P_{k-i,\ell-j}(\underline{x}_1, \underline{x}_2). \tag{3.3}$$

Now we claim that $\Gamma_{\underline{\omega}_1} P_{k-i,\ell-j}$ is an inner spherical monogenic of degree $(\ell-j)$ in $\underline{x}_2$ which is moreover homogeneous of degree $(k-i)$ in $\underline{x}_1$.

Indeed, as $\Gamma_{\underline{\omega}_1} = -\underline{x}_1 \wedge \partial_{\underline{x}_1} = -\lambda \underline{x}_1 \wedge \partial_{\lambda \underline{x}_1}$ for all $\lambda \in \mathbf{R} \backslash \{0\}$, $\Gamma_{\underline{\omega}_1} P_{k-i,\ell-j}(\underline{x}_1, \underline{x}_2)$ is clearly a homogeneous polynomial of degree $(k-i)$ in $\underline{x}_1$.

Furthermore, as $\Gamma_{\underline{\omega}_1}$ and $\Gamma_{\underline{\omega}_2}$ commute,

$$\begin{aligned}\Gamma_{\underline{\omega}_2}\Gamma_{\underline{\omega}_1}P_{k-i,\ell-j} &= \Gamma_{\underline{\omega}_1}\Gamma_{\underline{\omega}_2}P_{k-i,\ell-j}\\ &= -(\ell-j)\Gamma_{\underline{\omega}_1}P_{k-i,\ell-j}.\end{aligned}$$

From (3.1) - (3.3) we get

$$\begin{aligned}\Gamma_{\underline{\omega}_1}R_{k-i}(\underline{x}_1,\underline{x}_2) &= \sum_{j=0}^{\ell}\underline{x}_2^j\Gamma_{\underline{\omega}_1}P_{k-i,\ell-j}(\underline{x}_1,\underline{x}_2)\\ &= \sum_{j=0}^{\ell}\underline{x}_2^j(-(k-i))P_{k-i,\ell-j}(\underline{x}_1,\underline{x}_2),\end{aligned}$$

which expresses in two ways the Fischer decomposition of the homogeneous polynomial $\Gamma_{\underline{\omega}_1}P_{k-i}(\underline{x}_1,\underline{x}_2)$ in terms of spherical monogenics with respect to $\underline{x}_2$.

By the uniqueness of the Fischer decomposition we so obtain that

$$\Gamma_{\underline{\omega}_1}P_{k-i,\ell-j}(\underline{x}_1,\underline{x}_2) = -(k-i)P_{k-i,\ell-j}(\underline{x}_1,\underline{x}_2).$$

An analogous reasoning yields that

$$\Gamma_{\underline{\omega}_2}P_{k-i,\ell-j}(\underline{x}_1,\underline{x}_2) = -(\ell-j)P_{k-i,\ell-j}(\underline{x}_1,\underline{x}_2). \blacksquare$$

3.4 Separately monogenic functions in polyaxially symmetric domains

3.4.1 The case $G = \text{Spin}(m_1)\times\ldots\times\text{Spin}(m_n)$

In this subsection we assume that the domain $\underline{\Omega}\subset\mathbf{R}^m$ is invariant under the group $G = \text{Spin}(m_1)\times\ldots\times\text{Spin}(m_n)$.

If f is separately monogenic in $\underline{\Omega}$ then f, being thus monogenic too, admits a Laurent expansion given by (see Theorem 2.3.3)

$$f(\underline{x}) = \sum_{(k)=0}^{\infty}\Pi_{(k)}f(\underline{x}),$$

the convergence of this series being uniform on all compact subsets of $\underline{\Omega}$.

Moreover, as $\Pi_{(k)}f\in\mathcal{E}(\underline{\Omega};V)$ and $\Pi_{(k)}$ commutes with the operators $\partial_{r_j}, R_j, \Gamma_j$ and Ω_j (see Lemma 2.3.2), we have that for each $j=1,\ldots,n$,

$$\partial_{\underline{x}_j}\Pi_{(k)}f = \Pi_{(k)}\partial_{\underline{x}_j}f = 0$$

whence $\Pi_{(k)}f$ is separately monogenic in $\underline{\Omega}$.

Let us recall that (see (2.2))

$$\Pi_{(k)}f(\underline{x}) = \sum_{\sigma\in\{0,1\}^n}\underline{\omega}^\sigma A^\sigma_{(k)}(\underline{x})$$

where for $(r_1,\ldots,r_n)\in\underline{\tilde{\Omega}}$ fixed, $A^{\sigma}_{(k)}\in\mathcal{M}^0_{(k)}$.

Using the expressions (2.6) and (2.7), we thus obtain that for each $j=1,\ldots,n$

$$\begin{aligned}0 &= \partial_{\underline{x}_j}\Pi_{(k)}f\\ &= \sum_{\sigma\in\{0,1\}^n}\underline{\omega}^{\sigma}[(-1)^{\sigma_1+\ldots+\sigma_j+1}\\ &\qquad(\partial_{r_j}+\frac{1}{r_j}((-1)^{\sigma_j}k_j+(1-\sigma_j)(m_j-1)))A^{(\sigma_1,\ldots,1-\sigma_j,\ldots\sigma_n)}_{(k)}]\end{aligned}$$

and so for each $\sigma\in\{0,1\}^n$ and $j=1,\ldots,n$,

$$(\partial_{r_j}+\frac{1}{r_j}(-1)^{\sigma_j}k_j(1-\sigma_j)(m_j-1))A^{(\sigma_1,\ldots,1-\sigma_j,\ldots,\sigma_n)}_{(k)}=0. \tag{3.4}$$

Obviously, each solution of (3.4) is a solution of (2.9).

3.4.2 The case $G=\text{Spin}(2)\times\ldots\times\text{Spin}(2)$

Using the notations introduced in 2.4.3, the system (2.14) may now be written into the form

$$(\partial_{r_j}+\frac{\gamma_j+1}{r_j})A_{(\gamma_1,\ldots,-\gamma_j-1,\ldots,\gamma_n)}=0. \tag{3.5}$$

Replacing for each $j=1,\ldots,n$, $-\gamma_j-1$ by γ_j and calling $\gamma=(\gamma_1,\ldots,\gamma_n)$, we obtain the equations

$$(\partial_{r_j}-\frac{\gamma_j}{r_j})A_{\gamma}=0,\quad j=1,\ldots,n. \tag{3.6}$$

As is readily seen, a general solution of (3.6) is given by

$$A_{\gamma}=r_1^{\gamma_1}r_2^{\gamma_2}\ldots r_n^{\gamma_n}B_{\gamma}(\theta_1,\ldots,\theta_n) \tag{3.7}$$

where we put for each $j=1,\ldots,n$,

$$\underline{x}_j=(\underline{x}_j^{(1)},\underline{x}_j^{(2)})\text{ with }\underline{x}_j^{(1)}=r_j\cos\theta_j\ ,\ \underline{x}_j^{(2)}=r_j\sin\theta_j.$$

But, as for $(r_1,\ldots,r_n)\in\underline{\tilde{\Omega}}$ fixed, A_γ is a separately spherical monogenic function, we have that

$$\Gamma_{\underline{\omega}_j}B_{\gamma}=-k_jB_{\gamma}\quad ,j=1,\ldots,n.$$

Calling for each $j=1,\ldots,n$,

$$i_j=-e_{2j-1}e_{2j}$$

we have that $\Gamma_{\underline{\omega}_j} = i_j\partial_{\theta_j}$ whence a general solution $g(\theta_1,\ldots,\theta_n)$ of the equations

$$\Gamma_{\underline{\omega}_j} g = (-k_j)g, \quad j = 1,\ldots,n,$$

is of the form

$$g(\theta_1,\ldots,\theta_n) = (\prod_{j=1}^n e^{i_j\theta_j k_j})C \tag{3.8}$$

with $C \in V$ arbitrarily chosen.

Hence, by means of (3.7) and (3.8),

$$A_\gamma(r_1\theta_1,\ldots,r_n\theta_n) = (\prod_{j=1}^n r_j^{\gamma_j})(\prod_{j=1}^n e^{i_j\theta_j k_j})C_\gamma$$

with $C_\gamma \in V$ arbitrarily chosen.

We thus obtain that if f is separately monogenic in $\underline{\Omega} \subset \mathbf{R}^m$, $\underline{\Omega}$ being invariant under $\mathrm{Spin}(2) \times \ldots \times \mathrm{Spin}(2)$,

$$f(\underline{x}) = \sum_{(k)=0}^{\infty} \Pi_{(k)} f(\underline{x})$$

with

$$\Pi_{(k)} f(\underline{x}) = \sum_{\sigma\in\{0,1\}^n} (\prod_{j=1}^n r_j^{\gamma_j})\underline{\omega}^\sigma A_\gamma(\theta_1,\ldots,\theta_n)$$

and

$$A_\gamma(\theta_1,\ldots,\theta_n) = (\prod_{j=1}^n e^{i_j\theta_j k_j})C_\gamma.$$

But, as for each $j = 1,\ldots,n$,

$$\begin{aligned}
\underline{\omega}_j &= e_{2j-1}\cos\theta_j + e_{2j}\sin\theta_j \\
&= (\cos\theta_j + e_{2j-1}e_{2j}\sin\theta_j)e_{2j-1} \\
&= e^{-i_j\theta_j}\, e_{2j-1}
\end{aligned}$$

and $\underline{\omega}_j$ anticommutes with i_j,

$$\begin{aligned}
\underline{\omega}_j e^{i_j\theta_j k_j} &= e^{-i_j\theta_j k_j} e^{-i_j\theta_j}\, e_{2j-1} \\
&= e^{-(k_j+1)i_j\theta_j}\, e_{2j-1},
\end{aligned}$$

we obtain that

$$\begin{aligned}
&\underline{\omega}_1^{\sigma_1}\ldots\underline{\omega}_j^{\sigma_j}\ldots\underline{\omega}_n^{\sigma_n} e^{i_1\theta_1 k_1}\ldots e^{i_n\theta_n k_n} C_\gamma \\
&= (\exp(\sum_{j=1}^n i_j\theta_j((-1)^{\sigma_j}k_j - \sigma_j))e_1^{\sigma_1}e_3^{\sigma_2}\ldots e_{2n-1}^{\sigma_n} C_\gamma \\
&= (\exp(\sum_{j=1}^n i_j\theta_j\gamma_j))C'_\gamma
\end{aligned}$$

with $C'_\gamma = e_1^{\sigma_1} e_3^{\sigma_2} \dots e_{2n-1}^{\sigma_n} C_\gamma$.

So we finally obtain that

$$\Pi_{(k)} f(\underline{x}) = \sum_\gamma (\Pi_{j=1}^n r_j^{\gamma_j}) \exp(\Sigma_{j=1}^n i_j \theta_j \gamma_j) C'_\gamma.$$

Putting $z_j = \underline{x}_j^{(1)} + i_j \underline{x}_j^{(2)}$, we obtain

$$\Pi_{(k)} f(\underline{x}) = (\Pi_{j=1}^n z_j^{\gamma_j}) C'_\gamma \tag{3.9}$$

The expression (3.9) already suggests a possible relationship between separately monogenic functions and holomorphic functions in Spin(2) $\times \dots \times$ Spin(2)-invariant domains $\underline{\Omega}$.

That even a more general connection between the theory of separately monogenic functions and complex analysis exists will be shown in the next subsection.

3.5 Separately monogenic functions versus holomorphic functions

Let $\underline{\Omega} \subset \mathbf{R}^{2n}$ be open where $\mathbf{R}^{2n}$ is identified with $\mathbf{R}^{2n} = \mathbf{R}^2 \times \dots \times \mathbf{R}^2$.

Putting for each $j = 1, \dots, n$,

$$f_j = \frac{1}{2}(e_{2j-1} + i e_{2j})$$

and

$$\overline{f}_j = \frac{-1}{2}(e_{2j-1} - i e_{2j}),$$

we obtain that

$$W = \operatorname{span}_{\mathbf{C}}\{f_1, \dots, f_n\} \quad \text{and} \quad \overline{W} = \operatorname{span}_{\mathbf{C}}\{\overline{f}_1, \dots, \overline{f}_n\}$$

are maximally isotropic subspaces of $\mathbf{C}^{2n} = \mathbf{R}^{0,2n} \otimes_{\mathbf{R}} \mathbf{C}$ with $\mathbf{C}^{2n} = W \oplus \overline{W}$ (see also I. 4.7.2) .

Now let the function g be separately monogenic in $\underline{\Omega} \subset \mathbf{R}^{2n}$ open, i.e.

$$(e_{2j-1} \partial_{x_{2j-1}} + e_{2j} \partial_{x_{2j}}) g = 0 \quad \text{in} \quad \underline{\Omega}, \quad j = 1, \dots, n.$$

Then this system of equations is clearly equivalent to the system

$$(\partial_{x_{2j-1}} + i_j \partial_{x_{2j}}) g = 0 \quad \text{in} \quad \underline{\Omega}, \quad j = 1, \dots, n.$$

Now put

$$I_j^0 = \frac{1}{2}(1 - i i_j), \quad I_j^1 = \frac{1}{2}(1 + i i_j), \quad j = 1, \dots, n,$$

and

$$I^\sigma = I_1^{\sigma_1} \ldots I_n^{\sigma_n}, \quad \sigma \in \{0,1\}^n.$$

Then

$$\begin{aligned} I_j^{\sigma_j} I_k^{\sigma_k} &= I_k^{\sigma_k} I_j^{\sigma_j} \\ (I_j^{\sigma_j})^2 &= I_j^{\sigma_j} \\ I_j^{\sigma_j} I_j^{1-\sigma_j} &= 0. \end{aligned}$$

Consequently

$$(I^\sigma)^2 = I^\sigma$$

and

$$I^\sigma I^{\sigma'} = 0 \quad \text{if} \quad \sigma \neq \sigma'.$$

A straightforward calculation yields that

$$\mathbf{1} = \sum_{\sigma \in \{0,1\}^n} I^\sigma$$

whence the identity operator on $\mathbf{C}_{2n}$ is decomposed into mutually orthogonal projection operators.

Let now g be a V-valued function and put $f = I^\sigma g$. Then $i_j f = i_j I^\sigma g = (-1)^{\sigma_j} i\, I^\sigma g$.

This implies that the left multiplication operators $g \to I^\sigma g$, g being a V-valued function, are the projection operators onto the simultaneous eigenspaces of the left multiplication operators $g \to i_j g$.

Hence, if g is a separately monogenic function in $\underline{\Omega}$, then $g_\sigma = I^\sigma g$, $\sigma \in \{0,1\}^n$, will satisfy the equations

$$(\partial_{x_{2j-1}} + (-1)^{\sigma_j} i \partial_{x_{2j}}) g_\sigma = 0.$$

Putting

$$z_j^\sigma = x_{2j-1} + (-1)^{\sigma_j} i\, x_{2j}$$

we thus obtain that $g_\sigma(z_1^\sigma, \ldots, z_n^\sigma)$ is holomorphic in $\underline{\Omega}$.

Hence to each separately monogenic function g in $\underline{\Omega}$ there corresponds a family $(g_\sigma : \sigma \in \{0,1\}^n)$ of holomorphic functions in the variables $z_j^\sigma, j = 1, \ldots, n$.

Now the converse also holds.

Indeed, let $(g_\sigma : \sigma \in \{0,1\}^n)$ be a family of holomorphic functions in the variables $z_j^\sigma, j = 1, \ldots, n$.

Then the function

$$g(\underline{x}) = \sum_{\sigma\in\{0,1\}^n} I^\sigma g_\sigma(z_1^\sigma, \ldots, z_n^\sigma)$$

is separately monogenic.

Another expression of a separately monogenic function in $\underline{\Omega} \subset \mathbf{R}^{2n} = \mathbf{R}^2 \times \ldots \times \mathbf{R}^2$ may be obtained as follows.

Consider the nilpotent elements

$$\Phi^\sigma = \varphi_1^{\sigma_1} \ldots \varphi_n^{\sigma_n}, \sigma \in \{0,1\}^n,$$

where

$$\varphi_j^{\sigma_j} = \frac{1}{2}(e_{2j-1} + (-1)^{\sigma_j} i e_{2j}), j = 1, \ldots, n.$$

Then Φ^σ is the volume element of the maximally isotropic subspace

$$V^\sigma = \operatorname{span}_{\mathbf{C}}\{\varphi_1^{\sigma_1}, \ldots, \varphi_n^{\sigma_n}\}.$$

As $\varphi_j^{\sigma_j}\overline{\varphi}_j^{\sigma_j} = I_j^{\sigma_j}$ and $\varphi_k^{\sigma_k} I_j^{\sigma_j} = I_j^{\sigma_j}\varphi_k^{\sigma_k}$ for $k \neq j$, we find that $I^\sigma = \Phi^\sigma\overline{\Phi}^\sigma$ whence, by virtue of $\mathbf{1} = \sum_{\sigma\in\{0,1\}^n} I^\sigma$, each separately monogenic function g is of the form

$$g(x) = \sum_{\sigma\in\{0,1\}^n} \Phi^\sigma h_\sigma(z_1^{\sigma_1}, \ldots, z_n^{\sigma_n})$$

where

$$h_\sigma = \overline{\Phi}^\sigma I^\sigma g$$

is holomorphic in $z_1^\sigma, \ldots, z_n^\sigma$.

Now putting for $\underline{x} = \sum_{j=1}^{2n} e_j x_j \in \mathbf{R}^{2n}$,

$$\varphi_j^{\sigma_j} \cdot \underline{x} = -[\varphi_j^{\sigma_j}\underline{x}]_0$$

we obtain, as for each $j = 1, \ldots, n$,

$$z_j^\sigma = \varphi_j^{\sigma_j} . \underline{x},$$

the invariant expression

$$g(\underline{x}) = \sum_{\sigma\in\{0,1\}^n} \varphi_1^{\sigma_1} \ldots \varphi_n^{\sigma_n} h_\sigma(\varphi_1^{\sigma_1} . \underline{x}, \ldots, \varphi_n^{\sigma_n} . \underline{x})$$

of g in terms of the holomorphic functions h_σ.

4 Construction of orthogonal bases for $\mathcal{M}^+(m;s;\mathcal{C})$

In this section we indicate how an orthogonal basis may be constructed for the space $\mathcal{M}^+(m;s;\mathcal{C})$, consisting of the restrictions to S^{m-1} of all homogeneous monogenic $\mathcal{C}$-valued polynomials of degree s in $\mathbf{R}^m$.

This construction relies essentially upon the splitting of $\mathbf{R}^m$ into $\mathbf{R}^m = \mathbf{R}^{m_1} \times \mathbf{R}^{m_2}$ and the decomposition of $P_s \in M^+(m;s;\mathcal{C})$ into a sum $P_s = \sum_{k+\ell \leq s} M_{k,\ell}$, where the restriction of $M_{k,\ell}$ to $S^{m_1-1} \times S^{m_2-1}$ belongs to

$$\mathcal{M}^{00}_{(k,\ell)} \oplus \mathcal{M}^{10}_{(k,\ell)} \oplus \mathcal{M}^{0,1}_{(k,\ell)} \oplus \mathcal{M}^{11}_{(k,\ell)} = \mathcal{M}_{(k,\ell)},$$

the space of $\mathcal{C}$-valued separately spherical monogenic functions of degree (k,ℓ) in $(\underline{\omega}_1, \underline{\omega}_2) \in S^{m_1-1} \times S^{m_2-1}$.

By recurrence we are then led to the problem of constructing orthogonal bases in the lowest dimensional cases $m = 1$ and $m = 2$.

This method should be compared to the one worked out in [85] to obtain orthogonal bases for the space $\mathcal{H}_k(S^m; \mathbf{C})$ of spherical harmonics of degree k.

We start with some lemmas.

Lemma 4.1. *Let $\mathbf{R}^m = \mathbf{R}^{m_1} \times \mathbf{R}^{m_2}$ and let $k, \ell \in \mathbf{N}$. Then*

$$\dim \mathcal{M}^{00}_{(k,\ell)} = \dim \mathcal{M}^+(m_1; k; \mathcal{C}) \dim \mathcal{M}^+(m_2; \ell; \mathcal{C})$$

Proof. Put $\underline{x}_1 = (x_1, \underline{x}_1') \in \mathbf{R} \times \mathbf{R}^{m_1-1}$ and $\underline{x}_2 = (x_{m_1+1}, \underline{x}_2') \in \mathbf{R} \times \mathbf{R}^{m_2-1}$ and let $R_{k,\ell}(\underline{x}_1', \underline{x}_2') \in \mathcal{P}_{k,\ell}(m_1-1; m_2-1; \mathcal{C})$, the space of $\mathcal{C}$-valued polynomials homogeneous in $\underline{x}_1' \in \mathbf{R}^{m_1-1}$ and $\underline{x}_2' \in \mathbf{R}^{m_2-1}$ of degree k and ℓ respectively. Then, using twice a CK-extension procedure (see also Theorem 0.2.13), we have that

$$R^{\star\star}_{k,\ell}(\underline{x}_1, \underline{x}_2) = \sum_{j=0}^{k} \frac{1}{j!} (-x_1)^j (\bar{e}_1 \partial_{\underline{x}_1'})^j R^{\star}_{k,\ell}(\underline{x}_1', \underline{x}_2)$$

with

$$R^{\star}_{k,\ell}(\underline{x}_1', \underline{x}_2) = \sum_{p=0}^{\ell} \frac{1}{p!} (-x_{m_1+1})^p (\bar{e}_{m_1+1} \partial_{\underline{x}_2'})^p R(\underline{x}_1', \underline{x}_2')$$

belongs to $M^+_{(k,\ell)}$.

Conversely, given $P_{k,\ell} \in M^+_{(k,\ell)}$ then, denoting by $P'_{k,\ell}(\underline{x}_1', \underline{x}_2')$ its restriction to $\mathbf{R}^{m_1-1} \times \mathbf{R}^{m_2-1}$ obtained by putting $x_1 = x_{m_1+1} = 0$, we clearly have that

$$P'^{\star\star}_{k,\ell} = P_{k,\ell}.$$

Hence the spaces $M^+_{(k,\ell)}$ and $\mathcal{P}_{k,\ell}(m_1 - 1; m_2 - 1; \mathcal{C})$ are isomorphic.

But, as

$$\mathcal{P}_{k,\ell}(m_1-1;m_2-1;\mathcal{C}) \cong \mathcal{P}(m_1-1;k;\mathcal{C}) \otimes \mathcal{P}(m_2-1;\ell;\mathcal{C})$$

and moreover

$$\mathcal{P}(m_1-1;k;\mathcal{C}) \cong M^+(m_1;k;\mathcal{C})$$

and

$$\mathcal{P}(m_2-1;\ell;\mathcal{C}) \cong M^+(m_2;\ell;\mathcal{C})$$

we obtain the desired result. ■

Lemma 4.2. *Let $s \in \mathbf{N}$. Then*

$$\dim \mathcal{M}^+(m;s;\mathcal{C}) = \sum_{k+\ell\leq s} \dim \mathcal{M}^{00}_{(k,\ell)}.$$

Proof. By the Cauchy-Kovalevska Theorem,

$$\dim M^+(m;s;\mathcal{C}) = \dim \mathcal{P}(m-1;s;\mathcal{C})$$

and, as

$$\mathcal{P}(m-1;s;\mathcal{C}) = \sum_{k\leq s} \mathcal{P}(m_1-1;k;\mathcal{C}) \otimes \mathcal{P}(m_2;s-k;\mathcal{C})$$

we have that

$$\dim \mathcal{P}(m-1;s;\mathcal{C}) = \sum_{k\leq s} \dim \mathcal{P}(m_1-1;k;\mathcal{C}) \dim \mathcal{P}(m_2;s-k;\mathcal{C}).$$

By virtue of the Fisher decomposition (see 1.10) we now have that

$$\dim \mathcal{P}(m_2;s-k;\mathcal{C}) = \sum_{l\leq s-k} \dim M^+(m_2;\ell;\mathcal{C})$$

whence

$$\begin{aligned} \dim \mathcal{P}(m-1;s;\mathcal{C}) &= \sum_{k\leq s}\sum_{l\leq s-k} \dim \mathcal{P}(m_1-1;k;\mathcal{C}) \dim M^+(m_2;\ell;\mathcal{C}) \\ &= \sum_{k\leq s}\sum_{l\leq s-k} \dim M^+(m_1;k;\mathcal{C}) \dim M^+(m_2;\ell;\mathcal{C}) \end{aligned}$$

or, using Lemma 4.1.,

$$\dim \mathcal{P}(m-1;s;\mathcal{C}) = \sum_{k+\ell\leq s} \dim \mathcal{M}^{00}_{(k,\ell)}. \blacksquare$$

Now let $s \in \mathbf{N}$ and let $P_s \in \mathcal{M}^+(m;s;\mathcal{C})$.

Then we may write $P_s(\underline{x}) \in M^+(m;s;\mathcal{C})$ as

$$P_s(\underline{x}) = \sum_{\alpha+\beta=s} R_{\alpha,\beta}(\underline{x}_1, \underline{x}_2)$$

where $R_{\alpha,\beta}(\underline{x}_1, \underline{x}_2) \in \mathcal{P}_{\alpha,\beta}(m_1;m_2;\mathcal{C})$ are uniquely determined.

In view of Theorem 3.3.3, each $R_{\alpha,\beta}(\underline{x}_1, \underline{x}_2)$ admits a unique biaxial decomposition

$$R_{\alpha,\beta}(\underline{x}_1, \underline{x}_2) = \sum_{\substack{i+k=\alpha \\ j+\ell=\beta}} \underline{x}_1^i \underline{x}_2^j P_{k,\ell}^{i,j}(\underline{x}_1, \underline{x}_2)$$

where $P_{k,\ell}^{i,j}(\underline{x}_1, \underline{x}_2) \in M^+_{(\alpha-i,\beta-j)} = M^+_{k,\ell}$.

Putting for k, ℓ fixed with $k + \ell \leq s$

$$M_{k,\ell}(\underline{x}) = \sum_{i+j=s-k-\ell} \underline{x}_1^i \underline{x}_2^j P_{k,\ell}^{i,j}(\underline{x}_1, \underline{x}_2) \tag{4.1}$$

we thus obtain that

$$P_s(\underline{x}) = \sum_{k+\ell \leq s} M_{k,\ell}(\underline{x}). \tag{4.2}$$

Notice that the restriction of $M_{k,\ell}$ to $S^{m_1-1} \times S^{m_2-1}$ belongs to $\mathcal{M}_{(k,\ell)}$, the simultaneous eigenspace of $C(L_1)$ and $C(L_2)$.

Lemma 4.3. *Let $P_s \in M^+(m;s;\mathcal{C})$ and let for $k, \ell \in \mathbf{N}$ with $k + \ell \leq s$, $M_{k,\ell}$ be defined by* (4.1). *Then*

(i) $M_{k,\ell} \in M^+(m;s;\mathcal{C})$.

(ii) There exist $P_{k,\ell} \in M^+_{(k,\ell)}(m_1;m_2;\mathcal{C})$ fixed and constants $c_{ij} \in \mathbf{R}$ such that

$$M_{k,\ell}(\underline{x}) = \Big(\sum_{i+j=s-k-\ell} c_{ij} \underline{x}_1^i \underline{x}_2^j \Big) P_{k,\ell}(\underline{x}_1, \underline{x}_2).$$

Proof. As

$$\begin{aligned} \Gamma_{\underline{\omega}} &= -\underline{x} \wedge \partial_{\underline{x}} \\ &= -(\underline{x}_1 + \underline{x}_2) \wedge (\partial_{\underline{x}_1} + \partial_{\underline{x}_2}) \\ &= \Gamma_{\underline{\omega}_1} + \Gamma_{\underline{\omega}_2} + L^{(12)} \end{aligned}$$

with

$$\Gamma_{\underline{\omega}_j} = -\underline{x}_j \wedge \partial_{\underline{x}_j}, \quad j = 1, 2$$

and

$$\begin{aligned} L^{(12)} &= -(\underline{x}_1 \wedge \partial_{\underline{x}_2} + \underline{x}_2 \wedge \partial_{\underline{x}_1}) \\ &= -(\underline{x}_1 \partial_{\underline{x}_2} + \underline{x}_2 \partial_{\underline{x}_2}) \end{aligned}$$

we obtain in view of the commutation relations (see 1.12.2) that $\Gamma_{\underline{\omega}}$ commutes with both $C(L_1)$ and $C(L_2)$.

Consequently, as the spaces $\mathcal{M}_{(k,\ell)}$ are eigenspaces of $C(L_1)$ and $C(L_2)$, the same holds for $\Gamma_{\underline{\omega}}\mathcal{M}_{(k,\ell)}$.

By virtue of (4.2) we so get :

$$\begin{aligned} \Gamma_{\underline{\omega}} P_s(\underline{\omega}) &= \sum_{k+\ell \leq s} \Gamma_{\underline{\omega}} M_{k,\ell}(\underline{\omega}) \\ &= -\sum_{k+\ell \leq s} s M_{k,\ell}(\underline{\omega}) \end{aligned}$$

or

$$\sum_{k+\ell \leq s} (\Gamma_{\underline{\omega}} M_{k,\ell}(\underline{\omega}) + s M_{k,\ell}(\underline{\omega})) = 0,$$

whence for each pair k, ℓ with $k + \ell \leq s$,

$$\Gamma_{\underline{\omega}} M_{k,\ell}(\underline{\omega}) = -s M_{k,\ell}(\underline{\omega}).$$

As clearly $M_{k,\ell}(\underline{x}) \in \mathcal{P}(m; s; \mathcal{C})$ we thus obtain that $M_{k,\ell} \in M^+(m; s; \mathcal{C})$.

Furthermore, as (see also Proposition 3 in 1.12.1)

$$\Gamma_{\underline{\omega}_1}(\underline{x}_1^2 \underline{x}_2^j P_{k,\ell}^{i,j}(\underline{x}_1, \underline{x}_2) = \kappa(i) \underline{x}_1^i \underline{x}_2^j P_{k,\ell}^{i,j}(\underline{x}_1, \underline{x}_2)$$

with

$$\kappa(i) = \begin{cases} -k & \text{for } i \text{ even} \\ k + m_1 - 1 & \text{for } i \text{ odd} \end{cases} ;$$

$$\Gamma_{\underline{\omega}_2}(\underline{x}_1^i \underline{x}_2^j P_{k,\ell}^{i,j}(\underline{x}_1, \underline{x}_2)) = \lambda(j) \underline{x}_1^i \underline{x}_2^j P_{k,\ell}^{i,j}(\underline{x}_1, \underline{x}_2)$$

with

$$\lambda_{(j)} = \begin{cases} -\ell & \text{for } j \text{ even} \\ l + m_2 - 1 & \text{for } j \text{ odd} \end{cases} ;$$

$$\underline{x}_1 \partial_{\underline{x}_2}(\underline{x}_1^i \underline{x}_2^j P_{k,\ell}^{i,j}(\underline{x}_1, \underline{x}_2)) = \beta(j) \underline{x}_1^{i+1} \underline{x}_2^{j-1} P_{k,\ell}^{i,j}(\underline{x}_1, \underline{x}_2) \quad ;$$

$$\underline{x}_2 \partial_{\underline{x}_1}(\underline{x}_1^i \underline{x}_2^j P_{k,\ell}^{i,j}(\underline{x}_1, \underline{x}_2)) = \alpha(i) \underline{x}_1^{i-1} \underline{x}_2^{j+1} P_{k,\ell}^{i,j}(\underline{x}_1, \underline{x}_2) \quad ;$$

with

$$\alpha(i) = \begin{cases} (-1)^{i-1}(-i) & \text{for } i \text{ even} \\ (-1)^{i-1}(-(i+m_1+2k-1)) & \text{for } i \text{ odd} \end{cases}$$

and

$$\beta(j) = \begin{cases} (-1)^{i}(-j) & \text{for } j \text{ even} \\ (-1)^{i}(-(j+m_2+2\ell-1)) & \text{for } j \text{ odd,} \end{cases}$$

the relation $\Gamma_{\underline{\omega}} M_{k,\ell}(\underline{\omega}_1, \underline{x}_2) + sM_{k,\ell}(\underline{\omega}_1, \underline{x}_2) = 0$ turns into

$$\sum_{i+j=s-k-\ell} [(\kappa(i)+\lambda(j)+s)\underline{x}_1^i\underline{x}_2^j - \alpha(i)\underline{x}_1^{i-1}\underline{x}_2^{j+1} - \beta(j)\underline{x}_1^{i+1}\underline{x}_2^{j-1}]$$

$$P_{k,\ell}^{i,j}(\underline{x}_1, \underline{x}_2) = 0.$$

or

$$\sum_{i+j=s-k-\ell} \underline{x}_1^i\underline{x}_2^j[(\kappa(i)+\lambda(j)+s)P_{k,\ell}^{i,j}(\underline{x}_1, \underline{x}_2) - \alpha(i+1)P_{k,\ell}^{i+1,j-1}(\underline{x}_1, \underline{x}_2)$$

$$- \beta(j+1)P_{k,\ell}^{i-1,j+1}(\underline{x}_1, \underline{x}_2)] = 0.$$

Consequently, for k, ℓ fixed and $i, j \in \mathbf{N}$ with $i+j+k+\ell = s$, the following recurrence relation between $P_{k,\ell}^{i,j}, P_{k,\ell}^{i+1,j-1}$ and $P_{k,\ell}^{i-1,j+1}$ is obtained :

$$(\kappa(i)+\lambda(j)+s)P_{k,\ell}^{i,j} - \alpha(i+1)P_{k,\ell}^{i+1,j-1} - \beta(j+1)P_{k,\ell}^{i-1,j+1} = 0$$

where $\kappa(i)+\lambda(j)+s, \alpha(i+1)$ and $\beta(j+1)$ are real numbers.

This recurrence relation clearly implies that for all $i, j \in \mathbf{N}$ with $i+j+k+\ell = s$, $i \neq 0$, $j \neq s-k-\ell$, there ought to exist $c_{ij} \in \mathbf{R}$ such that

$$P_{k,\ell}^{i,j} = c_{ij}P_{k,\ell}^{0,s-k-\ell}.$$

Hence, putting $P_{k,\ell} = P_{k,\ell}^{0,s-k-\ell}$, we have

$$M_{k,\ell}(\underline{x}_1, \underline{x}_2) = \Big(\sum_{i+j=s-k-\ell} c_{ij}\underline{x}_1^i\underline{x}_2^j\Big)P_{k,\ell}. \blacksquare$$

Remarks.

(1) Notice that the constants $c_{ij} \in \mathbf{R}$ appearing in Lemma 4.3. satisfy the recurrence relation

$$(\kappa(i)+\lambda(j)+s)c_{ij} - \alpha(i+1)c_{i+1,j-1} - \beta(j+1)c_{i-1,j+1} = 0.$$

(2) Given $P_s \in M^+(m; s; C)$ and its decomposition $P_s(\underline{x}) = \sum_{k+\ell \leq s} M_{k,\ell}(\underline{x})$, we put in what follows for $k, \ell \in \mathbf{N}$ fixed,

$$S_{s,k,\ell}(\underline{x}_1, \underline{x}_2) = \sum_{i+j=s-k-\ell} c_{ij} \underline{x}_1^i \underline{x}_2^j.$$

By introducing polar coordinates $\underline{x}_1 = r_1\underline{\omega}_1$ and $\underline{x}_2 = r_2\underline{\omega}_2$ in $\mathbf{R}^{m_1}$ and $\mathbf{R}^{m_2}$ respectively we may thus write

$$P_s(r_1\underline{\omega}_1, r_2\underline{\omega}_2) = \sum_{k+\ell\leq s} M_{k,\ell}(r_1\underline{\omega}_1, r_2\underline{\omega}_2)$$

with

$$M_{k,\ell}(r_1\underline{\omega}_1, r_2\underline{\omega}_2) = S_{s,k,\ell}(r_1\underline{\omega}_1, r_2\underline{\omega}_2) r_1^k r_2^\ell P_{k,\ell}(\underline{\omega}_1, \underline{\omega}_2).$$

Denoting by $E_i(r_1, r_2)$, $i = 1, 2, 3, 4$, suitable polynomials in r_1, r_2 having real coefficients and in which only even powers of r_1 and r_2 are appearing, we may write $M_{k,\ell}$ in the following forms.

If $i + j = s - k - \ell$ is even,

$$M_{k,\ell}(r_1\underline{\omega}_1, r_2\underline{\omega}_2) = [A(r_1, r_2) + \underline{\omega}_1\underline{\omega}_2 D(r_1, r_2)] P_{k,\ell}(\underline{\omega}_1, \underline{\omega}_2)$$

where

$$A(r_1, r_2) = E_1(r_1, r_2) r_1^k r_2^\ell$$

and

$$D(r_1, r_2) = E_2(r_1, r_2) r_1^{k+1} r_2^{\ell+1}. \tag{4.3}$$

If $i + j = s - k - \ell$ is odd,

$$M_{k,\ell}(r_1\underline{\omega}_1, r_2\underline{\omega}_2) = [B(r_1, r_2)\underline{\omega}_1 + C(r_1, r_2)\underline{\omega}_2] P_{k,\ell}(\underline{\omega}_1, \underline{\omega}_2))$$

where

$$B(r_1, r_2) = E_3(r_1, r_2) r_1^{k+1} r_2^\ell$$

and

$$C(r_1, r_2) = E_4(r_1, r_2) r_1^k r_2^{l+1}. \tag{4.4}$$

Notice that A, B, C and D are homogeneous functions of degree s in $r = (r_1^2 + r_2^2)^{1/2}$ which moreover, by virtue of $\partial_{\underline{x}} M_{k,\ell} = 0$, satisfy Laurent equations of the type mentioned in 2.4.1, namely

$$(I)\begin{cases} (\partial_{r_1} - \frac{k}{r_1})A + (\partial_{r_2} + \frac{l+m_2-1}{r_2})D = 0 \\ (\partial_{r_1} + \frac{k+m_1-1}{r_1})D - (\partial_{r_2} - \frac{\ell}{r_2})A = 0 \end{cases}$$

and

$$(II)\begin{cases} (\partial_{r_1} + \frac{k+m_1-1}{r_1})B + (\partial_{r_2} + \frac{\ell+m_2-1}{r_2})C = 0 \\ (\partial_{r_1} - \frac{k}{r_1})C - (\partial_{r_2} - \frac{\ell}{r_2})B = 0 \end{cases}$$

(3) Finally observe that for (r_1, r_2) fixed, the expression (4.2) yields the decomposition of P_s with respect to $\sum_{k+\ell\leq s} \oplus\mathcal{M}_{(k,\ell)}$, i.e. its decomposition into separately spherical monogenics.

Now we give explicit forms of the polynomials $S_{s,k,\ell}(\underline{x}_1, \underline{x}_2)$ in terms of Jacobi polynomials by solving the equations (I) and (II).

Lemma 4.4. *Let $s, k, \ell \in \mathbb{N}$ be such that $k + \ell \neq s$.*

(i) If $s = k + \ell + 2n$, then

$$\begin{aligned} S_{s,k,\ell}(\underline{x}_1, \underline{x}_2) &= (\underline{x}_1^2 + \underline{x}_2^2)^{n-1}[(\underline{x}_1^2 + \underline{x}_2^2)P_n^{\ell+\frac{m_2}{2}-1,k+\frac{m_1}{2}-1}\left(\frac{\underline{x}_1^2 - \underline{x}_2^2}{\underline{x}_1^2 + \underline{x}_2^2}\right) \\ &\quad - \underline{x}_1\underline{x}_2 P_{n-1}^{\ell+\frac{m_2}{2},k+\frac{m_1}{2}}\left(\frac{\underline{x}_1^2 - \underline{x}_2^2}{\underline{x}_1^2 + \underline{x}_2^2}\right)]. \end{aligned}$$

(ii) If $s = k + \ell + 2n + 1$, then

$$\begin{aligned} S_{s,k,\ell}(\underline{x}_1\underline{x}_2) &= (\underline{x}_1^2 + \underline{x}_2^2)^n[(\ell - k - m_1 + 1 - s)\underline{x}_2 P_n^{\ell+\frac{m_2}{2}-1,k+\frac{m_1}{2}}\left(\frac{\underline{x}_1^2 - \underline{x}_2^2}{\underline{x}_1^2 + \underline{x}_2^2}\right) \\ &\quad - (k - \ell - m_2 + 1 - s)\underline{x}_1 P_n^{\ell+\frac{m_2}{2},k+\frac{m_1}{2}-1}\left(\frac{\underline{x}_1^2 - \underline{x}_2^2}{\underline{x}_1^2 + \underline{x}_2^2}\right)]. \end{aligned}$$

Proof. Put $r_1 = r\cos\theta$, $r_2 = r\sin\theta$, $\theta \in]0, \pi/2[$.
Then we have that

$$r\partial_r = r_1\partial_{r_2} + r_2\partial_{r_2} \quad , \quad \partial_\theta = r_1\partial_{r_2} - r_2\partial_{r_1}.$$

(i) Since A and D are homogeneous functions in r of degree s, we have

$$r\partial_r A = sA \quad \text{and} \quad r\partial_r D = sD.$$

Multiplying the first and second equations equations in (I) with, respectively, r_1 and r_2 and subtracting and adding the equations thus obtained, we find

$$(I') \quad \begin{cases} (\partial_\theta - (k + m_1 - 1)\tan\theta + (\ell + m_2 - 1)\cot\theta)D = (k + \ell - s)A \\ (\partial_\theta + k\tan\theta - \ell cot\theta)A = (k + \ell + s + m_1 + m_2 - 2)D \end{cases}$$

Putting $t = \cos 2\theta = \frac{r_1^2 - r_2^2}{r_1^2 + r_2^2}$ and, by taking into account (4.3),

$$\begin{aligned} A(r,\theta) &= r^s \cos^k\theta \sin^\ell\theta a(\cos 2\theta); \\ D(r,\theta) &= r^s \cos^{k+1}\theta \sin^{\ell+1}\theta d(\cos 2\theta). \end{aligned}$$

(I') then leads to the system

$$(I'')\quad \begin{cases} (t^2-1)d' + \frac{1}{2}[(2k+m_1)(t-1)+(2\ell+m_2)(t+1)]d = (k+\ell-s)a \\ a' = -\frac{1}{4}(k+\ell+s+m_1+m_2-2)d. \end{cases}$$

Deriving the first equation in (I''), d satisfies the differential equation for the classical Jacobi polynomials :

$$(1-t^2)d'' + [2(q-p) - 2(q+p)t]d' + (n-1)(n+2q+2p-2)d = 0$$

where

$$p = \frac{1}{4}(2\ell+m_2+2)\ ,\ q = \frac{1}{4}(2k+m_1+2)\ ,\ n = \frac{s-k-\ell}{2}$$

and so we may put

$$d(t) = P_{n-1}^{\ell+\frac{m_2}{2},k+\frac{m_1}{2}}(t).$$

Using the derivation formula for Jacobi polynomials we get from (I'') that

$$a(t) = -P_n^{\ell+\frac{m_2}{2}-1,k+\frac{m_1}{2}-1}(t).$$

Hence by (4.3) $S_{s,k\ell}(\underline{x}_1,\underline{x}_2)$ admits the form required.

(ii) Since B and C are homogeneous functions in r of degree s, we have

$$r\partial_r B = sB \quad \text{and} \quad r\partial_r C = sC.$$

Multiplying the first qnd second equations in (II) with, respectively, r_1 and r_2 and subtracting and adding the equations thus obtained, we get :

$$(II')\quad \begin{cases} (\partial_\theta + k\tan\theta + (\ell+m_2-1)\cot\theta)C &= (\ell-k-m_1+1-s)B \\ (\partial_\theta - \ell\cot\theta - (k+m_1-1)\tan\theta) &= -(k-\ell+m_2+1-s)C. \end{cases}$$

Putting again $t = \cos 2\theta = \frac{r_1^2-r_2^2}{r_1^2+r_2^2}$ and, by taking into account (4.4),

$$\begin{aligned} B(r,\theta) &= r^s\cos^{k+1}\theta\sin^\ell\theta b(\cos 2\theta) \\ C(r,\theta) &= r^s\cos^k\theta\sin^{\ell+1}\theta c(\cos 2\theta), \end{aligned}$$

(II') then leads to the system

$$(II'')\quad \begin{cases} 2(t-1)c' + (2\ell+m_2)c &= (\ell-k-m_1+1-s)b \\ 2(t+1)b' + (2k+m_1)b &= (k-\ell-m_2+1-s)c \end{cases}$$

which has solutions of the form

$$c(t) = (\ell-k-m_1+1-s)P_n^{\ell+\frac{m_2}{2}-1,k+\frac{m_1}{2}}(t)$$

and

$$b(t) = -(k - \ell - m_2 + 1 - s)P_n^{\ell+\frac{m_2}{2},k+\frac{m_1}{2}-1}(t).$$

Hence by (4.4) $S_{s,k,\ell}(\underline{x}_1, \underline{x}_2)$ admits the form required. ■

The following theorem tells us how, using the decomposition $\mathbf{R}^m = \mathbf{R}^{m_1} \times \mathbf{R}^{m_2}$ of $\mathbf{R}^m$, an orthogonal basis of $\mathcal{M}^+(m; s; \mathcal{C})$ may be constructed once orthogonal bases of $\mathcal{M}^+(m_1; k; \mathcal{C})$ and $\mathcal{M}^+(m_2; \ell; \mathcal{C})$ are known for all different pairs (k, ℓ) satisfying $k + \ell \leq s$.

The theorem is entirely formulated in terms of "orthogonal" polynomials in $(\underline{x}_1, \underline{x}_2)$, "orthogonality" here meaning that, when restricted to S^{m-1}, the polynomials considered are orthogonal in $L_2(S^{m-1}; \mathcal{C})$.

Notice also that, given a pair (k, ℓ) with $k + \ell \leq s$, we put for simplicity

$$d_1 = \dim \mathcal{M}^+(m_1; k; \mathcal{C})$$

and

$$d_2 = \dim \mathcal{M}^+(m_2; \ell; \mathcal{C})$$

where, obviously, these dimensions are depending on (m_1, k) and (m_2, ℓ) respectively.

Theorem 4.5. *Let $\mathbf{R}^m$ be split into $\mathbf{R}^m = \mathbf{R}^{m_1} \times \mathbf{R}^{m_2}$, let $s \in \mathbf{N}$ and let $k, \ell \in \mathbf{N}$ be such that $k + \ell \leq s$. Furthermore, let $S_{s,k,\ell}$ be as in Lemma 4.4., let*

$\{P_k^{(i)} : i = 1, \ldots, d_1\}$ be an orthogonal basis of $M^+(m_1; k; \mathcal{C})$ and let $\{P_\ell^{(j)} : j = 1, \ldots, d_2\}$ be an orthogonal basis of $M^+(m_2; \ell; \mathcal{C})$ where all $P_k^{(i)}$ and $P_\ell^{(j)}$ are, respectively, $\Gamma(0, m_1) \cup \{0\}$- and $\Gamma(0, m_2) \cup \{0\}$- valued. Then we have that, when putting

$$M_{k,\ell}^{(i,j)}(\underline{x}_1, \underline{x}_2) = S_{s,k,\ell}(\underline{x}_1, \underline{x}_2) P_k^{(i)}(\underline{x}_1) P_\ell^{(j)}(\underline{x}_2),$$

the set

$$\{M_{k,\ell}^{(i,j)}(\underline{x}_1, \underline{x}_2) : k + \ell \leq s; i = 1, \ldots, d_1, j = 1, \ldots, d_2\}$$

is an orthogonal basis of $M^+(m; s; \mathcal{C})$ consisting of $\Gamma(0, m) \cup \{0\}$-valued functions.

Proof. First of all notice that, as any two elements in $\Gamma(0, m_1)$ and $\Gamma(0, m_2)$ are commuting or anticommuting, for all i and j,

$$\Gamma_{\underline{\omega}_1}(P_k^{(i)} P_\ell^{(j)}) = (-k) P_k^{(i)} P_\ell^{(j)}$$

and

$$\Gamma_{\underline{\omega}_2}(P_k^{(i)} P_\ell^{(j)} = (-\ell) P_k^{(i)} P_\ell^{(j)}.$$

Hence each $P_k^{(i)}(\underline{\omega}_1)P_\ell^{(j)}(\underline{\omega}_2) \in \mathcal{M}^{0,0}_{(k,\ell)}$ so that by virtue of Lemma 4.1. and the hypotheses made, $\{P_k^{(i)}(\underline{\omega}_1)P_\ell^{(j)}(\underline{\omega}_2) : i = 1,\ldots,d_1; j = 1,\ldots,d_2\}$ is an orthogonal basis of $\mathcal{M}^{0,0}_{(k,\ell)}$, whence $\{P_k^{(i)}(\underline{x}_1)P_\ell^{(j)}(\underline{x}_2) : i = 1,\ldots,d_1, j = 1,\ldots,d_2\}$ is an orthogonal basis of $M^+_{(k,\ell)}(m_1;m_2;\mathcal{C})$.

Taking into account Lemma 4.3. and Lemma 4.4, we have by construction that each $M_{k,\ell}^{(i,j)} \in M^+(m;s;\mathcal{C})$.

Moreover, for $k,\ell \in \mathbf{N}$ fixed with $k+\ell \leq s$, we have that for any two different pairs (i,j) and (i',j') in $\{1,\ldots,d_1\} \times \{1,\ldots,d_2\}$, $M_{k,\ell}^{(i,j)}$ and $M_{k,\ell}^{(i',j')}$ are orthogonal in $L_2(S^{m-1};\mathcal{C})$.

Furthermore, for any two different pairs (k,ℓ) and (k',ℓ') satisfying $k+\ell \leq s$ and $k'+\ell' \leq s$, we have that $M_{k,\ell}^{(i,j)}$ and $M_{k',\ell'}^{(i',j')}$ are orthogonal in $L_2(S^{m-1};\mathcal{C})$ and this for all $(i,j) \in \{1,\ldots,d_1\} \times \{1,\ldots,d_2\}$ and $(i',j') \in \{1,\ldots,d_1'\} \times \{1,\ldots,d_2'\}$.

In view of Lemma 4.2. we may indeed conclude that the set of elements $M_{k,\ell}^{(i,j)}$ forms an orthogonal basis of $M^+(m;s;\mathcal{C})$.

Finally notice that if $P_k^{(i)}$ and $P_\ell^{(j)}$ are, respectively, $\Gamma(0,m_1) \cup \{0\}$- and $\Gamma(0,m_2) \cup \{0\}$-valued, then clearly $P_k^{(i)}P_\ell^{(j)}$ is $\Gamma(0,m) \cup \{0\}$-valued.

Now we claim that $S_{s,k,\ell}$ is also $\Gamma(0,m)$-valued.

Indeed, if $s-(k+\ell)$ is even we have by virtue of (4.3) that

$$\begin{aligned} S_{s,k,\ell}(r_1\underline{\omega}_1, r_2\underline{\omega}_2) &= E_1(r_1,r_2) + \underline{\omega}_1\underline{\omega}_2 E_2(r_1,r_2) \\ &= (-E_1(r_1,r_2)\underline{\omega}_2 + E_2(r_1,r_2)\underline{\omega}_1)\underline{\omega}_2 \end{aligned}$$

is $\Gamma(0,m) \cup \{0\}$-valued, while, if $s-(k+\ell)$ is odd, by means of (4.4),

$$S_{s,k,\ell}(r_1\underline{\omega}_1, r_2\underline{\omega}_2) = E_3(r_1,r_2)\underline{\omega}_1 + E_4(r_1,r_2)\underline{\omega}_2$$

is $\Gamma(0,m) \cup \{0\}$-valued too. ■

Since, starting from a splitting of $\mathbf{R}^m$ into $\mathbf{R}^{m_1} \times \mathbf{R}^{m_2}$, of $\mathbf{R}^{m_1}$ into $\mathbf{R}^{m_1^{(1)}} \times \mathbf{R}^{m_1^{(2)}}$ and of $\mathbf{R}^{m_2}$ into $\mathbf{R}^{m_2^{(1)}} \times \mathbf{R}^{m_2^{(2)}}$, etc ... , we finally arrive at components of the form $\mathbf{R}^1$ or $\mathbf{R}^2$, the foregoing Theorem 4.5. indicates how to obtain an orthogonal basis of $M^+(m;s;\mathcal{C})$ when knowing orthogonal bases of $M^+(1;k;\mathcal{C})$ and $M^+(2;\ell;\mathcal{C})$.

In such a way it is clear that the latter spaces are cornerstones in constructing an orthogonal basis for $M^+(m;s;\mathcal{C})$. Let us therefore have a closer look at them

(I) The space $M^+(1;k;\mathcal{C})$

Obviously, when $k \geq 1$, the set of homogeneous monogenic polynomials of degree k on the real line is empty, so that we only have to take into account the case $k=0$.

Clearly $\{1\}$ is an orthogonal basis for $M^+(1;0;\mathcal{C})$ consisting of a $\Gamma(0,1) \cup \{0\}$-valued function.

(II) The space $M^+(2;\ell;C)$

In the case of the plane, we already know that an orthogonal basis of $M^+(2;\ell;C)$ is given by $\{z_1^\ell\}$ where $z_1 = x_2 + x_1e_1e_2$ (see also section 1.5).

Clearly $z_1 = (-x_2e_2 + x_1e_1)e_2$ and hence also z_1^ℓ are $\Gamma(0,2) \cup \{0\}$-valued.

Notice that, by using polar coordinates, we may also write z_1^ℓ as

$$z_1^\ell = (re^{-e_1e_2\theta})^\ell (e_1e_2)^\ell.$$

5 Generalized Taylor and Laurent series

Let $\mathbf{R}^m$ be split into $\mathbf{R}^m = \mathbf{R}^p \oplus \mathbf{R}^q$. In this section, we want to obtain a suitable generalization of the Taylor series for monogenic functions, namely such that $\underline{y} \in \mathbf{R}^q$ is considered as a parameter and that we have a Taylor decomposition in the variable $\underline{x} \in \mathbf{R}^p$. Notice that monogenicity should be understood in terms of the **Dirac** operator.

The first idea coming into mind is to use functions of the form $P_k(\underline{x})A(\underline{y})$ as analogues of Taylor polynomials, where $P_k \in M^+(p,k;C)$ and $A(\underline{y})$ is analytic, and to express any monogenic function around $\mathbf{R}^q$ as a series in such polynomials. It does not work, however, since polynomials of the described type are usually not monogenic on $\mathbf{R}^m$. Nevertheless, there is a possibility to correct such a function by adding a series of functions of the type $\underline{x}^l P_k(\underline{x})A_l(\underline{y}), l > 0$, in such a way that the resulting function is monogenic in $\mathbf{R}^m$. As we shall see, there is exactly one way to do so. Note that all functions added to the original function $P_k(\underline{x})A(\underline{y})$ are also homogeneous polynomials in $\rho = |\underline{x}|$, but of higher order. As a consequence, the leading term of the sum "near" the axis $\mathbf{R}^q$ is the original function we started with. So such a sum indeed yields a good analogue of a Taylor polynomial. We shall show below that every monogenic function in a symmetric domain around $\mathbf{R}^q$ can be decomposed into functions of the described type and that these components are uniquely characterized by the corresponding "Taylor coefficients" $A(\underline{y})$, thus giving rise to a generalized Cauchy-Kovalevska extension property.

5.1 *Generalized Cauchy-Kovalevska extension*

Let us consider an SO(p)-invariant domain $\tilde{\Omega} \subset \mathbf{R}^m$ and let us assume that the intersections of $\tilde{\Omega}$ with all subspaces parallel to $\mathbf{R}^p$ are convex (and hence, if nonvoid, they contain a point in $\mathbf{R}^q$). Putting $\Omega = \tilde{\Omega} \cap \mathbf{R}^q$, we have

Theorem 5.1.1. *Let $P_k \in M^+(p,k;C)$ be fixed and let $A_0(\underline{y})$ be a (C-valued) analytic function in Ω. Then there exists a unique sequence $\left(A_l(\underline{y})\right)_{l>0}$ of analytic functions such that the series*

$$f(\underline{x},\underline{y}) = \sum_{l=0}^{\infty} \underline{x}^l P_k(\underline{x})A_l(\underline{y})$$

is convergent in a neighbourhood $U \subset \mathbf{R}^m$ of the domain Ω and its sum f is monogenic in U. The function $A_0(\underline{y})$ is determined by the relation

$$P_k(\underline{\xi})A_0(\underline{y}) = \lim_{\rho\to 0}\frac{1}{\rho^k} f(\underline{x},\underline{y}),\ \underline{x} = \rho\underline{\xi}.$$

Furthermore, the sum f is formally given by the expression

$$f(\underline{x},\underline{y}) = \Gamma\left(k+\frac{p}{2}\right)\left(\frac{\rho\sqrt{\Delta_{\underline{y}}}}{2}\right)^{-(k+\frac{p}{2})} \cdot$$

$$\cdot\left[\frac{\rho\sqrt{\Delta_{\underline{y}}}}{2}J_{k+\frac{p}{2}-1}\left(\rho\sqrt{\Delta_{\underline{y}}}\right)+\frac{\underline{x}\,\partial_{\underline{y}}}{2}J_{k+\frac{p}{2}}\left(\rho\sqrt{\Delta_{\underline{y}}}\right)\right](P_k(\underline{x})A_0(\underline{y})),$$

where $\sqrt{\Delta_{\underline{y}}}$ denotes the square root of the Laplacian $\Delta_{\underline{y}}$ (of which only even powers occur in the resulting series).

Proof. Let us state first the following basic identities:

$$\partial_{\underline{y}}\underline{x}^{\,l} = (-1)^l\underline{x}^{\,l}\partial_{\underline{y}},$$

where $\partial_{\underline{y}}$ and $\underline{x}^{\,l}$ are both considered as operators on functions, and

$$\partial_{\underline{x}}(\underline{x}^{\,2l}P_k(\underline{x})) = -2l\underline{x}^{\,2l-1}P_k(\underline{x}),$$

$$\partial_{\underline{x}}(\underline{x}^{\,2l+1}P_k(\underline{x})) = -(2l+2k+p)\underline{x}^{\,2l}P_k(\underline{x}).$$

The last two formulae can be found in 1.12.1, Proposition 3; they express the fact that the space $M^+(p,k;\mathcal{C})$ and its shifts of the form $\underline{x}^{\,l}\cdot M^+(p,k;\mathcal{C})$ are characterized exactly as subspace s of $\mathcal{C}$-valued polynomials of a given order such that the operations $\partial_{\underline{x}}$ and multiplication by $\underline{x}$ are inverses of each other (modulo a suitable constant).

If we apply now the operator $\partial_{\underline{x}} + \partial_{\underline{y}}$ to the (formal) sum f, we get, using the formulae above:

$$(2l+2k+p)P_k(\underline{x})A_{2l+1}(\underline{y}) = \partial_{\underline{y}}[P_k(\underline{x})A_{2l}(\underline{y})],$$

$$(2l+2)P_k(\underline{x})A_{2l+2}(\underline{y}) = -\partial_{\underline{y}}[P_k(\underline{x})A_{2l+1}(\underline{y})].$$

Hence the functions $A_l(\underline{y})$ are uniquely determined by the formulae

$$P_k(\underline{x})A_{2l}(\underline{y}) = \frac{(-1)^l\partial_{\underline{y}}^{2l}[P_k(\underline{x})A_0(\underline{y})]}{2l(2l-2)\ldots 2(2l+2k+p-2)\ldots(2k+p)}$$

$$= \frac{(-1)^l\Gamma(k+\frac{p}{2})\partial_{\underline{y}}^{2l}[P_k(\underline{x})A_0(\underline{y})]}{2^{2l}\,l!\,\Gamma(l+k+\frac{p}{2})},$$

and

$$P_k(\underline{x})A_{2l+1}(\underline{y}) = \frac{(-1)^l\partial_{\underline{y}}^{2l+1}[P_k(\underline{x})A_0(\underline{y})]}{2l(2l-2)\dots 2(2l+2k+p)\dots(2k+p)}$$

$$= \frac{(-1)^l\Gamma(k+\frac{p}{2})\partial_{\underline{y}}^{2l+1}[P_k(\underline{x})A_0(\underline{y})]}{2^{2l+1}\, l!\,\Gamma(l+k+\frac{p}{2}+1)}.$$

The last formula of the theorem follows from the expansion (for an indeterminate variable α)

$$\sum_{l=0}^{\infty}\frac{\alpha^{2l}}{2^{2l}\, l!\,\Gamma(l+k+\frac{p}{2})} = \Gamma\left(\frac{p}{2}\right)\left(\frac{\alpha}{2}\right)^{-k-\frac{p}{2}+1} J_{k+\frac{p}{2}-1}(\alpha).$$

Finally, we have to show that the series converges in a neighbourhood of Ω in $\mathbf{R}^m$. To this end, it is sufficient first to note that a function $f \in C_\infty(\Omega)$ is real analytic iff for every compact set $K \subset \Omega$ there exists $R_f(K) > 0$ such that the series

$$\sum_{n=0}^{\infty}\frac{M_n(f,K)}{(2n)!}z^n,\quad M_n(f,K) = \sup_{\underline{y}\in K}|\Delta_{\underline{y}}^n f(\underline{y})|,$$

has the radius of convergence equal to $R_f(K)$ (see [2]) and then to use Stirlings inequality to prove that, up to factors of slow growth in k, we have

$$2^{2l}\, l!\,\Gamma(l+k+\frac{p}{2}) \approx 2^{2l}\,(l!)^2 \approx \frac{(2l)^{2l}}{e^{2l}} \approx (2l)!. \blacksquare$$

Definition 5.1.2. *Let $k \in N$ be fixed and let Ω be a domain in $\mathbf{R}^q$. Then $\mathcal{T}_k(\Omega)$ stands for the space of all functions of the form $f = \sum_{l=0}^{\infty}\underline{x}^l A_l(\underline{x},\underline{y})$ such that*

(i) for any $\underline{y}$ fixed, $A_l(\underline{x},\underline{y})$ belongs to the space $M^+(p,k;C)$

(ii) the sum converges in a neighbourhood of Ω and f is monogenic there.

Functions belonging to $\mathcal{T}_k(\Omega)$ will be called generalized Taylor functions of order k.

The spaces $\mathcal{T}_k(\Omega)$ are direct generalizations of the one dimensional spaces

$$\{a_k z^k | a_k \in C\}, k \geq 0,$$

generated by z^k in complex analysis. We shall show now that every element of $\mathcal{T}_k(\Omega)$ can be characterized by an $M^+(p,k;C)$-valued function of $\underline{y}$, which is playing the role of a Taylor coefficient.

Theorem 5.1.3.

(i) For every $f \in \mathcal{T}_k(\Omega)$ and every $\underline{y} \in \Omega$,

$$\lim_{\rho\to 0}\frac{1}{\rho^k}f(\rho\underline{\xi},\underline{y})$$

exists. Moreover, putting

$$T_k(f)(r\underline{\xi},\underline{y}) = r^k\cdot\lim_{\rho\to 0}\frac{1}{\rho^k}f(\rho\underline{\xi},\underline{y})$$

we have that for every $\underline{y}\in\Omega$ *fixed,* $T_k(f)(\underline{x},\underline{y})$ *is inner monogenic, whence* $T_k(f)$ *can be considered to be an* $M^+(p,k;C)$*-valued analytic function on* Ω.

(ii) For every $M^+(p,k;C)$*-valued analytic function* A *on* Ω*, there exists a unique element* $f\in\mathcal{T}_k$ *such that*

$$T_k(f) = A.$$

Proof. Suppose first that $f\in\mathcal{T}_k(\Omega)$ is given by

$$f = \sum_{l=0}^{\infty}\underline{x}^l A_l(\underline{x},\underline{y}),$$

where $A_l(\underline{x},\underline{y})$ are $M^+(p,k;C)$-valued functions on Ω. Then, because of homogeneity in ρ,

$$\lim_{\rho\to 0}\frac{1}{\rho^k}f(\rho\underline{\xi},\underline{y}) = A_0(\underline{\xi},\underline{y}),$$

whence $T_k(f) = A_0$.

Suppose next that an $M^+(p,k;C)$-valued function $A(\underline{x},\underline{y})$ is given. If $(P_{k,\alpha})_{\alpha\in A_k}$ is any basis for the finite dimensional space $M^+(p,k;C)$, then $A(\underline{x},\underline{y})$ can be uniquely written in the form

$$A(\underline{x},\underline{y}) = \sum_{A_k}P_{k,\alpha}(\underline{\xi})t_{k,\alpha}(\underline{y}),$$

where $t_{k,\alpha}(\underline{y})$ are analytic functions on Ω. If $f_{k,\alpha}(\underline{x},\underline{y})$ are the monogenic functions corresponding to $t_{k,\alpha}(\underline{y})$ by Theorem 5.1, then their sum $f = \sum_{B_k} f_{k,\alpha}$ clearly belongs to $\mathcal{T}_k(\Omega)$ and $T_k(f) = A$. The unicity follows from the fact that if

$$f = \sum_{l=0}^{\infty}\underline{x}^l A_l(\underline{x},\underline{y})\in\mathcal{T}(\Omega)$$

with $A_0\equiv 0$, then the condition of monogenicity implies immediately (by induction) that all other functions $A_l, l>0$ are trivial. ■

Definition 5.1.4. *For any* $f\in\mathcal{T}_k(\Omega)$ *given, the associated function*

$$T_k(f)(r\underline{\xi},\underline{y}) = r^k\cdot\lim_{\rho\to 0}\frac{1}{\rho^k}f(\rho\underline{\xi},\underline{y})$$

is henceforth called the generalized Taylor coefficient of f *of order* k.

5.2 Generalized Taylor series

Let us recall (see 1.3.4) that any function on the sphere S^{p-1} admits an expansion

$$f(\underline{\xi}) = \sum_{k=0}^{\infty} \Pi(k) f(\underline{\xi}),$$

where

$$\Pi(k) f(\underline{\xi}) = \mathbf{P}(k) f(\underline{\xi}) - \underline{\xi}\mathbf{P}(k)(\underline{\xi} f)(\underline{\xi}),$$

$\mathbf{P}(k)$ being the projection onto $\mathcal{M}^+(p,k;\mathcal{C})$. The series converges uniformly.

For any function $f(\underline{x},\underline{y})$ on $\tilde{\Omega}$ and for any fixed pair $(\rho,\underline{y})$ such that $(\underline{x},\underline{y}) \in \tilde{\Omega}$, $\underline{x} = \rho\underline{\xi}$, we can thus decompose f as a function of $\underline{\xi}$ by the procedure above.

Definition 5.2.1. *The projection operator $\Pi(k)$ will be extended to the space of smooth $\mathcal{C}_m$-valued functions on $\tilde{\Omega}$ by the formula*

$$[\Pi(k) f](\rho\underline{\xi},\underline{y}) = \Pi(k)[f(\rho,\underline{\xi},\underline{y})],$$

($\Pi(k)$ being applied on the right hand side for $(\rho,\underline{y})$ fixed).

The projections $\Pi(k)$ commute with $\partial_{\underline{x}} + \partial_{\underline{y}}$, (see 1.12.2, Corollary 2), hence if f is monogenic in $\tilde{\Omega}$, then the functions $\Pi(k)f$ are monogenic.

The image of the space $M_l(\tilde{\Omega};\mathcal{C})$ of (left) monogenic functions under the map $\Pi(k)$ will be denoted by $\mathcal{T}_{l,k}(\tilde{\Omega})$.

Theorem 5.2.2. (Generalized Taylor series)

(i) Suppose that f is a monogenic function in $\tilde{\Omega}$. Then f can be written in a unique way as

$$f = \sum_{k=0}^{\infty} f_k,\ f_k \in \mathcal{T}_k(\tilde{\Omega}),$$

the series converging uniformly on compact sets in $\tilde{\Omega}$. The functions f_k are given by the formula

$$f_k = \mathbf{P}(k) f(\rho,\underline{\xi},\underline{y}) - \underline{\xi}\mathbf{P}(k)(\underline{\xi} f)(\rho,\underline{\xi},\underline{y}).$$

(ii) The functions f_k are generalized Taylor coefficients of f_k are given by

$$T_k(f_k)(r\underline{\xi},\underline{y}) = r^k \cdot \lim_{\rho\to 0} \frac{1}{\rho^k} \mathbf{P}(k) f(\rho,\underline{\xi},\underline{y}).$$

The function f is uniquely determined by its set of generalized Taylor coefficients.

Proof. The first part of the theorem follows immediately from the properties of the projection operator $\Pi(k)$. The uniform convergence on compact sets was proved in 2.3.3. The spaces $\mathcal{M}^+(p,k;\mathcal{C})$ being finite dimensional, each function $\Pi(k)f$ can be written as a (finite) sum of functions of the form

$$P_k(\underline{\xi})A(\rho,\underline{y}) + \underline{\xi}P_k(\underline{\xi})B(\rho,\underline{y}), \tag{5.1}$$

where $P_k \in \mathcal{M}^+(p,k;\mathcal{C})$ and A, B are analytic. But a function of the form (5.1) is analytic in points belonging to Ω iff it can be written in the form

$$P_k(\underline{x}\,)a(\rho^2,\underline{y}) + \underline{x}\,P_k(\underline{x}\,)b(\rho^2,\underline{y}),$$

where a and b are (for fixed $\underline{y}$) analytic near $\rho^2 = 0$ and $P_k(\underline{x}\,) = \rho^k P_k(\underline{\xi})$ belongs to $M^+(p,k;\mathcal{C})$. This implies that such functions belong to $\mathcal{T}_k(\Omega)$.

The uniqueness of the Taylor series follows from the fact that the spaces $\mathcal{T}_k(\Omega)$ are linearly independent. In the limit, only the part $P_k(\underline{x}\,)a(\rho^2,\underline{y})$ survives, which proves the rest of the theorem. ∎

Definition 5.2.3. *Let $f = \sum_k f_k$, $f_k \in \mathcal{T}(\Omega)$, be monogenic in $\tilde{\Omega}$ and let $T_k(f_k)$ be the generalized Taylor coefficient of f_k. Then $T_k(f_k)$ is called the generalized Taylor coefficient of f of order k; it will be denoted by $T_k(f)$.*

If a basis $(P_{k,\alpha})_{\alpha\in A_k}$ for the space $M^+(p,k;\mathcal{C})$ is given, then the generalized Taylor theorem can be written in a form which is closer to the standard Taylor theorem, the variable $\underline{y}$ playing the role of a parameter.

Theorem 5.2.4. *Let f be a monogenic function in $\tilde{\Omega}$. Then the generalized Taylor coefficient of f of order k can be decomposed in a unique way as*

$$T_k(f)(\underline{\xi},\underline{y}) = \sum_{\alpha\in A_k} P_{k,\alpha}(\underline{\xi})T_{k,\alpha}(f)(\underline{y}),$$

where $T_{k,\alpha}(f)(\underline{y})$ are real analytic functions on Ω.

If the generalized Taylor functions, corresponding to $P_{k,\alpha}(\underline{\xi})T_{k,\alpha}(f)(\underline{y})$, are denoted by $T_{k,\alpha}$, then

$$f = \sum_k \sum_{\alpha\in A_k} T_{k,\alpha}$$

in a neighbourhood of Ω.

5.3 Generalized Laurent series

Knowing what the generalized Taylor series for a monogenic function is, we now wish to define generalized Laurent series. The basic idea behind it is to use the duality between left and right monogenic functions g and f given by

$$\int_{\partial\Omega} f d\sigma g.$$

Moreover, the notion of a generalized Laurent coefficient is introduced which appears to be an $M_r^+(p,k;\mathcal{C})$-valued analytic functional (the subscript r referring to right monogenicity). It will be convenient to identify below the spaces $M_r^+(p,k;\mathcal{C})$ and $\mathcal{M}_r^+(p,k;\mathcal{C})$ (the isomorphism being the restriction to the unit sphere).

5.3.1 Taylor and Laurent part of the series

Before proceeding, let us first show how each monogenic function in $\tilde{\Omega}\setminus K, K\subset \mathbf{R}^q$ being compact, can be written as a sum of its Taylor and Laurent parts.

In what follows we are assuming that $\mathbf{R}^m = \mathbf{R}^p\times\mathbf{R}^q, \tilde{\Omega}\subset\mathbf{R}^m$ is an SO(p)-invariant domain such that the intersections of $\tilde{\Omega}$ with all subspaces parallel to $\mathbf{R}^p$ are convex, $\Omega=\tilde{\Omega}\cap\mathbf{R}^q$ and $K\subset\Omega$ is compact. Calling $M_{l,0}(\mathbf{R}^m\setminus K;\mathcal{C})$ the space of left monogenic functions in $\mathbf{R}^m\setminus K$ which vanish at infinity, we have

Theorem 5.3.1. *$M_l(\tilde{\Omega}\setminus K;\mathcal{C}) = M_{l,0}(\mathbf{R}^m\setminus K;\mathcal{C})\oplus M_l(\tilde{\Omega};\mathcal{C})$, i.e. each element $f\in M_l(\tilde{\Omega}\setminus K;\mathcal{C})$ admits a unique decomposition of the form $f = Tf + Lf$, where $Tf\in M_{l,0}(\mathbf{R}^m\setminus K;\mathcal{C})$ and $Lf\in M_l(\tilde{\Omega};\mathcal{C})$.*

Proof. Suppose first that $\tilde{\Omega}$ is a bounded set. We can exhaust $\tilde{\Omega}$ by a sequence $\tilde{\Omega}_n$ of open domains with a piecewise smooth boundary such that $\tilde{\Omega}_n\subset\subset\tilde{\Omega}_{n+1}\subset\tilde{\Omega}$. Indeed, taking a suitable sequence $\epsilon_n>0, \epsilon\downarrow 0$, we choose for every n a finite cover of the compact set $\partial\tilde{\Omega}$ by balls $B(p_i,\epsilon_n), i\in A$, of radius ϵ_n and we define a neighbourhood of $\partial\tilde{\Omega}$ by

$$U_n = \bigcup_{i\in A} B(p_i,\epsilon_n).$$

Analogously, we may define a neighbourhood of K by

$$U_n' = \bigcup_{i\in A'} B(p_i',\epsilon_n).$$

We then put

$$Tf = \int_{\partial U_n'} E(\underline{x}-\underline{y})d\sigma_{\underline{y}} f$$

on $\mathbf{R}^m\setminus U_n'$ and

$$L\,f = \int_{\partial U_n \cap \tilde{\Omega}} E(\underline{x} - \underline{y}) d\sigma_{\underline{y}} f$$

on $\tilde{\Omega} \setminus U_n$.

The first integral defines, in fact, a sequence of monogenic functions, vanishing at ∞, which coincide on the intersection of the domains of definition, whence it yields a monogenic function on $\mathbf{R}^m \setminus K$, vanishing at ∞. The function $L\,f$ is defined, in the same way, on $\tilde{\Omega}$. The fact that $f = T\,f + L\,f$ then follows from the Cauchy theorem.

In the case that $\tilde{\Omega}$ is not bounded, we exhaust it first by an increasing sequence of bounded domains and then apply the above procedure. ∎

5.3.2 Generalized Laurent series

The generalized Laurent series will be defined now by using the duality between the space $M_{l,0}(\mathbf{R}^m \setminus K; \mathcal{C})$ and the space $M_r(K; \mathcal{C})$ given by

$$\int_{\partial U} f\, d\sigma\, g,$$

where U is a suitable neighbourhood of K.

The spaces $M_r(K; \mathcal{C})$ are defined as the inductive limits

$$M_r(K; \mathcal{C}) = \operatorname{lim\,ind}_{U \supset K} M_r(U; \mathcal{C}).$$

The standard duality theory (see [10]) gives rise to the following

Theorem 5.3.2. *Let $K \subset \mathbf{R}^m$ be compact. Then $M_{l,0}(\mathbf{R}^m \setminus K; \mathcal{C}) \simeq M_r'(K; \mathcal{C})$ as topological vector spaces (the topology of uniform convergence on compact subsets is taken on $M_{l,0}(\mathbf{R}^m \setminus K; \mathcal{C})$ and the strong topology is taken on $M_r'(K; \mathcal{C})$). The corresponding isomorphism is given by the Cauchy transform*

$$\hat{T}(\underline{x}) = \frac{-1}{A_m} \langle T_{\underline{u}}, \frac{\underline{x} - \underline{u}}{|\underline{x} - \underline{u}|^m} \rangle,\ T \in M_r'(K; \mathcal{C}), \underline{x} \in \mathbf{R}^m \setminus K.$$

The map $B : M_{l,0}(\mathbf{R}^m \setminus K; \mathcal{C}) \to M_r'(K; \mathcal{C})$ inverse to the Cauchy transform can be described as follows:

$$B(f)[g] = \int_{\partial U} g\, d\sigma\, f,\ f \in M_{l,0}(\mathbf{R}^m \setminus K; \mathcal{C}),\ g \in M_r(K; \mathcal{C}),$$

where U is a suitable neighbourhood of K. The notation T_f instead of B_f is often used.

The spaces $M_{l,0}(\mathbf{R}^m \setminus K; \mathcal{C})$ and $M_r(K; \mathcal{C})$ thus form a dual pair. So, by duality the splitting of monogenic functions on open domains into general Taylor functions will lead to a splitting of $M_{l,0}(\mathbf{R}^m \setminus K; \mathcal{C})$.

Let us denote by $\mathcal{T}_{r,k}(K)$ the inductive limit of the spaces $\mathcal{T}_{r,k}(U), U \supset K$ (the subscript r is referring to right monogenicity). Then Theorem 5.2.1 implies that

$$M_r(K;\mathcal{C}) = \bigoplus_0^\infty \mathcal{T}_{r,k}(K). \tag{5.2}$$

The spaces $\mathcal{T}_{r,k}(K)$ being closed subspaces, by duality we have

$$M_r'(K;\mathcal{C}) = \bigoplus_0^\infty \mathcal{T}'_{r,k}(K), \tag{5.3}$$

where the spaces $\mathcal{T}'_{r,k}(K)$ are orthogonal to $\oplus_{k'\neq k}\mathcal{T}_{r,k'}(K)$. More precisely, the direct sum in (5.3) means that each element of $M_r'(K;\mathcal{C})$ can be written in a unique way as the sum of elements in $\mathcal{T}'_{r,k}(K)$, where the sum is taken in the weak topology. The Cauchy transform then translates the splitting to the space $\mathcal{L}_{l,k}(\mathbf{R}^m \setminus K)$, the definition of which is given in

Definition 5.3.3. *The space $\mathcal{L}_{l,k}(\mathbf{R}^m \setminus K)$ of (left) generalized Laurent functions of order k on $\mathbf{R}^m \setminus K$ is given by*

$$\mathcal{L}_{l,k}(\mathbf{R}^m \setminus K) = \{f \in M_{l,0}(\mathbf{R}^m \setminus K;\mathcal{C}) | B(f)[g] = 0 \text{ for all } g \in \mathcal{T}_{l,k'}, k' \neq k\}.$$

As an immediate consequence of the definition we get

Theorem 5.3.4. (Generalized Laurent series) *We have*

$$M_{l,0}(\mathbf{R}^m \setminus K;\mathcal{C}) = \bigoplus_0^\infty \mathcal{L}_{l,k}(\mathbf{R}^m \setminus K)$$

and

$$M_l(\tilde{\Omega} \setminus K;\mathcal{C}) = \left[\bigoplus_0^\infty \mathcal{T}_{l,k}(\tilde{\Omega})\right] \oplus \left[\bigoplus_0^\infty \mathcal{L}_{l,k}(\mathbf{R}^m \setminus K)\right],$$

both series being uniformly convergent on compact subsets.

Proof. The Cauchy transform gives (topological) isomorphism

$$\mathcal{L}_{l,k}(\mathbf{R}^m \setminus K) \simeq \mathcal{T}'_{r,k}(K).$$

The assertion then follows from the fact that the space $M_r'(K;\mathcal{C})$ is a Frechet-Montel space ([50]), whence the weak and strong sequential convergence coincide. ∎

5.4 Generalized Laurent coefficients

We have seen that generalized Taylor functions of order k can be characterized by their generalized Taylor coefficients of order k. We would like to do the same with generalized Laurent functions.

Let us denote by $\mathcal{A}(U, M_r^+(p,k;\mathcal{C}))$ the space of all $M_r^+(p,k;\mathcal{C})$-valued analytic functions on a domain U and by $\mathcal{A}(K, M_r^+(p,k;\mathcal{C}))$ the inductive limit over open neighbourhoods of K. Theorem 5.1.2 then implies that the spaces $\mathcal{A}(K, M_r^+(p,k;\mathcal{C}))$ and $\mathcal{T}_{r,k}(K)$ are isomorphic, the isomorphism being given by T_k. By duality, we then have the following diagram:

$$\begin{array}{ccc} \mathcal{T}_{r,k}(K) & \xrightarrow{T_k} & \mathcal{A}(K, M_r^+(p,k;\mathcal{C})) \\ \updownarrow & & \updownarrow \\ \mathcal{T}'_{r,k}(K) & \xleftarrow{T'_k} & \mathcal{A}'(K, M_r^+(p,k;\mathcal{C})) \\ \uparrow B & \nearrow L_k & \\ \mathcal{L}_{l,k}(\mathbf{R}^m \setminus K) & & \end{array}$$

Here B denotes the inverse to the Cauchy transform. The map T_k is an isomorphism, so the same is true for T_k'. The map L_k is then defined as $L_k = (T_k')^{-1} \circ B$ and so it is an isomorphism as well.

Definition 5.4.1. *For each $f \in \mathcal{L}_{l,k}(\mathbf{R}^m \setminus K)$, the functional*

$$L_k(f) \in \mathcal{A}'(K, M_r^+(p,k;\mathcal{C}))$$

is called the generalized Laurent coefficient of f of order k.

Note that the function $f \in \mathcal{L}_{l,k}(\mathbf{R}^m \setminus K)$ is uniquely determined by its Laurent coefficients $L_k(f)$ and, obviously, it would be useful to have a procedure available enabling us to compute the Laurent coefficients effectively. To this end, let us recall that the generalized Taylor coefficients of a generalized Taylor function f of order k were defined as

$$T_k(f)(r\underline{\xi}, \underline{y}) = r^k \cdot \lim_{\rho \to 0} \frac{1}{\rho^k} \mathbf{P}(k) f(\rho\underline{\xi}, \underline{y}),$$

the limit being pointwise.

We would like to have a similar formula for the computation of the generalized Laurent coefficients. Pointwise limits similar to that described above cannot be used anymore and should be substituted by a limit taken in a suitable topology on the corresponding function space. Let us therefore first describe how the space $\mathcal{A}(\overline{U}, \mathcal{M}_l^+(p,k;\mathcal{C}))$ of $\mathcal{M}_l^+(p,k;\mathcal{C})$-value d functions on $\overline{U}, U \subset \mathbf{R}^m$ open is embedded into the space $\mathcal{A}'(\overline{U}, \mathcal{M}_r^+(p,k;\mathcal{C}))$. The (finite dimensional) spaces $\mathcal{M}_l^+(p,k;\mathcal{C})$ and $\mathcal{M}_r^+(p,k;\mathcal{C})$ are dual to each other, the corresponding duality being

$$\langle f, g \rangle = \int_{S^{p-1}} f(\underline{\xi}) g(\underline{\xi}) d\Sigma(\underline{\xi}),\ f \in \mathcal{M}_r^+(p,k;\mathcal{C}),\ g \in \mathcal{M}_l^+(p,k;\mathcal{C}).$$

For $f \in \mathcal{A}(\overline{U}, \mathcal{M}_r^+(p,k;\mathcal{C}))$, $g \in \mathcal{A}(\overline{U}, \mathcal{M}_l^+(p,k;\mathcal{C}))$, the duality is given by

$$\langle f, g \rangle_U = \int_U \langle f(\underline{y}), g(\underline{y}) \rangle d\underline{y}.$$

The spaces $M_r^+(p,k;C)$ and $\mathcal{M}_r^+(p,k;C)$ are usually identified (the isomorphism being the restriction to the unit sphere).

Let us study now in more details the action of the Laurent coefficient $L_k(f)$ of a function $f \in \mathcal{L}_{l,k}(\mathbf{R}^m \setminus K)$ on an element $\phi \in \mathcal{A}(K, \mathcal{M}_r^+(p,k;C))$.

By definition,

$$\langle L_k(f), \varphi \rangle = \int_{\partial V} \phi \, d\sigma \, f, \tag{5.4}$$

where V is a suitable neighbourhood of K in $\mathbf{R}^m$ and $T_k(\phi) = \varphi, \phi \in \mathcal{T}_{r,k}(K)$.

It is possible to choose V in the form $V = B_\rho \times U$, where B_ρ is a ball in $\mathbf{R}^p$ of radius ρ and U is a neighbourhood of K in $\mathbf{R}^q$. We can suppose that $V \subset\subset \tilde{\Omega}$ and that ϕ is defined on a neighbourhood of $\overline{V}$. Putting $S_\rho = \partial B_\rho$ and denoting the restriction of ϕ to $S_\rho \times U$ by $\phi_\rho(\underline{\xi}, \underline{y})$, we know that we can write it as a sum

$$\phi_\rho = \phi_\rho^1 - \phi_\rho^2 \underline{\xi},$$

where ϕ_ρ^i are in $\mathcal{A}(\overline{U}, \mathcal{M}_r^+(p,k;C))$ and that $T_k(\phi) = \varphi$ is given by

$$T_k(\phi)(\underline{\xi}, \underline{y}) = \lim_{\rho \to 0} \frac{1}{\rho^k} \phi_\rho^1(\rho \underline{\xi}, \underline{y}). \tag{5.5}$$

The restriction f_ρ of the function f to $S_\rho \times U$ can be decomposed in the same way into the sum

$$f_\rho = f_\rho^1 - \underline{\xi} f_\rho^2,$$

where f_ρ^i are in $\mathcal{A}(U, \mathcal{M}_l^+(p,k;C))$. Moreover,

$$\lim_{\rho \to 0} \frac{1}{\rho^k} \phi_\rho^2(\underline{\xi}, \underline{y}) = 0. \tag{5.6}$$

Under the duality (5.4), only two of the four pieces survive and because of (5.6), only the piece

$$\int_{S_\rho \times U} \phi_\rho^1 \, d\sigma \, \underline{\xi} \, f_\rho^2$$

is expected to be important for us.

In analogy with the case of generalized Taylor functions of order k, an expression for the generalized Laurent coefficient $L_k(f)$ of a monogenic function f could be determined by

$$L_k(f)(\underline{\xi}, \underline{y}) = \lim_{\rho \to 0} \rho^{k+p-1} f_\rho^2(\underline{\xi}, \underline{y}), f_\rho^2(\underline{\xi}, \underline{y}) = \mathbf{P}_k(\underline{\xi} f_\rho)(\underline{\xi}, \underline{y}).$$

It is far from being clear, however, if such a limit exists and in what sense it should be taken. We are going to show now that for every neighbourhood U of K in Ω the limit exists in the weak topology of the space $\mathcal{A}'(U, \mathcal{M}_r^+(p,k;\mathcal{C}))$ and that it is equal to the corresponding Laurent coefficient. The proof of this fact will be done in several steps.

Theorem 5.4.2. *Suppose that* $f \in \mathcal{L}_{l,k}(\mathbf{R}^m \setminus K)$ *and that* $f_\rho = f_\rho^1 - \underline{\xi} f_\rho^2$, *where* f_ρ^i *are in* $\mathcal{A}(\mathbf{R}^q, \mathcal{M}_l^+(p,k;\mathcal{C}))$.

If the limits

$$f_0^i = \lim_{\rho\to 0} \rho^{k+p-1} f_\rho^i,\ i = 1,2$$

exist in the weak topology of the space $\mathcal{A}'(\overline{U}, \mathcal{M}_r^+(p,k;\mathcal{C}))$ *for every open neighbourhood* U *of* K *in* $\mathbf{R}^q$, *then the limit* f_0^2 *does not depend on* U *and*

$$L_k(f)(\underline{\xi},\underline{y}) = f_0^2(\underline{\xi},\underline{y}).$$

Proof. Let $f_0^2|_U$ denote the limit of $\rho^{k+p-1} f_\rho^2$ in the weak topology of the space $\mathcal{A}'(\overline{U}, \mathcal{M}_r^+(p,k;\mathcal{C}))$. It is sufficient to show that $\langle L_k(f), \varphi\rangle = \langle f_0^2|_U, \varphi\rangle|_U$ for all neighbourhoods U of K and for all $\varphi \in \mathcal{A}(\overline{U}, \mathcal{M}_r^+(p,k;\mathcal{C}))$.

So let a neighbourhood U and an element $\varphi \in \mathcal{A}(\overline{U}, \mathcal{M}_r^+(p,k;\mathcal{C}))$ be fixed and let us define again $\phi = T_k^{-1}(\varphi) \in \mathcal{T}_{r,k}(\overline{U})$. Then the open sets

$$V_n = B(0, 1/n) \times U$$

contain K and the element $\langle L_k(f), \varphi\rangle$ can be computed by

$$\langle L_k(f), \varphi\rangle = \int_{\partial V_n} \phi\, d\sigma\, f$$

for all $n \geq n_0$, n_o big enough.

Recalling that the restrictions of f and ϕ to the boundary of V_n define elements $f_{1/n}^i$ and $\phi_{1/n}^i$ in $\mathcal{A}(\overline{U}, \mathcal{M}_l^+(p,k;\mathcal{C}))$ and $\mathcal{A}(\overline{U}, \mathcal{M}_r^+(p,k;\mathcal{C}))$, respectively, and that $d\sigma = \rho^{p-1}\underline{\xi} d\Sigma(\underline{\xi}) d\underline{y}$ on $S_{1/n} \times U$, we get

$$\begin{aligned}\langle L_k(f), \varphi\rangle &= \lim_{n\to\infty} \int_U \rho^{p-1} \langle \phi_{1/n}^1(\underline{y}), f_{1/n}^2(\underline{y})\rangle d\underline{y} + \\ &\quad + \lim_{n\to\infty} \int_U \rho^{p-1} \langle \phi_{1/n}^2(\underline{y}), f_{1/n}^1(\underline{y})\rangle d\underline{y}.\end{aligned}$$

But $\rho^{k+p-1} f_{1/n}^i \rightharpoonup f_0^i|_U$ and $\rho^{-k}\phi_{1/n}^1 \to \varphi$, $\rho^{-k}\phi_{1/n}^2 \to 0$, whence

$$\langle L_k(f), \varphi\rangle = \lim_{n\to\infty} \langle \rho^{-k}\phi_{1/n}^1, \rho^{k+p-1} f_{1/n}^2\rangle_U = \langle f_0^2|_U, \varphi\rangle. \blacksquare$$

Using the theorem just proved, we are now able to compute some important examples of generalized Laurent coefficients.

Examples 5.4.3.

(1) Let us first consider the Cauchy kernel

$$E(\underline{x}+\underline{y}) = -\frac{1}{A_m}\frac{\underline{x}+\underline{y}}{|\underline{x}+\underline{y}|^m}.$$

It is easy to see that E belongs to $\mathcal{L}_{l,0}(\mathbf{R}^m\setminus\{0\})$ and its parts E^i in the decomposition $E = E^1 - \underline{\xi}E^2$ are given by

$$E^1 = -\frac{1}{A_m}\frac{\underline{y}}{[\rho^2+|\underline{y}|^2]^{m/2}},\ E^2 = \frac{1}{A_m}\frac{\rho}{[\rho^2+|\underline{y}|^2]^{m/2}}.$$

We now show that for every ball $U \subset \mathbf{R}^q$ with center in the origin the limit

$$\lim_{\rho\to 0}\rho^{p-1}(E^1_\rho - \underline{\xi}E^2_\rho) = \delta(\underline{y})$$

exists in the weak topology on the space $\mathcal{A}'(\overline{U}, \mathcal{M}^+_l(p,k;\mathcal{C}))$, whence

$$\mathcal{L}_0(E) = \delta(\underline{y}).$$

Indeed, for each φ analytic in a neighbourhood of $\overline{U}$ we have

$$\begin{aligned}
\lim_{\rho\to 0}\rho^{p-1}\langle(E^1_\rho - \underline{\xi}E^2_\rho),\varphi\rangle_U &= \lim_{\rho\to 0}\int_U\frac{\rho^{p-1}}{A_m}\int_{S^{p-1}}\frac{\varphi(\underline{y})(\rho-\underline{y})}{[\rho^2+|\underline{y}|^2]^{m/2}}d\Sigma d\underline{y}\\
&= \lim_{\rho\to 0}\int_U\frac{A_p\,\rho^{p-1}(\rho-\underline{y})}{A_m[\rho^2+|\underline{y}|^2]^{m/2}}\varphi(\underline{y})d\underline{y}\\
&= \lim_{\rho\to 0}\int_{\frac{1}{\rho}\cdot U}\frac{A_p\,\varphi(\rho\underline{u})(1-\underline{u})}{A_m[1+|\underline{u}|^2]^{m/2}}d\underline{u}\\
&= \frac{A_p}{A_m}\int_{\mathbf{R}^q}\frac{\varphi(0)}{[1+|\underline{u}|^2]^{m/2}}d\underline{u}\\
&= \frac{A_pA_q}{A_m}\int_0^\infty\frac{s^{q-1}}{(1+s^2)^{m/2}}ds\cdot\varphi(0)\\
&= \varphi(0).
\end{aligned}$$

(2) If $F \in \mathcal{A}'(K;\mathcal{C})$ is an analytic functional on a compact subset K in $\mathbf{R}^q$, then we define its Cauchy transform $\widehat{F}$ by the action of the functional F on E, i.e.

$$\widehat{F}(\underline{x}+\underline{y})=F_{\underline{v}}[E(\underline{x}+\underline{y}-\underline{v})],\ \underline{x}+\underline{y}\in\mathbf{R}^m\setminus K.$$

The Cauchy transform $\widehat{F}$ is a monogenic function in the domain $\mathbf{R}^m\setminus K$. Moreover, $L_0(\widehat{F}) = F$, i.e. the Cauchy transform is the inverse of the isomorphism L_0.

Indeed, for each φ analytic in a neighbourhood of $\overline{U}$ we have

$$\begin{aligned}
&\lim_{\rho\to 0}\rho^{p-1}\int_U F_{\underline{v}}[E_\rho(\rho\underline{\xi}+\underline{y}-\underline{v})]\varphi(\underline{y})d\underline{y} = \\
&= \lim_{\rho\to 0}F_{\underline{v}}[\rho^{p-1}\int_U E_\rho(\rho\underline{\xi}+\underline{y}-\underline{v})]\varphi(\underline{y})d\underline{y} \\
&= \lim_{\rho\to 0}F_{\underline{v}}[\int_{\frac{1}{\rho}\cdot U}\frac{A_p}{A_m}\frac{\varphi(\rho\underline{u}+\underline{v})(1-\underline{u})}{[1+|\underline{u}|^2]^{m/2}}d\underline{u}] \\
&= F_{\underline{v}}[\varphi(\underline{v})].
\end{aligned}$$

(3) Let $P_k(\underline{x})$ be a (left) inner spherical monogenic of degree k in $\mathbf{R}^p$. Then it is an inner spherical monogenic of degree k in $\mathbf{R}^m$ as well and its inverse in $\mathbf{R}^m$

$$\begin{aligned}
Q_k(\underline{x},\underline{y}) &= \frac{-1}{A_m}|\frac{\underline{x}+\underline{y}}{|\underline{x}+\underline{y}|^m}P_k\left(\frac{\underline{x}}{|\underline{x}+\underline{y}|^2}\right) \\
&= \frac{-1}{A_m}|\frac{\underline{x}+\underline{y}}{|\underline{x}+\underline{y}|^{2k+m}}P_k(\underline{x})
\end{aligned}$$

is monogenic in $\mathbf{R}^m\setminus\{0\}$ and vanishes at infinity. It is clear that it is a generalized Laurent function of degree k and its two parts Q_k^i are given by

$$Q^1_{k,\rho} = -\frac{1}{A_m}\frac{\rho^k\underline{y}P_k(\underline{\xi})}{[\rho^2+|\underline{y}|^2]^{k+m/2}},\quad Q^2_{k,\rho}=\frac{1}{A_m}\frac{\rho^{k+1}P_k(\underline{\xi})}{[\rho^2+|\underline{y}|^2]^{k+m/2}}.$$

Let us now compute the Laurent coefficient $L_k(Q_k)$. Let us consider again a test function $\varphi\in\mathcal{A}(\overline{U},\mathcal{M}_r^+(p,k;C))$, then

$$\begin{aligned}
&\lim_{\rho\to 0}\rho^{k+p-1}\langle Q_{k,\rho},\varphi\rangle_U \\
&= \lim_{\rho\to 0}\int_U\frac{\rho^{2k+p-1}}{A_m}\int_{S^{p-1}}\frac{\varphi(\underline{\xi},\underline{y})P_k(\underline{\xi})(\rho-\underline{y})}{[\rho^2+|\underline{y}|^2]^{k+m/2}}d\Sigma(\underline{\xi})d\underline{y} \\
&= \lim_{\rho\to 0}\int_{\frac{1}{\rho}\cdot U}\int_{S^{p-1}}\frac{\varphi(\underline{\xi},\rho\underline{u})P_k(\underline{\xi})(1-\underline{u})}{A_m[1+|\underline{u}|^2]^{k+m/2}}d\Sigma d\underline{u}
\end{aligned}$$

$$= \frac{A_q}{A_m}\int_0^\infty \frac{s^{q-1}}{(1+s^2)^{k+m/2}}ds \int_{S^{p-1}} \varphi(\underline{\xi},0)P_k(\underline{\xi})d\Sigma$$

$$= \frac{\Gamma(k+\frac{p}{2})\Gamma(\frac{m}{2})}{\Gamma(\frac{p}{2})\Gamma(k+\frac{m}{2})}\frac{1}{A_p}\int_{S^{p-1}} \varphi(\underline{\xi},0)P_k(\underline{\xi})d\Sigma.$$

Consequently

$$L_k(Q_k) = \frac{\Gamma(k+\frac{p}{2})\Gamma(\frac{m}{2})}{\Gamma(\frac{p}{2})\Gamma(k+\frac{m}{2})}\delta(\underline{y})P_k(\underline{\xi}).$$

The computation in the last example leads us to the following definition.

Theorem 5.4.4. *Let K be a compact subset of $\mathbf{R}^q$, let $F \in \mathcal{A}'(K;\mathcal{C})$ and let $P_k \in M^+(p,k;\mathcal{C})$. Then the Cauchy transform $C_k(F)$ of order k of F is defined by the convolution*

$$C_k(F)(\underline{x}+\underline{y}) = F_{\underline{v}}\left[Q_k(\underline{x}+\underline{y}-\underline{v})\right],\ \underline{x}+\underline{y}\in\mathbf{R}^m\setminus K,$$

where

$$Q_k(\underline{x}+\underline{y}-\underline{v}) = \frac{-1}{A_m}\frac{\underline{x}+\underline{y}-\underline{v}}{|\underline{x}+\underline{y}-\underline{v}|^{2k+m}}P_k(\underline{x}).$$

(Note that the kernel Q_k depends on the choice of P_k).

The Cauchy transform (of order k) $C_k(F)$ is a monogenic function in $\mathbf{R}^m\setminus K$. As in Example (2), the computation done in Example (3) yields, by convolution, that all weak limits of the corresponding restrictions exist and that the generalized Laurent coefficient is given by

$$L_k(C_k(F)) = \frac{\Gamma(k+\frac{p}{2})\Gamma(\frac{m}{2})}{\Gamma(\frac{p}{2})\Gamma(k+\frac{m}{2})}F\otimes P_k(\underline{\xi}),$$

where $L_k(C_k(F)) \in \mathcal{A}'(K;\mathcal{C})\otimes\mathcal{M}_l^+(p,k;\mathcal{C}) \simeq \mathcal{A}'(K,\mathcal{M}_r^+(p,k;\mathcal{C}))$. So the Cauchy transform of order k describes the inverse of the map L_k. Now we are finally able to prove that the Laurent coefficients can be computed by the corresponding limits in all cases.

Theorem 5.4.5. *Suppose that f belongs to $\mathcal{L}_{l,k}(\mathbf{R}^m\setminus K)$ and that $f_\rho = f_\rho^1 - \underline{\xi}f_\rho^2$, where f_ρ^i are in $\mathcal{A}(\mathbf{R}^q,\mathcal{M}_l^+(p,k;\mathcal{C}))$. Then*

$$f_0^i|_U = \lim_{\rho\to 0}\rho^{k+p-1}f_\rho^i,\ i=1,2$$

exists in the weak topology of the space $\mathcal{A}'(\overline{U},\mathcal{M}_r^+(p,k;\mathcal{C}))$ for every open neighbourhood U of K in $\mathbf{R}^q$ and

$$L_k(f)(\underline{\xi},\underline{y}) = f_0^2|_U(\underline{\xi},\underline{y}).$$

Proof. Take any $f \in \mathcal{L}_{l,k}(\mathbf{R}^m \setminus K)$ and choose a basis $(P_{k,\alpha})_{\alpha\in A_k}$ for the space $\mathcal{M}_l^+(p,k;\mathcal{C})$. Then $L_k(f) \in \mathcal{A}'(K,\mathcal{M}_r^+(p,k;\mathcal{C})) \simeq \mathcal{A}'(K;\mathcal{C}) \otimes \mathcal{M}_l^+(p,k;\mathcal{C})$ can be written as a sum

$$L_k(f) = \sum_{\alpha\in A_k} c_\alpha P_{k,\alpha},\ c_\alpha \in \mathcal{A}'(K;\mathcal{C}).$$

Putting

$$\tilde{f} = \left[\frac{\Gamma(k+\frac{p}{2})\Gamma(\frac{m}{2})}{\Gamma(\frac{p}{2})\Gamma(k+\frac{m}{2})}\right]^{-1} \sum_{\alpha\in A_k} C_k(c_\alpha),$$

$L_k(\tilde{f}) = \sum_{\alpha\in A_k} c_\alpha P_{k,\alpha} = L_k(f)$, whence $\tilde{f} = f$. Using all information on the limits computed in the examples, the proof is done. ■

Chapter III

Special functions and methods

In this chapter some methods are developed to defining special (monogenic) functions.

The classical orthogonal polynomials play a preponderant rôle in different areas of mathematics and physics. They are characterized by one of the following properties

(i) There is a weighted Hilbert space $L_2(\Omega, w)$ such that $\{\frac{d}{dx}p_n : n = 1, 2, \ldots\}$ is a sequence of orthogonal polynomials for this space.

(ii) Each polynomial p_n satisfies a differential equation

$$A(x)\frac{d^2 p_n(x)}{dx^2} + B(x)\frac{dp_n(x)}{dx} + \lambda_n p_n(x) = o$$

where A and B are independent of n and λ_n is independent of x.

(iii) The polynomials can be defined by Rodrigues' Formula

$$p_n(x) = \frac{1}{K_n w(x)}\frac{d^n}{dx^n}(w(x)X^n(x))$$

where K_n is a constant and X is a polynomial in x, independent of n.

As $i\frac{d}{dx}$ is nothing else but the Dirac operator on the real line, we may wonder whether or not such a scheme can be extended to the case of several variables, replacing derivation by the Dirac operator in $\mathbf{R}^m$. It turns out that the only candidate solutions are these with weight functions $e^{\underline{x}^2/2}$ and $(1 + \underline{x}^2)^\alpha$, corresponding with generalizations of the Hermite and the Gegenbauer polynomials (section 1). (For a more detailed exposition, see [18]).

Another way of getting special monogenic functions is by applying the Cauchy-Kovalevska extension principle to C-valued real analytic functions in $\mathbf{R}^m$ (section 2). As polynomials of the form $\underline{x}^s P_k(\underline{x})$, $s \in \mathbf{N}$, $P_k \in M^+(m; k; C)$, are building blocks for such functions (see also the Remark in II 1.10), it is natural to determine first the CK-extension of these polynomials. More generally, we also look at the CK-extension of functions of the form $\underline{x}^\alpha P_k(\underline{x})$ where $\alpha \in \mathbf{R}$. In such a way generalized power functions are introduced. Finally the CK-extension principle allows us to define axially exponential functions and to present another approach to Hermite polynomials.

A third way of generating special functions is by letting act a generalized Cauchy kernel on suitable classes of functions or distributions (section 3).

A combination of the method of plane waves developed by F. John in [45], the CK-extension principle and the Funk-Hecke Theorem gives rise to the introduction of so-called monogenic plane waves (section 4).

Finally a generalization of the classical Riesz potentials (see [34]) is obtained in section 5.

All methods exposed in §§2-5 were developed recently by F. Sommen in a series of papers (see e.g. [78] and [75]).

1 Gegenbauer and Hermite polynomials in $\mathbf{R}^m$

1.1 *Gegenbauer polynomials*

As is well known Gegenbauer polynomials on the real line are polynomials which form an orthonormal basis in the Hilbert space $L_2(]-1,1[;(1-x^2)^{\lambda-\frac{1}{2}})$.

The aim of this section is to generalize these polynomials to the Euclidean space $\mathbf{R}^m$, thus obtaining an orthonormal basis for the Hilbert module $L_2(B(1);(1+\underline{x}^2)^\alpha)$.

To this end consider the space $\mathbf{R}^{0,m}$ with orthogonal basis $e=\{e_1,\ldots,e_m\}$ and the associated complex Clifford algebra $\mathbf{C}_m=\mathbf{R}_{0,m}\otimes_{\mathbf{R}}\mathbf{C}$.

As usual an element $\underline{x}=(x_1,\ldots,x_m)$ in $\mathbf{R}^m$ is identified with $\underline{x}=\sum_{j=1}^m e_jx_j$ and the Dirac operator $\partial_{\underline{x}}$ in $\mathbf{R}^m$ is given by

$$\partial_{\underline{x}}=\sum_{j=1}^m e_j\partial_{x_j}.$$

Throughout this section $\partial_{\underline{x}}$ will act on V-valued functions f which are of the class C_1 where, as usual, V is the Clifford algebra $\mathbf{C}_m(=C)$ or the spinor space S.

1.1.1 The spaces ${}_n\mathrm{C}[\underline{x}]$, R_{m+2k} and $R_m(P_k)$

Consider the $\mathbf{C}_m$-valued "variable" $\underline{x}=\sum_{j=1}^m e_jx_j$.

Then for $s\in\mathbf{N}$,

$$\underline{x}^s=\begin{cases}(\underline{x}^2)^\ell & \text{if } s=2\ell\\ (\underline{x}^2)^\ell\underline{x} & \text{if } s=2\ell+1\end{cases}$$

whence $\underline{x}^s$ is an $\mathbf{R}$-valued or an $\mathbf{R}^1_{0,m}$-valued homogeneous polynomial of degree s in $x_1,\ldots,x_m$.

Putting

$$\mathrm{C}[\underline{x}]=\mathrm{span}_{\mathbf{C}}\{\underline{x}^s:s\in\mathbf{N}\}$$

and, for $n \in \mathbf{N}$ fixed,

$$_n\mathbf{C}[\underline{x}] = \text{span}_{\mathbf{C}}\{\underline{x}^s : s \in \mathbf{N}, \quad s \leq n\}$$

we obtain complex vector spaces called the *space of polynomials in* $\underline{x}$ and the *space of polynomials in* $\underline{x}$ *having degree less than or equal to* n. They have respective (algebraic) bases $\{\underline{x}^s : s \in \mathbf{N}\}$ and $\{\underline{x}^s : s \in \mathbf{N}, \quad s \leq n\}$.

Denoting henceforth by $p_j(\underline{x})$, $j \in \mathbf{N}$, an arbitrary element of $_j\mathbf{C}[\underline{x}]$, i.e.

$$p_j(\underline{x}) = \sum_{\ell=0}^{j} a_\ell \underline{x}^\ell, \quad a_\ell \in \mathbf{C},$$

we say that p_j has degree j if $a_j \neq 0$.

In the sequel when using the notation $p_j(\underline{x})$ it is understood that p_j has degree j.

Notice that we may write p_j as

$$p_j(\underline{x}) = \sum_{2\ell \leq j} (-1)^\ell r^{2\ell} a_{2\ell} + \underline{\xi} \sum_{2\ell+1 \leq j} (-1)^\ell r^{2\ell+1} a_{2\ell+1},$$

where we of course used polar coordinates $\underline{x} = r\underline{\xi}$.

Now let $(p_j)_{j=0}^{\infty}$ be a sequence of polynomials in $\underline{x}$. Then clearly $\{p_j(\underline{x}) : j \in \mathbf{N}\}$ is again an algebraic basis for $\mathbf{C}[\underline{x}]$ and, analogously, a sequence $(p_j)_{j=0}^{n}$ of polynomials in $\underline{x}$ is an algebraic basis for $_n\mathbf{C}[\underline{x}]$.

In what follows we shall meet regularly expressions of the form $p_j(\underline{x})P_k(\underline{x})$ where $P_k(x) \in M^+(k;V)$ is fixed and $p_j(\underline{x}) \in_j \mathbf{C}[\underline{x}]$.

In fact, as we already pointed out in 1.12, functions of this type are special cases of functions of the form $(f(r) + \underline{\xi} g(r))P_k(\underline{x}), r \in \mathbf{R}$, the space of which is henceforth denoted by $R_m(P_k)$. The space $R_m(P_k)$ may be described in an elegant way as follows.

Denoting an arbitrary element of $\mathbf{R}^{m+2k}$ by $\underline{b}$ and its polar form by $\underline{b} = \rho\underline{\beta}$, we call R_{m+2k} the space of functions of the form

$$H(\underline{b}) = f(\rho) + \underline{\beta} g(\rho)$$

where f and g are arbitrary radially symmetric functions of the class C_1 in $\mathbf{R}$.

Clearly the operators

$$\text{ext} : R_m(P_k) \to R_{m+2k}$$

and

$$\text{rest} : R_{m+2k} \to R_m(P_k)$$

with

$$(\text{ext}\, h)(\underline{b}) = f(\rho) + \underline{\beta} g(\rho)$$

and

$$(\text{rest}\, H)(\underline{x}) = (f(r) + \underline{\xi} g(r))P_k(\underline{x})$$

are bijections.

Henceforth putting $\text{ext}\, h = h^{\text{ext}}$ and $\text{rest}\, H = H^{\text{rest}}$, we clearly have that

$$(h^{\text{ext}})^{\text{rest}} = h, (H^{\text{rest}})^{\text{ext}} = H$$

for all $h \in R_m(P_k)$ and $H \in R_{m+2k}$.

In particular we have that

$$[p_j(\underline{x})P_k(\underline{x})]^{\text{ext}} = p_j(\underline{b})$$

and (1.1)

$$[p_j(\underline{b})]^{\text{rest}} = p_j(\underline{x})P_k(\underline{x})$$

for all $p_j(\underline{x}) \in {}_j\mathbf{C}[\underline{x}]$ and $p_j(\underline{b}) \in {}_j\mathbf{C}[\underline{b}]$.

We will also meet frequently expressions of the form $\partial_{\underline{x}}((f(r)+\underline{\xi} g(r))P_k(\underline{x}))$ which admit, as a particular case, expressions of the form $\partial_{\underline{x}}(p_j(\underline{x})P_k(\underline{x}))$, $p_j(\underline{x}) \in {}_j\mathbf{C}[\underline{x}]$.

For $\mathbf{R}$-valued smooth functions $f(r)$ in $\mathbf{R}$, we obtain by replacing m by $(m-1)$ in II.1.12, Proposition 2 and Proposition 3 :

$$\text{(i)}\ \partial_{\underline{x}}(f(r)P_k(\underline{x})) = \underline{\xi} f'(r)P_k(\underline{x})$$

$$\text{(ii)}\ \partial_{\underline{x}}(\underline{\xi} f(r)P_k(\underline{x})) = -(f'(r) + \tfrac{m-1+2k}{r} f(r))P_k(\underline{x}) \tag{1.2}$$

$$\text{(iii)}\ \partial_{\underline{x}}(\underline{x}^s P_k(\underline{x})) = \begin{cases} -s\underline{x}^{s-1}P_k(\underline{x}) & \text{for } s \text{ even} \\ -(s+2k+m-1)\underline{x}^{s-1}P_k(\underline{x}) & \text{for } s \text{ odd} \end{cases}$$

Denoting the Dirac operator in $\mathbf{R}^{m+2k}$ by $\partial_{\underline{b}}$, the relations (1.2) yield that for each $h \in R_m(P_k)$

$$\partial_{\underline{x}} h = \left[\partial_{\underline{b}} h^{\text{ext}}\right]^{\text{rest}}. \tag{1.3}$$

Moreover notice that, since $[\underline{x}^s P_k(\underline{x})]^{\text{ext}} = \underline{b}^s$ and

$$\partial_{\underline{b}}(\underline{b}^s) = \begin{cases} -s\underline{b}^{s-1} & \text{for } s \text{ even} \\ -(s-1+m+2k)\underline{b}^{s-1} & \text{for } s \text{ odd} \end{cases},$$

the relation (1.3) yields

$$\partial_{\underline{x}}(\underline{x}^s P_k(\underline{x})) = \begin{cases} -s\underline{x}^{s-1}P_k(\underline{x}) & \text{for } s \text{ even} \\ -(s-1+m+2k)\underline{x}^{s-1}P_k(\underline{x}) & \text{for } s \text{ odd} \end{cases},$$

thus showing once more the importance of the operators ext and rest.

Finally, from (1.2) (iii) it follows that for each $p_j(\underline{x}) \in {}_j\mathbf{C}[\underline{x}]$, there exists an element $p_{j-1}(\underline{x}) \in {}_{j-1}\mathbf{C}[\underline{x}]$ such that for all $P_k \in M^+(k;V)$

$$\partial_{\underline{x}}(p_j(\underline{x})P_k(\underline{x})) = p_{j-1}(\underline{x})P_k(\underline{x}). \tag{1.4}$$

1.1.2 The space $L_2(B(1);(1+\underline{x}^2)^\alpha)$

Let $\alpha \in \mathbf{R}$ with $\alpha > -1$. Then $L_2(B(1);(1+\underline{x}^2)^\alpha)$ stands for the space of V-valued measurable functions f in $B(1)$ such that

$$\int_{B(1)} |f(\underline{x})|^2(1+\underline{x}^2)^\alpha d\underline{x} < +\infty.$$

Provided with the V-valued inner product

$$(f,g)_\alpha = \int_{B(1)} \bar{f}(\underline{x})g(\underline{x})(1+\underline{x}^2)^\alpha d\underline{x},$$

it becomes a right Hilbert module over C if $V = C$ or a complex Hilbert space if $V = S$.

Proposition 1. *Let $k,\ell \in \mathbf{N}$ be such that $k \neq \ell$. Then for all $s,t \in \mathbf{N}$*

$$(\underline{x}^s P_k, \underline{x}^t P_\ell)_\alpha = 0.$$

Proof. Let us prove the statement in the cases where $s = 2i$ i s even and $t = 2j+1$ is odd, the other cases being similar.

We have, using the orthogonality relations between spherical monogenics, that

$$\begin{aligned}
&(\underline{x}^{2i} P_k, \underline{x}^{2j+1} P_\ell)_\alpha \\
&= (-1)^{i+j} \int_0^1 r^{2i+2j+1+k+\ell}(1-r^2)^\alpha dr \left[\int_{S^{m-1}} \overline{P_k(\underline{\xi})}\,\underline{\xi} P_\ell(\underline{\xi}) d\Sigma(\underline{\xi})\right] \\
&= 0. \blacksquare
\end{aligned}$$

Corollary. *Let $k,\ell \in \mathbf{N}$ be such that $k \neq \ell$. Then for all polynomials $p_n(\underline{x})$ and $p_t(\underline{x})$,*

$$(p_n P_k, p_t P_\ell)_\alpha = 0.$$

1.1.3 The operator D_α

In this subsection we introduce the differential operator D_α which will play a fundamental role in studying the space $L_2(B(1),(1+\underline{x}^2)^\alpha)$.

Definition. Let $\alpha \in \mathbf{R}$ with $\alpha > -1$. Then the operator D_α is defined on $C_1(\overline{B}(1);V)$ by

$$D_\alpha f(\underline{x}) = (1+\underline{x}^2)^{-\alpha}\partial_{\underline{x}}((1+\underline{x}^2)^{\alpha+1} f(\underline{x})). \tag{1.5}$$

Notice that, as for each $f \in C_1(\overline{B}(1);V)$,

$$\partial_{\underline{x}}((1+\underline{x}^2)^{\alpha+1}f(\underline{x}) = (-2(\alpha+1)\underline{x}(1+\underline{x}^2)^{\alpha} + (1+\underline{x}^2)^{\alpha+1}\partial_{\underline{x}})f(\underline{x})$$

we also have that

$$D_\alpha f(\underline{x}) = (-2(\alpha+1)\underline{x} + (1+\underline{x}^2)\partial_{\underline{x}})f(\underline{x}). \tag{1.6}$$

Proposition 1. *For all $f, g \in C_1(\overline{B}(1); V)$,*

$$(D_\alpha f, g)_\alpha = (f, \partial_{\underline{x}} g)_{\alpha+1}.$$

Proof. Using Stokes' Theorem we have :

$$\begin{aligned}
&(D_\alpha f, g)_\alpha \\
&= \int_{B(1)} (1+\underline{x}^2)^{-\alpha}\overline{\partial_{\underline{x}}((1+\underline{x}^2)^{\alpha+1}f(\underline{x}))}g(\underline{x})(1+\underline{x}^2)^{\alpha}d\underline{x} \\
&= \int_{B(1)} \overline{\partial_{\underline{x}}((1+\underline{x}^2)^{\alpha+1}f(\underline{x}))}g(\underline{x})d\underline{x} \\
&= -\int_{B(1)} ((\overline{(1+\underline{x}^2)^{\alpha+1}f(\underline{x})})\partial_{\underline{x}})g(\underline{x})d\underline{x} \\
&= -\int_{\partial B(1)} (1+\underline{x}^2)^{\alpha+1}\overline{f(\underline{x})}d\sigma_{\underline{x}}g(\underline{x}) + \int_{B(1)} ((1+\underline{x}^2)^{\alpha+1}\overline{f(\underline{x})}\partial_{\underline{x}}(g(\underline{x}))d\underline{x} \\
&= \int_{B(1)} \overline{f(\underline{x})}\partial_{\underline{x}}(g(\underline{x}))(1+\underline{x}^2)^{\alpha+1}d\underline{x} \\
&= (f, \partial_{\underline{x}} g)_{\alpha+1}
\end{aligned}$$

and this since $(1+\underline{x}^2)^{\alpha+1} = 0$ on $\partial B(1)$. ∎

Using the relations (1.2) we obtain straightforwardly

Proposition 2. *For each $p_n(\underline{x}) \in {}_n\mathbf{C}[\underline{x}]$, there exists a $p_{n+1}(\underline{x}) \in {}_{n+1}\mathbf{C}[\underline{x}]$ such that for all $P_k \in M^+(k; V)$,*

$$D_\alpha(p_n(\underline{x})P_k(\underline{x})) = p_{n+1}(\underline{x})P_k(\underline{x}).$$

1.1.4 The Gegenbauer polynomials $C^\alpha_{n,m}(P_k)$

Let again $\alpha \in \mathbf{R}$ be such that $\alpha > -1$, let $P_k \in M^+(k; V)$ be fixed and let $n \in \mathbf{N}$. Then we put

$$C^\alpha_{n,m}(P_k)(\underline{x}) = D_\alpha D_{\alpha+1} \cdots D_{\alpha+n-1}(P_k(\underline{x})) \tag{1.7}$$

Notice that by means of 1.1.3 Proposition 2,

$$C^{\alpha}_{n,m}(P_k)(\underline{x}) = C^{\alpha}_{n,m,k}(\underline{x})P_k(\underline{x}) \tag{1.8}$$

where $C^{\alpha}_{n,m,k}(\underline{x}) \in {}_n\mathbf{C}[x]$ has real coefficients depending on k whence it takes its values in $\mathbf{R}^0_{0,m} \oplus \mathbf{R}^1_{0,m}$.

Consequently $C^{\alpha}_{n,m}(P_k)(\underline{x})$ is a V-valued polynomial of degree $(n+k)$ in $x_1, \ldots, x_m$.

The polynomials $C^{\alpha}_{n,m}(P_k)(\underline{x})$ are called *Gegenbauer polynomials* in $\mathbf{R}^m$.

Theorem 1. (Rodrigues' Formula) *The Gegenbauer polynomials $C^{\alpha}_{n,m}(P_k)(\underline{x})$ are also determined by*

$$C^{\alpha}_{n,m}(P_k)(\underline{x}) = (1+\underline{x}^2)^{-\alpha}\partial^n_{\underline{x}}((1+\underline{x}^2)^{\alpha+n}P_k(\underline{x})) \tag{1.9}$$

Proof. We have

$$\begin{aligned}
&C^{\alpha}_{n,m}(P_k)(\underline{x}) \\
&= D_\alpha D_{\alpha+1}\cdots D_{\alpha+n-1}(P_k(\underline{x})) \\
&= (1+\underline{x}^2)^{-\alpha}\partial_{\underline{x}}((1+\underline{x}^2)^{\alpha+1}(D_{\alpha+1}\cdots D_{\alpha+n-1}P_k(\underline{x})) \\
&= (1+\underline{x}^2)^{-\alpha}\partial_{\underline{x}}((1+\underline{x}^2)^{\alpha+1}\cdot \\
&\quad \cdot\left[(1+\underline{x}^2)^{-(\alpha+1)}\partial_{\underline{x}}((1+\underline{x}^2)^{\alpha+2}D_{\alpha+2}\cdots D_{\alpha+n-1}P_k(x)\right] \\
&= (1+\underline{x}^2)^{-\alpha}\partial^2_{\underline{x}}((1+\underline{x}^2)^{\alpha+2}D_{\alpha+2}\cdots D_{\alpha+n-1}P_k(\underline{x})) \\
&= \ldots \\
&= (1+\underline{x}^2)^{-\alpha}\partial^n_{\underline{x}}((1+\underline{x}^2)^{\alpha+n}P_k(\underline{x})). \blacksquare
\end{aligned}$$

Corollary. $C^{\alpha}_{n,m}(P_k)(\underline{x}) = [C^{\alpha}_{n,m+2k,0}(\underline{b})]^{\mathrm{rest}}(\underline{x})$ (1.10)

Proof. Using Rodrigues' Formula (1.9) and the relations established in 1.1.1 we have :

$$\begin{aligned}
C^{\alpha}_{n,m}(P_k)(\underline{x}) &= \left[(1+\underline{b}^2)^{-\alpha}\partial^n_{\underline{b}}(1+\underline{b}^2)^{\alpha+n}\right]^{\mathrm{rest}} \\
&= \left[C^{\alpha}_{n,m+2k}(1)(\underline{b})\right]^{\mathrm{rest}} \\
&= \left[C^{\alpha}_{n,m+2k,0}(\underline{b})\right]^{\mathrm{rest}}. \blacksquare
\end{aligned}$$

Theorem 2. **(Recurrence relations for $C^{\alpha}_{n,m}(P_k)(\underline{x})$)**

$$\begin{aligned}&(i)\ C^{\alpha}_{n+1,m}(P_k)(\underline{x}) = D_{\alpha}C^{\alpha+1}_{n,m}(P_k)(\underline{x}).\\ &(ii)\ C^{\alpha}_{n+1,m}(P_k)(\underline{x}) = (-2(\alpha+1)\underline{x} + (1+\underline{x}^2)\partial_{\underline{x}})C^{\alpha+1}_{n,m}(P_k)(\underline{x}).\end{aligned} \tag{1.11}$$

Proof.

(i) By virtue of Rodrigues' Formula (1.9) we obtain

$$\begin{aligned}&C^{\alpha}_{n+1,m}(P_k)(\underline{x})\\ &= (1+\underline{x}^2)^{-\alpha}\partial^{n+1}_{\underline{x}}((1+\underline{x}^2)^{\alpha+n+1}P_k(\underline{x}))\\ &= (1+\underline{x}^2)^{-\alpha}\partial_{\underline{x}}\left[(1+\underline{x}^2)^{\alpha+1}((1+\underline{x}^2)^{-(\alpha+1)}\partial^{n}_{\underline{x}}(1+\underline{x}^2)^{\alpha+n+1}P_k(\underline{x}))\right]\\ &= D_{\alpha}C^{\alpha+1}_{n,m}(P_k)(\underline{x}).\end{aligned}$$

(ii) Apply (i) and (1.6). ■

Corollary. **(Recurrence relations for $C^{\alpha}_{n,m,0}(\underline{x})$)**

$$\begin{aligned}&(i)\ C^{\alpha}_{n+1,m,0}(\underline{x}) = D_{\alpha}C^{\alpha+1}_{n,m,0}(\underline{x})\\ &(ii)\ C^{\alpha}_{n+1,m,0}(\underline{x}) = (-2(\alpha+1)\underline{x} + (1+\underline{x}^2)\partial_{\underline{x}})C^{\alpha+1}_{n,m,0}(\underline{x}).\end{aligned} \tag{1.12}$$

Writing out $C^{\alpha+1}_{n,m,0}(\underline{x})$ and $C^{\alpha}_{n+1,m,0}(\underline{x})$ as

$$C^{\alpha+1}_{n,m,0}(\underline{x}) = a(n,m,\alpha+1,0)\underline{x}^n + \text{ lower order terms}$$

and

$$C^{\alpha}_{n+1,m,0}(\underline{x}) = a(n+1,m,\alpha,0)\underline{x}^{n+1} + \text{ lower order terms},$$

we obtain by (1.12) (ii) and (1.2) (iii) that

$$a(n+1,m,\alpha,0) = \begin{cases} -(2(\alpha+1)+n)a(n,m,\alpha+1,0) & \text{for } n \text{ even}\\ -(2(\alpha+1)+(n-m-1))a(n,m,\alpha+1,0) & \text{for } n \text{ odd.}\end{cases}$$

As $C^{\alpha}_{0,\alpha,0}(\underline{x}) = 1$ and so $a(0, m, \alpha, 0) = 1$ we thus obtain

(i) for $n = 2p$ even

$$a(2p, m, \alpha, 0) = 2^{2p}(\alpha + p + \frac{m}{2})_p(\alpha + p + 1)_p \tag{1.13}$$

(ii) for $n = 2p + 1$ odd

$$a(2p+1, m, \alpha, 0) = -2^{2p+1}(\alpha + p + 1 + \frac{m}{2})_p(\alpha + p + 1)_{p+1}.$$

Now writing out $C^{\alpha}_{n,m,k}(\underline{x})$ as

$$C^{\alpha}_{n,m,k}(\underline{x}) = a(n, m, \alpha, k)\underline{x}^n + \text{ lower order terms}$$

we get from (1.10) that

$$\begin{aligned}
&(a(n, m, \alpha, k)\underline{x}^n + \text{ lower order terms})P_k(\underline{x}) \\
&= \left[C^{\alpha}_{n,m+2k,0}(\underline{b})\right]^{\text{rest}}(\underline{x}) \\
&= \left[a(n, m+2k, \alpha, 0)\underline{b}^n + \text{ lower order terms}\right]^{\text{rest}}(\underline{x}) \\
&= (a(n, m+2k, \alpha, 0)\underline{x}^n + \text{ lower order terms})P_k(\underline{x})
\end{aligned}$$

whence

$$a(n, m, \alpha, k) = a(n, m+2k, \alpha, 0) \tag{1.14}$$

In virtue of (1.13) and (1.14) we so obtain that

(i) for $n = 2p$ even

$$a(2p, m, \alpha, k) = 2^{2p}(\alpha + p + \frac{m}{2} + k)_p(\alpha + p + 1)_p$$

(1.15)

(ii) for $n = 2p + 1$ odd

$$a(2p+1, m, \alpha, k) = -2^{2p+1}(\alpha + p + 1 + \frac{m}{2} + k)_p(\alpha + p + 1)_{p+1}.$$

Finally notice that by means of (1.12) (ii), we have explicitly that

$$\begin{aligned} C^{\alpha}_{0,m,0}(\underline{x}) &= 1 \\ C^{\alpha}_{1,m,0}(\underline{x}) &= -2(\alpha+1)\underline{x} \\ C^{\alpha}_{2,m,0}(\underline{x}) &= 2(\alpha+2)(2\alpha+2+m)\underline{x}^2 + 2(\alpha+2)m \end{aligned}$$

etc ...

As $\partial_{\underline{x}}$ maps an even power of $\underline{x}$ to an odd power of $\underline{x}$ and vice versa, we thus obtain by induction that for n even (resp. n odd), $C^{\alpha}_{n,m,0}(\underline{x})$ contains only even (resp. odd), powers in $\underline{x}$. By means of (1.10), the same holds for $C^{\alpha}_{n,m,k}(\underline{x})$.

Remark. The foregoing calculations clearly show that the coefficients of $C^{\alpha}_{n,m,k}(\underline{x})$ do not depend on the element $P_k \in M^+(k;V)$ chosen.

In the sequel we tacitly admit that $P_k \in M^+(k;V)$ has been taken in such a way that

$$\int_{S^{m-1}} \overline{P}_k(\underline{\xi})P_k(\underline{\xi})d\Sigma(\underline{\xi}) = \begin{cases} 1 & \text{if } V = C \\ I & \text{if } V = S \end{cases}.$$

This of course implies that

$$(C^{\alpha}_{n,m}(P_k), C^{\alpha}_{n,m}(P_k))_{\alpha} \in \begin{cases} \mathbf{R}^+ & \text{if } V = C \\ \mathbf{R}^+ I & \text{if } V = S \end{cases}.$$

Theorem 3. (Orthogonality relations) *Let $s, t, k, \ell \in \mathbf{N}$ be such that $s \neq t$ or $k \neq \ell$. Then for all $P_k \in M^+(k;V)$ and $P_\ell \in M^+(\ell;V)$*

$$(C^{\alpha}_{s,m}(P_k), C^{\alpha}_{t,m}(P_\ell))_{\alpha} = 0.$$

Proof. If $k \neq \ell$ then the result follows immediately from Proposition 1.2.1.

If $k = \ell$ and hence $s \neq t$, say $s > t$, we have by virtue of 1.3.1 Proposition 1 and (1.2) (iii) that

$$\begin{aligned} &(C^{\alpha}_{s,m}(P_k), C^{\alpha}_{t,m}(P_\ell))_{\alpha} \\ &= (D_{\alpha}D_{\alpha+1}\cdots D_{\alpha+s-1}(P_k), C^{\alpha}_{t,m}(P_\ell))_{\alpha} \\ &= (P_k, \partial^s_{\underline{x}}(C^{\alpha}_{t,m}(P_\ell))_{\alpha+s} \\ &= 0. \blacksquare \end{aligned}$$

Corollary. *Let $\alpha > -1$ and $n \in \mathbf{N}$ be fixed. Then for each $p_n[\underline{x}] \in {}_n\mathrm{C}[\underline{x}]$, there exists a (unique) sequence $(a_j)_{j=0}^n$ of real numbers such that for all $P_k \in M^+(k;V)$*

$$p_n(\underline{x})P_k(\underline{x}) = \sum_{j=0}^{n} a_j C^{\alpha}_{j,m}(P_k)(\underline{x}).$$

Proof. As each $C^{\alpha}_{j,m}(P_k)(\underline{x}) = C^{\alpha}_{j,m,k}(\underline{x})P_k(\underline{x})$ with $C^{\alpha}_{j.m,k}(\underline{x}) \in {}_j\mathbb{C}[\underline{x}]$ having real coefficients, the relation to be proved is meaningful.

Now let $V = \mathcal{C}$, then by means of Theorem 3 we find immediately that

$$a_j = \frac{1}{\|C^{\alpha}_{j,m}(P_k)\|^2_{\alpha}}(p_n(\underline{x})P_k(\underline{x}), C^{\alpha}_{j,m}(P_k)(\underline{x}))_{\alpha}.$$

The case $V = S$ is similar. ■

Theorem 4. (Differential equation for the Gegenbauer polynomials) *For all $n, k \in \mathbf{N}$ and $\alpha > -1$ there exists a real constant $C(\alpha, n, m, k)$ such that*

$$D_{\alpha}\partial_{\underline{x}}(C^{\alpha}_{n,m}(P_k)(\underline{x}) = C(\alpha, n, m, k)C^{\alpha}_{n,m}(P_k)(\underline{x})$$

or

$$(1 + \underline{x}^2)\partial^2_{\underline{x}}C^{\alpha}_{n,m}(P_k)(\underline{x}) - 2(\alpha + 1)\underline{x}\partial_{\underline{x}}C^{\alpha}_{n,m}(P_k)(\underline{x}) - C(\alpha, n, m, k)C^{\alpha}_{n,m}(P_k)(\underline{x}) = 0.$$

Proof. First notice that, as by definition,

$$D_{\alpha+1}\dots D_{\alpha+n-1}(P_k(\underline{x}) = C^{\alpha+1}_{n-1,m}(P_k)(\underline{x})$$

then

$$\begin{aligned} C^{\alpha}_{n,m}(P_k)(\underline{x}) &= D_{\alpha}(D_{\alpha+1}\dots D_{\alpha+n-1}(P_k(\underline{x})) \\ &= D_{\alpha}C^{\alpha+1}_{n-1,m}(P_k)(\underline{x}). \end{aligned} \tag{1.16}$$

Now we prove that there exists a real constant $C(\alpha, n, m, k)$ such that

$$\partial_{\underline{x}}(C^{\alpha}_{n,m}(P_k)(\underline{x}) = C(\alpha, n, m, k)C^{\alpha+1}_{n-1,m}(P_k)(\underline{x}) \tag{1.17}$$

As $C^{\alpha}_{n,m}(P_k)(\underline{x}) = C^{\alpha}_{n,m,k}(\underline{x})P_k(\underline{x})$ where $C^{\alpha}_{n,m,k}(\underline{x}) \in {}_n\mathbb{C}[\underline{x}]$, by virtue of (1.4) there ought to exist $p_{n-1}(\underline{x})$ such that

$$\partial_{\underline{x}}C^{\alpha}_{n,m}(P_k)(\underline{x}) = p_{n-1}(\underline{x})P_k(\underline{x}).$$

Furthermore, for all $t \in \mathbf{N}$ with $t < n - 1$, we have that

$$(C^{\alpha+1}_{t,m}(P_k), \partial_{\underline{x}}C^{\alpha}_{n,m}(P_k))_{\alpha+1} = 0.$$

Indeed, by means of Proposition 1 in 1.1.3, Theorem 2 and Theorem 3 we obtain :

$$\begin{aligned}
&(C^{\alpha+1}_{t,m}(P_k), \partial_{\underline{x}} C^{\alpha}_{n,m}(P_k))_{\alpha+1} \\
&= (D_\alpha C^{\alpha+1}_{t,m}(P_k), C^{\alpha}_{n,m}(P_k))_\alpha \\
&= (C^{\alpha}_{t+1,m}(P_k), C^{\alpha}_{n,m}(P_k))_\alpha \\
&= 0.
\end{aligned}$$

Now, as by the Corollary to Theorem 2, there ought to exist (real) constants a_j, $j = 0, \ldots, n-1$, such that

$$p_{n-1}(\underline{x}) P_k(\underline{x}) = \sum_{j=0}^{n-1} a_j C^{\alpha+1}_{j,m}(P_k)(\underline{x})$$

we find that for each $k < n-1$

$$\begin{aligned}
&a_t (C^{\alpha+1}_{k,m}(P_k), C^{\alpha+1}_{t,m}(P_k))_{\alpha+1} \\
&= (p_{n-1}(\underline{x}) P_k(\underline{x}), C^{\alpha+1}_{t,m}(P_k(\underline{x})) \\
&= 0
\end{aligned}$$

whence, as $(C^{\alpha+1}_{t,m}(P_k), C^{\alpha+1}_{t,m}(P_k))_{\alpha+1} \neq 0$, $a_t = 0$.

Consequently

$$p_{n-1}(\underline{x}) P_k(\underline{x}) = a_{n-1} C^{\alpha+1}_{n-1,m}(P_k)(\underline{x})$$

and so

$$\partial_{\underline{x}} C^{\alpha}_{n,m}(P_k)(\underline{x}) = a_{n-1} C^{\alpha+1}_{n-1,m}(P_k)(\underline{x}). \tag{1.18}$$

Hence (1.17) holds when taking $C(\alpha, n, m, k) = a_{n-1}$.

Letting act D_α on both sides of (1.18) and taking account into (1.17) we so obtain that

$$D_\alpha \partial_{\underline{x}} (C^{\alpha}_{n,m}(P_k)(\underline{x}) = C(\alpha, n, m, k) C^{\alpha}_{n,m}(P_k)(\underline{x}).$$

By the definition itself of D_α we finally have that

$$\begin{aligned}
D_\alpha \partial_{\underline{x}} C^{\alpha}_{n,m}(P_k)(\underline{x}) &= (1+\underline{x}^2)^{-\alpha} \partial_{\underline{x}}((1+\underline{x}^2)^{\alpha+1} \partial_{\underline{x}} C^{\alpha}_{n,m}(P_k)(\underline{x})) \\
&= (1+\underline{x}^2) \partial^2_{\underline{x}} C^{\alpha}_{n,m}(P_k)(\underline{x}) - 2(\alpha+1) \underline{x} \partial_{\underline{x}} C^{\alpha}_{n,m}(P_k)(\underline{x}). \blacksquare
\end{aligned} \tag{1.19}$$

Remarks.

(1) The equation (1.19) should be compared with the classical differential equation for the Gegenbauer polynomials on the real line (see e.g. [40])

$$(1+x^2)\frac{d^2}{dx^2}C_n^\gamma(x)-(2\gamma+1)x\frac{d}{dx}C_n^\gamma(x)+n(n+2\gamma)C_n^\gamma(x)=0$$

where it should be noticed that $\gamma=\alpha+\frac{1}{2}$.

(2) In the course of the foregoing proof we obtained the relation (1.18) :

$$\partial_{\underline{x}}C^\alpha_{n,m}(P_k)(\underline{x})=C(\alpha,n,m,k)C^{\alpha+1}_{n-1}(P_k)(\underline{x})$$

thus giving us the Dirac-derivative for a Gegenbauer polynomial of degree n. It is henceforth called the *Dirac derivation formula for Gegenbauer polynomials.*

Let us now determine explicitly the constant $C(\alpha,n,m,k)$ appearing in the differential equation (1.18).

To this end, first observe that (1.18) implies that, putting $k=0$,

$$\partial_{\underline{x}}C^\alpha_{n,m,0}(\underline{x})=C(\alpha,n,m,0)C^{\alpha+1}_{n-1,m,0}(\underline{x}).$$

Comparing the coefficients of $\underline{x}^{n-1}$ we obtain, using the relations (1.13), that

(i) for n even

$$-na(n,m,\alpha,0)=C(\alpha,n,m,0)a(n-1,\alpha+1)$$

and so

$$C(\alpha,n,,m,0)=n(2\alpha+n+m);$$

(1.20)

(ii) for n odd

$$-(n+m-1)a(n,m,\alpha,0)=C(m,\alpha,n,0)a(n-1,\alpha+1)$$

and so

$$C(\alpha,n,m,0)=(2\alpha+n+1)(n+m-1).$$

But, as on one hand, by virtue of (1.3) and (1.18)

$$\begin{aligned}\partial_{\underline{x}}C^\alpha_{n,m}(P_k)(\underline{x}) &= \left[\partial_{\underline{b}}C^\alpha_{n,m+2k,0}(\underline{b})\right]^{\text{rest}}\\ &= \left[C(\alpha,n,m+2k,0)C^{\alpha+1}_{n-1,m+2k,0}(\underline{b})\right]^{\text{rest}}\\ &= C(\alpha,n,m+2k,0)C^{\alpha+1}_{n-1,m}(P_k)(\underline{x})\end{aligned}$$

while on the other hand, again by (1.18),

$$\partial_{\underline{x}} C^{\alpha}_{n,m}(P_k)(\underline{x}) = C(\alpha, n, m, k) C^{\alpha+1}_{n-1,m}(P_k)(\underline{x}),$$

we find that

$$C(\alpha, n, m, k) = C(\alpha, nm + 2k, 0). \tag{1.21}$$

Consequently, by means of (1.20) and (1.21) we have

(i) for n even

$$C(\alpha, n, m, k) = n(2\alpha + n + m + 2k)$$

$$\tag{1.22}$$

(ii) for n odd

$$C(\alpha, n, m, k) = (2\alpha + n + 1)(n + m + 2k - 1).$$

As a final result of this section we establish a relationship between the polynomials $C^{\alpha}_{n,m,k}(\underline{x})$ and the classical Jacobi polynomials $P^{\beta,\alpha}_{[\frac{n}{2}]}(1 + 2\underline{x}^2)$ for β appropriate.

Theorem 5.

(i) For $n = 2p$ even

$$\begin{aligned} C^{\alpha}_{2p,m,0}(\underline{x}) &= 2^{2p}(\alpha + p + 1)_p(-1)^p p! P_p^{(\frac{m}{2}-1,\alpha)}(1 + 2\underline{x}^2) \\ &= 2^{2p}(\alpha + p + 1)_p(-1)^p(\frac{m}{2})_p F(-p, p + \alpha + \frac{m}{2}; \frac{m}{2}; -\underline{x}^2). \end{aligned}$$

(ii) For $n = 2p + 1$ odd

$$\begin{aligned} C^{\alpha}_{2p+1,m,0}(\underline{x}) &= \underline{x} 2^{2p+1}(\alpha + p + 1)_{p+1} p! (-1)^{p+1} P_p^{(\frac{m}{2},\alpha)}(1 + 2\underline{x}^2) \\ &= \underline{x} 2^{2p+1}(\alpha + p + 1)_{p+1}(\frac{m}{2} + 1)_p(-1)^{p+1} \cdot \\ &\quad \cdot F(-p, p + \alpha + \frac{m}{2} + 1; \frac{m}{2} + 1; -\underline{x}^2). \end{aligned}$$

Proof.

(i) Since for $n = 2p$ even, $C^{\alpha}_{n,m,0}(\underline{x})$ contains only even powers of $\underline{x}$, we may put $C^{\alpha}_{2p,m,0}(\underline{x}) = u_p(r^2)$ where u_p is a polynomial of degree p in r^2.

For any pair $C^{\alpha}_{2p,m,0}(\underline{x})$ and $C^{\alpha}_{2q,m,0}(\underline{x})$ with $p \neq q$ we so obtain that

$$\begin{aligned}(C^{\alpha}_{2p,m,0}(\underline{x}), C^{\alpha}_{2q,m,0}(\underline{x}))_{\alpha} &= \int_{B(1)} u_p(r^2)u_q(r^2)(1+\underline{x}^2)^{\alpha} d\underline{x} \\ &= \int_0^1 r^{m-1}(1-r^2)^{\alpha} u_p(r^2)u_q(r^2)A_m dr.\end{aligned}$$

As u_p is a polynomial of degree p in r^2, there exists a polynomial U_p such that

$$U_p(1-2r^2) = u_p(r^2).$$

By means of Theorem 3 we thus have

$$\begin{aligned}&(C^{\alpha}_{2p,m,0}(\underline{x}), C^{\alpha}_{2q,m,0}(\underline{x}))_{\alpha} \\ &= \int_{-1}^{1} (\frac{1+s}{2})^{\alpha}(\frac{1-s}{2})^{\frac{m}{2}-1}(\frac{A_m}{4})U_p(s)U_q(s)ds \\ &= 0.\end{aligned}$$

This orthogonality relation determines the polynomials U_p uniquely, up to a constant, namely :

$$u_p(\frac{1-s}{2}) = U_p(s) = c_p P_p^{(\frac{m}{2}-1,\alpha)}(s) \tag{1.23}$$

where $P_p^{(\frac{m}{2}-1,\alpha)}(s)$ is the p-th Jacobi polynomial corresponding to the weight function $(1-s)^{\frac{m}{2}-1}(1+s)^{\alpha}$.

The constants c_p can be easily determined by comparing the coefficients of s^p in (1.23).

Indeed, on the one hand (see also (1.13))

$$u_p(\frac{1-s}{2}) = 2^p(\alpha+p+\frac{m}{2})_p(\alpha+p+1)_p(-1)^p s^p + \text{ lower order terms}$$

while on the other hand (see e.g. [28])

$$P_p^{(\frac{m}{2}-1,\alpha)}(s) = 2^{-p}\frac{(p+\alpha+\frac{m}{2})_p}{p!}s^p + \text{ lower order terms.}$$

Consequently

$$c_p = 2^{2p}(\alpha+p+1)_p(-1)^p p!$$

Taking into account the expression of the Jacobi polynomials in terms of the hypergeometric function (see e.g. [28]), we thus obtain :

$$\begin{aligned}C^{\alpha}_{2p,m,0}(\underline{x}) &= 2^{2p}(\alpha+p+1)_p(-1)^p p! P_p^{(\frac{m}{2}-1,\alpha)}(1+2\underline{x}^2)^{\alpha} \\ &= 2^{2p}(\alpha+p+1)_p(-1)^p(\frac{m}{2})_p F(-p, p+\alpha+\frac{m}{2}; \frac{m}{2}; -\underline{x}^2).\end{aligned}$$

(ii) Since $C^{\alpha}_{2p+1,m,0}(\underline{x})$ cont ains only odd powers in $\underline{x}$, we may write it as

$$C^{\alpha}_{2p+1,m,0}(\underline{x}) = \underline{x}v_p(r^2)$$

where v_p is a polynomial of degree p in r^2.

For any pair $C^{\alpha}_{2p+1,m,0}(\underline{x})$ and $C^{\alpha}_{2q+1,m,0}(\underline{x})$ with $p \neq q$ we so obtain that

$$\begin{aligned}(C^{\alpha}_{2p+1,m,0}(\underline{x}), C^{\alpha}_{2q+1,m,0}(\underline{x}))_{\alpha} &= \int_{B(1)} v_p(r^2)v_q(r^2)\overline{\underline{x}}\underline{x}(1+\underline{x}^2)^{\alpha}d\underline{x} \\ &= \int_0^1 r^{m+1}(1-r^2)^{\alpha}v_p(r^2)v_q(r^2)A_m dr.\end{aligned}$$

Putting $V_p(1-2r^2) = v_p(r^2)$ we obtain, by virtue of Theorem 3, that

$$\begin{aligned}&(C^{\alpha}_{2p+1,m,0}(\underline{x}), C^{\alpha}_{2q+1,m,0}(\underline{x}))_{\alpha} \\ &= \int_{-1}^{1}(\frac{1+s}{2})^{\alpha}(\frac{1-s}{2})^{\frac{m}{2}}(\frac{A_m}{4})V_p(s)V_q(s)ds \\ &= 0.\end{aligned}$$

Hence for some suitable constant d_p

$$v_p(\frac{1-s}{2}) = d_p P_p^{(\frac{m}{2},\alpha)}(s). \tag{1.24}$$

Again comparing the coefficients of s^p in both sides of (1.24) and taking into account (1.13), we find that

$$\begin{aligned}C^{\alpha}_{2p+1,m,0}(\underline{x}) &= \underline{x}2^{2p+1}(\alpha+p+1)_{p+1}p!(-1)^{p+1}P_p^{(\frac{m}{2},\alpha)}(1+2\underline{x}^2) \\ &= \underline{x}2^{2p+1}(\alpha+p+1)_{p+1}(\frac{m}{2}+1)_p(-1)^{p+1}\cdot \\ &\quad \cdot F(-p, p+\alpha+\frac{m}{2}+1; \frac{m}{2}+1; -\underline{x}^2). \blacksquare\end{aligned}$$

Remarks.

(1) Replacing in the formulas appearing in Theorem 5, $\frac{m}{2}$ by $\frac{m}{2}+k$, we obtain the corresponding expressions of $C^{\alpha}_{n,m,k}(\underline{x})$ in terms of the Jacobi polynomials $P^{(\beta,\alpha)}_{[\frac{n}{2}]}(1+2\underline{x}^2)$.

(2) These expressions also yield :

$$C^{\alpha}_{2p+1,m,k}(\underline{x}) = -2\underline{x}(\alpha+2p+1)C^{\alpha}_{2p,m,k+1}(\underline{x}).$$

1.1.5 The space $L_2(B(1),(1+\underline{x}^2)^\alpha)$ revisited

In this subsection we take $V = C$.

Since for all $f,g \in L_2(B(1);(1+\underline{x}^2)^\alpha)$

$$\begin{aligned}(f,g)_\alpha &= \int_{B(1)} \bar{f}(\underline{x})g(\underline{x})(1+\underline{x}^2)^\alpha d\underline{x} \\ &= \int_0^1 r^{m-1}(1-r^2)^\alpha dr \int_{S^{m-1}} \bar{f}(r\xi)g(r\xi)d\Sigma(\xi) \\ &= \int_0^1 (f,g)_{S^{m-1}} w(r)dr\end{aligned}$$

with $w(r) = r^{m-1}(1-r^2)^\alpha$, we may identify $L_2(B(1);(1+\underline{x}^2)^\alpha)$ with the Hilbert module $L_2([0,1],w,L_2(S^{m-1};C))$ of $L_2(S^{m-1};C)$-valued functions f on $[0,1]$ such that $\|f(r\cdot)\|_{S^{m-1}} \in L_2([0,1],\mu)$ where $d\mu(r) = r^{m-1}(1-r^2)^\alpha dr$. Here the notation $f(r\cdot)$ means that f is considered as a function in the variable $r \in [0,1]$.

We may now extend the operator $\Gamma_{\underline{\xi}}$, originally defined on $L_2(S^{m-1};C)$ (see II. 1.3), to $L_2(B(1),(1+\underline{x}^2)^\alpha)$ in the following way : for $f(r\underline{\xi}) \in L_2(B(1);(1+\underline{x}^2)^\alpha)$ we put $\Gamma^e_{\underline{\xi}} f(r\underline{\xi}) = g(r\underline{\xi})$ if and only if

(i) $g(r\underline{\xi}) \in L_2(B(1);(1+\underline{x}^2)^\alpha)$

(ii) $\Gamma_{\underline{\xi}} f(r\underline{\xi}) = g(r\underline{\xi})$ for $\mu \cdot a \cdot e \cdot r$ in $[0,1]$

Since $\Gamma_{\underline{\xi}}$ is self adjoint on $L_2(S^{m-1};C)$ (see II. 1.3), $\Gamma^e_{\underline{\xi}}$ is self adjoint on the space $L_2(B(1);(1+\underline{x}^2)^\alpha)$ and it has the same pont spectrum as $\Gamma_{\underline{\xi}}$ (see also [84]).

Putting for convenience $L_2(B(1);(1+\underline{x}^2)^\alpha) = L_2^\alpha$, we introduce the modules

$$\begin{aligned}R_k^+ &= \{f : [0,1] \to L_2(S^{m-1},C) \mid f \quad \text{is measurable} \\ &\qquad \text{and } f(r\cdot) \in \mathcal{M}^+(k;C) \quad \text{for } a\cdot e\cdot r \quad \text{in} \quad [0,1]\}; \\ R_k^- &= \{f : [0,1] \to L_2(S^{m-1};C) \mid f \quad \text{is measurable} \\ &\qquad \text{and } f(r\cdot) \in \mathcal{M}^-(k;C) \quad \text{for } a\cdot e\cdot r \quad \text{in} \quad [0,1]\}; \\ R_k &= R_k^+ \cup R_k^-; \\ L_{2,k}^{\alpha+} &= L_2^\alpha \cap R_k^+; \\ L_{2,k}^{\alpha-} &= L_2^\alpha \cap R_k^-; \\ L_{2,k}^\alpha &= L_2^\alpha \cap R_k.\end{aligned}$$

The eigenspaces of $\Gamma^e_{\underline{\xi}}$ are then exactly the spaces $L_{2,k}^{\alpha+}$ and $L_{2,k}^{\alpha-}$ corresponding to, respectively, the eigenvalues $(-k)$ and $(k+m-1)$, while the spaces $L_{2,k}^\alpha$ are the eigenspaces of the self adjoint operator $\underline{\xi}\Gamma^e_{\underline{\xi}}\underline{\xi}\Gamma^e_{\underline{\xi}}$ corresponding to the eigenvalues $(-k)(k+m-1)$.

Moreover

$$L_2^\alpha = \sum_{k=0}^{\infty} \oplus_\perp L_{2,k}^\alpha.$$

Now take an orthonormal basis $\{P_k^{(i)} : i = 1, \ldots, K(m-1;k)\}$ of $\mathcal{M}^+(k;\mathcal{C})$ (see also II. 4) and define for $i = 1, \ldots, K(m-1,k)$:

$$\begin{aligned}
R_k^{(i)+} &= \{f : [0,1] \to L_2(S^{m-1}, \mathcal{C}) | f(r\underline{\xi}) = P_k^{(i)}(r\underline{\xi})a(r) \\
&\quad \text{with } a(r)\,\mathcal{C} - \text{valued}\}; \\
R_k^{(i)-} &= \{f\,[0,1] \to L_2(S^{m-1};\mathcal{C}) | f(r\cdot) = \underline{\xi}P_k^{(i)}(r\underline{\xi})b(r) \\
&\quad \text{with } b(r)\,\mathcal{C} - \text{valued}\}; \\
R_k^{(i)} &= R_k^{(i)+} \oplus_\perp R_k^{(i)-}; \\
L_{2,k}^{\alpha(i)+} &= L_2^\alpha \cap R_k^{(i)+}; \\
L_{2,k}^{\alpha(i)-} &= L_2^\alpha \cap R_k^{(i)-}; \\
L_{2,k}^{\alpha(i)} &= L_2^\alpha \cap R_k^{(i)}.
\end{aligned}$$

Then

$$L_2^\alpha = \sum_{k=0}^{\infty} \sum_{i=1}^{K(m-1;k)} \oplus_\perp L_{2,k}^{\alpha(i)}.$$

The importance of the spaces $R_k^{(i)}$ lies in the fact that their subspace, consisting of functions of the form

$$f(r\cdot) = P_k^{(i)}(r\underline{\xi})a(r) + \underline{\xi}P_k^{(i)}(r\underline{\xi})b(r)$$

where $a, b \in \mathcal{C}_\infty(]0,1[;\mathcal{C}) \cap L_2([0,1];\mathcal{C})$, are invariant dense subspaces for the Dirac operator.

Indeed, their density being obvious, by means of (1.2) (i) and (ii), the invariance property clearly holds when taking $a(r)$ and $b(r)$ C-valued whence, taking into account that each C-valued function $c(r)$ may be written as

$$c(r) = \sum_A c_A(r)e_A$$

with $c_A(r)$ C-valued, it remains true for $\mathcal{C}$-valued functions.

Notice that the Gegenbauer polynomials $C_{n,m}^\alpha(P_k^{(i)})(\underline{x})$ all belong to $R_k^{(i)}$ since

$$C^{\alpha}_{n,m}(P_k^{(i)})(\underline{x}) = C^{\alpha}_{n,m,k}(\underline{x})P_k^{(i)}(\underline{x})$$

with

$$C^{\alpha}_{n,m,k}(\underline{x}) = \begin{cases} u_p^{\star}(r^2) & \text{for } n = 2p \text{ even} \\ \underline{\xi} r v_p^{\star}(r^2) & \text{for } n = 2p+1 \text{ odd} \end{cases}$$

where $u_p^\star$ and $v_p^\star$ are $\mathbf{R}$-valued polynomials of degree p.

We have even more, namely

Theorem 1. *The set* $\{C^{\alpha}_{n,m}(P_k^{(i)})(\underline{x}) : n, k, i \in \mathbf{N},\ i \leq K(m-1;k)\}$ *is an orthogonal basis for* L_2^{α}.

Proof. The orthogonality follows from Theorem 3 in 1.1.4 and the fact that for $i \neq j$,

$$\begin{aligned} &(C^{\alpha}_{n,m}(P_k^{(i)}), C^{\alpha}_{n,m}(P_k^{(j)}))_{\alpha} \\ = &\int_{B(1)} \overline{P_k^{(i)}(\underline{x})}\,\overline{C^{\alpha}_{n,m,k}(\underline{x})} C^{\alpha}_{n,m,k}(\underline{x}) P_k^{(j)}(\underline{x})(1+\underline{x}^2)^{\alpha} d\underline{x} \\ = &\int_0^1 r^{m-1}(1-r^2)^{\alpha} Q(r^2) dr \Big(\int_{S^{m-1}} \overline{P_k^{(i)}(\underline{\xi})} P_k^{(j)}(\underline{\xi}) d\Sigma(\underline{\xi})\Big) \\ = &\ 0. \end{aligned}$$

Here Q is an $\mathbf{R}$-valued polynomial with

$$Q(r^2) = \begin{cases} r^{2k}(u_p^{\star}(r^2))^2 & \text{for } n = 2p \text{ even} \\ r^{2k+2}(v_p^{\star}(r^2))^2 & \text{for } n = 2p+1 \text{ odd} \end{cases}$$

Now we prove that for $k \in \mathbf{N}$ and $i \in \{1, \ldots, K(m-1;k)\}$ fixed, $\{C^{\alpha}_{n,m}(P_k^{(i)})(\underline{x}) : n \in \mathbf{N}\}$ is a basis for $L_{2,k}^{\alpha(i)}$.

To this end, take $f \in L_{2,k}^{\alpha(i)}$. Then

$$f = P_k^{(i)}(\underline{x})A(r) + \underline{x}P_k^{(i)}(\underline{x})B(r).$$

Defining the function a by

$$a(s) = A\left(\sqrt{\frac{1-s}{2}}\right),$$

then $a \in L_2([-1,+1];(1-s)^{\frac{m}{2}-1+k}(1+s)^{\alpha}ds)$ whence

$$a(s) = \sum_{p=0}^{\infty} P_p^{(\frac{m}{2}-1+k,\alpha)}(s)a_p$$

where $a_p \in \mathcal{C}$ for all $p \in \mathbf{N}$.

Hence, by virtue of the Corollary to Theorem 3 in 1.1.4, there ought to exist $A_p \in \mathcal{C}$ such that

$$P_k^{(i)}(\underline{x})A(r) = \sum_{p=0}^{\infty} C_{2p,m}^{\alpha}(P_k^{(i)})(\underline{x})A_p.$$

An analogous reasoning yields that for suitable $B_p \in \mathcal{C}$,

$$\underline{x}P_k^{(i)}(\underline{x})B(r) = \sum_{p=0}^{\infty} C_{2p+1,m}^{\alpha}(P_k^{(i)})(\underline{x})B_p.$$

Consequently f admits the expansion

$$f = \sum_{k=0}^{\infty} \sum_{i=1}^{K(m-1;k)} \sum_{p=0}^{\infty} (C_{2p,m}^{\alpha}(P_k^{(i)})(\underline{x})A_p + C_{2p+1,m}^{\alpha}(P_k^{(i)})(\underline{x})B_p). \blacksquare$$

Let us now calculate the normalizing constants

$$\gamma_{n,m,k}^{\alpha} = (C_{n,m}^{\alpha}(P_k^{(i)}), C_{n,m}^{\alpha}(P_k^{(i)}))_{\alpha}$$

We have

Theorem 2.

(i) For $n = 2p$

$$\gamma_{2p,m,k}^{\alpha} = \frac{2^{4p}p!(\Gamma(\alpha+2p+1))^2\pi^{\frac{m}{2}}\Gamma(\frac{m}{2}+k+p)}{\Gamma(\alpha+p+\frac{m}{2}+k)\Gamma(\alpha+p+1)\Gamma(\frac{m}{2})(\alpha+2p+\frac{m}{2}+k)}$$

(ii) For $n = 2p+1$

$$\gamma_{2p+1,m,k}^{\alpha} = \frac{2^{4p+2}p!(\Gamma(\alpha+2p+2))^2\pi^{\frac{m}{2}}\Gamma(\frac{m}{2}+k+p+1)}{\Gamma(\alpha+p+\frac{m}{2}+k+1)\Gamma(\alpha+p+1)\Gamma(\frac{m}{2})(\alpha+2p+\frac{m}{2}+k+1)}$$

Proof. From the differential equation (1.16), Proposition 1 i n (1.1.3) and the Dirac derivation formula (1.18) it follows that, for $n > 0$,

$$\begin{aligned}
\gamma_{n,m,k}^{\alpha}C(\alpha,n,mk) &= (D_{\alpha}\partial_{\underline{x}}C_{n,m}^{\alpha}(P_k^{(i)}), C_{n,m}^{\alpha}(P_k^{(i)}))_{\alpha} \\
&= (\partial_{\underline{x}}C_{n,m}^{\alpha}(P_k^{(i)}), \partial_{\underline{x}}C_{n,m}^{\alpha}(P_k^{(i)}))_{\alpha+1} \\
&= (C(\alpha,n,m,k))^2(C_{n-1,m}^{\alpha+1}(P_k^{(i)}), C_{n-1,m}^{\alpha+1}(P_k^{(i)}))_{\alpha+1} \\
&= (C(\alpha,n,m,k))^2\gamma_{n-1,m,k}^{\alpha+1}
\end{aligned}$$

or for $n > 1$,

$$\gamma^{\alpha}_{n,m,k} = C(\alpha, n, m, k)\gamma^{\alpha+1}_{n-1,m,k}. \tag{1.25}$$

For $n = 0$ we have

$$\begin{aligned}\gamma^{\alpha}_{0,m,k} &= \int_{B(1)} \overline{P^{(i)}_k(\underline{x})} P^{(i)}_k(\underline{x})(1+\underline{x}^2)^{\alpha} d\underline{x} \\ &= \int_0^1 r^{m+2k-1}(1-r^2)^{\alpha} dr \int_{S^{m-1}} \overline{P^{(i)}_k(\underline{\xi})} P^{(i)}_k(\underline{\xi}) d\Sigma(\underline{\xi})\end{aligned}$$

or

$$\gamma^{\alpha}_{0,m,k} = \frac{1}{2} A_m B(\frac{m}{2} + k, \alpha + 1) \tag{1.26}$$

where B is the classical Eulerian B-function.

By the recurrence relation (1.25) and the relations (1.22) and (1.26) we thus obtain that

(i) for $n = 2p$ even

$$\begin{aligned}\gamma^{\alpha}_{2p,m,k} &= C(\alpha, 2p, m, k)C(\alpha+1, 2p-1, m, k)\gamma^{\alpha+2}_{2p-2,m,k} \\ &= \ldots \\ &= C(\alpha, 2p, m, k)C(\alpha+1, 2p-1, m, k)\ldots \\ &\quad \ldots C(\alpha+2p-2, 2, m, k)C(\alpha+2p-1, 1, m, k)\gamma^{\alpha+2p}_{0,m,k} \\ &= (\Pi_{j=0}^{p-1} C(\alpha+2j, 2p-2j, m, k) \\ &\quad .(\Pi_{j=0}^{p-1} C(\alpha+2j+1, 2p-2j-1, m, k))\gamma^{\alpha+2p}_{0,m,k} \\ &= (\Pi_{j=0}^{p-1}(2p-2j)(2\alpha+2j+2p+m+2k). \\ &\quad .(\Pi_{j=0}^{p-1}(2\alpha+2j+2p+2)(2p-2j-2+m+2k))\gamma^{\alpha+2p}_{0,m,k} \\ &= 2^{4p} p!(\alpha+p+\tfrac{m}{2}+k)_p(\alpha+p+1)_p(\tfrac{m}{2}+k)_p \gamma^{\alpha+2p}_{0,m,k} \\ &= \frac{2^{4p} p!\Gamma(\alpha+2p+\frac{m}{2}+k)\Gamma(\alpha+2p+1)\Gamma(\frac{m}{2}+k+p)}{\Gamma(\alpha+p+\frac{m}{2}+k)\Gamma(\alpha+p+1)\Gamma(\frac{m}{2}+k)}\gamma^{\alpha+2p}_{0,m,k} \\ &= \frac{2^{4p} p!(\Gamma(\alpha+2p+1))^2 \pi^{\frac{m}{2}} \Gamma(\frac{m}{2}+k+p)}{\Gamma(\alpha+p+\frac{m}{2}+k)\Gamma(\alpha+p+1)\Gamma(\frac{m}{2})(\alpha+2p+\frac{m}{2}+k)}.\end{aligned}$$

(ii) for $n = 2p+1$ odd

$$\begin{aligned}\gamma^{\alpha}_{2p+1,m,k} &= (2\alpha+2p+2)(2p+m+2k)\gamma^{\alpha+1}_{2p,m,k} \\ &= \frac{2^{4p+2} p!(\Gamma(\alpha+2p+2))^2 \pi^{\frac{m}{2}} \Gamma(\frac{m}{2}+k+p+1)}{\Gamma(\alpha+p+\frac{m}{2}+k+1)\Gamma(\alpha+p+1)\Gamma(\frac{m}{2})(\alpha+2p+\frac{m}{2}+k+1)}. \quad \blacksquare\end{aligned}$$

Putting

$$T^{\alpha,i}_{n,m,k}(\underline{x}) = \frac{C^{\alpha}_{n,m}(P^{(i)}_k)(\underline{x})}{\sqrt{\gamma^{\alpha}_{n,m,k}}} \tag{1.27}$$

we thus get in view of Theorem 1 and Theorem 2 :

Theorem 3. *The set* $\{T^{\alpha,i}_{n,m,k}(\underline{x}) : n,k,i \in \mathbf{N},\ i \leq K(m-1,k)\}$ *is an orthonormal basis for* L^{α}_2.

Remarks.

(1) Considering the operator $\partial_{\underline{x}}$ as a linear operator from L^{α}_2 into $L^{\alpha+1}_2$ we have, since by (1.18)

$$\partial_{\underline{x}} T^{\alpha,i}_{n,m,k}(\underline{x}) = C(\alpha,n,m,k)\frac{\sqrt{\gamma^{\alpha+1}_{n-1,m,k}}}{\sqrt{\gamma^{\alpha}_{n,m,k}}} T^{\alpha+1,i}_{n-1,m,k}(\underline{x})$$

that $\partial_{\underline{x}}$ is densely defined from L^{α}_2 into $L^{\alpha+1}_2$.

As furthermore (see Proposition 1 in 1.1.3) for all $f,g \in C_1(\overline{B}(1);C)$ we have that

$$(\partial_{\underline{x}} f, g)_{\alpha+1} = (f, D_{\alpha} g)_{\alpha}$$

the adjoint of $\partial_{\underline{x}}$ is an extension of D_{α}, which, by Theorem 4 in 1.1.4, is thus densely defined from $L^{\alpha+1}_2$ into L^{α}_2. Consequently, $\partial_{\underline{x}}$ can be extended to a closed, densely defined operator from L^{α}_2 into $L^{\alpha+1}_2$ (see e.g. [89]), still denoted by $\partial_{\underline{x}}$. If D_{α} still stands for the adjoint of $\partial_{\underline{x}}$, then $D_{\alpha}\partial_{\underline{x}}$ is a non-negative self adjoint operator having pure point spectrum $\{C(\alpha,n,m,k) : n,k \in \mathbf{N}\}$.

Notice however that, in contrast with the one dimensional case, there are multiple eigenvalues in the more dimensional case. Obviously, for n and k fixed, there are several possible values of i, namely $i = 1,\ldots,K(m-1,k)$. But, on the other hand, it is also possible that the values $C(\alpha,n,m,k)$ coincide for different values of n and k. So we have e.g. that

$$C(\frac{3}{2},2,4,3) = C(\frac{3}{2},1,4,1) = 30.$$

It is easy to see that the multiplicity of each non-zero eigenvalue is finite. Indeed, let $\lambda > 0$ and $n > \lambda + 2$. Then for k arbitrary we have that $C(\alpha,n,m,k) > \lambda$, while for $n \geq 1$ fixed, $C(\alpha,n,m,k)$ is an increasing function of k with $\lim_{k\to\infty} C(\alpha,n,m,k) = \infty$. Consequently, for each non-zero eigenvalue $C(\alpha,n,m,k)$, the eigenspace is spanned by the corresponding Gegenbauer polynomials $C^{\alpha}_{n,m}(P^{(i)}_k(\underline{x})$.

The eigenvalue zero however has infinite multiplicity since its corresponding eigenspace coincides with $L_2^\alpha \cap M(B(1);C)$ and hence surely contains the set $\{P_k^{(i)}(\underline{x}) : k \in \mathbf{N},\, i \leq K(m-1;k)\}$.

(2) For $\alpha > 0$ the operator $\partial_{\underline{x}} D_{\alpha-1}$ is non-negative and selfadjoint in L_2^α ; it has pure point spectrum $\{C(\alpha-1;n+1;m;k) : n,k \in \mathbf{N}\}$.

Indeed, from the recurrence relation and the Dirac derivation formula we obtain that

$$\begin{aligned}
\partial_{\underline{x}} D_{\alpha-1} C_{n,m}^{\alpha}(P_k^{(i)})(\underline{x}) &= \partial_{\underline{x}} C_{n+1,m}^{\alpha-1}(P_k^{(i)})(\underline{x}) \\
&= C(\alpha-1,n+1,m,k) C_{n,m}^{\alpha}(P_k^{(i)})(\underline{x}).
\end{aligned}$$

Since the eigenspaces coincide, also the difference between the two operators $\partial_{\underline{x}} D_{\alpha-1}$ and $D_\alpha \partial_{\underline{x}}$ is selfadjoint.

For the one dimensional case this difference is the trivial operator $2\alpha\mathbf{1}$ while for the higher dimensional case we get on $C(\overline{B}(1);C)$, applying (1.6) and the Corollary to Proposition 1 in II. 1.12.1, that

$$\begin{aligned}
&(\partial_{\underline{x}} D_{\alpha-1} - D_\alpha \partial_{\underline{x}}) f \\
&= \left[\partial_{\underline{x}}((1+\underline{x}^2)\partial_{\underline{x}}) - 2\alpha\underline{x}) - D_\alpha \partial_{\underline{x}}\right] f \\
&= \left[(1+\underline{x}^2)\partial_{\underline{x}}^2 - 2\underline{x}\partial_{\underline{x}} - 2\alpha\partial_{\underline{x}}\underline{x} + 2(\alpha+1)\underline{x}\partial_{\underline{x}} - (1+\underline{x}^2)\partial_{\underline{x}}^2\right] f \\
&= 2\alpha(\underline{x}\partial_{\underline{x}} - \partial_{\underline{x}}\underline{x}) f \\
&= 2\alpha(\mathbf{1} - \Gamma_{\underline{\xi}} - \underline{\xi}\Gamma_{\underline{\xi}}\underline{\xi}) f
\end{aligned}$$

or

$$\partial_{\underline{x}} D_{\alpha-1} - D_\alpha \partial_{\underline{x}} = 2\alpha(\mathbf{1} - \Gamma_{\underline{\xi}} - \underline{\xi}\Gamma_{\underline{\xi}}\underline{\xi}).$$

1.2 Hermite polynomials

As is well known Hermite polynomials on the real line are polynomials which form an orthonormal basis in the Hilbert space $L_2(]-\infty,+\infty[;e^{\frac{-x^2}{2}})$.

The aim of this section is to generalize these polynomials to the Euclidean space $\mathbf{R}^m$, thus obtaining an orthonormal basis for the Hilbert module $L_2(\mathbf{R}^m; e^{\underline{x}^2/2})$.

As the construction of these polynomials is quite similar to the one made in the case of the Gegenbauer polynomials and their properties mentioned may be proved in an analogous way too, we restrict ourselves to giving a survey.

Just as in section 1.1, $(C =)\ \mathbf{C}_m = \mathbf{R}_{0,m} \otimes_{\mathbf{R}} \mathbf{C}$ stands for the complexification of the Clifford algebra $\mathbf{R}_{0,m}$, $\partial_{\underline{x}}$ is the Dirac operator in $\mathbf{R}^m$ and V is equal to C or an appropriate spinor space S.

1.2.1 The space $L_2(\mathbf{R}^m; e^{\underline{x}^2/2})$

By $L_2(\mathbf{R}^m; e^{\underline{x}^2/2})$ we mean the space of V-valued measurable functions f in $\mathbf{R}^m$ such that

$$\int_{\mathbf{R}^m} |f(\underline{x})|^2 e^{\underline{x}^2/2} d\underline{x} < +\infty.$$

Provided with the inner product

$$(f, g) = \int_{\mathbf{R}^m} \overline{f(\underline{x})} g(\underline{x}) e^{\underline{x}^2/2} d\underline{x}$$

it becomes a right Hilbert module over C if $V = C$ or a complex Hilbert space if $V = S$.

1.2.2 The operator D_+

Define the operator D_+ on the space $\mathcal{P}_\infty(\mathbf{R}^m; V)$ of V-valued infinitely differentiable functions in $\mathbf{R}^m$ which are of polynomial growth at ∞ by

$$D_+ f(\underline{x}) = e^{-\underline{x}^2/2} \partial_{\underline{x}}(e^{\frac{\underline{x}^2}{2}} f(\underline{x})).$$

Then by (1.2) we obtain that on $\mathcal{P}_\infty(\mathbf{R}^m; V)$

$$D_+ = \partial_{\underline{x}} - \underline{x}.$$

Moreover we have

Proposition 1. *For all $f, g \in \mathcal{D}(\mathbf{R}^m; V)$*

$$(D_+ f, g) = (f, \partial_{\underline{x}} g).$$

1.2.3 The Hermite polynomials $H_{n,m}(P_k(\underline{x}))$

Let $P_k \in M^+(k; V)$ and let $n \in \mathbf{N}$. Then we put

$$H_{n,m}(P_k)(\underline{x}) = ((-D_+)^n P_k)(\underline{x}) \tag{1.28}$$

By means of the relations (1.2) we may conclude that there exists a polynomial $H_{n,m,k}(\underline{x}) \in {}_n\mathrm{C}[\underline{x}]$ such that

$$H_{n,m}(P_k)(\underline{x}) = H_{n,m,k}(\underline{x}) P_k(\underline{x}).$$

Here $H_{n,m,k}(\underline{x})$ has real coefficients depending on k (not on $P_k \in M^+(k; V)$) and it contains only even (resp. odd) powers of $\underline{x}$ if n is even (resp. odd), whence it takes its values in $\mathbf{R}^0_{0,m} \oplus \mathbf{R}^1_{0,m}$. Consequently $H_{n,m}(P_k)(\underline{x})$ is a V-valued polynomial of degree $(n+k)$ in $x_1, \ldots, x_m$.

The polynomials $H_{n,m}(P_k)(\underline{x})$ are called *Hermite polynomials* in $\mathbf{R}^m$.

Theorem 1. (Rodrigues' Formula) *The Hermite polynomials $H_{n,m}(P_k)(\underline{x})$ are also determined by*

$$H_{n,m}(P_k)(\underline{x}) = e^{-\underline{x}^2/2}(-\partial_{\underline{x}})^n(e^{\underline{x}^2/2}P_k(\underline{x}))$$

Corollary. $H_{n,m}(P_k)(\underline{x}) = [H_{n,m+2k,0}(\underline{b})]^{\text{rest}}(\underline{x})$

Theorem 2. (Recurrence relations for $H_{n,m}(P_k)(\underline{x})$)

(i) $(-D_+)H_{n-1,m}(P_k)(\underline{x}) = H_{n,m}(P_k)(\underline{x})$

(ii) $(\underline{x} - \partial_{\underline{x}})H_{n-1,m}(P_k)(\underline{x}) = H_{n,m}(P_k)(\underline{x})$.

Writing out $H_{n,m,0}(\underline{x})$ as

$$H_{n,m,0}(\underline{x}) = b(n,m,0)\underline{x}^n + \quad \text{lower order terms}$$

it thus follows from Theorem 2 that $b(n,m,0) = 1$.
Corollary 1 to Theorem 1 then tells us that, if

$$H_{n,m,k}(\underline{x}) = b(n,m,k)\underline{x}^n + \quad \text{lower order terms,}$$

then $b(n,m,k) = 1$ as well.

Theorem 3. (Orthogonality relation) *Let $s,t,k,\ell \in \mathbf{N}$ be such that $s \neq t$ or $k \neq \ell$. Then for all $P_k \in M^+(k;V)$ and $P_\ell \in M^+(\ell;V)$*

$$(H_{s,m}(P_k), H_{t,m}(P_\ell)) = 0.$$

Corollary. *Let $n \in \mathbf{N}$ be fixed. Then for each $p_n(\underline{x}) \in {}_n\mathbf{C}[\underline{x}]$ there exists a (unique) sequence $(a_j)_{j=0}^n$ of real numbers such that for all $P_k \in M^+(k;V)$*

$$p_n(\underline{x})P_k(\underline{x}) = \sum_{j=0}^{n} a_j H_{n,m}(P_k)(\underline{x}).$$

Theorem 4. (Differential equation) *For all $n,k \in \mathbf{N}$ there exists a real constant $C(n,m,k)$ such that*

$$D_+\partial_{\underline{x}}H_{n,m}(P_k)(\underline{x}) = C(n,m,k)H_{n,m}(P_k)(\underline{x})$$

or

$$\partial_{\underline{x}}^2 H_{n,m}(P_k)(\underline{x}) - \underline{x}\partial_{\underline{x}}H_{n,m}(P_k)(\underline{x}) - C(n,m,k)H_{n,m}(P_k)(\underline{x}) = 0.$$

Remarks.

(1) The above differential equation should be compared with the classical differential equation for the Hermite polynomials on the real line [see e.g. [28]) :

$$\frac{d^2}{dx^2}H_n(x) - x\frac{d}{dx}H_n(x) + nH_n(x) = 0.$$

(2) The Dirac derivative of a Hermite polynomial is given by

$$\partial_{\underline{x}} H_{n,m}(P_k)(\underline{x}) = -C(n,m,k)H_{n-1,m}(P_k)(\underline{x}).$$

(3) The constant $C(n,m,k)$ is given by

$$C(n,m,k) = \begin{cases} n & \text{if } n \text{ is even} \\ (n+m+2k-1) & \text{if } n \text{ is odd} \end{cases}$$

Henceforth we restrict ourselves to the case $V = C$.

Theorem 5. *The set* $\{H_{n,m}(P_k^{(i)})(\underline{x})n, k \in \mathbf{N},\ i \leq K(m-1;k)\}$ *is an orthogonal basis for* $L_2(\mathbf{R}^m; e^{\underline{x}^2/2})$.

Let us now calculate the normalizing constants

$$\gamma_{n,m,k} = (H_{n,m}(P_k), H_{n,m}(P_k)).$$

We have

Theorem 6.

(i) For $n = 2p$

$$\gamma_{2p,m,k} = \frac{2^{2p+\frac{m}{2}+k} p! \pi^{\frac{m}{2}} \Gamma(\frac{m}{2}+k+p)}{\Gamma(\frac{m}{2})}$$

(ii) For $n = 2p+1$

$$\gamma_{2p+1,m,k} = \frac{2^{2p+\frac{m}{2}+k+1} p! \pi^{\frac{m}{2}} \Gamma(\frac{m}{2}+k+p+1)}{\Gamma(\frac{m}{2})}$$

Proof. First notice that for $n > 1$

$$\begin{aligned}\gamma_{n,m,k}C(n,m,k) &= (D_+\partial_{\underline{x}}H_{n,m}(P_k), H_{n,m}(P_k)) \\ &= (\partial_{\underline{x}}H_{n,m}(P_k), \partial_{\underline{x}}H_{n,m}(P_k)) \\ &= (C(n,m,k))^2(H_{n-1,m}(P_k), H_{n-1,m}(P_k)) \\ &= (C(n,m,k))^2\gamma_{n-1,m,k}.\end{aligned}$$

Furthermore

$$\begin{aligned}\gamma_{0,m,k} &= \int_0^{+\infty} r^{m-1+2k}e^{-r^2/2}A_m dr \\ &= 2^{\frac{m}{2}+k-1}A_m\int_0^{+\infty} v^{\frac{m}{2}+k-1}e^{-v}dv \\ &= 2^{\frac{m}{2}+k}\frac{\pi^{m/2}}{\Gamma(\frac{m}{2})}\Gamma(\frac{m}{2}+k).\end{aligned}$$

By iteration we get the desired results. ■

Putting

$$T^i_{n,m,k}(\underline{x}) = \frac{H_{n,m}(P_k^{(i)})(\underline{x})}{\sqrt{\gamma_{n,m,k}}}$$

we thus obtain

Theorem 6. *The set* $\{T^i_{n,m,k}(\underline{x}) : n, k, i\ in \mathbf{N},\ i \leq K(m-1,k)\}$ *is an orthonormal basis for* $L_2(\mathbf{R}^m; e^{\underline{x}^2/2})$.

Remark. Considering the operator $\partial_{\underline{x}}$ as a linear operator in $L_2(\mathbf{R}^m; e^{\underline{x}^2/2})$ we have, since

$$\partial_{\underline{x}}T^i_{n,m,k}(\underline{x}) = -C(n,m,k)\frac{\sqrt{\gamma_{n-1,m,k}}}{\sqrt{\gamma_{n,m,k}}}T^i_{n-1,m,k}(\underline{x}),$$

that $\partial_{\underline{x}}$ is densely defined.

As furthermore (see Proposition 1 in 1.2.2) for all $f, g \in \mathcal{D}(\mathbf{R}^m; \mathcal{C})$ we have that

$$(\partial_{\underline{x}}f, g) = (f, D_+g)$$

the adjoint of $\partial_{\underline{x}}$ is an extension of D_+, which, by Theorem 2 in 1.2.3, is thus densely defined. Consequently $\partial_{\underline{x}}$ can be extended to a closed densely defined operator, still denoted by $\partial_{\underline{x}}$ (see e.g. [89]). If D_+ still stands for the adjoint of $\partial_{\underline{x}}$, then $D_+\partial_{\underline{x}}$ is a non-negative selfadjoint operator in $L_2(\mathbf{R}^m; e^{\underline{x}^2/2})$ having pure point spectrum $\{C(n,m,k) : n, k \in \mathbf{N}\}$.

1.2.4 The relation between generalized Gegenbauer and Hermite polynomials

Clearly

$$\lim_{\alpha\to+\infty}(1+\frac{\underline{x}^2}{2\alpha})^{\alpha+i}=e^{\underline{x}^2/2}\quad\text{for all } i\in\mathbf{R}$$

and

$$\lim_{\alpha\to+\infty}(1+\frac{\underline{x}^2}{2\alpha})^{-\alpha}=e^{-\underline{x}^2/2}$$

where the convergence is uniform on compacta.

Moreover, taking $j\le n$, we have that, by Rodrigues' Formula for Gegenbauer polynomials,

$$C_{j,m}^{\alpha+n-j}(P_k)(\underline{x})=(1+\underline{x}^2)^{-(\alpha+n-j)}\partial_{\underline{x}}^j((1+\underline{x}^2)^{\alpha+n}P_k(\underline{x}))$$

or

$$\partial_{\underline{x}}^j((1+\underline{x}^2)^{\alpha+n}P_k(\underline{x}))=(1+\underline{x}^2)^{\alpha+n-j}C_{j,m}^{\alpha+n-j}(P_k)(\underline{x}).$$

Substituting $\underline{x}$ by $\underline{x}/\sqrt{2\alpha}$ we thus obtain :

$$\begin{aligned}&(\frac{1}{\sqrt{2\alpha}})^jC_{j,m,k}^{\alpha+n-j}(\underline{x}\sqrt{2\alpha})P_k(\underline{x})\\&=\ (1+(\underline{x}/\sqrt{2\alpha})^2)^{-(\alpha+n-j)}\partial_{\underline{x}}^j((1+(\underline{x}/\sqrt{2\alpha})^2)^{\alpha+n}P_k(\underline{x}))\end{aligned}\tag{1.29}$$

Using the explicit representation of the Gegenbauer polynomials in terms of hypergeometric series, one sees immediately that the coefficient of $\underline{y}^t$ in $C_{j,m,k}^{\alpha+n-j}(\underline{y})$ is a polynomial of degree $(j+t)$ in $\sqrt{\alpha}$.

Hence, uniformly on compacta,

$$\lim_{\alpha\to\infty}(\frac{1}{\sqrt{2\alpha}})^jC_{j,m,k}^{\alpha+n-j}(\underline{x}/\sqrt{2\alpha})=f_j(\underline{x})$$

exists.

Putting $j=n$ and taking the limit of both sides for $\alpha\to+\infty$, (1.29) yields

$$\begin{aligned}f_n(\underline{x})P_k(\underline{x})&=e^{-\underline{x}^2/2}\partial_{\underline{x}}^n(e^{\underline{x}^2/2}P_k(\underline{x}))\\&=(-1)^nH_{n,m}(P_k(\underline{x}))\end{aligned}$$

whence $f_n(\underline{x})=(-1)^nH_{n,m,k}(\underline{x})$.

We may thus conclude with

Theorem 1. *For all $n,k\in\mathbf{N}$ and $\alpha>-1$*

$$\lim_{\alpha\to+\infty}\left(\frac{-1}{\sqrt{2\alpha}}\right)^n C^{\alpha}_{n,m}(P_k)(\underline{x}) = H_{n,m}(P_k)(\underline{x})$$

In terms of the confluent hypergeometric series $F(a, b+d, c; \frac{z}{b})$, we find, by means of Theorem 1, and this by using the limit expression ${}_1F_1(a, c; z) = \lim_{b\to+\infty} F(a, b+d, c; \frac{z}{b})$ $(a, c \neq 0, c$ not a negative integer and d arbitrary) and Theorem 5 in 1.1.4, for the polynomials $H_{n,m,k}(\underline{x})$ that

Theorem 2.

(i) For $n = 2p$ *even*

$$H_{2p,m,k}(\underline{x}) = (-2)^p\left(\frac{m}{2}+k\right)_{p\,1}F_1\left(-p, \frac{m}{2}+k; -\underline{x}^2/2\right)$$

(ii) For $n = 2p+1$ *odd*

$$H_{2p+1,m,k}(\underline{x}) = \underline{x}(-2)^p\left(\frac{m}{2}+k+1\right)_{p\,1}F_1\left(-p, \frac{m}{2}+k+1; -\underline{x}^2/2\right).$$

Corollary. *For all* $p \in \mathbf{N}$,

(i) $H_{2p+1,m,k}(\underline{x}) = \underline{x}H_{2p,m,k+1}(\underline{x})$.

(ii) $H_{2p,m,k}(\underline{x}) = (-2)^p p! L^{\frac{m}{2}+k-1}_p(\underline{x}^2/2)$

(iii) $H_{2p+1,m,k}(\underline{x}) = (-2)^p p! \underline{x} L^{\frac{m}{2}+k}_p(\underline{x}^2/2)$.

The relations (ii) and (iii) in the above Corollary thus express the polynomials $H_{n,m,k}(\underline{x})$ in terms of the classical Laguerre polynomials $L^{\beta}_{[n/2]}$ for appropriate β.

2 The Cauchy-Kovalevska method

The aim of this section is to show how special monogenic functions in $\mathbf{R}^{m+1}$ may be obtained by applying the Cauchy-Kovalevska extension principle to special real analytic functions defined on some SO(m)-invariant open subset $\underline{\Omega} \subset \mathbf{R}^m$.

Throughout this section monogenicity in $\mathbf{R}^{m+1}$ should be considered in terms of the **Cauchy-Riemann** operator.

Let us therefore recall some conventions made in II. 0.1.3 and II. 0.1.4.

The Cauchy-Riemann operator in $\mathbf{R}^{m+1}$ is given by

$$D_x = \partial_{x_0} + \sum_{i=1}^{m} \varepsilon_i \partial_{x_i}$$

where $\varepsilon_i, i = 1, \ldots, m$, - although being bivectors in $\mathbf{C}_{m+1}$ - are regarded as vectors generating $\mathbf{R}^{0,m}$.

An element $x \in \mathbf{R}^{m+1}$ is then identified with $x = (x_0, \underline{x}) = x_0 + \underline{x} = x_0 + \sum_{i=1}^m x_i \varepsilon_i$ and $\underline{x} \in \mathbf{R}^m$ with $\underline{x} = \sum_{i=1}^m x_i \varepsilon_i$.

In what follows all functions under consideration are $\mathbf{C}_m$-valued. This is again in complete accordance with the convention made that in the case of the Weyl equation, we may take as an appropriate Clifford algebra C the algebra $C \cong \mathbf{C}_m$ generated by $\varepsilon_1, \ldots, \varepsilon_m$.

2.1 The CK-extension principle

Let $\underline{\Omega} \subset \mathbf{R}^m$ be an SO(m)-invariant open set and let $f(\underline{x})$ be a real analytic C-valued function in $\underline{\Omega}$. Then, with respect to the Cauchy-Riemann operator $D_x = \partial_{x_0} + \partial_{\underline{x}}$, $f(\underline{x})$ admits a CK-extension $f(x_0, \underline{x})$ given by (see 0.2.4)

$$f(x_0, \underline{x}) = \exp(-x_0 \partial_{\underline{x}}) f(\underline{x})$$

where $f(x_0, \underline{x})$ is monogenic in a suitable open SO(m)-invariant neighbourhood Ω of $\underline{\Omega}$ in $\mathbf{R}^{m+1}$

Notice that, by introducing spherical coordinates $\underline{x} = \rho\underline{\omega}$, the assumptions made upon $f(\underline{x})$ and $\underline{\Omega}$ lead to a decomposition of $f(\underline{x})$ in a series of the form

$$f(\underline{x}) = \sum_{k=0}^{\infty} (A_{k,\underline{\omega}}(\rho) + \underline{\omega} B_{k,\underline{\omega}}(\rho))$$

where for ρ fixed, $A_{k,\underline{\omega}}(\rho)$ and $B_{k,\underline{\omega}}(\rho)$ belong to $\mathcal{M}^+(m; k; V)$, i.e., with respect to the Dirac operator $\partial_{\underline{x}}$ in $\mathbf{R}^m$, they are inner spherical monogenics of degree k. The CK-extension of $f(\underline{x})$ then admits a decomposition of the form

$$f(x_0, \underline{x}) = \sum_{k=0}^{\infty} (A_{k,\underline{\omega}}(x_0, \rho) + \underline{\omega} B_{k,\underline{\omega}}(x_0, \rho))$$

where clearly $A_{k,\underline{\omega}}(x_0,\rho)$ and $B_{k,\underline{\omega}}(x_0,\rho)$ satisfy the initial conditions

$$\text{(I)}\ A_{k,\underline{\omega}}(x_0,\rho)|_{x_0=0} = A_{k,\underline{\omega}}(\rho);\ B_{k,\underline{\omega}}(x_0,\rho)|_{x_0=0} = B_{k,\underline{\omega}}(\rho)$$

and the plane elliptic system (see also I. 2.4.2)

$$\text{(II)} \left\{ \begin{array}{rcl} \partial_{x_0} A_{k,\underline{\omega}} - \partial_\rho B_{k,\underline{\omega}} & = & \frac{k+m-1}{\rho} B_{k,\underline{\omega}} \\ \partial_{x_0} B_{k,\underline{\omega}} + \partial_\rho A_{k,\underline{\omega}} & = & \frac{k}{\rho} A_{k,\underline{\omega}}. \end{array} \right.$$

Note that this system is similar to the system in I. 2.4.2 of Laurent equations for axially monogenic functions. Hence functions of the form $A_{k,\underline{\omega}} + \underline{\omega} B_{k,\underline{\omega}}$ satisfying the system (II) are also called *axially monogenics of degree k*.

In the sequel we restrict ourselves to solutions $A_{k,\underline{\omega}}(x_0,\rho)$ and $B_{k,\underline{\omega}}(x_0,\rho)$ to (I) and (II) of the form

$$A_{k,\underline{\omega}}(x_0,\rho) = A(x_0,\rho) P_k(\underline{\omega})$$

and

$$B_{k,\underline{\omega}}(x_0,\rho) = B(x_0,\rho) P_k(\underline{\omega})$$

where $A(x_0,\rho)$ and $B(x_0,\rho)$ are scalar valued and $P_k \in \mathcal{M}^+(m;k;V)$.

Notice that in view of the finite dimensionality of the space of inner spherical monogenics, this is no real restriction. Indeed, we may always decompose an expression of the form $A_{k,\underline{\omega}}(\rho) + \underline{\omega} B_{k,\underline{\omega}}(\rho)$ into a finite sum of functions of the form

$$(A(\rho) + \underline{\omega} B(\rho)) P_k(\underline{\omega}).$$

In order to obtain the CK-extension of a given real analytic function in an annular domain of $\mathbf{R}^m$, we may thus restrict ourselves to solving the following problem :

Given two real analytic functions $A(\rho)$ and $B(\rho)$, find two functions $A(x_0,\rho)$ and $B(x_0,\rho)$ satisfying the conditions (I)' and (II)' with

$$\text{(I)'}\ A(0,\rho) = A(\rho);\ B(0,\rho) = B(\rho)$$

and

$$\text{(II)'} \left\{ \begin{array}{rcl} \partial_{x_0} A - \partial_\rho B & = & \frac{k+m-1}{\rho} B \\ \partial_{x_0} B + \partial_\rho A & = & \frac{k}{\rho} A \end{array} \right.$$

To solve this problem, it may sometimes be convenient to look for solutions of the form

$$A(x_0,\rho) = \sum_{\ell=0}^{\infty} x_0^\ell A_\ell(\rho),\ B(x_0,\rho) = \sum_{\ell=0}^{\infty} x_0^\ell B_\ell(\rho).$$

The equations (II)' then lead to the recurrence relations

$$(\text{III}) \begin{cases} A_{\ell+1} &= \frac{1}{\ell+1}(B'_\ell + \frac{k+m-1}{\rho} B_\ell) \\ B_{\ell+1} &= \frac{1}{\ell+1}(-A'_\ell + \frac{k}{\rho} A_\ell) \end{cases}$$

which have to be solved taking into account the initial conditions

$$(\text{IV}) \ A_0(\rho) = A(\rho); B_0(\rho) = B(\rho).$$

2.2 The CK-extension of $\underline{x}^s P_k(\underline{x})(s \in \mathbf{N})$

Let $f(\underline{x})$ be a $\mathcal{C}$-valued real analytic function near the origin. Then, in a suitable open ball centred at the origin, $f(\underline{x})$ admits a decomposition of the form

$$f(\underline{x}) = \sum_{k=0}^{\infty} R_k(\underline{x})$$

where for each $k \in \mathbf{N}$, $R_k(\underline{x}) \in \mathcal{P}(m; k; \mathcal{C})$, i.e. R_k is a homogeneous $\mathcal{C}$-valued polynomial of degree k in $x_1, \ldots, x_m$.

So, homogeneous polynomials being building blocks for real analytic functions in $\mathbf{R}^m$, it is of fundamental importance to know their CK-extension and that is exactly what was done in Ch I when introducing inner spherical monogenics (see I. 1.2 (3)).

But, as we know from I. 1.10, each $R_k(\underline{x}) \in \mathcal{P}(m; k; \mathcal{C})$ admits the Fischer decomposition

$$R_k(\underline{x}) = \sum_{s=0}^{k} \underline{x}^s P_{k-s}(\underline{x})$$

where $P_{k-s}(\underline{x}) \in M^+(m; k; \mathcal{C})$.

Hence functions of the form $\underline{x}^s P_k(\underline{x})$ are building blocks for homogeneous polynomials in $\mathbf{R}^m$. Consequently, in order to characterize spaces of inner spherical monogenics in $\mathbf{R}^{m+1}$, it suffices to determine the CK-extension of polynomials of the form $\underline{x}^s P_k(\underline{x})$.

Theorem 2.2.1. *Let* $s \in \mathbf{N}$ *and* $P_k \in M^+(k; m; \mathcal{C})$. *Then the* CK*-extension of* $\underline{x}^s P_k(\underline{x})$ *has the form* $X_k^s(x_0, \underline{x}) P_k(\underline{x})$ *where*

$$X_k^s(x_0, \underline{x}) = \lambda_k^s r^s \left(C_s^{\frac{m-1}{2}+k}\left(\frac{x_0}{r}\right) + \frac{2k+m-1}{s+2k+m-1} C_{s-1}^{\frac{m+1}{2}+k}\left(\frac{x_0}{r}\right)\frac{\underline{x}}{r}\right)$$

Here

$$r^2 = x_0^2 + |\underline{x}|^2$$

and

$$\lambda_k^{2\ell} = (-1)^\ell (C_{2\ell}^{\frac{m-1}{2}+k}(0))^{-1}; \lambda_k^{2\ell+1} = (-1)^\ell \frac{2\ell+2k+m}{2k+m-1}(C_{2\ell}^{\frac{m+1}{2}+k}(0))^{-1}.$$

Proof. First notice that the CK-extension of $\underline{x}^s P_k(\underline{x})$ has the form

$$(A(x_0,\rho) + \underline{\omega} B(x_0,\rho))P_k(\underline{\omega}).$$

Now we prove we may take $A(x_0,\rho) = \rho^k a(x_0,\rho)$ and $B(x_0,\rho) = \rho^k b(x_0,\rho)$ where $\rho^k a(x_0,\rho)$ and $\rho^k b(x_0,\rho)$ satisfy the equation (II)'.

Notice here that, as the CK-extension of $\underline{x}^s P_k(\underline{x})$ is a homogeneous polynomial of degree $(s+k)$ in $\mathbf{R}^{m+1}$, we thus have that $a(x_0,\rho)$ and $b(x_0,\rho)$ should be homogeneous of degree s in $\mathbf{R}^{m+1}$.

Introducing polar coordinates $x_0 = r\cos\theta$ and $|\underline{x}| = \rho\sin\theta$ where $r^2 = x_0^2 + |\underline{x}|^2$ and $\theta \in [0,\pi]$, and putting $a(x_0,\rho)$ and $b(x_0,\rho)$ in the form

$$a(x_0,\rho) = r^s A(\cos\theta); b(x_0,\rho) = r^s \sin\theta B(\cos\theta),$$

we are thus looking for solutions to (II)' of the form

$$r^{s+k}(\sin^k\theta A(\cos\theta) + \underline{\omega}\sin^{k+1}\theta B(\cos\theta)).$$

Using the relations

$$r\partial_r = x_0\partial_{x_0} + \rho\partial_\rho \quad , \quad \partial_\theta = x_0\partial_\rho - \rho\partial_{x_0},$$

and putting $t = \cos\theta$, (II)' leads to the system

$$\begin{cases} (1-t^2)B' - (2k+m)tB &= sA \\ A' &= (2k+s+m-1)B \end{cases}$$

whence, eliminating B, we are led to the equation

$$(1-t^2)A'' - (2k+m)tA' - s(2k+s+m-1)A = 0.$$

Consequently

$$A(\cos\theta) = \lambda_k^s C_s^{\frac{m-1}{2}+k}(\cos\theta)$$

and

$$B(\cos\theta) = \lambda_k^s \frac{2k+m-1}{s+2k+m-1} C_{s-1}^{\frac{m+1}{2}+k}(\cos\theta).$$

The CK-extension of $\underline{x}^s P_k(\underline{x})$ may thus be written as

$$\lambda_k^s r^s (C_s^{\frac{m-1}{2}+k}(\frac{x_0}{r}) + \frac{2k+m-1}{s+2k+m-1} C_{s-1}^{\frac{m+1}{2}+k}(\frac{x_0}{r}) P_k(\underline{x}) \tag{2.1}$$

The scalar λ_k^s is here completely determined by the initial conditions (I)' which read as follows in our case.

Putting $x_0 = 0$ in (2.1) and taking into account that a Gegenbauer polynomial of odd index is zero at $t = 0$, we thus obtain that

(i) for $s = 2\ell$ even

$$\lambda_k^{2\ell} r^{2\ell} C_{2\ell}^{\frac{m-1}{2}+k}(0) = \underline{x}^{2\ell} = (-1)^\ell r^{2\ell}$$

(ii) for $s = 2\ell + 1$ odd

$$\lambda_k^{2\ell+1} r^{2\ell+1} \frac{2k+m-1}{2\ell+2k+m} C_{2\ell}^{\frac{m+1}{2}+k}(0)\frac{\underline{x}}{r} = \underline{x}^{2\ell+1}.$$

From these relations the desired values of λ_k^s indeed follow. ∎

Remarks 2.2.2.

(1) In view of [28] we have that

$$(C_{2\ell}^{\nu}(0))^{-1} = (-1)^\ell \frac{\ell!\Gamma(\nu)}{\Gamma(\nu+\ell)}.$$

Hence

$$\lambda_k^{2\ell} = \frac{\ell!\Gamma(k+\frac{m-1}{2})}{\Gamma(\ell+k+\frac{m-1}{2})}$$

and

$$\lambda_k^{2\ell+1} = \frac{2\ell+2k+m}{2k+m-1}\frac{\ell!\Gamma(k+\frac{m+1}{2})}{\Gamma(\ell+k+\frac{m+1}{2})}$$

(2) The polynomials $X_k^s(x_0,\underline{x})P_k(\underline{x})$ being the CK-extension of $\underline{x}^s P_k(\underline{x})$, where $P_k(\underline{x}) \in M^+(m;k;C)$ we clearly have that they are homogeneous monogenic polynomials of degree $(s+k)$ in $\mathbf{R}^{m+1}$. The space of such polynomials (resp. the space of their restrictions to S^m) is denoted by $X_k^s M^+(m;k;C)$ (resp. $X_k^s \mathcal{M}^+(m;k;C)$). We thus have that

$$X_k^s M^+(m;k;C) \subset M^+(m+1;s+k;C)$$

and

$$X_k^s \mathcal{M}^+(m;k;C) \subset \mathcal{M}^+(m+1;s+k;C).$$

(3) As

$$C_s^{\nu}(t) = \sum_{j=0}^{[s/2]} \frac{(-1)^j(\nu)_{s-j}}{j!(s-2j)!}(2t)^{s-2j}$$

it follows that

$$X_k^s(x) = \lambda_k^s r^s (C_s^{\frac{m-1}{2}+k}(\frac{x_0}{r}) + \frac{2k+m-1}{s+2k+m-1} C_{s-1}^{\frac{m+1}{2}+k}(\frac{x_0}{r})\frac{\underline{x}}{r})$$

is indeed a homogeneous polynomial of degree s in $(x_0, x_1, \ldots, x_m)$.

(4) Putting $k = 0$ we have that

$$X_0^s(x) = \lambda_0^s r^s (C_s^{\frac{m-1}{2}}(\frac{x_0}{r}) + \frac{m-1}{s+m-1} C_{s-1}^{\frac{m+1}{2}}(\frac{x_0}{r})\frac{\underline{x}}{r})$$

is the CK-extension of $\underline{x}^s$.

Theorem 2.2.3. *Let* $k \in \mathbb{N}$. *Then*

(i) $M^+(m+1;k;\mathcal{C}) = \oplus_{s=0}^k X_{k-s}^s M^+(m;k-s;\mathcal{C})$

(ii) $\mathcal{M}^+(m+1;k;\mathcal{C}) = \oplus_{s=0}^k X_{k-s}^s \mathcal{M}^+(m;k-s;\mathcal{C})$

the latter sum being orthogonal in $L_2(S^m;\mathcal{C})$.

Proof.

(i) As by definition $M^+(m+1;k;\mathcal{C}) = CK\mathcal{P}(m;k;\mathcal{C})$ and by the Fischer decomposition

$$\mathcal{P}(m;k;\mathcal{C}) = \sum_{s=0}^{k} \oplus \underline{x}^s M^+(m;k-s;\mathcal{C}),$$

we obtain that

$$\begin{aligned} M^+(m+1;k;\mathcal{C}) &= \sum_{s=0}^{k} \oplus CK(\underline{x}^s M^+(m;k-s;\mathcal{C})) \\ &= \sum_{s=0}^{k} \oplus X_{k-s}^s M^+(m;k-s;\mathcal{C}). \end{aligned}$$

(ii) It only remains to prove that the sum is orthogonal in $L_2(S^m;\mathcal{C})$. To this end it is sufficient to show that for $s \neq s'$, $X_{k-s}^s P_{k-s} \perp X_{k-s'}^{s'} P_{k-s'}$.

Now taking into account that on S^m,

$$\overline{X_{k-s}^s} X_{k-s'}^{s'} = \alpha + \underline{\omega}\beta$$

for some scalar valued functions α and β, where α and β are independent of $\underline{\omega} \in S^{m-1}$, we have :

$$\begin{aligned}
&(X^s_{k-s}P_{k-s}, X^{s'}_{k-s'}, P_{k-s'}) \\
&= \int_{S^m} \overline{P}_{k-s}\overline{X}^s_{k-s}X^{s'}_{k-s'}P_{k-s'}d\Sigma(\omega) \\
&= \int_{S^m} \overline{P}_{k-s}(\alpha + \underline{\omega}\beta)P_{k-s'}d\Sigma(\omega).
\end{aligned}$$

Putting

$$\omega = \cos\theta + \sin\theta\underline{\omega}, \quad \underline{\omega} \in S^{m-1}, \quad \theta \in [-\pi, \pi],$$

we find that

$$\begin{aligned}
&(X^s_{k-s}P_{k-s}, X^{s'}_{k-s'}P_{k-s'}) \\
&= \int_{-\pi}^{\pi} \sin^{m-1}\theta d\theta \left[\int_{S^{m-1}} \overline{P}_{k-s}(\sin\theta\underline{\omega})(\alpha + \underline{\omega}\beta)P_{k-s'}(\sin\theta\underline{\omega})d\Sigma(\underline{\omega})\right] \\
&= 0. \blacksquare
\end{aligned}$$

Theorem 2.2.4. (Abel's Lemma) *Let $f(\underline{x})$ be a real analytic C-valued function in the unit ball of $\mathbf{R}^m$ which is of the form $f(\underline{x}) = F(\underline{x})P_k(\underline{x})$ where $k \in \mathbf{N}$ is fixed and $F(\underline{x}) = \sum_{s=0}^{\infty} c_s\underline{x}^s$ for $|\underline{x}| < 1$ with $c_s \in \mathbf{C}$ for all $s \in \mathbf{N}$. Then the CK-extension of $f(\underline{x})$ is defined in the unit ball of $\mathbf{R}^{m+1}$ and is given there by*

$$f(x_0, \underline{x}) = \sum_{s=0}^{\infty} c_s X^s_k(x_0, \underline{x})P_k(\underline{x}).$$

Proof. As for $|t| < 1$,

$$|C^{\nu}_s(t)| \leq \frac{\Gamma(2\nu + s)}{s!\Gamma(2\nu)},$$

it easily follows from Stirling's inequality that for some constants α_k, depending on k,

$$|X^s_k(x_0, \underline{x}) \leq r^s(1 + s)^{\alpha_k}, \quad s \in \mathbf{N}. \blacksquare$$

2.3 Generalized power functions

The aim of this subsection is to introduce in Clifford analysis functions which are similar to the generalized power functions z^α in complex analysis. As z^α may be defined as the holomorphic extension of x^α, $x \in]0,+\infty[$,it seems natural to replace x^α on $]0,+\infty[$ by some kind of power function $\underline{x}^\alpha$, $\underline{x} \in \mathbf{R}^m\backslash\{0\}$ or, more generally, by functions of the form $\underline{x}^\alpha P_k(\underline{x})$ which for $\alpha = s \in \mathbf{N}$ reduce to $\underline{x}^s P_k(\underline{x})$ studied in 2.2.

The first problem of course consists in defining $\underline{x}^\alpha$ - and hence $\underline{x}^\alpha P_k(\underline{x})$ - in a convenient way ; the second problem is to determine the CK-extension of $\underline{x}^\alpha P_k(\underline{x})$.

I. The function $\underline{x}^\alpha$

Let $\alpha \in \mathbf{R}$. For a function $p_\alpha(\underline{x})$ that could play the role of $\underline{x}^\alpha$ it seems obvious to impose the following conditions :

$$(\mathrm{P}.1) : p_\alpha(\underline{x})p_\beta(\underline{x}) = p_{\alpha+\beta}(\underline{x})$$

$$(\mathrm{P}.2) : p_s(\underline{x}) = \underline{x}^s, \quad s \in \mathbf{Z}.$$

To this end first notice that, as $\underline{x} = |\underline{x}|\underline{\omega}$, $\underline{\omega} \in S^{m-1}$, where $\underline{\omega}^2 = -1$, the function p_α^1 with

$$\begin{aligned} p_\alpha^1(\underline{x}) &= |\underline{x}|^\alpha \exp\left(\frac{\alpha\pi}{2}\underline{\omega}\right) \\ &= \rho^\alpha\left(\cos\frac{\alpha\pi}{2} + \underline{\omega}\sin\frac{\alpha\pi}{2}\right) \end{aligned} \tag{2.2}$$

clearly satisfies the conditions (P.1) and (P.2).

But, as we may also write $\underline{x}$ as $\underline{x} = (-|\underline{x}|)(-\underline{\omega})$ where again $(-\underline{\omega})^2 = -1$, we might as well consider the function p_α^2 with

$$\begin{aligned} p_\alpha^2(\underline{x}) &= (-|\underline{x}|)^\alpha \exp\left(-\underline{\omega}\frac{\alpha\pi}{2}\right) \\ &= \rho^\alpha e^{i\pi\alpha}\left(\cos\frac{\alpha\pi}{2} - \underline{\omega}\sin\frac{\alpha\pi}{2}\right) \end{aligned} \tag{2.3}$$

and, as is easily seen, p_α^2 also satisfies the conditions (P.1) and (P.2).

Notice here that both expressions (2.2) and (2.3) may in fact be written into the form

$$p_\alpha(\underline{x}) = A_\alpha(\rho) + \underline{\omega}B_\alpha(\rho)$$

where A_α and B_α are scalar valued radial functions.

It thus seems natural to look for solutions p_α to (P.1) and (P.2) satisfying the extra condition

$$(\text{P.3}) : p_\alpha(\underline{x}) = A_\alpha(\rho) + \underline{\omega} B_\alpha(\rho)$$

with A_α and B_α C-valued radial functions.
Defining the functions Ω_+ and Ω_- with

$$\Omega_\pm = \frac{1}{2}(1 \mp i\underline{\omega}), \quad \underline{\omega} \in S^{m-1},$$

we have that on S^{m-1}

$$\begin{aligned} \Omega_+ + \Omega_- &= 1 \\ \Omega_+\Omega_- = \Omega_-\Omega_+ &= 0. \end{aligned}$$

So, introducing the C-valued radial functions $p_\alpha^\pm$ with

$$p_\alpha^\pm(\underline{x}) = A_\alpha(\rho) \pm i B_\alpha(\rho),$$

the function p_α may written in the following "complexified" form :

$$p_\alpha = \Omega_+ p_\alpha^+ + \Omega_- p_\alpha^-.$$

Moreover, for any P_α and p_β satisfying (P.3), we then have that

$$p_\alpha p_\beta = \Omega_+ p_\alpha^+ p_\beta^+ + \Omega_- p_\alpha^- p_\beta^-.$$

Hence the condition (P.1) leads to the scalar conditions

$$p_\alpha^\pm p_\beta^\pm = p_{\alpha+\beta}^\pm \tag{2.4}$$

while, for $S \in \mathbf{Z}$, in view of $\underline{x}^s = \Omega_+(i\rho)^s + \Omega_-(-i\rho)^s$, the condition (P.2) now reads

$$p_s^\pm(\underline{x}) = (\pm i\rho)^s. \tag{2.5}$$

From (2.4) and (2.5) it thus follows that

$$p_\alpha^\pm(\underline{x}) = \rho^\alpha(\exp(2j \pm \frac{1}{2})\pi i\alpha)$$

whence the general solution satisfying the conditions (P.1), (P.2) and (P.3) is given by

$$p_\alpha(\underline{x}) = \rho^\alpha e^{i\ell\pi\alpha}(\cos(\ell' + \frac{1}{2})\pi\alpha + \underline{\omega}\sin(\ell' + \frac{1}{2})\pi\alpha) \tag{2.6}$$

where $\ell \in \mathbf{Z}$ and $\ell' = \ell + 2j$, $j \in \mathbf{Z}$.

Observing that the function p_α defined by (2.6) is **C**-valued and that it becomes $\mathbf{R}_{0,m}$-valued if and only if $\ell = 0$ or, equivalently, $\ell' = 2j$, $j \in \mathbf{Z}$, we are finally led to consider the function $p_\alpha^{(j)}$ with

$$\begin{aligned} p_\alpha^{(j)}(\underline{x}) &= \rho^\alpha(\cos(2j+\frac{1}{2})\pi\alpha + \underline{\omega}\sin(2j+\frac{1}{2})\pi\alpha) \\ &= \rho^\alpha \exp\left((2j+\frac{1}{2})\pi\underline{\omega}\alpha\right) \\ &= (\exp(2j\pi\underline{\omega}\alpha))(\rho^\alpha \exp(\frac{\pi\alpha}{2}\underline{\omega})) \\ &= \exp(2j\pi\underline{\omega}\alpha)p_\alpha^1(\underline{x}). \end{aligned}$$

Henceforth we put

$$\begin{aligned} \underline{x}^\alpha &= p_\alpha^1(\underline{x}) \\ &= \rho^\alpha(\cos\frac{\pi\alpha}{2} + \underline{\omega}\sin\frac{\pi\alpha}{2}). \end{aligned} \tag{2.7}$$

whence

$$p_\alpha^{(j)}\underline{x} = (\exp(2j\pi\underline{\omega}\alpha))\underline{x}^\alpha.$$

In the sequel we also put

$$(-\underline{x})^\alpha = \rho^\alpha(\cos\frac{\pi\alpha}{2} - \underline{\omega}\sin\frac{\pi\alpha}{2}) \tag{2.8}$$

thus obtaining a special solution when the condition (P.2) is replaced by

$$\text{(P.2)'} : p_s(\underline{x}) = (-\underline{x})^s, \quad s \in \mathbf{Z}.$$

II. The CK-extension of $\underline{x}^\alpha P_k(\underline{x})$

Let again $\alpha \in \mathbf{R}$ and let $P_k \in M^+(m;k;C)$.

We now wish to determine the CK-extension of $\underline{x}^\alpha P_k(\underline{x})$, denoted by $X_k^\alpha(x)P_k(\underline{x})$, i.e.

$$\begin{cases} (\partial_{x_0} + \partial_{\underline{x}})(X_k^\alpha(x_0,\underline{x})P_k(\underline{x}) = 0 \\ X_k^\alpha(0,\underline{x}) = \underline{x}^\alpha. \end{cases}$$

By virtue of (2.7) we have that

$$\underline{x}^\alpha P_k(\underline{x}) = \rho^{\alpha+k}(\cos\frac{\pi\alpha}{2} + \underline{\omega}\sin\frac{\pi\alpha}{2})P_k(\underline{\omega}).$$

If thus suffices to determine the CK-extension of functions of the form $\rho^\alpha P_k(\underline{\omega})$ and $\rho^\alpha \underline{\omega} P_k(\underline{\omega})$, called, respectively, *inner* and *outer power functions*.

II.1 Inner power functions

The initial condition (I)' being here

$$A(0,\rho) = A_0(\rho) = \rho^\alpha$$

$$B(0,\rho) = B_0(\rho) = 0$$

we find, when trying a solution to (II)' of the form

$$A(x_0,\rho) = \sum_{\ell=0}^{\infty} x_0^\ell A_\ell(\rho), B(x_0,\rho) = \sum_{\ell=0}^{\infty} x_0^\ell B_\ell(\rho)$$

that the recurrence relations (III) become for $\ell \in \mathbf{N}$:

$$A_{2\ell+1}(\rho) = B_{2\ell}(\rho) = 0$$

and

$$A_{2\ell}(\rho) = \frac{(-1)^\ell 2^{2\ell}}{(2\ell)!} \frac{\Gamma(\ell+1-\frac{k+m+\alpha}{2})\Gamma(\ell+\frac{k-\alpha}{2})}{\Gamma(1-\frac{k+m+\alpha}{2})\Gamma(\frac{k-\alpha}{2})} \rho^{\alpha-2\ell},$$

$$B_{2\ell+1}(\rho) = \frac{(-1)^\ell 2^{2\ell+1}}{(2\ell+1)!} \frac{\Gamma(\ell+1-\frac{k+m+\alpha}{2})\Gamma(\ell+1+\frac{k-\alpha}{2})}{\Gamma(1-\frac{k+m+\alpha}{2})\Gamma(\frac{k-\alpha}{2})} \rho^{\alpha-2\ell-1}$$

where it is assumed that $1-\frac{k+m+\alpha}{2}$ and $\frac{k-\alpha}{2}$ are not belonging to $\{0,-1,-2,\ldots\}$.

The solution to (II)' subjected to the initial condition (I)' may then be expressed in terms of the hypergeometric function as

$$A(x_0,\rho) = \rho^\alpha F(1-\frac{k+m+\alpha}{2}; \frac{k-\alpha}{2}; \frac{1}{2}; -\frac{x_0^2}{\rho^2})$$

and

$$B(x_0,\rho) = (k-\alpha)x_0\rho^{\alpha-1} F(1-\frac{k+m+\alpha}{2}; \frac{k-\alpha}{2}+1; \frac{3}{2}; -\frac{x_0^2}{\rho^2}).$$

Hence the function

$$p_{\alpha,k,m}(x_0,\underline{x})P_k(\underline{\omega}) = (A(x_0,\rho) + \underline{\omega}B(x_0,\rho))P_k(\underline{\omega}) \tag{2.9}$$

is defined at least for $|x_0| < \rho$ and it there determines the CK-extension of $\rho^\alpha P_k(\underline{\omega})$.

It should be remarked that the restrictions $1-\frac{k+m+\alpha}{2}, \frac{k-\alpha}{2} \notin \mathbf{Z}\backslash\mathbf{N}_0$ may be easily removed.

Indeed, $\frac{k-\alpha}{2} \in \mathbf{Z}\backslash\mathbf{N}_0$ means that $\alpha = 2s+k$ for $s \in \mathbf{N}$ and in this case we already have that

$$p_{2s+k,k,m}(x)P_k(\underline{\omega}) = X_k^{2s}(x)P_k(\underline{x}),$$

i.e. the hypergeometric series breaks off.

If $1 - \frac{k+m+\alpha}{2} \in \mathbf{Z}\backslash\mathbf{N}_0$ we have that $\alpha = 2s + 2 - k - m$.

In this case the hypergeometric series breaks off to a polynomial of degree $2s+1$ in x_0 but it seldom becomes a polynomial in $(x_0, x_1, \ldots, x_m)$. This only happens for m even and $\alpha \geq k$, but this case was already treated in section 2.2.

II.2 Outer power functions

The initial condition (I)' being here

$$A(0,\rho) = A_0(\rho) = 0$$

$$B(0,\rho) = B_0(\rho) = \rho^\alpha$$

we find, when trying a solution to (II)' of the form

$$A(x_0,\rho) = \sum_{\ell=0}^{\infty} x_0^\ell A_\ell(\rho) \quad , \quad B(x_0,\rho) = \sum_{\ell=0}^{\infty} x_0^\ell B_\ell(\rho)$$

that the recurrence relations (III) become for $\ell \in \mathbf{N}$.

$$A_{2\ell}(\rho) = B_{2\ell+1}(\rho) = 0$$

and

$$B_{2\ell}(\rho) = \frac{(-1)^\ell 2^{2\ell}}{(2\ell)!}\frac{\Gamma(\ell + \frac{1-\alpha-k-m}{2})\Gamma(\ell + \frac{k-\alpha+1}{2})}{\Gamma(\frac{1-\alpha-k-m}{2})\Gamma(\frac{k-\alpha+1}{2})}\rho^{\alpha-2\ell},$$

$$A_{2\ell+1}(\rho) = \frac{(-1)^{\ell+1} 2^{2\ell+1}}{(2\ell+1)!}\frac{\Gamma(\ell + 1 + \frac{1-\alpha-k-m}{2})\Gamma(\ell + \frac{k-\alpha+1}{2})}{\Gamma(\frac{1-\alpha-k-m}{2})\Gamma(\frac{k-\alpha+1}{2})}\rho^{\alpha-2\ell-1}$$

where it is assumed that $\frac{1-\alpha-k-m}{2}, \frac{k-\alpha+1}{2} \notin \{0,-1,-2,\ldots\}$.

The solution to (II)' subjected to the initial condition (I)' may then be expressed in terms of the hypergeometric function as

$$B(x_0,\rho) = \rho^\alpha F(\frac{k-\alpha+1}{2}; \frac{1-\alpha-k-m}{2}; \frac{1}{2}; -\frac{x_0^2}{\rho^2})$$

and

$$A(x_0,\rho) = (\alpha+k+m-1)x_0\rho^{\alpha-1} F(\frac{k-\alpha+1}{2}; \frac{1-\alpha-k-m}{2}; \frac{3}{2}; -\frac{x_0^2}{\rho^2})$$

whence the function

$$q_{\alpha,k,m}(x_0,\underline{x})P_k(\underline{\omega}) = (A(x_0,\rho)+\underline{\omega}B(x_0,\rho))P_k(\underline{\omega}) \qquad (2.10)$$

is defined at least for $|x_0| < \rho$ and it there determines the CK-extension of $\rho^\alpha \underline{\omega} P_k(\underline{\omega})$.

Again it should be remarked that the restrictions

$$\frac{1-\alpha-k-m}{2}, \frac{k-\alpha+1}{2} \notin \{0,-1,-2,\ldots\}$$

may be easily removed.

Indeed, if $\alpha = 2s+k+1$, $s \in \mathbf{N}$, the hypergeometric series breaks off and we get

$$q_{2s+k+1.k.m}(x)P_k(\underline{\omega}) = X_k^{2s+1}(x)P_k(\underline{x}).$$

If $\alpha = 2s+1-k-m$, $s \in \mathbf{N}$, the hypergeometric series breaks off to a polynomial of degree $2s$ in x_0, but it only becomes a polynomial in $(x_0, x_1, \ldots, x_m)$ if m is even and $\alpha \geq k+1$, a case which was treated in section 2.2.

The discussion held may thus be summarized as follows.

Let for $\alpha \in \mathbf{R}$, $\underline{x}^\alpha$ be defined by (2.7), i.e.

$$\underline{x}^\alpha = \rho^\alpha(\cos\frac{\pi\alpha}{2} + \underline{\omega}\sin\frac{\pi\alpha}{2}),$$

let $P_k(\underline{x}) \in M^+(m;k;C)$ and let $p_{\alpha,k,m}(x)$ and $q_{\alpha,k,m}(x)$ be determined by (2.9) and (2.10) respectively. Then we have

Theorem 2.3.1. *The CK-extension of $\underline{x}^\alpha P_k(\underline{x})$ is given by the function $X_k^\alpha(x)P_k(\underline{x})$ where*

$$X_k^\alpha(x)P_k(\underline{x}) = (p_{\alpha+k,k,m}(x)\cos\frac{\pi\alpha}{2} + q_{\alpha+k,k,m}(x)\sin\frac{\pi\alpha}{2})P_k(\underline{\omega})$$

Remarks 2.3.2.

(1) The function $X_k^\alpha(x)P_k(\underline{x})$ is called a *generalized power function.*

(2) The functions $p_{\alpha,k,m}(x)P_k(\underline{\omega})$ and $q_{\alpha,k,m}(x)P_k(\underline{\omega})$ defined in (2.8) and (2.9) respectively, are clearly monogenic and homogeneous of degree α. Moreover, as their A and B components satisfy the Vekua system (II)', these functions are in fact axial monogenics of degree k. We therefore call them *axially monogenic generalized powers of degree k.*

(3) Introducing the inversion $\mathbf{J}$ with

$$\mathbf{J}f(x) = \frac{\overline{x}}{|x|^{m+1}} f(-\frac{\overline{x}}{|x|^2})$$

we have that **J** transforms monogenic functions into monogenic functions (see also I. 0.2.5)

Putting

$$\mathbf{J}(p_{\alpha,k,m}(x)P_k(\underline{\omega})) = \tilde{p}_{\alpha,k,m}(x)P_k(\underline{\omega})$$

and

$$\mathbf{J}(q_{\alpha,k,m}(x)P_k(\underline{\omega})) = \tilde{q}_{\alpha,k,m}(x)P_k(\underline{\omega})$$

we have , in view of the homogenicity of degree α of the functions involved, that

$$\tilde{p}_{\alpha,k,m}(x) = \frac{\overline{x}}{|x|^{2\alpha+m+1}} p_{\alpha,k,m}(-\overline{x})$$

and

$$\tilde{q}_{\alpha,k,m}(x) = \frac{\overline{x}}{|x|^{2\alpha+m+1}} q_{\alpha,k,m}(-\overline{x}).$$

From these relations we obtain that

$$\tilde{p}_{\alpha,k,m}(\underline{x}) = -\underline{\omega}\rho^{-m-\alpha}$$

and

$$\tilde{q}_{\alpha,k,m}(\underline{x}) = \rho^{-m-\alpha}.$$

By virtue of the uniqueness of monogenic extensions we thus find

$$\begin{aligned} q_{-m-\alpha,k,m}(x) &= -\tilde{p}_{\alpha,k,m}(x) = \tfrac{-\overline{x}}{|x|^{2\alpha+m+1}} p_{\alpha,k,m}(-\overline{x}) \\ \text{and} \quad p_{-m-\alpha,k,m}(x) &= \tilde{q}_{\alpha,k,m}(x) = \tfrac{\overline{x}}{|x|^{2\alpha+m+1}} q_{\alpha,k,m}(-\overline{x}) \end{aligned} \tag{2.11}$$

(4) In (2.8) we introduced the function

$$(-\underline{x})^\alpha = \rho^\alpha(\cos\frac{\pi\alpha}{2} - \underline{\omega}\sin\frac{\pi\alpha}{2})$$

whence we could as well look for the CK-extension of $(-\underline{x})^\alpha P_k(\underline{x})$.

To this end notice first that, if $F(x) = A(x_0,\rho) + \underline{\omega}B(x_0,\rho)$ satisfies the Vekua system (II)', then so does $\overline{F}(-\overline{x}) = A(-x_0,\rho) - \underline{\omega}B(-x_0,\rho)$.

Applying this remark to the function

$$F(x) = |x|^k X_k^\alpha(x)$$

we thus get that $\overline{X_k^\alpha}(-\overline{x})P_k(\underline{x})$ is monogenic.

As the restriction of the latter function to $\mathbf{R}^m$ yields $(-\underline{x})^\alpha P_k(\underline{x})$, we so proved

$$\overline{X_k^\alpha}(-\overline{x})P_k(\underline{x}) = CK((-\underline{x})^\alpha P_k(\underline{x})).$$

We may thus conclude that one single function, namely X_k^α, is sufficient to determine the CK-extensions of both $\underline{x}^\alpha P_k(\underline{x})$ and $(-\underline{x})^\alpha P_k(\underline{x})$.

(5) Let $s \in \mathbf{N}$. Then, as we saw in Theorem 2.2.3, the functions $p_{s,k,m}(x)P_k(\underline{\omega})$, $s = 2\ell+k$ and $q_{s,k,m}(x)P_k(\underline{\omega})$, $s = (2\ell+1)+k$, are inner spherical monogenics of degree s in $\mathbf{R}^{m+1}$. Moreover, when restricted to S^m, they are orthogonal in $L_2(S^m;\mathcal{C})$ and they generate the space $\mathcal{M}^+(m+1;s;\mathcal{C})$. Consequently, by inversion, it follows that the functions $q_{-m-s,k,m}(x)P_k(\underline{\omega})$, $s = 2\ell+k$ and $p_{-m-s,k,m}(x)P_k(\underline{\omega})$, $s = (2\ell+1)+k$, generate the space $M^-(m+1;s;\mathcal{C})$ and, when restricted to S^m, thus generate the space $\mathcal{M}^-(m+1;s;\mathcal{C})$ and are orthogonal in $L_2(S^m;\mathcal{C})$.

(6) The CK-extension $X_k^\alpha(x)P_k(\underline{x}$ of $\underline{x}^\alpha P_k(\underline{x})$ appearing in Theorem 2.3.1 is given in terms of the hypergeometric function.

It is conceptually nicer to use simply the expression mentioned in I. 0.2.13, namely

$$X_k^\alpha(x)P_k(\underline{x}) = \exp(-x_0\partial_{\underline{x}})(\underline{x}^\alpha P_k(\underline{x})).$$

Analogously

$$\overline{X_k^\alpha}(-\overline{x})P_k(\underline{x}) = \exp(-x_0\partial_{\underline{x}})((-\underline{x})^\alpha P_k(\underline{x})).$$

To evaluate these expressions, first notice that

$$\begin{aligned}\partial_{\underline{x}}(\underline{x}^\alpha P_k(\underline{x})) &= -\alpha\underline{x}^{\alpha-1}P_k(\underline{x}) - (2k+m-1)\cos\frac{\pi}{2}(\alpha-1)\rho^{\alpha-1}P_k(\underline{x})\\ &= \left[-\alpha\underline{x}^{\alpha-1} - (k+\frac{m-1}{2})(\underline{x}^{\alpha-1}+(-\underline{x})^{\alpha-1})\right]P_k(\underline{x})\end{aligned}$$

and, analogously,

$$\partial_{\underline{x}}((-\underline{x})^\alpha P_k(\underline{x})) = \left[\alpha(-\underline{x})^{\alpha-1} + (k+\frac{m-1}{2})(\underline{x}^{\alpha-1}+(-\underline{x})^{\alpha-1}\right]P_k(\underline{x})$$

whence

$$\partial_{\underline{x}}\left[\begin{pmatrix}\underline{x}^\alpha\\(-\underline{x})^\alpha\end{pmatrix}P_k(\underline{x})\right] = \begin{pmatrix}-(\alpha+k+\frac{m-1}{2}) & -(k+\frac{m-1}{2})\\ k+\frac{m-1}{2} & \alpha+k+\frac{m-1}{2}\end{pmatrix}\begin{pmatrix}\underline{x}^{\alpha-1}\\(-\underline{x})^{\alpha-1}\end{pmatrix}P_k(\underline{x})$$

or, putting $I = \begin{pmatrix}1 & \\ & -1\end{pmatrix}$ and $J = \begin{pmatrix} & 1\\ -1 & \end{pmatrix}$, we get

$$\begin{aligned}\partial_{\underline{x}}[(I\underline{x})^\alpha P_k(\underline{x})] &= -\left[(\alpha+k+\frac{m-1}{2})I + (k+\frac{m-1}{2})J\right](I\underline{x})^{\alpha-1}P_k(\underline{x})\\ &= -G_k(\alpha)(I\underline{x})^{\alpha-1}P_k(\underline{x})\end{aligned}$$

where $G_k(\alpha) = \alpha I + (k+\frac{m-1}{2})(I+J)$.

By iteration we obtain

$$\partial_{\underline{x}}^j \left[(I\underline{x})^\alpha P_k(\underline{x})\right] = (-1)^j G_k(\alpha)G_k(\alpha-1)\dots G_k(\alpha-j+1)(I\underline{x})^{\alpha-j}P_k(\underline{x})$$

whence

$$\begin{aligned} & CK((I\underline{x})^\alpha P_k(\underline{x})) \\ &= \exp(-x_0\partial_{\underline{x}})\left[(I\underline{x})^\alpha P_k(\underline{x})\right] \\ &= \sum_{j=0}^{\infty} \frac{G_k(\alpha)\dots G_k(\alpha-j+1)}{j!} x_0^j (I\underline{x})^{\alpha-j} P_k(\underline{x}). \end{aligned} \tag{2.12}$$

The expression (2.12) shows how the generalized power functions $X_k^\alpha(x)P_k(\underline{x})$ and $\overline{X}_k^\alpha(-\overline{x})P_k(\underline{x})$ may be written out in terms of $\underline{x}^\alpha$ and $(-\underline{x})^\alpha$.

Although $\underline{x}^\alpha$ and $(-\underline{x})^\alpha$ are very important from the conceptual point of view since they generalize the function x^α, $x \in]0,+\infty[$, it reveals to be easier for computation to work directly with the CK-extensions $p_{\alpha+k,k,m}(x)P_k(\underline{\omega})$ and $q_{\alpha+k,k,m}(x)P_k(\underline{\omega})$ of, respectively, $\rho^{\alpha+k}P_k(\underline{\omega})$ and $\rho^{\alpha+k}\underline{\omega}P_k(\underline{\omega})$.

Let us finish this subsection by showing how, given $\alpha \in \mathbf{R}$ fixed, the set V_α determined by

$$V_\alpha = \left\{p_{\alpha,k,m}(x)P_k(\underline{\omega}), q_{\alpha,k,m}(x)P_k(\underline{\omega}) : k \in \mathbf{N},\ P_k \in \mathcal{M}^+(m;k;\mathcal{C})\right\}$$

generates all monogenic functions homogeneous of degree α.

Theorem 2.3.3. *Let $f(\underline{x})$ be a $\mathcal{C}$-valued real analytic function in $\mathbf{R}^m\backslash\{0\}$ which is homogeneous of degree α, i.e. $f(\lambda\underline{x}) = \lambda^\alpha f(\underline{x})$, $\lambda \in \mathbf{R}^+$. Then the CK-extension $f(x_0,\underline{x})$ of f is expressible as a series of functions, all of them belonging to V_α.*

Proof. The function $f(\underline{x})$ may be decomposed as a s eries of the form

$$f(\underline{x}) = \sum_{k=0}^{\infty}\sum_{i=1}^{K(m;k)} (A_k^{(i)}(\rho) + \underline{\omega}B_k^{(i)}(\rho))P_k^{(i)}(\underline{\omega})$$

where for each $k \in \mathbf{N}$, $\{P_k^{(i)}(\underline{\omega}) : i = 1,\dots,K(m;k)\}$ is an orthonormal basis of $\mathcal{M}^+(m+1;k;\mathcal{C})$.

As the homogenicity condition $f(\lambda\underline{x}) = \lambda^\alpha f(\underline{x})$ is then clearly inherited by the functions $A_k^{(i)}(\rho)$ and $B_k^{(i)}(\rho)$, the latter have to be of the form $A_k^{(i)}(\rho) = \rho^\alpha a_k^{(i)}$ and $B_k^{(i)}(\rho) = \rho^\alpha b_k^{(i)}$ for some suitable scalars $a_k^{(i)}$, $b_k^{(i)}$.

Applying the CK-extension to both sides of (2.13) then yields the desired result.

■

2.4 Axially exponential functions

In this subsection we introduce so-called *axially exponential functions of degree k*. These are by definition axially monogenic functions of degree k of the form

$$\mathcal{E}_k(x_0, \underline{x})P_k(\underline{x}) = (A(x_0,\rho) + \underline{\omega}B(x_0,\rho))P_k(\underline{\omega})$$

satisfying the conditions

$$A(x_0,\rho) = e^{x_0}a(\rho);\, B(x_0,\rho) = e^{x_0}b(\rho)$$

The Vekua system (II)' then leads to the system of differential equations

$$\begin{cases} a(\rho) - b'(\rho) &= \frac{k+m-1}{\rho}b(\rho) \\ b(\rho) + a'(\rho) &= \frac{k}{\rho}a(\rho) \end{cases}.$$

Eliminating b from this system leads to Bessel's equation

$$a''(\rho) + \frac{m-1}{\rho}a'(\rho) + (1 - \frac{k(k+m-2)}{\rho^2})a(\rho) = 0$$

which has solutions of the form

$$a(\rho) = C\rho^{1-\frac{m}{2}}J_{k+\frac{m}{2}-1}(\rho).$$

Notice that $a(\rho)$ is of the form $a(\rho) = \rho^k A(\rho)$ as expected where

$$A(0) = C\frac{2^{1-k-\frac{m}{2}}}{\Gamma(k+\frac{m}{2})}.$$

Imposing the initial condition $A(0) = 1$ we get

$$a(\rho) = 2^{k+\frac{m}{2}-1}\Gamma(k+\frac{m}{2})\rho^{1-\frac{m}{2}}J_{k+\frac{m}{2}-1}(\rho).$$

From the above system we so obtain

$$b(\rho) = 2^{k+\frac{m}{2}-1}\Gamma(k+\frac{m}{2})\rho^{1-\frac{m}{2}}J_{k+\frac{m}{2}}(\rho)$$

whence finally, putting $\nu = k + \frac{m}{2}$,

$$\begin{aligned} \mathcal{E}_k(x) &= 2^{\nu-1}\Gamma(v)e^{x_0}\rho^{1-\nu}(J_{\nu-1}(\rho) + \underline{\omega}J_\nu(\rho)) \\ &= e^{x_0}\Gamma(\nu)\sum_{\ell=0}^{\infty}\frac{(\underline{x}/2)^{2\ell}}{\ell!\Gamma(\nu+\ell)}(1 + \frac{\underline{x}}{2(\nu+\ell)}). \end{aligned}$$

Notice that at $\underline{x} = 0$,

$$\mathcal{E}_k(x_0, 0) = e^{x_0}.$$

Furthermore from

$$\partial_{x_0}(\mathcal{E}_k(x_0, \underline{x})P_k(\underline{x})) = \mathcal{E}_k(x_0, \underline{x})P_k(\underline{x})$$

it follows that, as $(\partial_{x_0} + \partial_{\underline{x}})(\mathcal{E}_k(x_0, \underline{x})P_k(\underline{x})) = 0$,

$$(\partial_{\underline{x}} + 1)\mathcal{E}_k(\underline{x})P_k(\underline{x}) = 0$$

and hence for $\lambda \in \mathbf{C}$,

$$(\partial_{\underline{x}} - \lambda)\mathcal{E}_k(-\lambda\underline{x})P_k(\underline{x}) = 0.$$

This means that $\mathcal{E}_k(-\lambda\underline{x})P_k(\underline{x})$ is a global eigenfunction of the Dirac operator $\partial_{\underline{x}}$ in $\mathbf{R}^m$ corresponding to the eigenvalue $\lambda \in \mathbf{C}\backslash\{0\}$.

It thus follows that all real analytic eigenfunctions $f_\lambda(\underline{x})$ of $\partial_{\underline{x}}$ corresponding to the eigenvalue $\lambda \in \mathbf{C}\backslash\{0\}$ have the form

$$f_\lambda(\underline{x}) = \sum_{k=0}^{\infty} \mathcal{E}_k(-\lambda\underline{x})P_k(\underline{x}).$$

Indeed, it suffices to decompose $f_\lambda(\underline{x})$ into a series of spherical monogenics.

For further details concerning the function theory of the operator

$$\partial_{\underline{x}} - \lambda,\ \lambda \in \mathbf{C}\backslash\{0\},$$

we refer the reader to [93].

2.5 Hermite polynomials revisited

In (1.28) we introduced the Hermite polynomials $H_{n,m}(P_k)(\underline{x})$ by

$$H_{n,m}(P_k)(\underline{x}) = ((-D_+)^n P_k)(\underline{x})$$

where $D_+ = \underline{x} - \partial_{\underline{x}}$ and $P_k(\underline{x}) \in M^+(m; k; C)$.

It was shown that

$$H_{n,m}(P_k)(\underline{x}) = H_{n,m,k}(\underline{x})P_k(\underline{x})$$

where $H_{n,m,k}(\underline{x}) \in {}_n\mathbf{C}[\underline{x}]$ has real coefficients depending on k (not on $P_k(\underline{x})$).

In this section the Hermite polynomials are introduced by means of a generating function which is the CK-extension of the C-valued real analytic function

$$G_k(\underline{x}) = e^{\underline{x}^2/2}P_k(\underline{x}).$$

Notice that for $k = 0$ this function turns into $G_0(\underline{x}) = e^{-|\underline{x}|^2/2}$, which is the Gauss-distribution in $\mathbf{R}^m$.

Although the main properties of the Hermite polynomials $H_{n,m}(P_k)(\underline{x})$ are already stated in section 1.2.1, we reformulate some of them here, giving at the same time more detailed proofs.

Further notice that the case $k = 0$ is of particular interest since, as will be seen, it shows quite a lot of similarity with the way the Hermite polynomials H_n may be defined on the real line by the generating function

$$e^{(x+it)^2/2} = e^{x^2/2} \sum_{k=0}^{\infty} \frac{H_n(ix)t^n}{n!}$$

In the case $k = 0$ the polynomials thus obtained are called ***radial Hermite polynomials*** while for $k \neq 0$, they are called *(general) Hermite polynomials.*

I. Radial Hermite polynomials

Consider the radial function $G_0(\underline{x}) = e^{\underline{x}^2/2} = e^{-|\underline{x}|^2/2}$ and let

$$G_0(x_0, \underline{x}) = \exp(-x_0 \partial_{\underline{x}}) G_0(\underline{x})$$

be its CK-extension. Then putting

$$G_0(x_0, \underline{x}) = e^{\underline{x}^2/2} \sum_{n=0}^{\infty} \frac{x_0^n}{n!} H_{n,m}(\underline{x})$$

we claim that the functions $H_{n,m}(\underline{x})$ are polynomials in $\underline{x}$ of degree n having real coefficients.

Indeed, from the equation $(\partial_{x_0} + \partial_{\underline{x}}) G_0(x_0, \underline{x}) = 0$ we immediately obtain the recurrence formula

$$H_{n+1,m}(\underline{x}) = (\underline{x} - \partial_{\underline{x}}) H_{n,m}(\underline{x}) \tag{2.13}$$

whence, since $H_{0,m}(\underline{x}) = 1$, $H_{n,m}(\underline{x}) \in {}_n\mathbf{C}[x]$.

Comparing the relation (2.13) with Theorem 2 (ii) in 1.2.1, we find that $H_{n,m}(\underline{x})$ is nothing else but the Hermite polynomial $H_{n,m,0}(\underline{x})$ introduced in section 1.2.

The first radial Hermite polynomials are thus given by

$$\begin{aligned} H_{0,m}(\underline{x}) &= 1 \\ H_{1,m}(\underline{x}) &= \underline{x} \\ H_{2,m}(\underline{x}) &= \underline{x}^2 + m \\ H_{3,m}(\underline{x}) &= \underline{x}^3 + (m+2)\underline{x} \end{aligned}$$

etc ...

The relation (2.13) again shows that $H_{2p,m}(\underline{x})$ contains only even powers of $\underline{x}$ while $H_{2p+1,m}(\underline{x})$ contains only odd ones.

We also have that $H_{n,m}(\underline{x})$ is determined by the Rodrigues' Formula

$$H_{n,m}(\underline{x}) = e^{-\underline{x}^2/2}(-\partial_{\underline{x}})^n e^{\underline{x}^2/2}. \tag{2.14}$$

Indeed, putting

$$f_n(\underline{x}) = e^{-\underline{x}^2/2}(-\partial_{\underline{x}})^n e^{\underline{x}^2/2}$$

we clearly have that $f_0(\underline{x}) = 1$ and $f_{n+1}(\underline{x}) = (\underline{x} - \partial_{\underline{x}})f_n(\underline{x})$, thus defining the sequence $(f_n)_{n\in\mathbf{N}}$ completely with $f_n(\underline{x}) = H_{n,m}(\underline{x})$.

Furthermore the polynomials $H_{n,m}(\underline{x})$ satisfy the orthogonality relations

$$\int_{\mathbf{R}^m} \overline{H}_{n',m}(\underline{x})H_{n,m}(\underline{x})e^{\underline{x}^2/2}d\underline{x} = 0, \quad n \neq n'. \tag{2.15}$$

To prove this, it clearly suffices to show that for $n' < n$,

$$\int_{\mathbf{R}^m} \underline{x}^{n'} H_{n,m}(\underline{x})e^{\underline{x}^2/2}d\underline{x} = 0. \tag{2.16}$$

Rodrigues' Formula (2.14) together with Cauchy's Theorem yield the identity

$$\int_{\mathbf{R}^m} \underline{x}^{n'} H_{n,m}(\underline{x})e^{\underline{x}^2/2}d\underline{x} = \int_{\mathbf{R}^m} (\underline{x}^{n'}\partial_{\underline{x}} H_{n-1,m}(\underline{x})d\underline{x}.$$

Now for $n' = 0$, $(\underline{x}^{n'})\partial_{\underline{x}} = 0$, while in general $(\underline{x}^{n'})\partial_{\underline{x}}$ is proportional to $\underline{x}^{n'-1}$ (see also II. 1.12.1, Corollary 1 to Proposition 3).

The orthogonality relation (2.16) thus follows by induction on n'.

Next we wish to establish the analogue of the differential equation

$$H_n'' + (-x)H_n' + nH_n = 0$$

valid for the classical Hermite polynomials on $\mathbf{R}$.

To this end first notice that for each $n \in \mathbf{N}$

$$\overline{H}_{n,m}(\underline{x})H_{n,m}(\underline{x}) = (-1)^n H_{n,m}^2(\underline{x}) > 0$$

whence $\alpha_{n,m} > 0$ where

$$\alpha_{n,m} = (-1)^n \int_{\mathbf{R}^m} H_{n,m}^2(\underline{x})e^{\underline{x}^2/2}d\underline{x}.$$

As

$$\begin{aligned} \partial_{\underline{x}}(e^{\underline{x}^2/2}\partial_{\underline{x}} H_{n,m}(\underline{x})) &= e^{\underline{x}^2/2}(\partial_{\underline{x}}^2 - \underline{x}\partial_{\underline{x}})H_{n,m}(\underline{x}) \\ &= p_n(\underline{x})e^{\underline{x}^2/2} \end{aligned} \tag{2.17}$$

for some $p_n \in {}_n\mathbf{C}[\underline{x}]$ having real coefficients, there ought to exist real constants c_s, $s = 0, \ldots, n$, such that

$$(\partial_{\underline{x}}^2 - \underline{x}\partial_{\underline{x}})H_{n,m}(\underline{x}) = \sum_{s=0}^{n} c_s H_{s,m}(\underline{x}). \tag{2.18}$$

We here used the fact that $(H_{s,m}(\underline{x}))_{s=0}^n$, being a sequence of mutually orthogonal polynomials, spans ${}_n\,\mathbf{C}[x]$.

From (2.15), (2.17) and (2.18) it thus follows that for each $j < n$,

$$\begin{aligned} c_j \int_{\mathbf{R}^m} \overline{H}_{j,m}(\underline{x}) e^{\underline{x}^2/2} H_{j,m}(\underline{x}) d\underline{x} &= \int_{\mathbf{R}^m} \overline{H}_{j,m}(\underline{x}) e^{\underline{x}^2/2} (\partial_{\underline{x}}^2 - \underline{x}\partial_{\underline{x}}) H_{n,m}(\underline{x}) d\underline{x} \\ &= \int_{\mathbf{R}^m} \overline{H}_{j,m}(\underline{x}) p_n(\underline{x}) e^{\underline{x}^2/2} d\underline{x} \\ &= 0 \end{aligned}$$

whence

$$(\partial_{\underline{x}}^2 - \underline{x}\partial_{\underline{x}}) H_{n,m}(\underline{x}) = c_n H_{n,m}(\underline{x}). \tag{2.19}$$

Let us now determine the constant c_n.

To this end first recall that $H_{2\ell,m}(\underline{x})$ has the form $f(\rho)$ with $f(\rho) = \sum_{s=0}^{\ell} A_{2s}\rho^{2s}$, while $H_{2\ell+1,m}(\underline{x})$ has the form $\underline{x}g(\rho)$ with $g(\rho) = \sum_{s=0}^{\ell} B_{2s}\rho^{2s}$.

Now the equation

$$(\partial_{\underline{x}}^2 - \underline{x}\partial_{\underline{x}} - \lambda) f(\rho) = 0$$

leads to the differential equation

$$f'' + (\frac{m-1}{\rho} - \rho) f' + \lambda f = 0 \tag{2.20}$$

and hence to

$$(2s+2)(2s+m)A_{2s+2} = (2s-\lambda)A_{2s}$$

so that polynomial solutions of degree 2ℓ only exist for $\lambda = 2\ell$.

Consequently $c_{2\ell} = 2\ell$, while furthermore it follows from (2.20) that

$$f(\rho) = C L_{\ell}^{\frac{m}{2}-1)}(\frac{\rho^2}{2}) >$$

As to the equation

$$(\partial_{\underline{x}}^2 - \underline{x}\partial_{\underline{x}} - \lambda)(\underline{x}g(\rho)) = 0$$

using the identities (see also II 1.2.13)

$$\partial_{\underline{x}}(\underline{x}g(\rho)) = -\rho g'(\rho) - m g(\rho)$$

and

$$\partial_{\underline{x}}^2(\underline{x}g(\rho)) = -\frac{\underline{x}}{\rho}(g'(\rho) + \rho g''(\rho) + m g'(\rho))$$

we are led to the equation

$$g'' + (\frac{m+1}{\rho} - \rho)g' + (\lambda - m)g = 0 \tag{2.21}$$

Polynomial solutions of degree 2ℓ to the equation (2.21) only exist for $\lambda = 2\ell + m$ whence $c_{2\ell+1} = 2\ell + m$.

Moreover (2.20) then implies that $g(\rho) = C' L_\ell^{\frac{m}{2}}(\frac{\rho^2}{2})$

Now comparing the coefficient of $\underline{x}^n$ in both sides of (2.19) finally leads us to

$$H_{2\ell,m}(\underline{x}) = (-2)^\ell \ell! L_\ell^{\frac{m}{2}-1}(\frac{\rho^2}{2})$$

and

$$H_{2\ell+1,m}(\underline{x}) = (-2)^\ell \ell! \underline{x} L_\ell^{\frac{m}{2}}(\frac{\rho^2}{2}).$$

Hence

$$G_0(x_0, \underline{x}) = e^{\underline{x}^2/2} \sum_{\ell=0}^{\infty} \frac{(-2x_0^2)^\ell}{(2\ell)!} \ell! \left[L_\ell^{\frac{m}{2}-1}(\frac{\rho^2}{2}) + \frac{x_0 \underline{x}}{2\ell+1} L_\ell^{\frac{m}{2}}(\frac{\rho^2}{2}) \right]$$

thus determining the CK-extension of the Gauss distribution $G_0(\underline{x}) = e^{\underline{x}^2/2}$ explicitly in terms of the classical Laguerre polynomials.

II. General Hermite polynomials

Let $P_k(\underline{x}) \in M^+(m; k; C)$ and let $G_k(\underline{x}) = e^{\underline{x}^2/2} P_k(\underline{x})$ be the corresponding generalized Gauss distribution, the CK-extension of which is given by $G_k(x_0, \underline{x})$.

Putting

$$G_k(x_0, \underline{x}) = e^{\underline{x}^2/2} \sum_{n=0}^{\infty} \frac{x_0^n}{n!} H_{n,m,k}(\underline{x}) P_k(\underline{x})$$

we claim that the functions $H_{n,m,k}(\underline{x})$ are polynomials in $\underline{x}$ of degree n, having real coefficients depending on k.

Indeed, from the equation $(\partial_{x_0} + \partial_{\underline{x}})G_k(x_0, \underline{x}) = 0$ we can immediately derive the recurrence relation

$$H_{n+1,m,k}(\underline{x}) P_k(\underline{x}) = (\underline{x} - \partial_{\underline{x}})(H_{n,m,k}(\underline{x}) P_k(\underline{x})) \tag{2.22}$$

whence

$$\begin{aligned} H_{2\ell+1,m,k}(\underline{x}) &= (\underline{x} - \partial_{\underline{x}}) H_{2\ell,m,k}(\underline{x}) \\ \text{and} \qquad & \\ H_{2\ell+2,m,k}(\underline{x}) &= (\underline{x} + \tfrac{2k}{\underline{x}} - \partial_{\underline{x}}) H_{2\ell+1,m,k}(\underline{x}). \end{aligned} \tag{2.23}$$

But, as $H_{0,m,k}(\underline{x}) = 1$, we thus obtain from the relation (2.23) that

$$H_{n,m,k}(\underline{x}) \in_n \mathbf{C}[\underline{x}].$$

Moreover, comparing the relation (2.22) with Theorem 2 (ii) in 1.2.1, we may conclude that the polynomials $H_{n,m,k}(\underline{x})P_k(\underline{x})$ are indeed nothing else but the (general) Hermite polynomials $H_{n,m}(P_k)(\underline{x})$ introduced in 1.2.

Notice also that

$$\begin{aligned} H_{0,m,k}(\underline{x}) &= 1 \\ H_{1,m,k}(\underline{x}) &= \underline{x} \\ H_{2,m,k}(\underline{x}) &= \underline{x}^2 + (2k+m) \\ H_{3,m,k}(\underline{x}) &= \underline{x}^3 + (2k+m+2)\underline{x} \\ &\text{etc ..} \end{aligned}$$

thus showing that for all $n, k \in \mathbf{N}$,

$$H_{n,m,k}(\underline{x}) = H_{n,m+2k}(\underline{x}).$$

This implies that is suffices to know the radial Hermite polynomials.

The Rodrigues' Formula

$$H_{n,m,k}(\underline{x})P_k(\underline{x}) = e^{-\underline{x}^2/2}(-\partial_{\underline{x}})^n(e^{\underline{x}^2/2}P_k(\underline{x})$$

leads to the orthogonality relations

$$0 = \int_{\mathbf{R}^m} \overline{P}_{k'}(\underline{x})\overline{H}_{n',m,k'}(\underline{x})H_{n,m,k}(\underline{x})P_k(\underline{x})e^{\underline{x}^2/2}d\underline{x}$$

for $k' \neq k$ or $n' \neq n$.

Finally, let us recall that the functions $H_{n,m,k}(\underline{x})P_k(\underline{x})$ are solutions of the differential equation

$$(\partial_{\underline{x}}^2 - \underline{x}\partial_{\underline{x}} - \mu_n)f(\underline{x}) = 0$$

where $\mu_{2n} = 2n$ and $\mu_{2n+1} = 2(n+k)+m$.

Explicitly we have

$$H_{2\ell,m,k}(\underline{x}) = (-2)^\ell \ell! L_\ell^{k+\frac{m}{2}-1}(\frac{\rho^2}{2})$$

and

$$H_{2\ell+1,m,k}(\underline{x}) = (-2)^\ell \ell! \underline{x} L_\ell^{k+\frac{m}{2}}(\frac{\rho^2}{2})$$

so that

$$\begin{aligned} &CK(e^{\underline{x}^2/2}P_k(\underline{x})) \\ &= e^{\underline{x}^2/2}\sum_{\ell=0}^{\infty}\frac{(-2x_0^2)^\ell}{(2\ell)!}\ell!\left[L_\ell^{k+\frac{m}{2}-1}(\frac{\rho^2}{2}) + \frac{x_0\underline{x}}{2\ell+1}L_\ell^{k+\frac{m}{2}}(\frac{\rho^2}{2})\right]P_k(\underline{x}). \end{aligned}$$

3 Cauchy type integrals

The aim of this section is to generate monogenic functions by means of the generalized Cauchy kernel $\frac{\overline{x}\pm t}{|\overline{x}\pm t|^{m+1+2k}}$ acting on suitable classes of functions $f(t)$ or distributions T_t defined on the half line $[0,+\infty[$.

Throughout this section, monogenicity in $\mathbf{R}^{m+1}$ should again be interpreted in terms of the **Cauchy-Riemann** operator (see also section 2 for the notations used).

3.1 A Cauchy type integral

Let $k \in \mathbf{N}$ and let $P_k \in M^+(m;k;C)$, the latter space being associated to the Dirac operator $\partial_{\underline{x}}$ in $\mathbf{R}^m$. Then the function

$$(x_0, \underline{x}) \rightarrow P_k(\underline{x})$$

is monogenic in $\mathbf{R}^{m+1}$ whence, by applying the inversion operator, the function

$$(x_0, \underline{x}) \rightarrow \frac{x_0 - \underline{x}}{|x_0 - \underline{x}|^{m+1+2k}} P_k(\underline{x}) = \frac{\overline{x}}{|\overline{x}|^{m+1+2k}} P_k(\underline{x})$$

belongs to $M^-(m+1;k;C)$.

Notice that, up to the constant $\frac{1}{A_{m+1}}$, it coincides for $k=0$ with the fundamental solution $E(x)$ of the Cauchy-Riemann operator D_x.

Now let $f(t)$ (resp. T_t) belong to a suitable class of functions (resp. distributions) defined on the half line $[0,+\infty[$. Then we define

$$\Lambda_k^{\pm} f(x_0, \underline{x}) = \frac{\pm 1}{A_{m+1}} \int_0^{+\infty} \frac{\overline{x} \pm t}{|\overline{x} \pm t|^{m+1+2k}} f(t) dt,$$

resp.

$$\Lambda_k^{\pm} T_t(x_0, \underline{x}) = < T_t, \frac{\pm 1}{A_{m+1}} \frac{\overline{x} \pm t}{|\overline{x} \pm t|^{m+1+2k}} > .$$

Then clearly $\Lambda_k^{\pm} f(x_0, \underline{x}) P_k(\underline{x})$ (resp. $\Lambda_k^{\pm} T_t(x_0, \underline{x}) P_k(\underline{x})$) is monogenic in $\mathbf{R}^{m+1} \backslash \mathbf{R}^{\mp}$ and, e.g. for f it is of the form, putting $\underline{x} = \rho \underline{\omega}$,

$$\Lambda_k^{\pm} f(x_0, \underline{x}) = A_k^{\pm} f(x_0, \rho) + \underline{\omega} B_k^{\pm} f(x_0, \rho). \tag{3.1}$$

Assuming f to be differentiable on $[0,+\infty[$ and its derivative f' to belong to a suitable class of functions too, we moreover have that

$$\Lambda_k^{\pm} f' = \mp \partial_{x_0} \Lambda_k^{\pm} f$$

so that, under the assumptions made,

$$\partial_{\underline{x}}(\Lambda_k^{\pm} f P_k(\underline{x})) = (\pm \Lambda_k^{\pm} f'(x_0, \underline{x})) P_k(\underline{x}). \tag{3.2}$$

Using the expression $\partial_{\underline{x}} = \underline{\omega}(\partial_\rho + \frac{1}{\rho}\Gamma_{\underline{\omega}})$ and the relations $\Gamma_{\underline{\omega}} P_k(\underline{\omega}) = (-k)P_k(\underline{\omega})$, $\Gamma_{\underline{\omega}}(\underline{\omega} P_k(\underline{\omega})) = (k+m-1)\underline{\omega} P_k(\underline{\omega})$, it easily follows from (3.1) and (3.2) that

$$\underline{\omega}\partial_\rho \Lambda_k^{\pm} f - \frac{2k+m-1}{\rho} B_k^{\pm} f = \pm \Lambda_k^{\pm} f'. \tag{3.3}$$

Analogous relations may obviously be established when f is replaced by a suitable distribution T.

In particular, observe that, taking $T_t = Y(t)$, the Heaviside function, and putting $k = 0$, (3.2) implies that

$$\begin{aligned}(\partial_{\underline{x}} \Lambda_0^{\pm} Y)(x_0, \underline{x}) &= (\pm \Lambda_0^{\pm} \delta)(x_0, \underline{x}) \\ &= \frac{1}{A_{m+1}} \frac{\overline{x}}{|x|^{m+1}}.\end{aligned}$$

Hence $\Lambda_0^{\pm} Y$ generalizes the classical logarithm on the real line.

3.2 Integral representation for generalized power functions

Let $\alpha \in \mathbf{C}$ be such that $-1 < Re\alpha < 2k + m - 1$.

Then we put

$$\begin{aligned}\Lambda_{\alpha,k}^{\pm}(x_0, \underline{x}) &= \Lambda_k^{\pm}(t^\alpha)(x_0, \underline{x}) \\ &= \frac{\pm 1}{A_{m+1}} \int_0^{+\infty} \frac{(x_0 \pm t - \underline{x}) t^\alpha}{|x_0 \pm t - \underline{x}|^{m+1+2k}} dt \\ &= A_{\alpha,k}^{\pm}(x_0, \rho) + \underline{\omega} B_{\alpha,k}^{\pm}(x_0, \underline{x}).\end{aligned} \tag{3.4}$$

If moreover $Re\alpha + 1 < 2k + m - 1$, we obtain in view of (3.2) the recursion formula

$$\Lambda_{\alpha,k}^{\pm}(x) P_k(\underline{x}) = \pm \frac{1}{\alpha+1} \partial_{\underline{x}} (\Lambda_{\alpha+1,k}^{\pm}(x) P_k(\underline{x}))$$

or, by means of (3.3),

$$\Lambda_{\alpha,k}^{\pm}(x) = \pm \frac{1}{\alpha+1} (\underline{\omega}\partial_\rho \Lambda_{\alpha+1,k}^{\pm} - \frac{2k+m-1}{\rho} B_{\alpha+1,k}^{\pm}) \tag{3.5}$$

Clearly the expression (3.5) may thus be used to extend $\Lambda_{\alpha,k}^{\pm}(x)$ holomorphically for all values of α for which $\alpha \notin \{-1, -2, \ldots,\}$ and $Re\alpha < 2k + m - 1$.

Let us now calculate $\Lambda_{\alpha,k}^{\pm}(x)$.

To this end, first notice that for $-1 < Re\alpha < 2k + m - 1$ and $x_0 = 0$,

$$\begin{aligned}\Lambda_{\alpha,k}^{\pm}(\underline{x}) &= \frac{1}{A_{m+1}} \int_0^{+\infty} \frac{t^{\alpha+1} \mp \underline{x} t^\alpha}{||\underline{x}|^2 + t^2|^{k+\frac{m+1}{2}}} dt \\ &= |\underline{x}|^{\alpha-m+1-2k} (\sigma_{m,k}(\alpha+1) \mp \underline{\omega} \sigma_{m,k}(\alpha))\end{aligned}$$

where

$$\begin{aligned}\sigma_{m,k}(\alpha) &= \frac{1}{A_{m+1}}\int_0^{+\infty}\frac{t^\alpha dt}{(1+t^2)^{k+\frac{m+1}{2}}}\\ &= \frac{1}{2A_{m+1}}B(\frac{2k+m-\alpha}{2},\frac{\alpha+1}{2})\\ &= \sigma_{m,k}(2k+m-1-\alpha).\end{aligned} \tag{3.6}$$

So, after multiplying $\Lambda^{\pm}_{\alpha,k}(\underline{x})$ with $P_k(\underline{x})$ and taking its CK-extension, we find that

$$\begin{aligned}&\Lambda^{\pm}_{\alpha,k}(x)P_k(\underline{x})\\ &= [\sigma_{m,k}(\alpha+1)p_{\alpha-m-k+1,k,m}(x) \mp \sigma_{m,k}(\alpha)q_{\alpha-m-k+1,k,m}(x)]\,P_k(\underline{\omega})\end{aligned} \tag{3.7}$$

where for the definition of $p_{\alpha-m-k+1,k,m}(x)$ and $q_{\alpha-m-k+1,k,m}(x)$ we refer to (2.9) and (2.10) respectively.

From (3.6) and (3.7) it clearly follows that, as holomorphic functions in α, $\sigma_{m,k}(\alpha)$ and $\Lambda^{\pm}_{\alpha,k}(x)P_k(\underline{x})$ have first order poles belonging to, respectively, the sets

$$\{-1,-3,-5,\ldots\}\cup\{2k+m,2k+m+2,\ldots\}$$

and

$$\{-1,-2,\ldots\}\cup\{2k+m-1,2k+m,\ldots\}.$$

Putting

$$\begin{aligned}\sigma_\ell &= \mathrm{Res}(\sigma_{m,k}(\alpha),-2\ell-1)\\ &= \mathrm{Res}(\sigma_{m,k}(\alpha),2(k+\ell)+m)\\ &= \frac{1}{2A_{m+1}}\frac{\Gamma(\frac{2(k+\ell)+m+1}{2})}{\Gamma(\frac{2k+m+1}{2})}\mathrm{Res}(\Gamma,-\ell)\\ &= \frac{1}{2A_{m+1}}\frac{(-1)^\ell}{\ell!}\frac{\Gamma(\frac{2(k+\ell)+m+1)}{2}}{\Gamma(\frac{2k+m+1}{2})},\end{aligned}$$

we thus obtain the following integral representations for the inner and outer power functions :

$$\begin{aligned}\mp\sigma_\ell\, q_{-m-k-2\ell,k,m}(x)P_k(\underline{\omega}) &= \mathrm{Res}(\Lambda^{\pm}_{\alpha,k}(x)P_k(\underline{x}),-2\ell-1);\\ \sigma_\ell\, p_{-m-k-2\ell-1,k,m}(x)P_k(\underline{\omega}) &= \mathrm{Res}(\Lambda^{\pm}_{\alpha,k}(x)P_k(\underline{x}),-2\ell-2);\\ \sigma_\ell\, p_{2\ell+k,k,m}(x)P_k(\underline{\omega}) &= \mathrm{Res}(\Lambda^{\pm}_{\alpha,k}(x)P_k(\underline{x}),m+2k+2\ell-1);\\ \mp\sigma_\ell\, q_{2\ell+k+1,k,m}(x)P_k(\underline{\omega}) &= \mathrm{Res}(\Lambda^{\pm}_{\alpha,k}(x)P_k(\underline{x}),m+2k+2\ell).\end{aligned}$$

Finally notice that by combining the expression (3.7) with the relations obtained in (2.11), we get that

$$\Lambda^{\pm}_{m-2+2k-\alpha,k}(x)P_k(\underline{x}) = \mp\mathbf{J}(\Lambda^{\mp}_{\alpha,k}(x)P_k(\underline{x})).$$

4 PLANE WAVE INTEGRALS

The aim of this section is to investigate two examples of so-called monogenic plane wave integrals, namely the exponential plane wave integral and the generalized power plane wave integrals and to relate them to, respectively, the axially exponential function and the generalized power functions studied in 2.4 and 2.3.

Throughout this section monogenicity should be understood in terms of the **Cauchy-Riemann** operator (see also section 2 for the notations used).

4.1 Monogenic plane waves

The notion of a plane wave, i.e. a function of the scalar product $<\underline{x},\underline{t}>$, $\underline{x},\underline{t} \in \mathbf{R}^m$, was introduced by F. John in [45].

Let g be a **C**-valued plane wave. Then formally, for each $\underline{t} \in \mathbf{R}^m$ fixed, its CK-extension may be written in the form

$$G(x_0,\underline{x},\underline{t}) = g_1(<\underline{x},\underline{t}>, x_0|\underline{t}|) - \frac{\underline{t}}{|\underline{t}|} g_2(<\underline{x},\underline{t}>, x_0|\underline{t}|)$$

where g_1 and g_2 are **C**-valued functions of the two real variables $x = <\underline{x},\underline{t}>$ and $y = x_0|\underline{t}|$ satisfying the Cauchy-Riemann system

$$\left\{ \begin{array}{lcl} \partial_x g_1 - \partial_y g_2 & = & 0 \\ \partial_x g_2 + \partial_y g_1 & = & 0 \end{array} \right. \tag{4.1}$$

Conversely, let g_1 and g_2 be **C**-valued functions in $\tilde{\Omega} \subset \mathbf{C}$ satisfying the Cauchy-Riemann system (4.1). Then for each $\underline{t} \in \mathbf{R}^m$, the function

$$G(x_0,\underline{x},\underline{t}) = g_1(<\underline{x},\underline{t}>, x_0|\underline{t}|) - \frac{\underline{t}}{|\underline{t}|} g_2(<\underline{x},\underline{t}>, x_0|\underline{t}|) \tag{4.2}$$

is clearly monogenic in $\Omega_{\underline{t}} = \{(x_0,\underline{x}) : (<\underline{x},\underline{t}>, x_0|\underline{t}|) \in \tilde{\Omega}\}$.

So CK-extensions of **C**-valued plane waves and pairs of **C**-valued functions (g_1, g_2) satisfying the Cauchy-Riemann system (4.1) are related to each other.

A function of the type (4.2) is henceforth called a *monogenic plane wave.*

In particular, notice that, if f is a holomorphic function in $\tilde{\Omega} \subset \mathbf{C}$, we have that, writing f as $f = f_1 + i f_2$, for each $\underline{t} \in \mathbf{R}^m$, the function

$$F(x_0,\underline{x},\underline{t}) = f_1(<\underline{x},\underline{t}>, x_0|\underline{t}|) - \frac{\underline{t}}{|\underline{t}|} f_2(<\underline{x},\underline{t}>, x_0|\underline{t}|)$$

is a monogenic plane wave.

Now let g be a **C**-valued function on **R** and let h be a $\mathbf{C}_m$-valued function on S^{m-1}. Then we consider integrals of the form

$$\int_{S^{m-1}} g(<\underline{x},\underline{\omega}>) h(\underline{\omega}) d\Sigma(\underline{\omega}) \tag{4.3}$$

To evaluate such integrals, we take the expansion of h into spherical monogenics (see II. 1.3.4)

$$h(\underline{\omega}) = \sum_{k=0}^{\infty}(\mathbf{P}(k)h(\underline{\omega}) + \mathbf{Q}(k)h(\underline{\omega})).$$

Using the relation (see II. 1.9)

$$\mathbf{Q}(k) = -\mathbf{I}^*\mathbf{P}(k)\mathbf{I}^*$$

we thus have that, as $\mathbf{P}(k)h$ and $\mathbf{P}(k)(\underline{\nu}h)$ both belong to $\mathcal{M}^+(m;k;\mathcal{C})$, the expression (4.3) leads us to consider integrals of the form

$$\int_{S^{m-1}} g(<\underline{x},\underline{\omega}>)P_k(\underline{\omega})d\Sigma(\omega)$$

and

$$\int_{S^{m-1}} g(<\underline{x},\underline{\omega}>)\underline{\omega}P_k(\underline{\omega})d\Sigma(\underline{\omega})$$

where $P_k \in \mathcal{M}^+(m;k;\mathcal{C})$.

Observing that $P_k(\underline{\omega})$ and $\underline{\omega}P_k(\underline{\omega})$ are spherical harmonics of degree k and $k+1$, respectively, we so obtain by virtue of the Funk-Hecke Theorem that (see [40])

$$\int_{S^{m-1}} g(<\underline{x},\underline{\omega}>)P_k(\underline{\omega})d\Sigma(\underline{\omega}) = A_{m-1}I_k g(\rho)P_k(\underline{\xi}) \tag{4.4}$$

and

$$\int_{S^{m-1}} g(<\underline{x},\underline{\omega}>)\underline{\omega}P_k(\underline{\omega})d\Sigma(\underline{\omega}) = A_{m-1}I_{k+1}g(\rho)\underline{\xi}P_k(\underline{\xi})$$

where $\underline{x} = \rho\underline{\xi}$ and

$$I_k g(\rho) = \int_{-1}^{1} g(\rho s)P_{k,m}(s)(1-s^2)^{\frac{m-3}{2}}ds,$$

$P_{k,m}(s)$ being the Legendre polynomial of degree k in $\mathbf{R}^m$.

Now returning to a pair of C-valued functions (g_1, g_2) satisfying (4.1) and constructing the associated monogenic function $G(x_0, \underline{x}, \underline{t})$ given by (4.2) we may thus consider integrals of the form

$$\int_{S^{m-1}} G(x_0,\underline{x},\underline{\omega})P_k(\underline{\omega})d\Sigma(\underline{\omega})$$

$$= \int_{S^{m-1}}(g_1(<\underline{x},\underline{\omega}>,x_0) - \underline{\omega}g_2(<\underline{x},\underline{\omega}>,x_0))P_k(\underline{\omega})d\Sigma(\underline{\omega}) \tag{4.5}$$

where again $P_k(\underline{\omega}) \in \mathcal{M}^+(m;l;\mathcal{C})$.

An integral of the type (4.5) is called a *monogenic plane wave integral.* It clearly defines a monogenic function in $\Omega = \bigcap_{\underline{t}\in S^{m-1}}\Omega_{\underline{t}}$ and, in view of the relations (4.4) and (4.5), it is the CK-extension of

$$A_{m-1}(I_k(g_1|_{x_0=0})(\rho) - \underline{\xi}I_{k+1}(g_2|_{x_0=0})(\rho))P_k(\underline{\xi}).$$

Hence it is of the form $(A(x_0,\rho) + \underline{\xi}B(x_0,\rho))P_k(\underline{\xi})$ where A and B satisfy the Vekua system (II)' and so it is an axial monogenic of degree k.

In such a way the Funk-Hecke Theorem establishes a link between solutions of, respectively, the Cauchy-Riemann and the Vekua systems.

4.2 Exponential plane wave integrals

Consider the plane wave $g(<\underline{x},\underline{t}>) = e^{i<\underline{x},\underline{t}>}$. Then its CK-extension has the form (see also II. 0.2.4)

$$\mathrm{Exp}(x_0,\underline{x},\underline{t}) = e^{i<\underline{x},\underline{t}>}(ch(x_0|\underline{t}|) - i\frac{\underline{t}}{|\underline{t}|}sh(x_0|\underline{t}|)), \tag{4.6}$$

i.e. $g_1(x,y) = e^{ix}chy, g_2(x,y) = ie^{ix}shy$.

The function $\mathrm{Exp}(x_0,\underline{x},\underline{t})$ is called the *exponential plane wave.*

Putting

$$\mathcal{E}_\pm(x_0,\underline{x},\underline{t}) = \frac{1}{2}(1 \mp i\frac{\underline{t}}{|\underline{t}|})e^{i<\underline{x},\underline{t}>\mp x_0|\underline{t}|}$$

we may also write (4.6) into the form

$$\mathrm{Exp}(x_0,\underline{x},\underline{t}) = \mathcal{E}_+(x_0,\underline{x},\underline{t}) + \mathcal{E}_-(x_0,\underline{x},\underline{t}).$$

Notice here that $\mathcal{E}_\pm$ is again of the type

$$\mathcal{E}_\pm(x_0,\underline{x},\underline{t}) = g_1^\pm(<\underline{x},\underline{t}>, x_0|\underline{t}|) - \frac{\underline{t}}{|\underline{t}|}g_2^\pm(<\underline{x},\underline{t}>, x_0|\underline{t}|)$$

where $(g_1^\pm, g_2^\pm)$ is a pair of C-valued functions satisfying (4.1), i.e. $\mathcal{E}_\pm$ is a monogenic plane wave.

Consequently, in order to evaluate $\int_{S^{m-1}} \mathrm{Exp}(x_0,\underline{x},\underline{\omega})P_k(\underline{\omega})d\Sigma(\underline{\omega})$ it suffices to evaluate $\int_{S^{m-1}} \mathcal{E}_\pm(x_0,\underline{x},\underline{\omega})P_k(\underline{\omega})d\Sigma(\underline{\omega})$.

But, as

$$\int_{S^{m-1}} \mathcal{E}_-(x_0,\underline{x},\underline{\omega})P_k(\underline{\omega})d\Sigma(\underline{\omega}) = (-1)^k \int_{S^{m-1}} \mathcal{E}_+(-x_0,-\underline{x},\underline{\omega})P_k(\underline{\omega})d\Sigma(\underline{\omega})$$

we are led to calculate $\int_{S^{m-1}} \mathcal{E}_+(x_0,\underline{x},\underline{\omega})P_k(\underline{\omega})d\Sigma(\underline{\omega})$.

A direct application of the Funk-Hecke Theorem yields :

$$\begin{aligned}&\int_{S^{m-1}} \mathcal{E}_+(x_0,\underline{x},\underline{\omega})P_k(\underline{\omega})d\Sigma(\underline{\omega})\\ &= \frac{e^{x_0}}{2}A_{m-1}(I_k(e^{is})(\rho) - i\underline{\xi}I_{k+1}(e^{is})(\rho))P_k(\underline{\xi})\end{aligned}$$

where (see also [40])

$$\begin{aligned}A_{m-1}I_k(e^{is})(\rho) &= A_{m-1}\int_{-1}^{1} e^{i\rho t}(1-t^2)^{\frac{m-1}{2}}P_{k,m}(t)dt\\ &= C\rho^{1-\frac{m}{2}}J_{k+\frac{m}{2}-1}(\rho)\end{aligned}$$

with $C = (2\pi)^{\frac{m}{2}}i^k$.

Hence we find that

$$\int_{S^{m-1}} \mathcal{E}_+(x_0, \underline{x}, \underline{\omega}) P_k(\underline{\omega}) d\Sigma(\underline{\omega}) = C_k \mathcal{E}_k(x_0, \underline{x}) P_k(\underline{x})$$

where $\mathcal{E}_k(x_0, \underline{x}) P_k(\underline{x})$ is the axially exponential function of degree k introduced in section 2.4 and

$$C_k = \frac{\pi^{k+\frac{m}{2}}}{\Gamma(k+\frac{m}{2})} \frac{1}{(-2\pi i)^k}.$$

4.3 Generalized power plane wave integrals

Let $\alpha \in \mathbf{C}$ and consider the (multivalued) holomorphic function

$$z^\alpha = p^\alpha(z) = |z|^\alpha(\cos(\alpha \arg z) + i \sin(\alpha \arg z)).$$

Putting $f_1 = |z|^\alpha \cos(\alpha \arg z)$ and $f_2 = |z|^\alpha \sin(\alpha \arg z))$ we may thus define the (multivalued) monogenic plane wave

$$p^\alpha(o, \underline{x}, \underline{t}) = (<\underline{x}, \underline{t}>^2 + x_0^2|\underline{t}|^2)^{\alpha/2}(\cos(\alpha \arg(x, \underline{t}) - \frac{\underline{t}}{|\underline{t}|} \sin(\alpha \arg(x, \underline{t}))$$

where

$$\arg(x, \underline{t}) = \arg(<\underline{x}, \underline{t}> + i x_0 |\underline{t}|).$$

To obtain "single valued" monogenic plane waves, it suffices to "cut" the argument function as follows.

Let $\theta \in \mathbf{R}$; then classically $\arg_\theta z$ is the branch of $\arg z$ for which $\arg_\theta z \in [\theta - \pi, \theta + \pi[$, $z \in \mathbf{C}$.

Putting

$$\arg_\theta(x, \underline{t}) = \arg_\theta(<\underline{x}, \underline{t}> + i x_0 |\underline{t}|)$$

we introduce the plane wave generalized powers p_θ^α by

$$\begin{aligned} & p_\theta^\alpha(x_0, \underline{x}, \underline{t}) \\ = \; & (<\underline{x}, \underline{t}>^2 + x_0^2|\underline{t}|^2)^{\alpha/2}(\cos(\alpha \arg_\theta(x, \underline{t}) - \frac{\underline{t}}{|\underline{t}|} \sin(\alpha \arg_\theta(x, \underline{t})). \end{aligned} \qquad (4.7)$$

The expression thus defines a monogenic plane wave where, for $\underline{t} \in \mathbf{R}^m$ fixed, the monogenicity holds in $\Omega_{\underline{t}} = \{x \in \mathbf{R}^{m+1} : \arg_\theta(x, \underline{t}) \neq \theta - \pi\}$.

Notice that, as for $k \in \mathbf{N}$, $(<\underline{x}, \underline{t}>)^k$ admits the CK-extension $p_k(x, \underline{t})$ with (see also II. 0.2.13)

$$p_k(x, \underline{t}) = (<\underline{x}, \underline{t}> - x_0 \underline{t})^k$$

the function $p_k(x, \underline{t})$ is a monogenic extension of p_θ^k to the whole of $\mathbf{R}^{m+1}$.

Taking $\theta = \pm\pi/2$ we define the integrals

$$\begin{aligned} \mathcal{C}_\alpha^\pm(P_k)(x) &= \int_{S^{m-1}} p_{\pm\pi/2}^\alpha(x, \underline{\omega}) P_k(\underline{\omega}) d\Sigma(\underline{\omega}) \\ \text{and} & \\ \mathcal{S}_\alpha^\pm(P_k)(x) &= \int_{S^{m-1}} p_{\pm\pi/2}^\alpha(x, \underline{\omega}) \underline{\omega} P_k(\underline{\omega}) d\Sigma(\underline{\omega}) \end{aligned} \tag{4.8}$$

which are well defined for $Re\alpha > -1$ and monogenic at least in

$$\mathbf{R}_\pm^{m+1} = \{x = (x_0, \underline{x}) : x_0 \gtrless 0\}.$$

To evaluate $\mathcal{C}_\alpha^\pm(P_k)(x)$ and $\mathcal{S}_\alpha^\pm(P_k)(x)$, we first restrict them to $\mathbf{R}^m$ (by putting $x_0 = 0$) and then extend monogenically.

As

$$\begin{aligned} p_{\pm\pi/2}^\alpha(<\underline{x}, \underline{t}>) &= |<\underline{x}, \underline{t}>|^\alpha (\cos \pi\alpha Y(-<\underline{x}, \underline{t}>) \\ &\qquad \mp \frac{\underline{t}}{|\underline{t}|} \sin \pi\alpha Y(-<\underline{x}, \underline{t}>)) \end{aligned}$$

we obtain by applying the Funk-Hecke Theorem that

$$\begin{aligned} \mathcal{C}_\alpha^\pm(P_k)(\underline{x}) &= A_{m-1}|\underline{x}|^\alpha (C_k(\alpha) \mp S_k(\alpha)\underline{\xi}) P_k(\underline{\xi})) \\ \text{and} & \\ \mathcal{S}_\alpha^\pm(P_k)(\underline{x}) &= A_{m-1}|\underline{x}|^\alpha (C_{k+1}(\alpha)\underline{\xi} \pm S_{k-1}(\alpha)) P_k(\underline{\xi}) \end{aligned} \tag{4.9}$$

where

$$\begin{aligned} C_k(\alpha) &= \int_{-1}^1 |s|^\alpha \cos(\pi\alpha Y(-s)) P_{k,m}(s)(1-s^2)^{\frac{m-3}{2}} ds \\ \text{and} & \\ S_k(\alpha) &= \sin \pi\alpha \int_{-1}^0 |s|^\alpha P_{k+1,m}(s)(1-s^2)^{\frac{m-3}{2}} ds. \end{aligned} \tag{4.10}$$

Moreover, as $\mathcal{C}_\alpha^+(P_k)$ and $\mathcal{S}_\alpha^+(P_k)$ (resp. $\mathcal{C}_\alpha^-(P_k)$ and $\mathcal{S}_\alpha^-(P_k)$) are monogenic in $\mathbf{R}_+^{m+1}$ (resp. $\mathbf{R}_-^{m+1}$) and admit distributional boundary values for $x_0 \to 0+$ (resp. $x_0 \to 0-$), these functions are completely determined by their limits for $x_0 \to 0+$ (resp. $x_0 \to 0-$).

In view of the definition of the axially monogenic generalized powers of degree k (see 2.3) we thus obtain that

$$\begin{aligned} \mathcal{C}_\alpha^\pm(P_k)(x) &= A_{m-1}(C_k(\alpha) p_{\alpha,k,m}(x) \mp S_k(\alpha) q_{\alpha,k,m}(x)) P_k(\underline{\xi}) \\ \text{and} & \\ \mathcal{S}_\alpha^\pm(P_k)(x) &= A_{m-1}(C_{k+1}(\alpha) q_{\alpha,k,m}(x) \pm S_{k-1}(\alpha) p_{\alpha,k,m}(x)) P_k(\underline{\xi}). \end{aligned} \tag{4.11}$$

The monogenic plane wave integrals $\mathcal{C}_\alpha^\pm(P_k)$ are related to the Cauchy type integrals (see section 3) in the following way.

Theorem 4.3.1. *For $-1 < Re\alpha < 0$ we have that*

$$\mathcal{C}_\alpha^\pm(P_k)(x) = \frac{A_{m-1}C_k(\alpha)}{\sigma_{m,k}(\alpha+k+m)} \Lambda_{\alpha+k+m-1,k,m}^\pm(x) P_k(\underline{x}) \tag{4.12}$$

Proof. From the expressions (4.11) of $C_\alpha^\pm(P_k)$ and the relations (see (3.7))

$$\Lambda^\pm_{\alpha+k+m-1,k,m}(x)$$
$$= (\sigma_{m,k}(\alpha+k+m)p_{\alpha,k,m}(x) \mp \sigma_{m,k}(\alpha+k+m-1)q_{\alpha,k,m}(x))P_k(\underline{\xi})$$

and by putting $x_0 = 0$ into the expression

$$\frac{1}{\sigma_{m,k}(\alpha+k+m)}(\Lambda^+_{\alpha+k+m-1,k,m}(x) + \Lambda^-_{\alpha+k+m-1,k,m}(x))P_k(\underline{x})$$

it follows that

$$\frac{1}{\sigma_{m,k}(\alpha+k+m)}(\Lambda^+_{\alpha+k+m-1,k,m}(\underline{x}) + \Lambda^-_{\alpha+k+m-1,k,m}(\underline{x}))P_k(\underline{\xi})$$
$$= p_{\alpha,k,m}(\underline{x})P_k(\underline{\xi})$$
$$= \rho^\alpha P_k(\underline{\xi}).$$

But, as by virtue of (4.9)

$$\rho^\alpha P_k(\underline{\xi}) = \frac{1}{A_{m-1}C_k(\alpha)}(C_\alpha^+(P_k)(\underline{x}) + C_\alpha^-(P_k)(\underline{x})),$$

we so find that

$$\frac{1}{\sigma_{m,k}(\alpha+k+m)}(\Lambda^+_{\alpha+k+m-1,k,m}(\underline{x}) + \Lambda^-_{\alpha+k+m-1,k,m}(\underline{x}))P_k(\underline{x})$$
$$= \frac{1}{A_{m-1}C_k(\alpha)}(C_\alpha^+(P_k)(\underline{x}) + C_\alpha^-(P_k)(\underline{x}))$$

so that also, since $Re\alpha > -1$,

$$C_\alpha^+(P_k)(\underline{x}) - \frac{A_{m-1}C_k(\alpha)}{\sigma_{m,k}(\alpha+k+m)}\Lambda^+_{\alpha+k+m-1,k,m}(\underline{x})P_k(\underline{x})$$
$$= -C_\alpha^-(P_k)(\underline{x}) + \frac{A_{m-1}C_k(\alpha)}{\sigma_{m,k}(\alpha+k+m)}\Lambda^-_{\alpha+k+m-1,k,m}(\underline{x})P_k(\underline{x}). \tag{4.13}$$

Denoting both members of (4.13) by $f_\alpha(\underline{x})$, we thus have that, since the left hand side of (4.13) is the boundary value for $x_0 \to 0+$ of a monogenic function in $\mathbf{R}^{m+1}_+$, while the right hand side of (4.13) is the boundary value for $x_0 \to 0-$ of a monogenic function in $\mathbf{R}^{m+1}_-$, Painlevé's Theorem tells us that $f_\alpha(\underline{x})$ is extensible to a monogenic function $f_\alpha(x)$ in $\mathbf{R}^{m+1}$ (see also II. 0.2.12). Moreover, as in $\mathbf{R}^{m+1}$,

$$f_\alpha(x) = \frac{1}{A_{m-1}C_k(\alpha)}\int_{S^{m-1}} p^\alpha_{\pi/2}(x,\underline{\omega})P_k(\underline{\omega})d\Sigma(\underline{\omega})$$
$$-\frac{1}{\sigma_{m,k}(\alpha+k+m)}\Lambda^+_\alpha(x)P_k(\underline{x})$$

we have that, as $Re\alpha < 0$, $\lim_{x_0 \to \infty} f_\alpha(x) = 0$.

So by Liouville's Theorem (see II. 1.6.3, Theorem 2) $f_\alpha(x) \equiv 0$ in $\mathbf{R}^{m+1}$ whence also $f_\alpha(\underline{x}) = 0$. The relations (4.13) then lead to the desired results. ■

Remarks 4.3.2.

(1) Noticing that from (4.11) and the relations (3.7) it also follows that

$$\begin{aligned} q_{\alpha,k,m}(\underline{x})P_k(\underline{\xi}) &= \frac{1}{\sigma_{m,k}(\alpha+k+m-1)} \\ &\qquad (\Lambda^+_{\alpha+k+m-1,k,m}(\underline{x}) - \Lambda^-_{\alpha+k+m-1,k,m}(\underline{x}))P_k(\underline{x}) \\ &= \frac{1}{A_{m-1}S_k(\alpha)}(\mathcal{C}^+_\alpha(P_k)(\underline{x}) - \mathcal{C}^-_\alpha(P_k)(\underline{x})) \end{aligned}$$

a similar reasoning as made in the proof of the foregoing theorem will lead to another relation between the monogenic plane wave integrals $\mathcal{C}^\pm_\alpha(P_k)$ and the Cauchy type integrals, namely

$$\mathcal{C}^\pm_\alpha(P_k)(x) = \frac{A_{m-1}S_k(\alpha)}{\sigma_{m,k}(\alpha+k+m-1)}\Lambda^\pm_{\alpha+k+m-1,k,m}(x)P_k(\underline{x}) \tag{4.14}$$

A combination of both identities (4.11) and (4.13) leads to the relation

$$\frac{C_k(\alpha)}{S_k(\alpha)} = \frac{\sigma_{m,k}(\alpha+k+m)}{\sigma_{m,k}(\alpha+k+m-1)}. \tag{4.15}$$

(2) As well in (4.12) as in (4.14), the left hand and the right hand sides are monogenic extensions of linear combinations of $\rho^\alpha P_k(\underline{\xi})$ and $\rho^\alpha \underline{\xi} P_k(\underline{\xi})$. The identities thus established determine a relationship between these monogenic extensions.

Theorem 4.3.3. (Recurrence Formulae) *The transforms $\mathcal{C}^\pm_\alpha(P_k)$ and $\mathcal{S}^\pm_\alpha(P_k)$ satisfy the recurrence relations*

$$\begin{aligned} \mathcal{S}^\pm_\alpha(P_k)(x) &= \tfrac{1}{\alpha+1}\partial_{\underline{x}}\mathcal{C}^\pm_{\alpha+1}(P_k)(x) \\ &\textit{and} \\ \mathcal{C}^\pm_\alpha(P_k)(x) &= \tfrac{1}{\alpha+1}\partial_{\underline{x}}\mathcal{S}^\pm_{\alpha+1}(P_k)(x) \end{aligned} \tag{4.16}$$

Proof. As $(\partial_{x_0} + \partial_{\underline{x}})(p^{\alpha+1}_{\pm\pi/2}(x,\underline{t})) = 0$, the expressions (4.8) yield :

$$\begin{aligned} \partial_{\underline{x}} p^{\alpha+1}_{\pm\pi/2}(x,\underline{t}) &= -\partial_{x_0} p^{\alpha+1}_{\pm\pi/2}(x,\underline{t}) \\ &= \underline{t}(\alpha+1)p^\alpha_{\pm\pi/2}(x,\underline{t}) \\ &= (\alpha+1)p^\alpha_{\pm\pi/2}(x,\underline{t})\underline{t} \end{aligned}$$

whence the desired results follow immediately. ∎

Corollary 4.3.4. *The constants $C_k(\alpha)$ and $S_k(\alpha)$ satisfy the recurrence relations*

$$C_{k+1}(\alpha) = \frac{\alpha + 1 - k}{\alpha + 1} C_k(\alpha + 1)$$

and

$$S_{k-1}(\alpha) = \frac{\alpha + k + m}{\alpha + 1} S_k(\alpha + 1).$$

Proof. Putting $x_0 = 0$ in (4.11) we find on the one hand that

$$S_\alpha^\pm(P_k)(\underline{x}) = A_{m-1}\rho^\alpha(C_{k+1}(\alpha)\underline{\xi} \pm S_{k-1}(\alpha))P_k(\underline{\xi}),$$

while, on the other hand, putting $x_0 = 0$ in (4.15), we obtain

$$\begin{aligned} & S_\alpha^\pm(P_k)(\underline{x}) \\ = & \frac{A_{m-1}}{\alpha+1}\left[C_k(\alpha+1)\partial_{\underline{x}}\rho^{\alpha+1-k}P_k(\underline{x}) \mp S_k(\alpha+1)\partial_{\underline{x}}\rho^{\alpha+m+k}\frac{\underline{x}}{\rho^m}P_k(\frac{\underline{x}}{\rho^2})\right] \\ = & \frac{A_{m-1}}{\alpha+1}\left[(\alpha+1-k)C_k(\alpha+1)\underline{\xi} \pm (\alpha+m-k)S_k(\alpha+1)\right]\rho^\alpha P_k(\underline{\xi}) \end{aligned}$$

The desired result then follows straightforwardly. ∎

Direct calculation of $S_0(\alpha)$ and $C_0(\alpha)$ now yield

$$\begin{aligned} S_0(\alpha) &= -\int_{-1}^{0} |s|^{\alpha+1}\sin\pi\alpha(1-s^2)^{\frac{m-3}{2}}\,ds \\ &= -\frac{\sin\pi\alpha}{2}B(\frac{\alpha}{2}+1, \frac{m-1}{2}) \end{aligned}$$

and

$$\begin{aligned} C_0(\alpha) &= (1+\cos\pi\alpha)\int_0^1 s^\alpha(1-s^2)^{\frac{m-3}{2}}\,dt \\ &= \frac{1+\cos\pi\alpha}{2}B(\frac{\alpha+1}{2}, \frac{m-1}{2}). \end{aligned}$$

The recurrence relations from Corollary 4.3.4 then lead to

$$\begin{aligned} C_k(\alpha) &= \frac{1+\cos\pi(\alpha-k)}{2^{k+1}}\frac{\Gamma(\alpha+1)\Gamma(\frac{\alpha-k+1}{2})\Gamma(\frac{m-1}{2})}{\Gamma(\alpha-k+1)\Gamma(\frac{\alpha+k+m}{2})} \\ \text{and} & \\ S_k(\alpha) &= \frac{\sin\pi(k-\alpha)}{2^{k+1}}\frac{\Gamma(\alpha+1)\Gamma(\frac{\alpha-k}{2}+1)\Gamma(\frac{m-1}{2})}{\Gamma(\alpha-k+1)\Gamma(\frac{\alpha+k+m+1}{2})}. \end{aligned} \tag{4.17}$$

Notice that, as

$$\frac{1+\cos\pi(\alpha-k)}{-\sin\pi(\alpha-k)} = \frac{\Gamma(\frac{k-\alpha}{2})\Gamma(1+\frac{\alpha-k}{2})}{\Gamma(\frac{1+k-\alpha}{2})\Gamma(\frac{1-k+\alpha}{2})}$$

we thus find back the expression (4.15) :

$$\frac{C_k(\alpha)}{S_k(\alpha)} = \frac{\Gamma(\frac{k-\alpha}{2})\Gamma(\frac{k+\alpha+m-1}{2})}{\Gamma(\frac{1+k-\alpha}{2})\Gamma(\frac{\alpha+k+m}{2})} = \frac{\sigma_{m,k}(\alpha+k+m)}{\sigma_{m,k}(\alpha+k+m-1)}.$$

The recurrence relations for $C_k(\alpha)$ and $S_k(\alpha)$ also enable us to define $C_\alpha^\pm(P_k)$ and $S_\alpha^\pm(P_k)$ as holomorphic functions of α at least in $\mathbf{C}\backslash\{-1,-2,\ldots\}$.

But, since for $x_0 \gtrless 0$, $p^\alpha_{\pm\pi/2}(x,\underline{t})$ has no singularities, the defining relations (4.9) tell us that, at least for $x \in \mathbf{R}^{m+1}_\pm$, the functions $C_\alpha^\pm(P_k)$ and $S_\alpha^\pm(P_k)$ are entire functions. Hence also $C_k(\alpha)$ and $S_k(\alpha)$ are entire functions, which of course also follows directly from the expression (4.17).

Let us now study $C_k(\alpha)$ and $S_k(\alpha)$ for integer values of α.

First consider the case $\alpha = \ell \in \mathbf{N}$. From the definition of $C_k(\alpha)$ and $S_k(\alpha)$ and the orthogonality properties of the Legendre polynomials $P_{k,m}$, we obtain immediately

Lemma 4.3.5. *Let $\ell \in \mathbf{N}$. Then*

(i) $S_k(\ell) = 0$;

(ii) $C_k(\ell) = 0$ *for* $k+\ell$ *odd ;*

(iii) $C_k(\ell) = 0$ *for* $k+\ell$ *even and* $\ell < k$.

It thus follows that we only have to consider $C_k(k+2s)$, $s \in \mathbf{N}$. Taking into account Lemma 4.3.5 and the relations (4.11) we obtain

Theorem 4.3.6. *Let $s \in \mathbf{N}$. Then*

(i) For $\ell = k+2s$,

$$\begin{aligned} C_\ell^\pm(P_k)(x) &= \int_{S^{m-1}} (<\underline{x},\underline{\omega}> - x_0\underline{\omega})^{k+2s} P_k(\underline{\omega})d\Sigma(\underline{\omega}) \\ &= A_{m-1}C_k(k+2s)p_{k+2s,k;m}(x)P_k(\underline{\xi}) \\ &= (-1)^s A_{m-1}C_k(k+2s)X_k^{2s}(x)P_k(\underline{x}). \end{aligned}$$

(ii) For $\ell = k+2s+1$,

$$\begin{aligned} \mathcal{S}_\ell^\pm(P_k)(x) &= \int_{S^{m-1}} (<\underline{x},\underline{\omega}> - x_0\underline{\omega})^{k+2s+1}\underline{\omega}P_k(\underline{\omega})d\Sigma(\underline{\omega}) \\ &= A_{m-1}C_{k+1}(k+2s+1)q_{k+2s+1,k,m}(x)P_k(\underline{\xi}) \\ &= (-1)^s A_{m-1}C_{k+1}(k+2s+1)X_k^{2s+1}(x)P_k(\underline{x}). \end{aligned}$$

Notice that by means of Theorem 4.3.6, the CK-extensions $X_k^{2s}(x)P_k(\underline{x})$ and $X_k^{2s+1}(x)P_k(\underline{x})$ of, respectively , $\underline{x}^{2s}P_k(\underline{x})$ and $\underline{x}^{2s+1}P_k(\underline{x})$ are now expressed in terms of plane wave integrals.

Now let $\alpha = -\ell$, $\ell \in \mathbf{N}$.

Lemma 4.3.7. *Let $\ell \in \mathbf{N}$. Then*

(i) $S_k(-\ell) = C_k(-\ell) = 0$ for $k+\ell$ odd ;

(ii) $S_k(-\ell) = 0$ for $k+\ell$ even, m odd and $\ell > m+k$;

(iii) $C_k(-\ell) = 0$ for $k+\ell$ even, m even and $\ell \geq m+k$.

Proof.

(i) Follows directly from the definitions since then $\mathcal{C}_\ell^\pm(P_k) = \mathcal{S}_{\ell-1}^\pm(P_k) = 0$.

(ii) We have that

$$S_k(-\ell) = \frac{\pi\cos\frac{\pi}{2}(k+\ell)}{2^k}\frac{\Gamma(1-\ell)\Gamma(\frac{m-1}{2})}{\Gamma(1-\ell-k)\Gamma(\frac{k+\ell}{2})\Gamma(\frac{m+k-\ell+1}{2})}.$$

Hence for $k+\ell$ even, m odd and $\ell > m+k$, $\Gamma(1-\lambda)$, $\Gamma(1-\lambda-k)$, $\Gamma(\frac{k+m-\lambda+1}{2})$ have first order poles at $\lambda = \ell$, so that $S_k(-\ell) = 0$.

(iii) We have that

$$C_k(-\ell) = \frac{1+\cos\pi(k+\ell)}{2^{k+1}}\frac{\Gamma(1-\ell)\Gamma(\frac{1-\ell-k}{2})\Gamma(\frac{m-1}{2})}{\Gamma(1-\ell-k)\Gamma(\frac{k+m-\ell}{2})}.$$

Hence for $k+\ell$ even, m even and $\ell \geq m+k$, $\Gamma(1-\lambda)$, $\Gamma(1-\lambda-k)$, $\Gamma(\frac{k+m-\lambda}{2})$ have first order poles at $\lambda = \ell$ so that $C_k(-\ell) = 0$. ∎

Lemma 4.3.7 immediately implies

Theorem 4.3.8. *Let $s \in \mathbf{N}$. Then*

(i) For m even and $\ell = k+m+2s$,

$$C^{\pm}_{-\ell}(P_k)(x) = \mp A_{m-1} S_k(-\ell) q_{-\ell,k,m}(x) P_k(\underline{\xi})$$

and

$$S^{\pm}_{-\ell-1}(P_k)(x) = A_{m-1} S_{k-1}(-\ell-1) p_{-\ell-1,k,m}(x) P_k(\underline{\xi})$$

(ii) For m odd and $\ell = k+m+2s$,

$$S^{\pm}_{-\ell}(P_k)(x) = A_{m-1} C_{k+1}(-\ell) q_{-\ell,k,m}(x) P_k(\underline{\xi})$$

and

$$C^{\pm}_{-\ell-1}(P_k)(x) = A_{m-1} C_k(-\ell-1) p_{-\ell-1,k;m}(x) P_k(\underline{\xi}).$$

Observing that

$$q_{-m,0,m}(x) = \frac{x_0 - \underline{x}}{|x_0 - \underline{x}|^{m+1}}$$

the foregoing theorem thus yields the plane wave decomposition of the Cauchy kernel when putting $k = s = 0$, namely

Theorem 4.3.9. *Let $m \in \mathbf{N}$. Then*

(i) For m even

$$\frac{1}{A_{m+1}} \frac{x_0 - \underline{x}}{|x_0 - \underline{x}|^{m+1}} = \pm \frac{(-1)^{m/2}(m-1)!}{2(2\pi)^m} \int_{S^{m-1}} (< \underline{x}, \underline{\omega} > - x_0 \underline{\omega})^{-m} d\Sigma(\underline{\omega})$$

(ii) For m odd

$$\frac{1}{A_{m+1}} \frac{x_0 - \underline{x}}{|x_0 - \underline{x}|^{m+1}} = \frac{(-1)^{\frac{m+1}{2}}(m-1)!}{2(2\pi)^m} \int_{S^{m-1}} (< \underline{x}, \underline{\omega} > - x_0 \underline{\omega})^{-m} \underline{\omega} d\Sigma(\underline{\omega}).$$

These results were used to define a generalization of the Radon transform in Clifford analysis. For further information we refer to [73], [77], [74], [72], [76].

5 RIESZ POTENTIALS

The aim of this section is to define generalized powers of the Dirac operator, i.e. given $\alpha \in \mathbf{C}$ and the Dirac operator $\partial_{\underline{x}}$ in $\mathbf{R}^m$, is there any meaning for an expression of the form $\partial_{\underline{x}}^{\alpha}$. Of course this problem is somewhat related to defining generalized powers of the Laplacian $\Delta_{\underline{x}}$ in $\mathbf{R}^m$ since $\partial_{\underline{x}}^2 = -\Delta_{\underline{x}}$ but, as we shall see, in the case of the Dirac operator we really obtain results which differ from the case of the Laplacian. For the latter we refer to [34].

5.1 Statement of the problem

Let $\partial_{\underline{x}}$ be the Dirac operator in $\mathbf{R}^m$ and let $\alpha \in \mathbf{C}$. Then we wish to introduce the operator $\partial_{\underline{x}}^{\alpha}$ as a convolution operator acting on the space $\mathcal{S}'(\mathbf{R}^m;\mathcal{C})$ of $\mathcal{C}$-valued distributions in such a way that, when putting

$$R_m^{(\alpha)} = \partial_{\underline{x}}^{\alpha}\delta, \tag{5.1}$$

we have that

$$R_m^{(\alpha)} \star R_m^{(\beta)} = R_m^{(\alpha+\beta)}. \tag{5.2}$$

In particular we are thus requiring that

$$R_m^{(1)} \star R_m^{(\alpha)} = \partial_{\underline{x}} R_m^{(\alpha)} = R_m^{(\alpha+1)} \tag{5.3}$$

whence the function $E_m^{[\ell]} = R_m^{(-\ell)}$ ($l \in \mathbf{N}$) should be a fundamental solution of $\partial_{\underline{x}}^{\ell}$, i.e.

$$\partial_{\underline{x}}^{\ell} E_m^{[\ell]} = \delta$$

To solve this problem it is natural to use the Fourier transform $\mathcal{F}$ given by

$$\mathcal{F}f(\underline{y}) = \int_{\mathbf{R}^m} e^{i<\underline{x},\underline{y}>} f(\underline{x})d\underline{x}$$

where f belongs to $\mathcal{S}(\mathbf{R}^m;\mathcal{C})$.

In a formal way, the definition (5.1) implies that

$$\mathcal{F}(\partial_{\underline{x}}^{\alpha}\delta) = (-i\underline{y})^{\alpha}$$

and so, once a meaning could be given to $(-i\underline{y})^{\alpha}$ - in the sense of distributions - it would be natural to put

$$R_m^{(\alpha)} = \mathcal{F}^{-1}(-i\underline{y})^{\alpha} \tag{5.4}$$

As to the requirement (5.2) it suffices to observe that, whenever for $T, S \in \mathcal{S}'(\mathbf{R}^m;\mathcal{C})$, the product $\mathcal{F}T \cdot \mathcal{F}S$ is defined in $\mathcal{S}'(\mathbf{R}^m;\mathcal{C})$, we then have that

$$T \star S = \mathcal{F}^{-1}(\mathcal{F}T \cdot \mathcal{F}S).$$

Hence $(-i\underline{y})^{\alpha}$ should be determined in such a way that

$$(-i\underline{y})^{\alpha}(-i\underline{y})^{\beta} = (-i\underline{y})^{\alpha+\beta}. \tag{5.5}$$

5.2 The distribution $(-i\underline{y})^\alpha$

Let us again consider the functions $\Omega_\pm$ given by (see also §2.3)

$$\Omega\pm = \frac{1}{2}(1 \mp \underline{\omega}), \underline{\omega} \in S^{m-1}.$$

Then, putting $\underline{y} = \rho\underline{\omega}$, we have that $-i\underline{\omega}\Omega_\pm = \pm\Omega_\pm$, whence symbolically

$$\begin{aligned}(-i\underline{y})^\alpha\Omega_\pm &= (-i\underline{\omega})^\alpha\rho^\alpha\Omega_\pm \\ &= (\pm 1)^\alpha\rho^\alpha\Omega_\pm\end{aligned}$$

Moreover, as $\Omega_+ + \Omega_- = \mathbf{1}$, we also get symbolically that

$$\begin{aligned}(-i\underline{y})^\alpha &= \frac{1}{2}(1 - i\underline{\omega})1^\alpha\rho^\alpha + \frac{1}{2}(1 + i\underline{\omega})(-1)^\alpha\rho^\alpha \\ &= \frac{1}{2}(1^\alpha + (-1)^\alpha)\rho^\alpha + \frac{1}{2}(1^\alpha - (-1)^\alpha)(-i\underline{\omega})\rho^\alpha \\ &= \frac{1}{2}(1^\alpha + (-1)^\alpha)\rho^\alpha + \frac{1}{2}(1^\alpha - (-1)^\alpha)(-i\underline{y})\rho^{\alpha-1}.\end{aligned}$$

Consequently, once a suitable meaning is given to 1^α, $(-1)^\alpha$ and ρ^α, the same would hold for $(-i\underline{y})^\alpha$.

As to 1^α and $(-1)^\alpha$ we admit the standard definitions

$$1^\alpha = e^{2\ell\pi i\alpha} \quad ; \quad (-1)^\alpha = e^{(2\ell+1)\pi i\alpha} \quad , \ell \in \mathbf{Z}.$$

As far as the distribution ρ^α is concerned, we rely upon the standard definition of the distribution x_+^α (see e.g. [30]). For completeness, let us recall some properties of this distribution.

Let $\alpha \in \mathbf{C}$ with $Re\alpha > -1$; then on the real line the action of x_+^α is defined by

$$< x_+^\alpha, \varphi >= \int_0^{+\infty} x^\alpha\varphi(x)dx, \quad \varphi \in \mathcal{S}(\mathbf{R}).$$

Using the recurrence formula

$$x_+^\alpha = \frac{1}{\alpha+1}\frac{d}{dx}x_+^{\alpha+1}$$

valid for $Re\alpha > -1$, a holomorphic extension of x_+^α may be obtained for $\alpha \in \mathbf{C}\backslash\{-1, -2, \ldots\}$.

The residues of x_+^α are given by

$$\mathrm{Res}(x_+^\alpha; -n) = \frac{(-1)^{n-1}\delta^{(n-1)}}{(n-1)!} \quad , \quad n \in \mathbf{N}_0.$$

Now let $\alpha \in \mathbf{C}$ be such that $Re\alpha > -m$. Then we put

$$< \rho^\alpha, \varphi > = \int_{\mathbf{R}^m} |\underline{y}|^\alpha \varphi(\underline{y}) d\underline{y}, \quad \varphi \in \mathcal{S}(\mathbf{R}^m; \mathcal{C}). \tag{5.6}$$

Where defined, we have that, putting $\underline{y} = \rho\underline{\omega}$,

$$\begin{aligned} < \rho^\alpha, \varphi > &= \int_0^\infty \rho^{\alpha+m-1} d\rho \left[\int_{S^{m-1}} \varphi(\rho\underline{\omega}) d\Sigma(\underline{\omega}) \right] \\ &= A_m < x_+^{\alpha+m-1}, M\varphi(0,x) > . \end{aligned} \tag{5.7}$$

Here

$$M\varphi(\underline{x}, \rho) = \frac{1}{A_m} \int_{S^{m-1}} \varphi(\rho\underline{\omega} + \underline{x}) d\Sigma(\underline{\omega}) \quad , \rho \in \mathbf{R},$$

is nothing else but the spherical mean operator (see e.g. [45] and [30]).

By means of (5.6) and (5.7) we may extend ρ^α holomorphically to

$$\mathbf{C} \setminus \{-m, -m-1, \ldots\}.$$

We so obtain that

$$\begin{aligned} < \text{Res}(\rho^\alpha; -m-j), \varphi > &= A_m < \text{Res}(x_+^{\alpha+m-1}; -m-j), M\varphi(0,x) > \\ &= \frac{A_m}{j!} (-1)^j < \delta^{(j)}, M\varphi(0,x) > \\ &= \frac{A_m}{j!} \frac{d^j}{dx^j} M\varphi(0,x)|_{x=0}. \end{aligned}$$

As x_+^α has first order poles at $\{-1, -2, \ldots\}$ and as $M\varphi(0,\rho) = M\varphi(0,-\rho)$, M being an even function, we have thus proved that ρ^α has first order poles at $-m-2j$ and removable singularities at $m-2j-1$, $j \in \mathbf{N}_0$.

Furthermore, in view of the formula (see [45])

$$M\varphi(0,\rho) = \Gamma(\frac{m}{2}) \sum_{\ell=0}^\infty \frac{\rho^{2\ell} \Delta_{\underline{x}}^\ell \varphi(0)}{2^{2\ell} \ell! \Gamma(\frac{m}{2} + \ell)}$$

we in fact obtain that

$$\text{Res}(\rho^\alpha, -m-2j) = \frac{A_m \Gamma(\frac{m}{2})}{2^{2j} j! \Gamma(\frac{m}{2} + j)} \Delta_{\underline{y}}^j \delta. \tag{5.8}$$

We are now in a position to define $(-i\underline{y})^\alpha$ and this by means of

$$(-i\underline{y})^\alpha = F(\alpha)\rho^\alpha + G(\alpha)(-i\underline{y})\rho^{\alpha-1} \tag{5.9}$$

where

$$F(\alpha) = \frac{1}{2}(1^\alpha + (-1)^\alpha); G(\alpha) = \frac{1}{2}(1^\alpha - (-1)^\alpha).$$

Notice that $F(\alpha)$ has first order zeros for $\alpha = 2\ell + 1$, $\ell \in \mathbf{Z}$, while $G(\alpha)$ has first order zeros for $\alpha = 2\ell$, $\ell \in \mathbf{Z}$.

Furthermore ρ^α (resp. $\rho^{\alpha-1}$) has first order poles at points $\alpha = -m - 2\ell$ (resp. $\alpha = -m - 2\ell + 1$), $\ell \in \mathbf{N}$.

Hence for m odd, all singularities of the expression defined by (5.8) may be removed and so, in that case, $(-i\underline{y})^\alpha$ is an $\mathcal{S}'(\mathbf{R}^m; \mathcal{C})$-valued entire function.

However, for m even, first order poles are to be expected at $\alpha = -m - \ell + 1$, $\ell \in \mathbf{N}$.

By virtue of (5.8) we get :

$$\text{Res}((-i\underline{y})^\alpha; -m - 2j) = \frac{A_m \Gamma(\frac{m}{2}}{2^{2j} j' \Gamma(\frac{m}{2} + j)} \Delta_{\underline{y}}^j \delta$$

and (5.10)

$$\text{Res}((-i\underline{y})^\alpha; -m - 2j + 1) = \frac{A_m \Gamma(\frac{m}{2})}{2^{2j} j! \Gamma(\frac{m}{2} + j)} (-i\underline{y}) \Delta_{\underline{y}}^j \delta.$$

Putting $j = 0$ in the second formula of (5.10) and taking into account that $\underline{y}\delta = 0$, we have that $\text{Res}((-i\underline{y})^\alpha; -m + 1) = 0$ so that $(-i\underline{y})^\alpha$ has a removable singularity at $-m + 1$ and first order poles at $\alpha = -m - \ell$, $\ell \in \mathbf{N}$.

Furthermore, as $(-i\partial_{\underline{y}})^2 = \Delta_{\underline{y}}$, we obtain that

$$\Delta_{\underline{y}}^j \delta = (-i\partial_{\underline{y}})^{2j} \delta \tag{5.11}$$

while

$$\begin{aligned} (-i\underline{y}) \Delta_{\underline{y}}^j \delta &= (-i\underline{y})(-i\partial_{\underline{y}})(-i\partial_{\underline{y}})^{2j-1} \delta \\ &= (E + \Gamma_{\underline{\omega}})(-i\partial_{\underline{y}})^{2j-1} \delta \end{aligned}$$

the latter expression following from

$$\begin{aligned} (-i\underline{y})(-i\partial_{\underline{y}}) &= \sum_{j=1}^m y_j \partial_{y_j} - \underline{y} \wedge \partial_{\underline{y}} \\ &= E + \Gamma_{\underline{\omega}} \end{aligned}$$

where $E = \rho\partial_\rho = \sum_{j=1}^m y_j \partial_{y_j}$ is the Euler operator.

But, as δ is homogeneous of degree $-m$ and so $(-i\partial_{\underline{y}})^{2j-1}\delta$ is homogeneous of degree $-m + 2j - 1$, we have that $E(-i\partial_{\underline{y}})^{2j-1}\delta = -(2j + m - 1)(-i\partial_{\underline{y}})^{2j-1}\delta$.

On the other hand, as any smooth vector-valued function $F(\underline{y})$ may be written as $F(\underline{y}) = F(|\underline{y}|)\underline{\omega}$ whence $\Gamma_{\underline{\omega}} F(\underline{y}) = (m - 1)F(|\underline{y}|)\underline{\omega}$ because of $\Gamma_{\underline{\omega}}\underline{\omega} = (m-1)\underline{\omega}$, we obtain by approximating the vector-valued $\mathcal{S}'(\mathbf{R}^m; \mathcal{C})$ distribution $(-i\partial_{\underline{y}})^{2j-1}\delta$ by a sequence of vector-valued smooth functions, that $\Gamma_{\underline{\omega}}(-i\partial_{\underline{y}})^{2j-1}\delta = (m-1)(-i\partial_{\underline{y}})^{2j-1}\delta$.

Consequently

$$\begin{aligned}(-i\underline{y})\Delta_{\underline{y}}^{j}\delta &= (-(2j+m-1)+m-1)(-i\partial_{\underline{y}})^{2j-1}\delta \\ &= (-2j)(-i\partial_{\underline{y}})^{2j-1}\delta \qquad (5.12)\end{aligned}$$

From (5.10) - (5.12) it thus follows that for m even, the residues at $\alpha = -m-\ell$, $\ell \in \mathbf{N}$, may be expressed as Dirac-derivatives of δ.

Finally observe that the formula (5.12) may also be written as

$$(-i\underline{y})(-i\partial_{\underline{y}})^{2j}\delta = (-2j)(-i\partial_{\underline{y}})^{2j-1}\delta$$

thus yielding an m-dimensional analogue of the well known formula

$$x\delta^{(j)} = c(j)\delta^{(j-1)} \quad , \quad c(j) \in \mathbf{R},$$

valid on the real line.

After having defined the distribution $(-i\underline{y})^{\alpha}$, we now wish to check if it indeed satisfies the requirement (5.5), i.e. $(-i\underline{y})^{\alpha}(-i\underline{y})^{\beta} = (-i\underline{y})^{\alpha+\beta}$.

As we already observed, for m odd all three functions involved are entire (in $\mathbf{C}^2$). Hence, if the identity (5.5) holds in a neighbourhood of say $(\alpha,\beta) = (1,1)$, it holds everywhere in $\mathbf{C}^2$ because of analytic continuation. For m even, the identity (5.5) will be valid where all three functions involved are holomorphic, i.e. where the conditions $\alpha \notin m-\mathbf{N}$, $\beta \notin m-\mathbf{N}$ and $\alpha+\beta \notin m-\mathbf{N}$ are satisfied. The region where they hold being connected in $\mathbf{C}^2$, the identity will also follow from analytic continuation.

Now let us have a closer look at the case m odd.

Around the points $\alpha = 1$, $\beta = 1$, $\gamma = \alpha+\beta = 2$, the functions $(-i\underline{y})^{\alpha}$, $(-i\underline{y})^{\beta}$ and $(-i\underline{y})^{\gamma}$ are everywhere defined and continuous as functions of $\underline{y}$ and we have that for $\underline{y} \neq 0$,

$$\begin{aligned}(-i\underline{y})^{\alpha}(-i\underline{y})^{\beta} &= (F(\alpha)\rho^{\alpha} + G(\alpha)(-i\underline{y})\rho^{\alpha-1})(F(\beta)\rho^{\beta} + G(\beta)(-i\underline{y})\rho^{\beta-1}) \\ &= \rho^{\alpha}\rho^{\beta}(F(\alpha)F(\beta)) + \rho^{\alpha-1}\rho^{\beta-1}(-i\underline{y})^2(G(\alpha)G(\beta)) \\ &\quad + \rho^{\alpha}\rho^{\beta-1}(-i\underline{y})(F(\alpha)G(\beta)) + (-i\underline{y})\rho^{\alpha-1}\rho^{\beta}(G(\alpha)F(\beta)) \\ &= \rho^{\alpha+\beta}(F(\alpha)F(\beta) + G(\alpha)G(\beta)) \\ &\quad + (-i\underline{y})\rho^{\alpha+\beta-1}(F(\alpha)G(\beta) + G(\alpha)F(\beta)).\end{aligned}$$

Now

$$\begin{aligned}F(\alpha)F(\beta) + G(\alpha)G(\beta) &= \frac{1}{4}(1^{\alpha} + (-1)^{\alpha})(1^{\beta} + (-1)^{\beta}) \\ &\quad + \frac{1}{4}(1^{\alpha} - (-1)^{\alpha})(1^{\beta} - (-1)^{\beta}) \\ &= \frac{1}{2}(1^{\alpha}1^{\beta} + (-1)^{\alpha}(-1)^{\beta})\end{aligned}$$

and

$$\begin{aligned} F(\alpha)G(\beta)+G(\alpha)F(\beta) &= \frac{1}{4}(1^\alpha+(-1)^\alpha)(1^\beta-(-1)^\beta) \\ &\quad +\frac{1}{4}(1^\alpha-(-1)^\alpha)(1^\beta-(-1)^\beta) \\ &= \frac{1}{2}(1^\alpha 1^\beta-(-1)^\alpha(-1)^\beta). \end{aligned}$$

It thus suffices to put $1^\alpha = \ell^{2\ell\pi i\alpha}$, $1^\beta = e^{2n\pi i\beta}$, $(-1)^\alpha = e^{(2\ell'+1)\pi i\alpha}$, $(-1)^\beta = e^{(2n'+1)\pi i\beta}$ to arrive at

$$\begin{aligned} (-i\underline{y})^\alpha(-i\underline{y})^\beta &= F(\alpha+\beta)\rho^{\alpha+\beta}+(-i\underline{y})G(\alpha+\beta)\rho^{\alpha+\beta-1} \\ &= (-i\underline{y})^{\alpha+\beta}. \end{aligned}$$

We thus obtain the desired property and so the identity holds everywhere in the case m odd.

For m even this is not yet the case since $(-i\underline{y})^\alpha$, $(-i\underline{y})^\beta$, $(-i\underline{y})^{\alpha+\beta}$ are not everywhere defined. Of course the identity (5.5) holds in the intersection of the domains of definition of the three functions involved and, as will be shown, it remains valid by extending these functions in a suitable way.

To this end, let us first extend the definition of $(-i\underline{y})^\alpha$ to the whole complex plane. For $\alpha \notin -m-\mathbf{N}$, we just introduced a definition. Taking into account the Laurent series of $(-i\underline{y})^\alpha$ near $\alpha = -m-\ell$, namely

$$(-i\underline{y})^\alpha = \frac{c_{-1}}{\alpha+m+\ell}+\sum_{j=0}^{\infty} c_j(\alpha+m+\ell)^j,$$

we put for $\alpha = -m-\ell$, $\ell \in \mathbf{N}$,

$$\begin{aligned} (-i\underline{y})^{-m-\ell} &= c_0 \\ &= \lim_{\alpha\to -m-\ell}\frac{d}{d\alpha}((\alpha+m+\ell)(-i\underline{y})^\alpha) \\ &= \oint \frac{(-i\underline{y})^\alpha}{\alpha+m+\ell}d\alpha \\ &= \frac{1}{2\pi}\int_0^{2\pi}(-i\underline{y})^{-m-\ell+\varepsilon e^{i\theta}}\,d\theta \\ &= Mf(-m-\ell;\varepsilon). \end{aligned} \tag{5.13}$$

Next assume that either α, β or $(\alpha+\beta)$ belongs to $-m-\mathbf{N}$.

If $\alpha = -m - \ell$ and $\beta \notin -m - \mathbf{N}$, we put

$$\begin{aligned}(-i\underline{y})^{-m-\ell}(-i\underline{y})^{\beta} &= (\frac{1}{2\pi}\int_0^{2\pi}(-i\underline{y})^{-m-\ell+\varepsilon e^{i\theta}}d\theta)(-i\underline{y})^{\beta} \\ &= I^{(-m-\ell,\beta)}.\end{aligned}$$

If $\alpha + \beta \notin -m - \mathbf{N}$, then for $\varepsilon > 0$ sufficiently small, $-m - \ell + \varepsilon e^{i\theta} + \beta \notin -m - \mathbf{N}$ and so in that case

$$\begin{aligned}I^{(-m-\ell,\beta)} &= \frac{1}{2\pi}\int_0^{2\pi}(-i\underline{y})^{-m-\ell+\beta+\varepsilon e^{i\theta}}d\theta \\ &= (-i\underline{y})^{-m-\ell+\beta} \\ &= (-i\underline{y})^{\alpha+\beta}\end{aligned}$$

Also for $\alpha + \beta \in -m - \mathbf{N}$, we have that for $\varepsilon > 0$ small enough, $-m - \ell + \beta + \varepsilon e^{i\theta} \notin -m - \mathbf{N}$ and so, by virtue of the definition of $(-i\underline{y})^{\gamma}$ in that case,

$$I^{(-m-\ell,\beta)} = (-i\underline{y})^{\alpha+\beta}.$$

We are thus left with the case $\alpha = -m - \ell$ and $\beta = -m - \ell'$ where $\ell, \ell' \in \mathbf{N}$, whence also $\alpha + \beta \in -m - \mathbf{N}$. Then by virtue of (5.13)

$$\begin{aligned}&(-i\underline{y})^{-m-\ell}(-i\underline{y})^{-m-\ell'} \\ &= \frac{1}{(2\pi)^2}\int_0^{2\pi}\int_0^{2\pi}(-i\underline{y})^{-m-\ell+\varepsilon e^{i\theta}}(-i\underline{y})^{-m-\ell'+\eta e^{i\psi}}d\theta d\psi \\ &= I^{(-m-\ell,-m-\ell')}.\end{aligned}$$

Now let $\varepsilon > 0$ and $\eta > 0$ be sufficiently small and such that $\varepsilon \neq \eta$ (take e.g. $\eta < \varepsilon$). Then for no $(\theta, \psi) \in [0, 2\pi] \times [0, 2\pi]$, the element

$$-m - \ell - m - \ell' + \varepsilon e^{i\theta} + \eta e^{i\psi} = \alpha + \beta + \varepsilon e^{i\theta} + \eta e^{i\psi}$$

belongs to $-m - \mathbf{N}$ and we thus have that

$$I^{(-m-\ell,-m-\ell')} = \frac{1}{(2\pi)^2}\int_0^{2\pi}\int_0^{2\pi}(-i\underline{y})^{\alpha+\beta+\varepsilon e^{i\theta}+\eta e^{i\psi}}d\theta d\psi.$$

Now this integral is nothing else but the iterated spherical mean $Mf(\alpha + \beta, \varepsilon, \eta)$ in the sense of F. John of the function $f(z) = (-i\underline{y})^z$ (see [45]). It is proved in [45] that such iterated spherical mean may be expressed in spherical means by a formula of the form

$$\begin{aligned}I^{(-m-\ell,-m-\ell')} &= Mf(\alpha + \beta, \varepsilon, \eta) \\ &= \int_{\varepsilon-\eta}^{\varepsilon+\eta} G(s)Mf(\alpha + \beta, s)ds\end{aligned}$$

where $G(s)$ is a weight function with total weight 1 on $[\varepsilon-\eta,\varepsilon+\eta]\subset]0,1[$.

Consequently, as by definition, for $\alpha+\beta\in -m-\mathbf{N}$,

$$Mf(\alpha+\beta,s)=(-i\underline{y})^{\alpha+\beta}$$

and as moreover

$$\int_{\varepsilon-\eta}^{\varepsilon+\eta} G(s)ds=1,$$

we arrive at

$$\begin{aligned} I^{(-m-\ell,-m-\ell')} &= \int_{\varepsilon-\eta}^{\varepsilon+\eta} G(s)(-i\underline{y})^{\alpha+\beta}ds \\ &= (-i\underline{y})^{\alpha+\beta}. \end{aligned}$$

Hence we proved that also in the case m even, the identity

$$(-i\underline{y})^{\alpha}(-i\underline{y})^{\beta}=(-i\underline{y})^{\alpha+\beta}$$

propagates by analytic continuation through the singularities.

5.3 *The distribution* $R_m^{(\alpha)}(\underline{x})$

Using the inverse Fourier transform

$$\mathcal{F}^{-1}\varphi(\underline{x})=\frac{1}{(2\pi)^m}\int_{\mathbf{R}^m} e^{-i<\underline{x},\underline{y}>}\varphi(\underline{y})d\underline{y},$$

the following expression is valid for the classical Riesz potential (see [34]) :

$$\mathcal{F}^{-1}\rho^{\alpha}=2^{\alpha}\pi^{-m/2}\frac{\Gamma(\frac{m+\alpha}{2})}{\Gamma(-\frac{\alpha}{2})}|\underline{x}|^{-\alpha-m}.$$

This formula holds normally only for α, $-\alpha-m\notin 2\mathbf{N}$ but it may be extended through the singularities if we define the value of a holomorphic function in a singular point by its coefficient of z^0 in the corresponding Laurent series (see also the foregoing subsection).

Applying the inverse Fourier transform to the function

$$(-i\underline{y})^{\alpha}=F(\alpha)\rho^{\alpha}+G(\alpha)(-i\underline{y})\rho^{\alpha-1}$$

the relation (5.4) and the remark just made lead to the formula

$$R_m^{(\alpha)}(\underline{x})=F(\alpha)\mathcal{F}^{-1}\rho^{\alpha}+G(\alpha)\mathcal{F}^{-1}(-i\underline{y})\rho^{\alpha-1}.$$

Using the fact that in the sense of distributions

$$\mathcal{F}^{-1}((-i\underline{y})f)=\partial_{\underline{x}}\mathcal{F}^{-1}f,$$

we thus obtain that

$$R_m^{(\alpha)}(\underline{x}) = 2^\alpha \pi^{-m/2} \left[\frac{F(\alpha)\Gamma(\frac{m+\alpha}{2})}{\Gamma(-\frac{\alpha}{2})} |\underline{x}|^{-\alpha-m} + \frac{G(\alpha)\Gamma(\frac{m+\alpha-1}{2})}{2\Gamma(\frac{1-\alpha}{2})} \partial_{\underline{x}} |\underline{x}|^{1-\alpha-m} \right].$$

But, as in the sense of homogeneous distributions of degree $(-\alpha - m)$,

$$\partial_{\underline{x}} |\underline{x}|^{1-\alpha-m} = -(m+\alpha-1)|\underline{x}|^{-\alpha-m} \frac{\underline{x}}{|\underline{x}|},$$

we also have that

$$R_m^{(\alpha)}(\underline{x}) = 2^\alpha \pi^{-m/2} |\underline{x}|^{\alpha-m} \left[\frac{F(\alpha)\Gamma(\frac{m+\alpha}{2})}{\Gamma(-\frac{\alpha}{2})} - \frac{G(\alpha)\Gamma(\frac{m+\alpha+1}{2})}{\Gamma(\frac{1-\alpha}{2})} \frac{\underline{x}}{|\underline{x}|} \right].$$

Chapter IV

Monogenic differential forms and residues

After a comprehensive study of the properties of spinor-valued solutions to the Dirac equation, we extend this study to spinor-valued differential forms.

As we saw in Chapter II, the theory of monogenic functions is a generalization of complex function theory to higher dimensions. Holomorphic differential forms were studied from the very beginning of complex analysis and form an important part of it. Holomorphic functions (i.e. 0-forms) and holomorphic 1-forms are the only cases which occur in dimension 2. In this chapter we want to describe a generalization of the concept of holomorphic forms to higher dimensions, called monogenic forms. The analogues of the operators $d' \sim \partial$ and $d'' \sim \bar{\partial}$ will be introduced; monogenic forms will be defined as the kernel of the d'' - operator, while the d' - operator will map one space of monogenic forms into another. We shall meet not only monogenic functions and 1-forms, but also monogenic forms of higher degree (up to $(m-1)$-forms). They will form a complex (the operator d' mapping monogenic k-forms into $(k+1)$-forms). We shall show that (as in the complex plane) the homology of the complex will describe faithfully the topology of the domain $\Omega \subset \mathbf{R}^m$ on which the forms are defined. A suitable analogue of the Cauchy theorem will be proved for monogenic forms.

There are several possibilities for a generalization of holomorphic forms to higher dimensions depending on the properties which we want to be preserved under such a generalization. The guiding principle behind the generalization presented here is the requirement of invariance with respect to the group Spin(m). Although this requirement is somewhat hidden, it emerges naturally from the role played below by the special invariant Clifford-valued differential form $d\underline{x}$ and the Dirac operator $\partial_{\underline{x}}$.

Generally speaking, there is a deep connection between Clifford analysis and the representation theory of the Spin group. The use of the structure of finite dimensional representations, their classification and their tensor products might hence bring a much better understanding and organization of facts in Clifford analysis. Unfortunately, as the theory of finite dimensional representations of compact groups still does not belong to the standard part of education in analysis, we decided not to use this language here (adding a summary and a reasonable explanation of the necessary facts from representation theory would indeed enlarge the book too much).

In the sections 1-5 we study monogenic differential forms. The structure of spinor valued differential forms can be described nicely from the representation theoretical point of view, showing at the same time that the differential operators used to define monogenic forms are Spin invariant (for more details see [25]). As a consequence, the structure of spinor valued forms is preserved and monogenic differential forms are well defined and can be studied on any manifold with a given Spin structure, but we are not going to describe this topic here.

In the second part of the chapter (§.6), the classical theory of residues for monogenic functions is generalized. It seems natural to consider monogenic functions with singularities on higher dimensional surfaces. In general, it is possible to study monogenic functions with singularities on any compact set K and to develop a residue theory for them. In the case that K is a compact manifold, however, the information on the behaviour of a function around the singularity set can be described nicely by differential forms defined on K and it turns out that a residue theory is also available for monogenic differential forms. The main tool needed for this generalized residue theory is the Leray-Norguet theory of residues for closed differential forms having a singularity on a submanifold of any codimension. The Leray residue was originally introduced in a complex situation (singularities considered were on a complex hypersurface, see [53]) and most of its generalizations were formulated in the complex case. The real version of the theory, worked out by Norguet in [59] is exactly the theory needed in Clifford analysis. The definition and properties of the Leray-Norguet residue, needed for applications in Clifford analysis, can be found in Appendix A.

1 SPINOR-VALUED FORMS

The basic objects of our study are spaces of spinor-valued differential forms. Let S be a left module over the Clifford algebra $\mathbf{R}_{0,m}$ denoted henceforth by C for simplicity. Here any module (real or complex) is taken into consideration and we do not restrict ourselves to irreducible representations only. Basic examples to keep in mind are, of course, any irreducible left module S and the real and complex Clifford algebras $\mathbf{R}_{0,m}$ and $\mathbf{C}_m$ themselves.

Occasionally, we also need right modules over $\mathbf{R}_{0,m}$. When necessary, we thus denote the corresponding modules by S_ℓ and S_r, respectively.

The space of S-valued forms is denoted by

$$\mathcal{E}^k = \mathcal{E}^k(\Omega, S) = \mathcal{E}^k(\Omega) \otimes_{\mathbf{R}} S = \{f : \Omega \mapsto \Lambda^k((\mathbf{R}^n)^*) \otimes_{\mathbf{R}} S\}.$$

The space $\mathcal{E}^0$ is the space of S-valued C_∞-functions and an arbitrary element in $\mathcal{E}^k$ can be written in the form

$$\omega = \sum_{|I|=k} \omega_I dx_I,\ \omega_I \in \mathcal{E}^0(\Omega, S),$$

where the meaning of the symbol $\omega_I dx_I$ is clear.

The space of Clifford-valued forms $\mathcal{E}^k(\Omega, \mathbf{R}_{0,m})$ plays a special role. It is a tensor product of two algebras, whence it is again an algebra and the multiplication into it is the combination of the Clifford and outer multiplications. Moreover, it acts on the space $\mathcal{E}^k(\Omega, S)$ of spinor-valued forms. Indeed, if $\tau = \sum_{|I|=k} \tau_I dx_I \in \mathcal{E}^k(\Omega, \mathbf{R}_{0,m})$ and $\omega = \sum_{|J|=l} s_J dx_J \in \mathcal{E}^k(\Omega, S)$, then

$$\Big(\sum_{|I|=k} \tau_I dx_I\Big) \wedge \Big(\sum_{|J|=l} \omega_J dx_J\Big) = \sum_{|I|=k,|J|=l} \tau_I\omega_J dx_I \wedge dx_J,$$

where the multiplication $\tau_I\omega_J$ denotes the action of the Clifford algebra on the representation space S (we shall use systematically the language of left multiplication for the left action). In the sense described above, the space $\mathcal{E}^*(\Omega, S)$ is a left module over the algebra $\mathcal{E}^*(\Omega, \mathbf{R}_{0,m})$.

In the case of a right module S_r over $\mathbf{R}_{0,m}$, we write a general form as

$$\omega = \sum_{|I|=k} dx_I\omega_I,\ \omega_I \in \mathcal{E}^0(\Omega, S_r),$$

and the space $\mathcal{E}^*(\Omega, S_r)$ is a right module over the algebra $\mathcal{E}^*(\Omega, \mathbf{R}_{0,m})$.

The most important example of a 1-form is the form

$$d\underline{x} = \sum_{i=1}^{m} e_i dx_i.$$

Let us warn the reader that the 1-form $d\underline{x}$ should not be confounded with the notation $d\underline{x}$ we used in earlier chapters to indicate the oriented volume element $d\underline{x} = dx_1 \wedge dx_2 \wedge \ldots \wedge dx_m$ in $\mathbf{R}^m$. However, as will be seen in a few lines, both notions are connected to each other.

Using the invariant form $d\underline{x}$, the space of spinor-valued forms will be decomposed into suitable subspaces. The main property of the form $d\underline{x}$ is that its powers $(d\underline{x})^k$ are nowhere vanishing for all $k = 1, \ldots, m$, since its coefficients are Clifford numbers which do not commute. Hence they provide nice examples of (Clifford-valued) k-forms. In particular,

$$(d\underline{x})^m = m!\, e.dx_1 \wedge \ldots \wedge dx_m,$$

where $e = e_1 \ldots e_m$ and

$$\begin{aligned}(d\underline{x})^{m-1} &= (m-1)!\{\sum_{J,|J|=m-1} e_J dx_J\} \\ &= (m-1)!\,\{\sum_{i=1}^{m}(-1)^i e_i d\hat{x}_i\}.e \\ &= -(m-1)!\, d\sigma.e\,,\end{aligned}$$

where $d\sigma = \sum_{i=1}^m (-1)^{i+1} e_i d\hat{x}_i$ is the form introduced in II.0.2.1.

Due to the nontriviality of $(d\underline{x})^k$, we can thus define the distinguished subspaces

$$\mathcal{E}^{k,0} = \{\omega \in \mathcal{E}^k | \omega = (d\underline{x})^k s, s \in \mathcal{E}^0\}$$

of the space $\mathcal{E}^k$ of k-forms. In the case of zero forms and m-forms they exhaust the space of all such forms, but for $k = 1, \ldots, m-1$ they are proper subspaces.

There is a natural complement of $\mathcal{E}^{1,0}$ in $\mathcal{E}^1$, namely

$$\mathcal{E}^{1,1} = \{\omega \in \mathcal{E}^1 | (d\underline{x})^{m-1} \wedge \omega = 0\}.$$

The fact that $(d\underline{x})^m \neq 0$ everywhere implies that $\mathcal{E}^1 \cong \mathcal{E}^{1,0} \oplus \mathcal{E}^{1,1}$. Indeed, it is clear that $\mathcal{E}^{1,0} \cap \mathcal{E}^{1,1} = \{0\}$ and as any $\omega \in \mathcal{E}^1$ can be written as

$$\omega = d\underline{x}\, s + [\omega - d\underline{x}\, s]$$

where $s \in \mathcal{E}^0$ is determined by the relation $(d\underline{x})^{m-1} \wedge \omega = (d\underline{x})^m s$, the sum is direct indeed.

The described splitting of 1-forms can be shifted by multiplication with $d\underline{x}$ to higher order forms, namely

$$\mathcal{E}^{k,1} = \{\omega \in \mathcal{E}^k | \omega = (d\underline{x})^{k-1} \wedge \tau, \tau \in \mathcal{E}^{1,1}\}, k = 1, \ldots, m-1.$$

In such a way we have introduced another distinguished subspace of k-forms, but we do not know a priori, if it is a nontrivial subspace. That this is indeed the case follows from an additional important information concerning the multiplication by $d\underline{x}$, namely that $\mathcal{E}^1$ is mapped isomorphically onto $\mathcal{E}^{m-1}$ under the mupltiplication by $(d\underline{x})^{m-2}$.

Lemma 1.1. *If* $(d\underline{x})^{m-2} \wedge \omega = 0, \omega \in \mathcal{E}^1$, *then* $\omega = 0$.

Proof. Let $\omega = \sum_{i=1}^m \omega_i \, dx_i \in \mathcal{E}^1$ be such that $(d\underline{x})^{m-2} \wedge \omega = 0$. Then

$$\begin{aligned}
(d\underline{x})^{m-2} \wedge \omega &= (m-2)!(\sum_{|B|=m-2} e_B \, dx_B) \wedge (\sum_{i=1}^m \omega_i \, dx_i) \\
&= \sum_{i=1}^m \sum_{|B|=m-2} [-e_B e_i e_i \omega_i] dx_B \wedge dx_i \\
&= -\sum_{k=1}^m \hat{e}_k [\sum_{i \neq k} e_i \omega_i] d\hat{x}_k \\
&= 0,
\end{aligned}$$

where $\hat{e}_k = e_1 \dots e_{k-1} e_{k+1} \dots e_m$.

It means that the Clifford-valued functions $\alpha_j = e_j \omega_j$ are solutions of a linear system of equations, the matrix of which is

$$A_1 = \begin{bmatrix} 1 & 1 & \dots & 1 & 1 & 0 \\ 1 & 1 & \dots & 1 & 0 & 1 \\ \vdots & \vdots & & \vdots & \vdots & \vdots \\ 0 & 1 & \vdots & 1 & 1 & 1 \end{bmatrix}.$$

It is easy to see that A_1 is a regular matrix, hence all α_i, and consequently all ω_i, are identically zero. ■

So we have found that 1-forms and $m-1$-forms are decomposed into a direct sum of two subspaces and that there are two distinguished subspaces of the space of k-forms for $k = 2, \dots, m-2$. It is more or less clear now how the multiplication by $d\underline{x}$ can be used to find further nice subspace s of k-forms.

Definition 1.2. *Let $j \leq m/2$. The form $\omega \in \mathcal{E}^j$ will be called a primitive j-form if $(d\underline{x})^{m-2j+1} \wedge \omega = 0$. The space of all primitive j-forms in $\mathcal{E}^j$ will be denoted by $\mathcal{E}^{j,j}$.*

For all j such that $0 \leq j \leq k, j \leq m-k$, we put

$$\mathcal{E}^{k,j} = \{\omega \in \mathcal{E}^k | \, \omega = (d\underline{x})^{k-j} \wedge (\omega'), \omega' \in \mathcal{E}^{j,j}\}.$$

Let us illustrate the subspaces thus obtained in a low dimensional case, say $m = 6$:

$$\begin{array}{ccccccc} \mathcal{E}^{0,0} & \mathcal{E}^{1,0} & \mathcal{E}^{2,0} & \mathcal{E}^{3,0} & \mathcal{E}^{4,0} & \mathcal{E}^{5,0} & \mathcal{E}^{6,0} \\ & \mathcal{E}^{1,1} & \mathcal{E}^{2,1} & \mathcal{E}^{3,1} & \mathcal{E}^{4,1} & \mathcal{E}^{5,1} & \\ & & \mathcal{E}^{2,2} & \mathcal{E}^{3,2} & \mathcal{E}^{4,2} & & \\ & & & \mathcal{E}^{3,3} & & & \end{array}$$

Notice that multiplication by $d\underline{x}$ is acting on the rows from the left (starting with primitive forms) to the right.

We want to prove now that the subspaces (appearing in each column) yield a decomposition of the space of k-forms into a direct sum of subspaces. The main ingredient in the proof is the fact that the multiplication by $d\underline{x}$ in individual lines is an isomorphism between the corresponding spaces.

Lemma 1.3. *The map $\omega \mapsto (d\underline{x})^{m-2j} \wedge \omega$ is an isomorphism between $\mathcal{E}^j$ and $\mathcal{E}^{m-j}$.*

Proof. It is sufficient to prove that the map is injective. So let us consider a form $\omega = \sum_{|A|=j} \omega_A dx_A$ such that $(d\underline{x})^{m-2j} \wedge \omega = 0$. Then

$$\begin{aligned} &\Big(\sum_{|B|=m-2j} e_B dx_B\Big) \wedge \Big(\sum_{|A|=j} \omega_A dx_A\Big) \\ &= (-1)^{j(j+1)/2} \sum_{A,B} (e_B e_A e_A \omega_A) dx_B \wedge dx_A \\ &= (-1)^{j(j+1)} \sum_{|C|=m-j} e_C \Big[\sum_{|A|=j} e_A \omega_A\Big] dx_C \\ &= 0. \end{aligned}$$

We thus obtain $\binom{m}{m-j}$ equations for $\binom{m}{j}$ unknowns $\alpha_A = e_A \omega_A$.. As before, we have to prove that the matrix A_j of this system of equations is regular. We have proved it before in the case $j = 1$. The general case will be proved by induction with respect to j. So suppose now that the corresponding matrix is regular for all m and for all $j' < j$. Then we show that any solution $(\alpha_A, |A| = j)$ of the linear system of equations determined by the matrix A_j is trivial. Take any set $I, I \subset \{1, \ldots, m-2\}, |I| = j-1$. The two equations corresponding to the sets $C = I \cup \{m-1\}$ and $C = I \cup \{m\}$ imply that

$$\sum_{J \subset I, |J| = j-1} \alpha_{J \cup \{m-1\}} = \sum_{J \subset I, |J| = j-1} \alpha_{J \cup \{m\}},$$

which means that the $\binom{m-2}{j-1}$ unknowns $\alpha_{J\cup\{m-1\}}$ as well as $\alpha_{J\cup\{m\}}$ satisfy the same linear system of $\binom{m-2}{m-j+1}$ equations. The matrix of this system is of the type considered above, but with $j-1$ instead of j, so the induction hypothesis implies that

$\alpha_{J\cup\{m-1\}} = \alpha_{J\cup\{m\}}$ for all J considered. It thus follows that all α_A in the original system of equations are equal, hence they are all zero. ∎

Theorem 1.4. *For all $k \leq m/2$, we have:*

$$\begin{aligned} \mathcal{E}^k &\cong \mathcal{E}^{k,k} \oplus \mathcal{E}^{k,k-1} \oplus \mathcal{E}^{k,k-2} \ldots\ldots \oplus \mathcal{E}^{k,0}; \\ \mathcal{E}^{m-k} &\cong \mathcal{E}^{m-k,k} \oplus \mathcal{E}^{m-k,k-1} \oplus \mathcal{E}^{m-k,k-2} \oplus \ldots \oplus \mathcal{E}^{m-k,0}. \end{aligned}$$

Proof. We have proved the claim already in the case $k = 1$. Take $j \leq m/2$ and suppose, by induction on j, that the theorem holds for $j-1$. The multiplication by $d\underline{x}$ being an iso morphism (whenever acting between nontrivial spaces), we thus have that the image $d\underline{x} \wedge (\mathcal{E}^{j-1})$ is isomorphic to $\mathcal{E}^{j,j-1} \oplus \ldots \oplus \mathcal{E}^{j,0}$.

Let us consider first $\omega \in (d\underline{x} \wedge \mathcal{E}^{j-1}) \cap \mathcal{E}^{j,j}$. Writing ω as $d\underline{x} \wedge \omega', \omega' \in \mathcal{E}^{j-1}$, we get $(d\underline{x})^{m-2j+2} \wedge \omega' = 0$, whence by Lemma 1.3 $\omega' = 0$ and so the sum of $d\underline{x} \wedge \mathcal{E}^{j-1}$ and $\mathcal{E}^{j,j}$ is direct.

If, on the other hand, $\omega \in \mathcal{E}^j$, then

$$(d\underline{x})^{m-2j+1} \wedge \omega \in \mathcal{E}^{m-j+1},$$

whence there exists $\tau \in \mathcal{E}^{j-1}$ such that

$$(d\underline{x})^{m-2j+2} \wedge \tau = (d\underline{x})^{m-2j+1} \wedge \omega.$$

But the decomposition

$$\omega = d\underline{x} \wedge \tau + [\omega - d\underline{x} \wedge \tau]$$

shows that $\mathcal{E}^j \cong d\underline{x} \wedge (\mathcal{E}^{j-1}) \oplus \mathcal{E}^{j,j}$. Consequently the proof is done for $j \leq m/2$. The proof of the statement for $j \geq m/2$ then follows immediately from Lemma 1.1. ∎

There is, in fact, a dual way how to describe the same splitting. As we saw, the multiplication by $d\underline{x}$ shifts subspaces of forms from the left to the right in the diagram. There is another natural linear operator going from the right to the left; it is a Clifford version of the contraction of a form by a vector field, although it is somewhat more general, since as well the forms as the vector fields can be spinor and C-valued.

Definition 1.5. *Let t be a C-valued vector field, i.e. $t = \sum_1^m t_j \partial_{x_j}, t_j \in \mathbf{R}_{0,m}$, and let $\omega = \sum_{|A|=k} \omega_A dx_A \in \mathcal{E}^k$. Then we define the contraction of ω by t as*

$$t \rfloor \omega = \sum_{j,A} t_j \omega_A (\partial_{x_j} \rfloor dx_A).$$

The notation for the contraction from the right (the form ω taking values in a right module) will be

$$\omega\lfloor t = \sum_{j,A} \omega_A t_j (-1)^{k-1}(\partial_{x_j}\rfloor dx_A).$$

Notice that contraction by a vector field defines a linear operator on the space of all forms; applying the contraction operator k times, we denote shortly

$$(t\rfloor)^k\omega = \underbrace{(t\rfloor(t\rfloor\ldots(t\rfloor}_{k}\omega)\ldots))$$

One of the vector fields which will be constantly used in the sequel is given by the Dirac operator

$$\partial_{\underline{x}} = \sum_{j=1}^{m} e_j \partial_{x_j}.$$

Examples.

(1) It is easily seen that

$$\partial_{\underline{x}}\rfloor d\underline{x} = -m.$$

We shall indicate in a moment how to compute quickly $\partial_{\underline{x}}\rfloor(d\underline{x})^k$. For $k = m$ we have

$$\partial_{\underline{x}}\rfloor(dx_1 \wedge \ldots \wedge dx_m) = d\sigma\,,$$

so $\partial_{\underline{x}}\rfloor((d\underline{x})^m f) = (-m)(d\underline{x})^{m-1} f, f \in \mathcal{E}^0$.

(2) We have

$$(\partial_{\underline{x}}\rfloor)^m dx_1 \wedge \ldots \wedge dx_m f = m!\, e\, f,\; f \in \mathcal{E}^0\,.$$

More generally, it is clear that contraction with $(\partial_{\underline{x}})^k$ maps $\mathcal{E}^m$ into $\mathcal{E}^{m-k,0}$. The question hence arises, if the two maps - the multiplication by $d\underline{x}$ and the contraction by $\partial_{\underline{x}}$ - are inverse one to another. We shall see that this is only partially true and that the structure of the decomposition of the space of forms is closely connected to that question.

The spaces $\mathcal{E}^{k,j}$ were defined above by purely using the properties of the left multiplication by $d\underline{x}$; starting on the left hand side of the picture, we could go to the right by multiplication. As may be expected, starting with the definition of the spaces on the right hand side of the triangle we may go to the left by the contraction with $\partial_{\underline{x}}$.

Theorem 1.6. *For $k \leq m/2$*

$$\mathcal{E}^{m-k,k} = \{\omega \in \mathcal{E}^k | (\partial_{\underline{x}})^{m-2k-1} \rfloor \omega = 0\}$$

and for $j = 0, \ldots, \min(k, m-k)$

$$\mathcal{E}^{k,j} = \{\omega \in \mathcal{E}^k | \omega = (\partial_{\underline{x}})^{m-j-k} \rfloor \omega', \omega' \in \mathcal{E}^{m-j,j}\}.$$

It is clear that the subspaces, defined above using the contraction, form a decomposition of $\mathcal{E}^k$ similar to the one defined using the multiplication by $d\underline{x}$ (a possible proof is just a mirror image of t he proofs before). Theorem 1.6 states that both decompositions coincide.

In order to prove Theorem 1.6, let us first state some basic lemmas giving important information on the commutator $[d\underline{x}, \partial_{\underline{x}}]$ between $d\underline{x}$ and $\partial_{\underline{x}}$.

Let us prove first the following basic lemma giving an important information on the commutator of both maps.

Lemma 1.7. *For each $\omega \in \mathcal{E}^k$*

$$\partial_{\underline{x}} \rfloor (d\underline{x} \wedge \omega) - d\underline{x} \wedge (\partial_{\underline{x}} \rfloor \omega) = -(m-2k)\omega.$$

Proof. Put $\omega = \sum_{|A|=k} \omega_A dx_A$; then

$$\partial_{\underline{x}} \rfloor (d\underline{x} \wedge \omega) = \sum_{i,j,A;\, j \notin A} e_i e_j \omega_A [\partial_{x_i} \rfloor (dx_j \wedge dx_A)]$$

$$= \sum_{i,j,A;\, i=j;\, j \notin A} e_i e_j \omega_A [\partial_{x_i} \rfloor (dx_j \wedge dx_A)]$$

$$+ \sum_{i,j,A;\, i \neq j;\, j \notin A} e_i e_j \omega_A [\partial_{x_i} \rfloor (dx_j \wedge dx_A)]$$

$$= -(m-k) \sum_A \omega_A dx_A - \sum_{i,j,A;\, i \neq j;\, j \notin A} e_i e_j \omega_A [dx_j \wedge dx_{A \setminus \{i\}} \mathrm{sgn}(i, A)]$$

where $\mathrm{sgn}(i, A)$ denotes the sign of the permutation changing the set $\{i, A \setminus \{i\}\}$ into the ordered set A.

Similarly

$$d\underline{x} \wedge (\partial_{\underline{x}} \rfloor \omega) = \sum_{i,j,A} e_i e_j \omega_A [dx_j \wedge (\partial_{x_i} \rfloor dx_A)]$$

$$= \sum_{i,j,A;\, i=j} e_i e_j \omega_A [dx_j \wedge (\partial_{x_i} \rfloor dx_A)] + \sum_{i,j,A;\, i \neq j;\, j \notin A} e_i e_j \omega_A [dx_j \wedge (\partial_{x_i} \rfloor dx_A)]$$

$$= -k \sum_A \omega_A dx_A - \sum_{i,j,A;\, i \neq j;\, j \notin A} e_i e_j \omega_A [dx_j \wedge dx_{A \setminus \{i\}} \mathrm{sgn}(i, A)]. \blacksquare$$

As a consequence we get

Lemma 1.8. *For all* $\omega \in \mathcal{E}^{k,j}, k = j, \ldots, m-j$,

$$d\underline{x} \wedge (\partial_{\underline{x}} \rfloor \omega) = (-m+k+j-1)(k-j)\omega$$

$$\partial_{\underline{x}} \rfloor (d\underline{x} \wedge \omega) = (-m+k+j)(k-j+1)\omega.$$

Proof. Using Lemma 1.7 we have for $k = m - j$ that

$$\partial_{\underline{x}} \rfloor (d\underline{x} \wedge \omega) = 0$$

and

$$d\underline{x} \wedge (\partial_{\underline{x}} \rfloor \omega) = -(m-2j)\omega.$$

This implies immediately that for $k = m - j - 1$

$$\partial_{\underline{x}} \rfloor (d\underline{x} \wedge \omega) = -(m-2j)\omega.$$

The rest is done by induction, taking into account that

$$(-m+2j) + \ldots + (-m+2k) = (-m+k+j)(k-j+1). \quad \blacksquare$$

Theorem 1.6 is then a consequence of Lemma 1.8 just proved.

It is clear now that there are a lot of possibilities to characterize the spaces $\mathcal{E}^{k,j}$ by the property that the composition of a suitable number of contractions by $\partial_{\underline{x}}$ and a suitable number of multiplications by $d\underline{x}$ is equal to zero. We shall use often the fact that

$$\mathcal{E}^{j,j} = \{\omega \in \mathcal{E}^j | \partial_{\underline{x}} \rfloor \omega = 0\}$$

for $j \leq m/2$.

Notice that for the direct sum decompositions obtained in Theorem 1.4, we may write out the projections onto $\mathcal{E}^{k,j}$ explicitly.

For example, it is easy to see that

$$\omega = -\frac{1}{m} d\underline{x} \wedge (\partial_{\underline{x}} \rfloor \omega) + [\omega + \frac{1}{m} d\underline{x} \wedge (\partial_{\underline{x}} \rfloor \omega)]$$

is the decomposition of an 1-form into its corresponding pieces.

The general form of the projection $\pi^{k,j}$ onto $\mathcal{E}^{k,j}$ follows by induction:

$$\begin{aligned}
\pi^{k,0}\omega &= \frac{1}{(k-m-1)k} d\underline{x} \wedge \pi^{k-1,0}(\partial_{\underline{x}} \rfloor \omega) \\
\pi^{k,1}\omega &= \frac{1}{(-m+k)(k-1)} d\underline{x} \wedge \pi^{k-1,1}(\partial_{\underline{x}} \rfloor \omega) \\
&\vdots \\
\pi^{k,k-1}\omega &= \frac{1}{(-m+2k-2)} d\underline{x} \wedge \pi^{k-1,k-1}(\partial_{\underline{x}} \rfloor \omega) \\
\pi^{k,k}\omega &= \omega - \pi^{k,0}\omega - \ldots - \pi^{k,k-1}\omega.
\end{aligned}$$

The splitting can be described by a triangle-shaped diagram which will be worked out explicitly in a few low dimensional cases. The arrows will indicate the maps $d\underline{x}\wedge.$ and $\partial_{\underline{x}} \rfloor .$. Their composition is a multiple of the identity and the factor is indicated above the arrow.

(i) $k = 4$

$$\begin{array}{ccccccccc}
\mathcal{E}^0 & \overset{-4}{\longleftrightarrow} & \mathcal{E}^{1,0} & \overset{-6}{\longleftrightarrow} & \mathcal{E}^{2,0} & \overset{-6}{\longleftrightarrow} & \mathcal{E}^{3,0} & \overset{-4}{\longleftrightarrow} & \mathcal{E}^4 \\
 & & \oplus & & \oplus & & \oplus & & \\
 & & \mathcal{E}^{1,1} & \overset{-2}{\longleftrightarrow} & \mathcal{E}^{2,1} & \overset{-2}{\longleftrightarrow} & \mathcal{E}^{3,1} & & \\
 & & & & \oplus & & & & \\
 & & & & \mathcal{E}^{2,2} & & & &
\end{array}$$

(ii) $k = 5$

$$\begin{array}{ccccccccccc}
\mathcal{E}^0 & \overset{-5}{\longleftrightarrow} & \mathcal{E}^{1,0} & \overset{-8}{\longleftrightarrow} & \mathcal{E}^{2,0} & \overset{-9}{\longleftrightarrow} & \mathcal{E}^{3,0} & \overset{-8}{\longleftrightarrow} & \mathcal{E}^{4,0} & \overset{-5}{\longleftrightarrow} & \mathcal{E}^5 \\
 & & \oplus & & \oplus & & \oplus & & \oplus & & \\
 & & \mathcal{E}^{1,1} & \overset{-3}{\longleftrightarrow} & \mathcal{E}^{2,1} & \overset{-4}{\longleftrightarrow} & \mathcal{E}^{3,1} & \overset{-3}{\longleftrightarrow} & \mathcal{E}^{4,1} & & \\
 & & & & \oplus & & \oplus & & & & \\
 & & & & \mathcal{E}^{2,2} & \overset{-1}{\longleftrightarrow} & \mathcal{E}^{3,2} & & & &
\end{array}$$

The spaces in the rows are of the same size: they have the same dimension as $\mathcal{E}^0$ -modules. The number of pieces increases by 1 up to the middle and then decreases by 1. The longest column appears in the middle for even dimensions, while for odd dimensions there are two columns of equal length there.

There is an alternative approach to the description of the above spaces based on Spin(m)-invariance. It was described in [25] that if S is one of the basic (irreducible) Spin(m)-modules and if $\mathcal{E}^{k,j}$ are irreducible components of $\Lambda^k((\mathbf{R}^n)^*) \otimes_{\mathbf{R}} S$, the latter being considered as a Spin(m)-module, then $\mathcal{E}^{k,j} \simeq \mathcal{E}^0(\Omega, \Lambda^k((\mathbf{R}^n)^*) \otimes_{\mathbf{R}} S)$. If S is a sum of such irreducible modules (which is always true for a finite-dimensional $\mathbf{R}_{0,m}$-module), then there will be some pieces repeated, corresponding to the multiplicities of the basic spinor modules inside S. The spaces $\mathcal{E}^{k,j}$ however remain invariant subspaces and the highest weights of their irreducible parts are the same as given in [25] for the case of an irreducible module S.

2 Invariant operators on forms

The monogenicity condition introduced below will be expressed by a combination of first order partial differential operators given by a composition of the de Rham differential d with projections onto the pieces in the decomposition of spinor valued forms.

Definition 2.1. *Let again $\pi^{k,j}$ stand for the projection*

$$\pi^{k,j} : \mathcal{E}^k \to \mathcal{E}^{k,j}, \ j \leq k, l \leq m-k.$$

Then we define the operators $d^{(j)}, \partial_{\underline{x}}^{(j)}$ by

$$d^{(j)} = \pi^{k,j} \circ d, \ \ \partial_{\underline{x}}^{(j)} = \pi^{k,j} \circ \partial_{\underline{x}} \, .$$

Notice that the operators $d^{(j)}$ and $\partial_{\underline{x}}^{(j)}$ are Spin(m)-invariant (see [25]). The operators $d^{(i)}$ can be considered as building bloks ("atoms") for Spin(m)-invariant first-order differential operators from $\mathcal{E}^k$ to $\mathcal{E}^{k+1}$ since each operator of this type can be written as linear combinations of them.

Not all the operators $d^{(i)}$ are nontrivial on $\mathcal{E}^k$; it is somewhat surprising that they are nontrivial only for a few exceptional cases. As may be expected, we define the action of $\partial_{\underline{x}}$ on an S-valued differential form as follows.

Definition 2.2. *Let $k = 0, \ldots, m$ and let $\omega \in \mathcal{E}^k$ with $\omega = \sum_{|A|=k} \omega_A dx_A$. Then we put*

$$\partial_{\underline{x}} \, \omega = \sum_{|A|=k} (\partial_{\underline{x}} \, \omega_A) dx_A.$$

We so have

Lemma 2.3. *For each $\omega \in \mathcal{E}^k$, $\partial_{\underline{x}} \, \omega = d(\partial_{\underline{x}} \rfloor \omega) + \partial_{\underline{x}} \rfloor (d\omega)$.*

Proof. If $\omega = \sum_{|A|=k} \omega_A dx_A$, then

$$\begin{aligned} \partial_{\underline{x}} \rfloor d\omega &= \sum_{i,j,A} e_i \partial_{x_j} \omega_A [\partial_{x_i} \rfloor (dx_j \wedge dx_A)] \\ &= \sum_{i,A,i \notin A} e_i \partial_{x_i} \omega_A dx_A - \sum_{i,j,A,i \neq j} e_i \partial_{x_j} \omega_A dx_j \wedge (\partial_{x_i} \rfloor dx_A) \end{aligned}$$

and

$$\begin{aligned} d(\partial_{\underline{x}} \rfloor \omega) &= \sum_{i,j,A} e_i \partial_{x_j} \omega_A [dx_j \wedge (\partial_{x_i} \rfloor dx_A)] \\ &= \sum_{i,A,i \in A} e_i \partial_{x_i} \omega_A dx_A + \sum_{i,j,A,i \neq j} e_i \partial_{x_j} \omega_A dx_j \wedge (\partial_{x_i} \rfloor dx_A). \end{aligned}$$

The sum of the two terms gives exactly $\partial_{\underline{x}} \, \omega$. ∎

The following lemma provides important information about the action of the Dirac operator on forms and, as a consequence, about the behaviour of the de Rham operator d.

Lemma 2.4. *If* $\partial_{\underline{x}} \rfloor \omega = 0$, *then* $\partial_{\underline{x}} \rfloor (\partial_{\underline{x}} \rfloor \partial_{\underline{x}} \omega) = 0$.

Proof. Let us write $\omega \in \mathcal{E}^k$ as

$$\omega = \sum_{i_1,\ldots,i_k=1}^{m} \omega_{i_1\ldots i_k} dx_{i_1} \wedge \ldots \wedge dx_{i_k},$$

where the components $\omega_{i_1\ldots i_k}$ are antisymmetric in their indices. The condition $\partial_{\underline{x}} \rfloor \omega = 0$ implies that

$$0 = \sum_{i_1,\ldots,i_k} [e_{i_1}\omega_{i_1\ldots i_k} dx_{i_2} \wedge \ldots \wedge dx_{i_k} - e_{i_2}\omega_{i_1\ldots i_k} dx_{i_1} \wedge dx_{i_3} \wedge \ldots \wedge dx_{i_k} + \ldots]$$

$$= k \sum_{i_2,\ldots,i_k} [\sum_j e_j \omega_{j i_1\ldots i_k}] dx_{i_2} \wedge \ldots \wedge dx_{i_k},$$

whence $\sum_j e_j \omega_{j i_1\ldots i_k} = 0$ for all $i_2, \ldots, i_k = 1, \ldots, m$.

Consequently

$$\partial_{\underline{x}} \rfloor (\partial_{\underline{x}} \rfloor \partial_{\underline{x}} \omega) = \sum_{i,j,l} \sum_{i_3,\ldots,i_k} k(k-1)[e_i e_j e_l \partial_{x_l} \omega_{i j i_3\ldots i_k}] dx_{i_3} \wedge \ldots \wedge dx_{i_k}$$

$$= 2 \sum_{i,j,l;\, j=l} \ldots + \sum_{i,j,l;\, j\neq l} \ldots - \sum_{i,j,l;\, j=l} \ldots$$

$$= -2k(k-1) \sum_{i_3,\ldots,i_k} [\sum_l \partial_{x_l} (\sum_i e_i \omega_{i l i_3\ldots i_k})] dx_{i_3} \wedge \ldots \wedge dx_{i_k}$$

$$-k(k-1) \sum_{i,j,l} \sum_{i_3,\ldots,i_k} e_i e_l \partial_{x_l} (\sum_j (e_j \omega_{i j i_3\ldots i_k}) dx_{i_3} \wedge \ldots \wedge dx_{i_k})$$

$$= 0. \blacksquare$$

Notice that for $j = k, k < m/2$ (i.e. $\omega \in \mathcal{E}^{k,k}$) we have:

$$\partial_{\underline{x}} \omega = \partial_{\underline{x}} \rfloor d\omega$$

and

$$\partial_{\underline{x}}^{(i)} \omega = \partial_{\underline{x}} \rfloor d^{(i)} \omega, \; i = k-1, k, k+1.$$

Moreover $\partial_{\underline{x}}^{(k+1)} \omega = 0$ for each $\omega \in \mathcal{E}^{k,k}$, while $d^{(k+1)} \omega$ need not be zero.

Theorem 2.5. *Let $\omega \in \mathcal{E}^{k,j}$. Then*

$$d\omega = d^{(j-1)}\omega + d^{(j)}\omega + d^{(j+1)}\omega$$

and

$$\partial_{\underline{x}}\,\omega = \partial_{\underline{x}}^{(j-1)}\omega + \partial_{\underline{x}}^{(j)}\omega + \partial_{\underline{x}}^{(j+1)}\omega.$$

Proof. If $\omega \in \mathcal{E}^{j,j}, j \leq m/2$, (i.e. $\partial_{\underline{x}} \rfloor \omega = 0$), then Lemma 2.4 implies that $\partial_{\underline{x}}\,\omega$ belongs to $\mathcal{E}^{j,j-1} \oplus \mathcal{E}^{j,j}$. Moreover $\partial_{\underline{x}}\,\omega = \partial_{\underline{x}} \rfloor\, d\omega$, so the theorem is true in this case.

In the general case it is sufficient to note that if $\omega = (d\underline{x})^{k-j} \wedge \omega', \omega' \in \mathcal{E}^{j,j}$, then (applying the projection $\pi^{k,l}$)

$$d^{(l)}\omega = (-1)^{k-j}(d\underline{x})^{k-j} \wedge d^{(l)}\omega'.$$

By the definition of the action of $\partial_{\underline{x}}$ on forms and properties already proved, the following diagram illustrates the case of the Dirac operator.

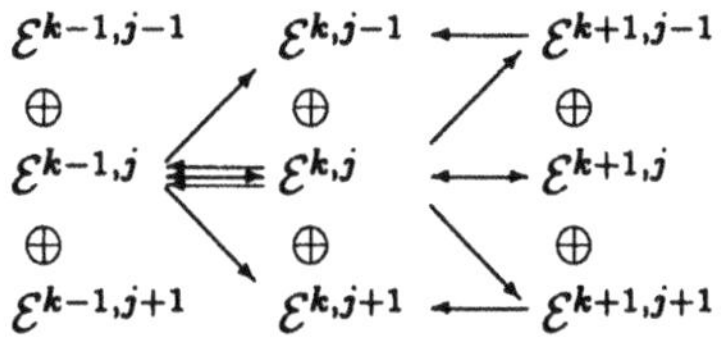

■

Detailed information on the action of the differential operators $d^{(i)}$ is summarized in the following

Lemma 2.6. *Let $\omega \in \mathcal{E}^{k,j}$, $j < k$, $j \leq m - k$, and suppose that $\omega = (d\underline{x})^{k-j} \wedge \omega'$, $\omega' \in \mathcal{E}^{j,j}$. Then for $i = j-1, j, j+1$,*

$$\begin{aligned} d^{(i)}(\partial_{\underline{x}} \rfloor \omega) &= (-m+k+j-1)(k-j)(-1)^{k-j-1}(d\underline{x})^{k-j-1} \wedge d^{(i)}\omega', \\ \partial_{\underline{x}} \rfloor (d^{(i)}\omega) &= (-m+k+i)(k-i+1)(-1)^{k-j}(d\underline{x})^{k-j-1} \wedge d^{(i)}\omega' \end{aligned}$$

and

$$\begin{aligned} \partial_{\underline{x}}^{(j-1)}\omega &= (-1)^{k-j}(j-1+k-m)(d\underline{x})^{k-j-1} \wedge d^{(j-1)}\omega'. \\ \partial_{\underline{x}}^{(j)}\omega &= (-1)^{k-j}(-m+2k)(d\underline{x})^{k-j-1} \wedge d^{(j)}\omega'. \\ \partial_{\underline{x}}^{(j+1)}\omega &= (-1)^{k-j}2(k-j)(d\underline{x})^{k-j-1} \wedge d^{(j+1)}\omega'. \end{aligned}$$

Proof. We have

$$\begin{aligned}\partial_{\underline{x}} \rfloor \omega &= \partial_{\underline{x}} \rfloor \{ d\underline{x} \wedge [(d\underline{x})^{k-j-1} \wedge \omega'] \} \\ &= (-m+k+j-1)(k-j)(d\underline{x})^{k-j-1} \wedge \omega',\end{aligned}$$

whence

$$d^{(i)}(\partial_{\underline{x}} \rfloor \omega) = (-1)^{k-j-1}(-m+k+j-1)(k-j)(d\underline{x})^{k-j-1} \wedge d^{(i)}\omega'.$$

On the other hand

$$\begin{aligned}\partial_{\underline{x}} \rfloor \omega &= (-1)^{k-j} \partial_{\underline{x}} \rfloor (d\underline{x} \wedge [(d\underline{x})^{k-j-1} \wedge d^{(i)}\omega']) \\ &= (-1)^{k-j}(-m+k+i)(k-i+1)(d\underline{x})^{k-j-1} \wedge d^{(i)}\omega'. \quad \blacksquare\end{aligned}$$

Theorem 2.7. *Let $\omega \in \mathcal{E}^{k,j}$. Then the operators $\omega \mapsto d\underline{x} \wedge \partial_{\underline{x}} \omega$ and $\omega \mapsto \partial_{\underline{x}} (d\underline{x} \wedge \omega)$ are determined by*

$$\begin{aligned}& d\underline{x} \wedge \partial_{\underline{x}} \omega \\ &= 2(j-1+k-m)d^{(j-1)}\omega + (2k-m)d^{(j)}\omega + 2(k-j)d^{(j+1)}\omega,\end{aligned}$$

$$\begin{aligned}& \partial_{\underline{x}} (d\underline{x} \wedge \omega) \\ &= -2(j+k-m)d^{(j-1)}\omega - (2k-m+2)d^{(j)}\omega - 2(k-j+1)d^{(j+1)}\omega.\end{aligned}$$

Moreover, the following relation is valid:

$$2d\omega = -\partial_{\underline{x}} (d\underline{x} \wedge \omega) - d\underline{x} \wedge \partial_{\underline{x}} \omega.$$

Proof. The first relation follows from Lemma 2.6 by multiplication by $d\underline{x}$. To prove the second one it is sufficient to use Lemma 2.6 for $d\underline{x} \wedge \omega$. The last relation then follows immediately, but it is also possible to prove it directly by straithforward computation using coordinates. ■

Let us recall finally that, by letting act the operators from the right, i.e. by replacing the maps $\omega \mapsto d\underline{x} \wedge \omega$ and $\omega \mapsto \partial_{\underline{x}} \rfloor \omega$ by their right versions $\omega \mapsto \omega \wedge d\underline{x}$ and $\omega \mapsto \omega \lfloor \partial_{\underline{x}}$ and by defining the pieces $\mathcal{E}_r^{k,j}$ by

$$\mathcal{E}_r^{k,j} = \{ \omega = \omega' \wedge (d\underline{x})^{k-j} \mid \omega' \in \mathcal{E}^j{}_r, \omega' \lfloor \partial_{\underline{x}} = 0 \},$$

we may as well obtain decompositions of the spaces $\mathcal{E}^k$. When both types of forms are used at the same time, the pieces in the decomposition described above (the left version) will be denoted by $\mathcal{E}_\ell^{k,j}$.

3 MONOGENIC DIFFERENTIAL FORMS

The differential operators $d^{(i)}$ introduced in the last section will be used now to define monogenic differential forms.

To motivate the definition, let us recall the definition of holomorphic forms in the plane. The space of 1-forms is split there into a direct sum of two pieces

$$\mathcal{E}^1 = \mathcal{E}^{1,0} \oplus \mathcal{E}^{0,1}$$

and the value of the $\bar{\partial}$ -operator on a function f is defined to be the composition of the de Rham operator d and the projection onto the (0,1)-part of df. We have the diagram

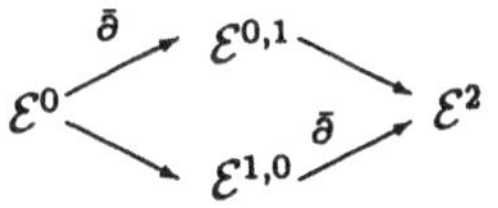

and holomorphic functions and holomorphic 1-forms are elements of the kernels of the map $\bar{\partial}$.

A similar procedure in higher dimension is based on a splitting

$$\mathcal{E}^k = \mathcal{E}^{k'} \oplus \mathcal{E}^{k''}.$$

It leads to the diagram

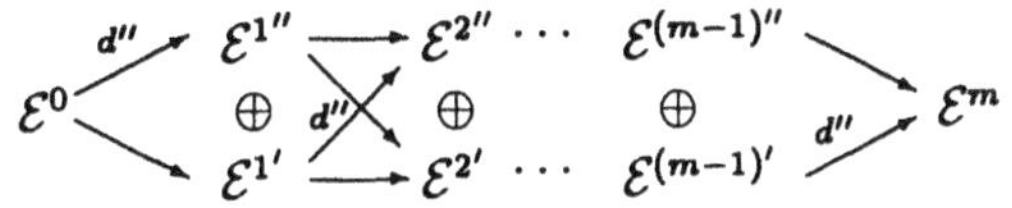

where the operators d'' are defined as the composition of d with the projection onto $\mathcal{E}^{k''}$. Monogenic forms could then be defined as elements of the kernel of the operators d'', the main question being however how to define the spaces $\mathcal{E}^{k'}$ and $\mathcal{E}^{k''}$.

The requirement of invariance tells us that we have to choose for the primed and double primed parts sums of pieces in the decomposition described above. One such possibility was worked out in [79]. It was shown there that if we take $\mathcal{E}^{k''} = \mathcal{E}^{k,0}$, then the homology of the monogenic forms describes correctly the homology groups of the domain. The disadvantage of such a choice is the fact that a suitable analogue of the Cauchy theorem is not available. To have such an analogue, it is necessary to make all $\mathcal{E}^{k''}$ much bigger. But choosing them bigger, there is a danger that homology will not reproduce the topology of the domain anymore.

Let us describe now a suitable splitting of $\mathcal{E}^k$ having the property that monogenic forms describe correctly homological properties of the domains of definition of forms and that the Cauchy theorem holds for them.

To describe the splitting, we have to distinguish between the cases of odd and even dimensions.

I. The case m = 2l+1.

Definition 3.1. *Put for* $1 \le k \le l$

$$\mathcal{E}^{k'} = \bigoplus_{\substack{0 \le j \le k \\ j \text{ even}}} \mathcal{E}^{k,k-j}; \quad \mathcal{E}^{k''} = \bigoplus_{\substack{0 \le j \le k \\ j \text{ odd}}} \mathcal{E}^{k,k-j} \tag{3.1}$$

and

$$\mathcal{E}^{(m-k)'} = \bigoplus_{\substack{0 \le j \le k \\ j \text{ odd}}} \mathcal{E}^{m-k,k-j}; \quad \mathcal{E}^{(m-k)''} = \bigoplus_{\substack{0 \le j \le k \\ j \text{ even}}} \mathcal{E}^{m-k,k-j}. \tag{3.2}$$

Let us illustrate the definition in the case $m = 9$. All spaces involved are finite dimensional $\mathcal{E}^0$ - modules, so the individual subspaces $\mathcal{E}^{k,j}$ are denoted by their dimensions in the diagram. The pieces, belonging to $\mathcal{E}^{k'}$, are indicated by boxes. In the top row the dimension of the full spaces $\mathcal{E}^k$ is written.

1	9	36	84	126	126	84	36	9	1
$\boxed{1}$	1	$\boxed{1}$	1	$\boxed{1}$	1	$\boxed{1}$	1	$\boxed{1}$	1
	$\boxed{8}$	8	$\boxed{8}$	8	$\boxed{8}$	8	$\boxed{8}$	8	
		$\boxed{27}$	27	$\boxed{27}$	27	$\boxed{27}$	27		
			$\boxed{48}$	48	$\boxed{48}$	48			
				$\boxed{42}$	42				

As to the dimensions of the irreducible pieces and of $\mathcal{E}^{k'}$ we find

Lemma 3.2. *Let* $m = 2l + 1$ *and let* $1 \le k \le l$. *Then*

$$\dim \mathcal{E}^{k,j} = \dim \mathcal{E}^{j,j} = \left\{ \binom{m}{j} - \binom{m}{j-1} \right\} \dim\ S;$$

$$\dim \mathcal{E}^{k'} = \binom{m-1}{k}.$$

Proof. The relation $d\underline{x} \wedge \mathcal{E}^{j-1} = \mathcal{E}^{j,0} \oplus \ldots \oplus \mathcal{E}^{j,j-1}$ implies that

$$\dim \mathcal{E}^{j,j} = \dim \mathcal{E}^{j} - \dim \mathcal{E}^{j-1}.$$

To prove the second part of the Lemma, it is sufficient to use the relation

$$\sum_{j=0}^{k}(-1)^{k-j}\binom{m}{j} = \binom{m-1}{k}. \quad \blacksquare$$

II. The case m = 2l.

The basic reason why the even-dimensional case is special is the fact that the l-forms split into self-dual and antiself-dual forms under the Hodge $*$- map. It can be shown that the two choices possible in the middle dimension correspond to the selfdual and antiselfdual forms.

Definition 3.3.

(i) Put for $k \leq l$

$$\mathcal{E}^{k'} = \bigoplus_{\substack{0 \leq j \leq k \\ j \text{ even}}} \mathcal{E}^{k,k-j}; \quad \mathcal{E}^{k''} = \bigoplus_{\substack{0 \leq j \leq k \\ j \text{ odd}}} \mathcal{E}^{k,k-j} \tag{3.3}$$

and for $k < l$

$$\mathcal{E}^{(m-k)'} = \bigoplus_{\substack{0 \leq j \leq k \\ j \text{ odd}}} \mathcal{E}^{m-k,k-j}; \quad \mathcal{E}^{(m-k)''} = \bigoplus_{\substack{0 \leq j \leq k \\ j \text{ even}}} \mathcal{E}^{m-k,k-j} \tag{3.4}$$

(ii) Define $\widetilde{\mathcal{E}^{k'}}$ and $\widetilde{\mathcal{E}^{k''}}$ by (3.3) for $k < l$ and by (3.4) for $k \leq l$.

Note that both possibilities differ only in the middle dimension. We have $\mathcal{E}^{k'} = \widetilde{\mathcal{E}^{k'}}$, $\mathcal{E}^{k''} = \widetilde{\mathcal{E}^{k''}}$ for all $k \neq l$ and $\mathcal{E}^{l'} = \widetilde{\mathcal{E}^{l''}}, \mathcal{E}^{l''} = \widetilde{\mathcal{E}^{l'}}$. The dimensions of the spaces $\mathcal{E}^{l'}$ and $\mathcal{E}^{l''}$ are the same and they are equal to $\binom{m}{l}$.

Let us illustrate both possibilities explicitly in the case $m = 8$.

(i) The spaces $\mathcal{M}$

1	8	28	56	70	56	28	8	1
$\boxed{1}$	1	$\boxed{1}$	1	$\boxed{1}$	$\boxed{1}$	1	$\boxed{1}$	1
	$\boxed{7}$	7	$\boxed{7}$	7	7	$\boxed{7}$	7	
		$\boxed{20}$	20	$\boxed{20}$	$\boxed{20}$	20		
			$\boxed{28}$	28	28			
				$\boxed{14}$	42			

(ii) The spaces $\widetilde{\mathcal{M}}$

1	8	28	56	70	56	28	8	1
$\boxed{1}$	1	$\boxed{1}$	1	1	$\boxed{1}$	1	$\boxed{1}$	1
	$\boxed{7}$	7	$\boxed{7}$	$\boxed{7}$	7	$\boxed{7}$	7	
		$\boxed{20}$	20	20	$\boxed{20}$	20		
			$\boxed{28}$	$\boxed{28}$	28			
				14				

We are now in position to define monogenic differential forms as well in the even as in the odd dimensional cases.

Definition 3.4. *Put*

$$\mathcal{M}^k{}_\ell = \{\omega \in \mathcal{E}^{k'}_\ell \mid d''\omega = 0\}$$

and

$$\widetilde{\mathcal{M}^k{}_\ell} = \{\omega \in \tilde{\mathcal{E}}^{k'}_\ell \mid d''\omega = 0\}.$$

The forms $\omega \in \mathcal{M}^k{}_\ell$ (resp. $\omega \in \widetilde{\mathcal{M}^k{}_\ell}$) will be called (left) monogenic k-forms.

Right monogenic k-forms are defined analogously and the corresponding spaces are denoted by, respectively, $\mathcal{M}^k{}_r$ and $\widetilde{\mathcal{M}^k{}_r}$.

Note that the definition of monogenicity can reformulated in the following way:

$$\omega \in \widetilde{\mathcal{M}^k{}_\ell} \iff d\omega \in \mathcal{E}^{k'}_\ell.$$

Moreover, it is clear that the space $\mathcal{M}^0{}_\ell$ is just the usual space of (left) monogenic functions and that the space $\mathcal{M}^{m-1}{}_\ell$ of monogenic $(m-1)$-forms consists of all forms $\omega = d\sigma f$, where f is a left monogenic function.

The next section will be devoted to the description of the basic properties of monogenic forms.

4 Homology of monogenic differential forms

In this section we discuss the complex of monogenic differential forms on an open subset $\Omega \subset \mathbf{R}^m$. We want to show that the homology of the complex is describing faithfully the topology of the domain Ω.

In the holomorphic case (n=2), it is well known that the number of holes in a domain in the complex plane can be counted using holomorphic functions, defined on the domain and having no primitive there. More conceptually, the same fact can be described by saying that the homology of the sequence

$$0 \to C \to \Gamma(\Omega, \mathcal{O}) \xrightarrow{d} \Gamma(\Omega, \mathcal{O}^{1,0}) \to 0, \tag{4.1}$$

– where $\Gamma(\Omega, \mathcal{O})$ denotes the space of holomorphic functions on $\Omega \subset \mathbf{R}^2$ and $\Gamma(\Omega, \mathcal{O}^{1,0})$ denotes the space of holomorphic 1-forms on Ω – coincides with the corresponding de Rham cohomology groups. So, for example, as for the number of holes in Ω, we have

$$H_1(\Omega, C) \cong \Gamma(\Omega, \mathcal{O}^{1,0}) / \operatorname{Im} d.$$

Going back to general dimensions, let us consider first the following abstract situation. Let

$$\begin{array}{ccccccccccc} 0 & \to & A_{-1} & \xrightarrow{i} & A_0 & \xrightarrow{d} & A_1 & \xrightarrow{d} & \dots & \xrightarrow{d} & A_n & \to & 0 \\ & & = & & \cup & & \cup & & & & \cup & & \\ 0 & \to & B_{-1} & \xrightarrow{i} & B_0 & \xrightarrow{d} & B_1 & \xrightarrow{d} & \dots & \xrightarrow{d} & B_n & \to & 0 \end{array}$$

be a complex of vector spaces (i.e. $d^2 = 0$) and its subcomplex.

If H_j^A and H_j^B are homologies of (A_i, d), resp. (B_i, d), we determine first a condition under which the natural inclusion

$$i : H_j^B \mapsto H_j^A$$

is an isomorphism.

Lemma 4.1. *Suppose that for all $j = 0, \dots, n$ and for every $\omega \in A_j$ such that $d\omega \in B_{j+1}$, there is an element $\tau \in A_{j-1}$ such that $\omega - d\tau \in B^j$. Then*

$$H_j^B \simeq H_j^A$$

for all $j = 0, \dots, n$.

Proof. Take $j = 0, \dots, n$ fixed. Then we first prove that the map i is onto. Suppose that $\omega \in A_j$ with $d\omega = 0$. Then there exists an element $\tau \in A_{j-1}$ such that $\omega - d\tau \in B_j$. It means that the class of $\omega - d\tau$ in H_j^B is mapped by i to the class of ω in H_j^A.

Next we show that the map i is injective. Suppose that the class of $\omega \in B_j$ with $d\omega = 0$ is mapped to zero in H_j^A. It means that there exists an element $\tau \in A_{j-1}$

such that $\omega = d\tau$, whence there is $\nu \in A_{j-2}$ such that $\tau - d\nu \in B_{j-1}$. But then $d(\tau - d\nu) = d\tau = \omega$. So ω is trivial in H_j^B. ■

The condition in Lemma 4.1 is expressing exactly the fact that the quotient complex A_i/B_i (with the inherited, well-defined differential) is exact. It means that no homology is lost by the restriction to B_i.

Let us apply now the procedure described above to the complex of monogenic differential forms. To this end consider again the splitting $\mathcal{E}^k = \mathcal{E}^{k'} \oplus \mathcal{E}^{k''}$ together with the projections $\pi' : \mathcal{E}^k \mapsto \mathcal{E}^{k'}$, $\pi'' \mapsto \mathcal{E}^k \mapsto \mathcal{E}^{k''}$. Recall that $d' = \pi' od$, $d'' = \pi'' od$.

Theorem 4.2. *The map $d'' : \mathcal{E}^k \mapsto \mathcal{E}^{k+1''}$, $k = 0, \ldots, m-1$, is onto.*

The main step in the proof of Theorem 4.2 is the following property, which is interesting for its own. It shows that horizontal arrows in the diagram describe surjective maps.

Lemma 4.3. *The map $d^{(j)} : \mathcal{E}^{k,j} \mapsto \mathcal{E}^{k+1,j}$, $j = 0, \ldots, \min\{k, m-k-1\}$, is onto.*

Proof. It is sufficient to prove the lemma for the case $k = j$, since the other cases may be easily deduced from it by using properties of the shifts $\omega \mapsto d\underline{x} \wedge \omega$ (see Lemma 2.6).

The basic fact which we shall use is that the map $\partial_{\underline{x}} : \mathcal{E}^k \mapsto \mathcal{E}^k$ is surjective, which follows immediately from the surjectivity of the Dirac operator on functions. As a consequence, the image of $\mathcal{E}^k$ under the map $\omega \mapsto d\underline{x} \wedge \partial_{\underline{x}} \omega$ is the space $\mathcal{E}^{k+1,0} \oplus \ldots \oplus \mathcal{E}^{k+1,k}$ (for $k < (m-1)/2$), resp. the whole $\mathcal{E}^{k+1}$ (for $k \geq (m-1)/2$).

Let us consider now $\omega \in \mathcal{E}^{j+1,j}$ fixed. Then there is an element $\tau \in \mathcal{E}^k$ such that $d\underline{x} \wedge \partial_{\underline{x}} \tau = \omega$. The form τ can be decomposed into $\tau = \sum_{i=0}^{j} \tau_i, \tau_i \in \mathcal{E}^{j,i}$. Theorem 2.7 implies then that $d^{(j)}(\tau_{j-1} + \tau_j) = \omega$. Repeating the procedure once more, we get that there is an element $\nu = \nu_{j-2} + \nu_{j-1} \in \mathcal{E}^{j-1,j-2} \oplus \mathcal{E}^{j-1,j-1}$ such that $d^{(j-1)}\nu = \tau_{j-1}$. Then $d^{(j)}(\tau - d\nu) = \omega$, whence the component of $\tau - d\nu$ in $\mathcal{E}^{j,j-1}$ is zero. By virtue of Theorem 2.7 $d^{(j)}(\tau_j - d^{(j)}\nu) = d^{(j)}(\tau - d\nu) = \omega$. ■

Proof of Theorem 4.2 Let us first discuss the case $k < m/2$. As

$$\mathcal{E}^{(k+1)''} = \bigoplus_{\substack{0 \leq j \leq k+1 \\ j \text{ odd}}} \mathcal{E}^{k+1,k+1-j}, \tag{4.2}$$

we have that $\mathcal{E}^{(k+1)''} \subset \mathcal{E}^{k+1,0} \oplus \ldots \oplus \mathcal{E}^{k+1,k}$.

Now let ω be an element of $\mathcal{E}^{(k+1)''}$ and let ω_j be its component in $\mathcal{E}^{k+1,j}$ where $1 \leq j \leq k$, j odd. It follows from Lemma 4.3 that for every $j, 1 \leq j \leq k$, j odd there is an element $\tau_j \in \mathcal{E}^{k,j}$ such that $d^{(j)}(\tau_j) = \omega_j$. In view of Theorem 2.5 we get:

$$d''(\sum \tau_j) = \sum d^{(j)}(\tau_j) = \sum \omega_j = \omega.$$

The case $k \geq m/2$ is similar to the one described above, even somewhat simpler.

In the case that m is even, we have still to discuss the case $k+1 = m/2$. There are now two possible definitions of the spaces $\mathcal{E}^{(k+1)''}$.

If

$$\mathcal{E}^{(k+1)''} = \bigoplus_{\substack{0 \leq j \leq k+1 \\ j \text{ odd}}} \mathcal{E}^{k+1,k+1-j}, \tag{4.3}$$

then the procedure above can be applied without any change. In the other case, we have to show (in addition to the procedure above) that the map $d^{k+1} : \mathcal{E}^{k,k} \mapsto \mathcal{E}^{k+1,k+1}$ is onto. To this end recall that $\partial_{\underline{x}}\,\omega = \partial_{\underline{x}} \rfloor d\omega + d(\partial_{\underline{x}} \rfloor \omega)$ and that $\partial_{\underline{x}} : \mathcal{E}^{k+1} \mapsto \mathcal{E}^{k+1}$ is onto. Moreover, if $\omega \in \mathcal{E}^{k+1,k+1}$, then $\partial_{\underline{x}} \rfloor \omega = 0$, whence $\partial_{\underline{x}}\,\omega = \partial_{\underline{x}} \rfloor d^{(k)}\omega = \partial_{\underline{x}}^{(k)}\omega$ and $\partial_{\underline{x}}^{(k)}\omega = 0$. The fact that the map $\partial_{\underline{x}} : \mathcal{E}^{k+1,k} \oplus \mathcal{E}^{k+1,k+1} \mapsto \mathcal{E}^{k+1,k+1}$ is onto implies that $\partial_{\underline{x}}^{(k)} : \mathcal{E}^{k+1,k} \mapsto \mathcal{E}^{k+1,k+1}$ is onto. But if $\omega \in \mathcal{E}^{k+1,k}$, then $\partial_{\underline{x}}^{(k+1)}\omega = d^{(k+1)}(\partial_{\underline{x}} \rfloor \omega)$, whence $d^{(k+1)}$ is also onto. ■

Summarizing we have

Theorem 4.4. *Let Ω be a domain in $\mathbf{R}^m$. Then d maps $\mathcal{M}^k(\Omega)$ into $\mathcal{M}^{k+1}(\Omega)$. Moreover, if d_k stands for the restriction of d to the space $\mathcal{M}^k(\Omega)$, then*

$$H^k(\Omega, S) \simeq \operatorname{Ker} d_{k+1} / \operatorname{Im} d_k.$$

Proof. The complex $(\mathcal{M}^k(\Omega), d_k)$ of monogenic differential forms is a subcomplex of the de Rham complex of differential forms with coefficients in S. The cohomology groups $H^k(\Omega, S)$ being given by the homology groups of the de Rham complex, it is sufficient to use Lemma 4.1. and Theorem 4.2. ■

5 The Cauchy theorem

The Cauchy Theorem for monogenic functions was given in Chap.II,§.0. It tells us that if f is a left monogenic function and g is a right monogenic function, then the form $f\,d\sigma\,g$ is closed. In order to obtain a possible generalization of this property to monogenic forms, observe that we can express the Cauchy Theorem in the following way. Recalling that an $(m-1)$-form $\omega = d\sigma\,g$ is left monogenic iff the function g is left monogenic, the Cauchy Theorem says that the wedge product of any monogenic 0-form with any monogenic $(m-1)$-form is a closed form.

A natural generalization of the Cauchy Theorem might hence be that the wedge product of monogenic forms having complementary degrees is closed. That this is indeed true is proved in

Theorem 5.1. (Cauchy Theorem)

(i) Let $m = 2l+1$ and $k = 0, \ldots, m-1$. Then for all $\omega \in \mathcal{M}_r^k$ and $\tau \in \mathcal{M}_l^{m-k-1}$

$$d(\omega \wedge \tau) = 0.$$

(ii) Let $m = 2l$ and $k = 0, \ldots, m-1$. Then for all $\omega \in \mathcal{M}_r^k$ and $\tau \in \widetilde{\mathcal{M}}_l^{m-k-1}$ or, respectively, for all $\omega \in \widetilde{\mathcal{M}}_r^k$ and $\tau \in \mathcal{M}_l^{m-k-1}$

$$d(\omega \wedge \tau) = 0.$$

Let us first prove some properties concerning wedge products of forms having complementary degrees.

Lemma 5.2. *If $i \neq j$ and $k \in \{1, \ldots, m-1\}$, then $\omega \wedge \tau = 0$ for all $\omega \in \mathcal{E}_r^{k,i}$ and $\tau \in \mathcal{E}_l^{m-k,j}$.*

Proof. If $\omega \in \mathcal{E}_r^{k,i}$, $\tau \in \mathcal{E}_l^{m-k,j}$, then we can write these forms as $\omega = \omega' \wedge (d\underline{x})^{k-i}$, $\omega' \in \mathcal{E}_r^{i,i}$ and $\tau = (d\underline{x})^{m-k-j} \wedge \tau'$, $\tau' \in \mathcal{E}_l^{j,j}$. Hence $\omega \wedge \tau = \omega' \wedge (d\underline{x})^{m-i-j} \wedge \tau'$.

Now if $i < j$,, then $(d\underline{x})^{m-i-j} \wedge \tau' \in \mathcal{E}_l^{m-i,j} = \{0\}$, since $m-i-j > m-2j$. Similarly, if $i > j$, then $\omega' \wedge (d\underline{x})^{m-i-j} \in \mathcal{E}_r^{m-j,i} = \{0\}$, since $m-i-j > m-2i$. ∎

Lemma 5.3.

(i) Let $m = 2l+1$ and $k = 0, \ldots, m-1$. Then for all $\omega \in \mathcal{E}_r^{k'}$ and $\tau \in \mathcal{E}_l^{(m-k)'}$

$$\omega \wedge \tau = 0.$$

(ii) Let $m = 2l$ and $k = 0, \ldots, m-1$. Then for all $\omega \in \mathcal{E}_r^{k'}$ and $\tau \in \tilde{\mathcal{E}}_l^{(m-k)'}$, or, respectively, for all $\omega \in \tilde{\mathcal{E}}_r^{k'}$ and $\tau \in \mathcal{E}_l^{(m-k)'}$

$$\omega \wedge \tau = 0.$$

Proof. Using the definition of primed and double primed spaces, it is sufficient to check that both spaces are never on the same row (see also the diagrams for low dimensions given in §.3). So we can apply Lemma 5.2. ∎

Proof of Theorem 5.1. As

$$d(\omega \wedge \tau) = d\omega \wedge \tau + (-1)^{\deg \omega} \omega \wedge d\tau,$$

it suffices to observe that the monogenicity condition is equivalent to the condition $d\omega \in \mathcal{E}_r^{(k+1)'}$ and $d\tau \in \mathcal{E}_l^{(m-k)'}$ (resp. $d\tau \in \tilde{\mathcal{E}}_l^{(m-k)'}$). ∎

6 The Residue Theorem

In this section we shall introduce in two different ways the notion of residues for monogenic functions having singularities on submanifolds of dimension k, $k = 1, \ldots, m-1$ (see Fig.1)

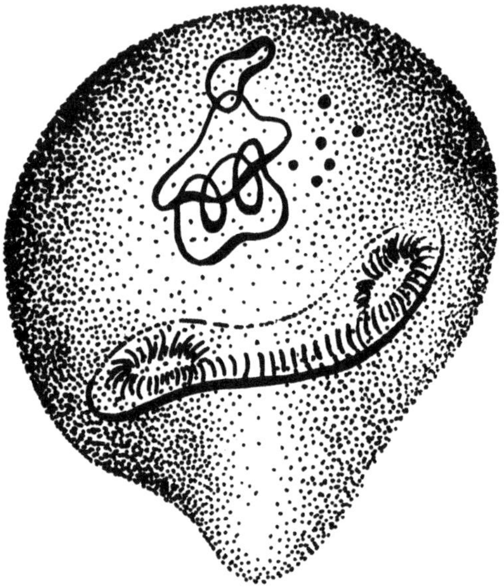

Figure 1.

For $k = 0$, a residue theory was already developed in [10] and this by using the Laurent series for monogenic functions. Notice that the case $k = m-1$ is somewhat peculiar because the surface under consideration then splits the domain into different connected pieces and it corresponds to considering singularities of holomorphic functions on curves.

A new tool should be developed in the higher dimensional situation. Leray introduced in [53] a new type of residue for complex hypersurfaces; it was generalized by Norguet (see [59]) to the case of real submanifolds of any codimension. It is exactly this Leray-Norguet residue which will be used to describe residues of monogenic functions on submanifolds. The case of pointwise singularities naturally fits into the scheme since it corresponds to the case of dimension 0. The Leray-Norguet residue is a cohomology class on the submanifold considered (where the original form has singularities) and it provides important topological information on the behaviour of the form near its singularity set.

It appears that it is also possible to introduce a more general notion of residue for monogenic forms, if the form has its singularities on a manifold Σ, and this by considering the form as a functional acting on forms of complementary degree, which are monogenic in a neighbourhood of Σ. The analogue in the complex plane is then given when considering a residue at z for a product fg of holomorphic functions, f having a singularity in z and g being holomorphic at z. Considering g as a test function, it is clear that the corresponding functional carries information on all coefficients in the Laurent series of g. A residue theory of such type is described below for any monogenic form with singularities on submanifolds of any codimension.

6.1 Residues as numbers

Let us first consider the simplest case of monogenic $(m-1)$-forms, i.e. the case of a form $\omega = d\sigma\, g$ where g is a left monogenic function. We know that ω is a closed form.

Definition 6.1.1. *Let Ω be a domain in $\mathbf{R}^m$ and let $\Sigma \subset \Omega$ be a compact (oriented) q-dimensional submanifold ($q = 0, \ldots, m-2$). If ω is a (left) monogenic $(m-1)$-form in $\Omega \setminus \Sigma$ and* $\operatorname{Res} \omega \in H^q(\Sigma)$ *is its Leray-Norguet residue, then the number*

$$res_\Sigma \omega = \int_\Sigma \operatorname{Res} \omega$$

is called the residue of the form ω at the submanifold Σ.

The residue $res_\Sigma \omega$ does not differ too much from the Leray-Norguet residue $\operatorname{Res} \omega$ itself; it is just the evaluation of the corresponding cohomology class on a generator of the top dimensional homology of Σ, represented by Σ itself. Because of the fact that the vector space $H^q(\Sigma)$ is one-dimensional, the information carried by both notions is the same.

Now, just as in complex analysis, the importance of the residue relies on the fact that the result of an integration of ω over a boundary $\partial\Omega$ of a "big" domain Ω is computable in terms of the local information on the behaviour of ω near the singularities, encoded in the value of the residue. This is exactly the information offered by

Theorem 6.1.2. **(Residue Theorem)** *Suppose that Ω is a domain in $\mathbf{R}^m$ and $\Omega' \subset\subset \Omega$ is a relatively compact subdomain with a smooth boundary (oriented by its outer normal). Furthermore, suppose that Σ_i, $i \in I$, is a finite family of pairwise disjoint compact submanifolds of Ω', the dimensions of which may vary from 0 to $m-2$, and let $\omega = d\sigma\, f$, $f \in M_l(\Omega \setminus (\cup_{i\in I} \Sigma_i))$. Then*

$$\int_{\partial\Omega'} \omega = \sum_{i\in I} res_{\Sigma_i} \omega.$$

Proof. For each $i \in I$, we choose an (oriented) tubular neighbourhood U_i of Σ_i in Ω' in such a way that their respective closures are pairwise disjoint and we assume their boundaries ∂U_i to be oriented by the outer normal (in each fiber). Then the boundary of the domain $\Omega'' = \Omega' \setminus (\cup_{i\in I}\bar{U}_i)$ (together with the orientation given by the outer normal) will split into the union of $\partial\Omega'$ (oriented by the outer normal) and of all ∂U_i, $i \in I$ (with the opposite orientation to that described above).

By the Cauchy Theorem applied to the domain Ω'', we get

$$\int_{\partial\Omega'} \omega = \sum_{i\in I} \int_{\partial U_i} \omega.$$

But the Leray theorem tells us that

$$\int_{\partial U_i} \omega = \int_{\Sigma_i} \operatorname{Res} \omega = res_{\Sigma}\omega. \blacksquare$$

6.2 *Residues as functionals*

Let us discuss now the more general case of differential forms of the type $\omega = \mu \wedge \nu$ with $\mu \in \mathcal{M}^k(\Omega \setminus \Sigma)$ and $\nu \in \mathcal{M}^{m-k-1}(\Sigma)$.

The Cauchy Theorem tells us that ω is a closed form in $\Omega \setminus \Sigma$. So the residue

$$res\,\omega = \int_{\delta\Sigma} \omega = \int_{\Sigma} \operatorname{Res} \omega$$

is well-defined and it carries comprehensive information concerning the behaviour of the form ω near Σ.

Definition 6.2.1. *Let Σ be a compact q-dimensional manifold in $\Omega \subset \mathbf{R}^m$, where $q = 0, \ldots, m-2$. The space $\mathcal{M}_r^j(\Sigma)$ of right monogenic j-forms on Σ is defined as the direct limit*

$$\mathcal{M}_r^j(\Sigma) = \lim\operatorname{ind}_{U\supset\Sigma}\mathcal{M}_r^j(U),$$

where the limit is taken over all neighbourhoods U of Σ in $\mathbf{R}^m$. The space $\tilde{\mathcal{M}}_r^j$ is defined in a similar way.

In order to enable us to formulate the statements below independently of the parity of m, let us admit that for the odd dimensional case the symbol $\tilde{\mathcal{M}}^k$ denotes the same space as $\mathcal{M}^k$.

Definition 6.2.2. *Let Σ be a compact q-dimensional manifold in $\Omega \subset \mathbf{R}^m$, where $q = 0, \ldots, m-2$ and let $\mu \in \mathcal{M}_\ell^k$ be a left monogenic k-form on $\Omega \setminus \Sigma$. Then the functional $Rs_\Sigma\,\mu \in \tilde{\mathcal{M}}_r^{m-k-1}(\Sigma)'$ given by*

$$Rs_\Sigma\,\mu\,[\nu] = \int_\Sigma \operatorname{Res}_\Sigma(\mu \wedge \nu),\ \nu \in \tilde{\mathcal{M}}_r^{m-k-1}(\Sigma)$$

will be called the grand residue of μ at Σ.

The grand residue may be used to compute integrals of the form $\mu \wedge \nu$ over $(m-1)$-dimensional surfaces for all monogenic forms μ and ν.

Theorem 6.2.3. *Let $\Omega' \subset\subset \Omega$ be a relatively compact subdomain with a smooth boundary (oriented by its outer normal) and let Σ_i, $i \in I$, be a finite family of pairwise disjoint compact submanifolds of Ω', the dimensions of which may vary from 0 to $m-2$. Suppose further that $\mu \in \mathcal{M}_r^k(\Omega \setminus (\cup_{i\in I}\Sigma_i))$ and that $\nu \in \mathcal{M}_\ell^{m-k-1}(\Omega)$. Then*

$$\int_{\partial\Omega'} \mu \wedge \nu = \sum_{i\in I}(Rs_{\Sigma_i}\mu)[\nu].$$

6.3 Computation of residues

Let us discuss first the case of pointwise singularities studied in [94]. We wish to show how this case fits into the scheme described above.

Notice that this is exactly the case when the dimension of the manifold Σ is zero. In such case, the Leray-Norguet residue is a number and an integral of it over a point is equal to this number.

Furthermore, the Leray cobord of a point p is represented by any sphere S containing p in its interior and the integral along fibres of an $(n-1)$-form ω is the number given by $\int_S \omega$. Finally, just as in the complex case, the Leray-Norguet residue of the form $f\, d\sigma\, g$ can be computed using the Taylor and Laurent coefficients of the functions f and g.

Theorem 6.3.1. *Let $\Omega \subset \mathbf{R}^m$ be open and let $p \in \Omega$. Furthermore, let $g \in \mathcal{M}_\ell(\Omega)$ and $f \in \mathcal{M}_r(\Omega \setminus \{p\})$ and let (c_β) be the Taylor coefficients of g at p with respect to a chosen basis (P_β) of inner spherical monogenics and let (d_β) be the Laurent coefficients of f at p with respect to the dual basis (Q_β) of outer spherical monogenics. Then*

$$res_{\{p\}}\, f d\sigma g = \sum_\beta d_\beta c_\beta.$$

Proof. Writing out the Taylor and Laurent series for f and g in terms of the basis chosen, it is sufficient to use the orthogonality relations for spherical monogenics to obtain the desired result. ∎

If Σ is a compact submanifold of $\mathbf{R}^m$ of dimension q, then any q-form ω on Σ defines, by convolution with the Cauchy kernel, a monogenic function $\hat{\omega}$ outside Σ, called *the indicatrix of* ω. As may be expected, the Leray-Norguet residue of $\hat{\omega}$ is, as a cohomology class, represented by ω (see Theorem 6.3.2 below).

At the same time it shows that the theory of residues is rich enough, in the sense that it enables us to construct for any class of cohomology a suitable monogenic function f such that the residue of $d\sigma\, f$ is represented by ω.

More generally, let us recall that the cohomology groups of Σ can also be represented by closed distributional forms in $\mathcal{E}'(\Sigma, \Lambda^q)$ (see e.g. [69] or [24]. Taking such a distributional form $\omega \in \mathcal{E}'(\Sigma, \Lambda^q)$, the corresponding indicatrix $\hat{\omega}$ can then be defined by the same procedure, namely

$$\hat{\omega}(\underline{x}) = \int_{\Sigma_{\underline{y}}} E(\underline{x} - \underline{y})\omega(\underline{y}),$$

and we have again that $\operatorname{Res}_\Sigma (d\sigma\, \hat{\omega})$ is represented by the distributional form ω in $H^q(\Sigma)$

As a typical example, the residue $\operatorname{Res}_\Sigma E_{\underline{y}}$ of the Cauchy kernel in a point $\underline{y} \in \Sigma$ is represented in $H^q(\Sigma)$ by the distributional q-form $\delta_{\underline{y}}$.

We would like to stress that, when considering the residue as a functional, much more can even be proved. As we already know, the grand residue $Rs_\Sigma(\omega)$ is a functional in $\mathcal{M}'_r(\Sigma)$ acting on test functions φ which are restrictions of monogenic functions on a neighbourhood of Σ. For some monogenic forms, it may happen that the grand residue belongs to the subspace $\mathcal{E}'(\Sigma, \Lambda^q)$ of distributional q-forms. It then means that the action of $Rs_\Sigma(d\sigma\hat{\omega})$ can be extended to the space of all smooth functions. This is exactly what occurs in the case described above. Note that, in particular, we thus have that $Rs_\Sigma\, E_{\underline{y}} = \delta_{\underline{y}}$.

Theorem 6.3.2. *Let Σ be a compact submanifold of $\mathbf{R}^m$ of dimension q and let $\omega \in \mathcal{E}'(\Sigma, \Lambda^q)$. Putting*

$$\hat{\omega}(\underline{x}) = \int_{\Sigma_{\underline{y}}} E(\underline{x} - \underline{y})\omega(\underline{y}),$$

where E is the Cauchy kernel, we have that

$$Rs_\Sigma (d\sigma\, \hat{\omega}) = \omega.$$

Proof. To show that $Rs_\Sigma (d\sigma\, \hat{\omega})$ is equal to ω, it is sufficient to prove that for all $f \in \overset{0}{\widetilde{\mathcal{M}}_r}(\Sigma)$

$$\int_{\delta\Sigma} f d\sigma\hat{\omega} = \int_\Sigma f\, \omega.$$

But

$$\begin{aligned} \int_{\delta\Sigma} f d\sigma\hat{\omega} &= \int_{\delta\Sigma_{\underline{x}}} f(\underline{x}) d\sigma_{\underline{x}} \int_{\Sigma_{\underline{y}}} E(\underline{x} - \underline{y})\omega(\underline{y}), \\ &= \int_{\Sigma_{\underline{y}}} \left[\int_{\delta\Sigma_{\underline{x}}} f(\underline{x}) d\sigma(\underline{x}) E(\underline{x} - \underline{y}) \right] \omega(\underline{y}), \\ &= \int_{\Sigma_{\underline{x}}} f(\underline{y})\omega(\underline{y}). \blacksquare \end{aligned}$$

The question of course arises how to compute residues in higher dimensions using Taylor and Laurent coefficients, but it is even not easy to find in general a suitable definition of Taylor and Laurent coefficients. A theory of Taylor and Laurent series was developed in Sect.II.5 in the case when the singularity set is part of the flat space $\mathbf{R}^q \subset \mathbf{R}^m$. The results obtained there may now be used to compute residues.

Theorem 6.3.4. *Let Σ be a compact submanifold dimension q in $\tilde{\Omega} \subset \mathbf{R}^m$ and let $\omega \in \mathcal{E}^q(\Sigma)$. Furthermore, let $K \subset \Omega$ be a compact subset x such that there is an open subset Ω of an affine subspace A_q of dimension q containing K. Then for all functions $f \in M_r(\tilde{\Omega} \setminus K)$ and $g \in M_l(\tilde{\Omega} \setminus K)$, we have*

$$res_\Sigma [f\, d\sigma\, g] = \sum_{k=0}^{\infty} (L_k(f)[T_k(g)] + L_k(g)[T_k(f)]),$$

where $L_k(f) \in \mathcal{A}'(K, \mathcal{M}_{l,+,k})$, $L_k(g) \in \mathcal{A}'(K, \mathcal{M}_{r,+,k})$, $T_k(f) \in \mathcal{A}(K, \mathcal{M}_{r,+,k})$, $T_k(g) \in \mathcal{A}(K, \mathcal{M}_{l,+,k})$.

Proof. After a suitable transformation of coordinates, we may suppose that $K \subset \Omega \subset \mathbf{R}^q$.

Taking into account that $K \subset \Omega$ and that f and g are monogenic in $\tilde{\Omega} \setminus K$, the Cauchy Theorem implies that for a sufficiently small neighbourhood U of K in $\mathbf{R}^m$

$$\int_{\delta\Sigma} f\, d\sigma\, g = \int_{\partial U} f\, d\sigma\, g.$$

Writing out now the generalized Laurent series for f and g and using the orthogonality relations for spherical monogenics in the dimension $m - q$, the desired result follows immediately. ■

Note that if the set K is a point, then we can embed it into an affine space of dimension 0 and Theorem 6.3.4 reduces to Theorem 6.3.1 (see [94]). Note also that it is necessary to have a flat piece of dimension q around K available to apply the decomposition of f and g into generalized Laurent series.

Chapter V

Clifford analysis and the Penrose transform

The Penrose transform has its origin in a contour integral formula established by R. Penrose; a holomorphic function with suitable singularities on the twistor space was used in the formula to produce a solution of the massless field equations (i.e. the Laplace and Dirac equations and similar equations for higher spins) on Minkowski space (see [61]). This procedure quickly developed into a full mathematical theory, where the Penrose transform is presented as a 1-1 map between holomorphic solutions of the massless field equation on a domain in the complexified Minkowski space and certain cohomology groups on the corresponding region in the twistor space (see [27]). A systematic description of the Penrose transform in this setting can be found in the book by Ward and Wells ([88]).

Meanwhile, several approaches have been developed to describe the Penrose transform. For the description of cohomology groups on the twistor space, it is possible to use either the Čech cohomology with coefficients in sheaves (see [88]) or the Dolbeault cohomology (see [31] or [92]). There are two ways how to consider the Minkowski and the twistor spaces can be considered as homogeneous spaces, the corresponding group being either $SL(4, \mathbf{C})$ or $SO(6, \mathbf{C})$. The symmetry given by the action of these groups can be interpreted as the symmetry with respect to the conformal group acting on the Minkowski space. In dimension 4, most of all the group $SL(4, \mathbf{C})$ is chosen (exception is made in e.g. [42]). In the Riemannian case, i.e. for the twistor spaces corresponding to Riemannian manifolds, it is even possible to extend the Penrose transform to the curved case, namely to self-dual Riemannian manifolds (see [39], [92]).

A major generalization of the Penrose transform was worked out recently by Baston and Eastwood (see [3], [4]). They are extending it not only to the case of conformal symmetry in higher dimensions, but they are also describing the Penrose transform corresponding to the symmetry given by any complex semisimple Lie group G. Of course, such general situation demands quite advanced and sophisticated tools. From representation theory two basic results, namely the Bott-Borel-Weil theorem and the Berstein-Gelfand-Gelfand resolution, are needed, while from algebraic topology the Leray spectral sequence is a basic tool used for the computation of the cohomology with coefficients in sheaves on fiber bundles.

In this chapter, we discuss a special case of the general Penrose transform described in [4] and its relation to Clifford analysis, described in previous chapters. Such a relation should be expected in the case of the series $B_n = \mathrm{so}(2n+1, \mathbf{C})$ and $D_n = \mathrm{so}(2n, \mathbf{C})$ of simple Lie algebras; we are discussing here only the latter case. The Dirac equation is one of the conformally invariant partial differential equations for which the Penrose transform can be used and as the whole book is devoted to the Dirac equation, we restrict ourselves to the case of the Dirac equation (for further information, see [92]).

The Penrose transform is presented here from the analyst's point of view. The main idea behind it can be described as follows. Extend first the Euclidean scalar product on $\mathbf{R}^{2n}$ to a bilinear nondegenerate form $(.,.)$ on $\mathbf{C}^{2n}$. As described in Ch.I.4.9, there are two families of isotropic subspaces in $\mathbf{C}^{2n}$. For any subspace W in a family chosen (denoted by IG_n), a corresponding pure spinor s may be found which is characterized by the property that for each $f \in W$ we have that $fs = 0$. If $[f_1, \ldots, f_n]$ is a basis of W such that $[\bar{f}_1, \ldots, \bar{f}_n]$ is the dual basis in $\bar{W}$, i.e. if $(f_i, \bar{f}_j) = \frac{1}{2}\delta_{ij}$, then $[f_1, \ldots, f_n, \bar{f}_1, \ldots, \bar{f}_n]$ is a basis of $\mathbf{C}^{2n}$. In the coordinates $(P_1, \ldots, P_{2n})$ corresponding to this basis, the (complexified) Dirac operator is given by

$$D = \sum_{j=1}^{n} (f_j \partial_{P_{n+j}} + \bar{f}_j \partial_{P_j}).$$

It is then possible to construct a family of solutions to the Dirac equation on $\mathbf{C}^{2n}$ depending on n parameters in the following way. Let F be a holomorphic function in $\mathbf{C}^{2n}$ which depends only on the variables $P_{n+1}, \ldots, P_{2n}$, i.e. the function F is constant along W and all affine subspaces in $\mathbf{C}^{2n}$ parallel to W. Then the spinor field $\phi(P) = sF(P)$ is a solution to the Dirac equation. Indeed, we have

$$D\phi = \sum_{j=1}^{n} f_j s \frac{\partial F}{\partial P_{n+j}} = 0.$$

The same can be done for any isotropic subspace of the given family, whence for a given holomorphic function on $\mathbf{T} \simeq \mathbf{C}^n \times IG_n$, we may construct the corresponding solutions for $W \in IG_n$ fixed and then superpose these solutions by integrating them over IG_n. Again a solution to the Dirac equation is thus obtained. Moreover, quite a lot solutions can be obtained in such a way; it may happen, of course, that the resulting superposition is trivial. Roughly speaking, it is exactly what the Penrose transform is doing.

To be more precise, we have to consider $\bar{\partial}$-closed forms on $\mathbf{T}$ of suitable degree and having a certain homogeneity with respect to the homogeneous coordinates on IG_n (see section 3.5), those leading to trivial solutions being neatly described by exact forms of the same type. This is exactly the Dolbeault description of the cohomology groups with coefficients in the sheaf of holomorphic sections of a line bundle used in the mathematical description of the Penrose transform.

We h .ve tried to keep the description of the Penrose transform as elementary as possible. As the elements of cohomology groups can be represented in homogeneous coordinates by ordinary differential forms, the Penrose transform is then described by means of some canonical invariant forms on the twistor space IG_n. Clifford analysis is entering at the moment when we show that each solution can be obtained in such a way, i.e. that the Penrose transform is surjective. The characterization of the image is the most important problem in all types of generalized Radon transforms (the Penrose transform being one of them). Integral formulae of hyperbolic type developed in Clifford analysis form the basic tool used here to prove the surjectivity. As for the injectivity of the Penrose transform, we refer to the standard results. Moreover, as we are describing the Penrose transform locally, we restrict ourselves to special types of domains, namely to Lie balls.

In section 1 we discuss first integral formulae of elliptic and hyperbolic type. The first one is an analytic continuation of the standard Cauchy Integral Formula in Clifford analysis to a complex domain and it expresses the value of a solution by an integral along a cycle of a suitable dimension having no intersection with the complex null cone.

A substantial step is made, when the Leray residue is used to prove an integral formula of hyperbolic type. Here the value of the solution is determined by an integral along a cycle *inside* the complex null cone of the point. Such types of integral formulae are discussed in section 2. Taking the cycle inside the Minkowski space, we obtain an integral formula for solutions of the Dirac equation on Minkowski space (see [60]) which gave the inspiration for the twistor description of solutions. It can be easily deduced, together with its generalization to higher dimensions (see [14]), from the hyperbolic integral formula proved in section 2.

In section 3 we describe the isotropic flag manifolds used for the Penrose transform and we construct as well some invariant forms on them as some line bundles and forms with values in them. The geometry of the basic twistor correspondence, including suitable coordinates used in the description of the Penrose transform, is discussed in section 4. Finally in section 5, we then study the Penrose transform for solutions to the Dirac equation.

1 Elliptic integral formulae

In this section, we shall study properties of analytic continuations of solutions of the Dirac equation to $\mathbf{C}^m$. It is a quite natural procedure because the Dirac operator on Euclidean space is elliptic and all solutions of the Dirac equation are hence real analytic. So their analytic continuations to a neighbourhood in $\mathbf{C}^m$ always exist.

This leads to the study of holomorphic solutions of the complexified Dirac equation on domains in $\mathbf{C}^m$. Although the analytic continuation works in any dimension, the situation is substantially different for even and odd dimensions. We shall restrict ourselves here to the case of even dimensions and we shall suppose henceforth that $m = 2n, n > 1$.

Let S be a left module (not necessarily irreducible) over the complex Clifford algebra $\mathbf{C}_{2n}$ and let $\Omega \subset \mathbf{C}^{2n}$ be a domain. Then we shall study holomorphic maps f from Ω into S satisfying the (complexified) Dirac equation

$$\partial_{\underline{z}} f = 0, \; \partial_{\underline{z}} = \sum_{i=1}^{2n} e_i \frac{\partial}{\partial z_i}. \tag{1.1}$$

The Cauchy Integral Formula for monogenic functions is one of the most important theorems in Clifford analysis. The main aim of this section is to show, how it can be extended to holomorphic solutions of the Dirac equation in $\mathbf{C}^m$. To do so, we first wish to describe what will be the contours of integration in the complexified situation.

In Euclidean space, the standard choice is to integrate over the (smooth) boundary of a domain in $\mathbf{R}^m$. Although in the complex situation there is much more freedom, it turns out that we have to consider manifolds of real dimension $m-1$ in $\mathbf{C}^m \simeq \mathbf{R}^{2m}$. A lack of geometrical intuition in such situation can be substituted well by the use of the language of homology classes. So to integrate p-forms defined on a domain Ω, we shall use chains in the p-th singular homology group $H_p(\Omega, \mathbf{Z})$ as contours of integration (for a suitable description of these groups see [56]). To realize how many times a surface is wrapped around a point, we introduce first the notion of index of a point with respect to a cycle.

1.1 Index of a point

In standard complex analysis, the notion of index of a point with respect to a closed curve which does not contain this point is well known. A quite similar definition works as well in Clifford analysis in $\mathbf{R}^m$. Indeed, let Σ be an $(m-1)$-dimensional cycle which does not contain the point $\underline{x}$. Then by means of the Cauchy Integral Formula we put

$$\mathrm{Ind}_\Sigma \underline{x} = \frac{1}{A_m} \int_\Sigma \frac{-(\underline{y} - \underline{x})}{|\underline{y} - \underline{x}|^{2n}} d\sigma_{\underline{y}}.$$

Just as in complex analysis in one variable, the role of the Cauchy kernel appears to be fundamental in defining the index of a point in real Clifford analysis. In the case of the complexified Dirac operator $\partial_{\underline{z}}$ in $\mathbf{C}^{2n}$, the complexification of the fundamental solution is clearly given by $E(z) = \frac{1}{A_m} \frac{\overline{z}}{|z|^{2n}}$ and, obviously, the singularity set of E consists of the null cone in $\mathbf{C}^{2n}$. The definition of the index of a point therefore demands more care.

To distinguish the complex situation from the real one, we shall use letters $P, Q, \ldots$ for points in $\mathbf{C}^{2n}$. We extend the scalar product on $\mathbf{R}^{2n}$ to a bilinear form $(.,.)$ on $\mathbf{C}^{2n}$ and the norm is defined as $|P|^2 = (P,P) = -\sum_1^{2n} P_i^2$. If $P \in \mathbf{C}^{2n}$, the symbol CN_P will denote the complex null cone at P, namely

$$CN_P = \{Q \in \mathbf{C}^{2n} \,|\, |Q-P|^2 = 0\}.$$

We shall keep the convention that the vector space $\mathbf{C}^{2n}$ is embedded into the complex Clifford algebra $\mathbf{C}_{2n}$ and that all multiplications met in formulae mean Clifford multiplication, respectively left multiplication of a spinor by a Clifford number.

Definition 1.1.1. *Let $P \in \mathbf{C}^{2n}$ be given and let Σ be a $(2n-1)$-dimensional cycle in $\mathbf{C}^{2n} \setminus CN_P$. Then we define the index of P with respect to Σ by*

$$\mathrm{Ind}_\Sigma P = \frac{1}{A_m} \int_\Sigma \frac{-(Q-P)}{|Q-P|^{2n}} DQ,$$

where $DQ = \sum_{j=1}^{2n} (-1)^{j+1} e_j d\hat{Q}_j$, $d\hat{Q}_j = dQ_1 \wedge \ldots \wedge \widehat{dQ_j} \wedge \ldots \wedge dQ_{2n}$.

Example 1.1.2. Consider a sphere $S^{2n-1} \subset \mathbf{R}^{2n} \subset \mathbf{C}^{2n} \setminus CN_0$ oriented by the outer normal in $\mathbf{R}^{2n}$. Then, by a direct corollary of the Cauchy integral formula in Euclidean space applied to a constant function, we have that $\mathrm{Ind}_{S^{2n-1}} 0 = 1$.

Basic properties of the index quite similar to the properties of the index in complex analysis, are summarized in the following

Theorem 1.1.3. *Let Σ be a $(2n-1)$-dimensional cycle in $\mathbf{C}^{2n}$. Then*

(i) $\mathrm{Ind}_\Sigma P$ *does not depend on the choice of a cycle representing an element in $H_{2n-1}(\mathbf{C}^{2n} \setminus CN_P, \mathbf{Z})$.*

(ii) $\mathrm{Ind}_\Sigma P$ *is an integer.*

(iii) The function $\mathrm{Ind}_\Sigma P$ *(with Σ fixed) is locally constant on its domain of definition.*

Proof. Suppose first that there exists $k \in \mathbf{Z}$ such that $\Sigma \sim k \cdot S^{2n-1}$.

Then the first and the second part of the theorem follow from Stokes' Theorem, because the form under the integral sign in the definition of the index is closed. For the last part, notice that the function $\mathrm{Ind}_\Sigma P$ is continuous and has its values in $\mathbf{Z}$, whence it is locally constant.

It thus remains to prove that every class of homology in $H_{2n-1}(\mathbf{C}^{2n} \setminus CN_P, \mathbf{Z})$ contains a representative of the form $k \cdot S^{2n-1}$, $k \in \mathbf{Z}$. This is exactly the content of Theorem 1.2.4 given below. ■

Theorem 1.2.4. *Let $S^{2n-1} \subset \mathbf{R}^{2n}$ denote a sphere with center at the origin. Then*

$$H_{2n-1}(\mathbf{C}^{2n} \setminus CN_P, \mathbf{Z}) \simeq \mathbf{Z}$$

and the sphere $S^{2n-1} \subset \mathbf{R}^{2n}$ is a generator of this group.

Proof. Take $P = 0$ and consider the sphere

$$S^{4n-1} = \{Q = \underline{x} + \underline{y} \in \mathbf{C}^{2n} | \sum_{j=1}^{2n} \underline{x}_j^2 + \sum_{j=1}^{2n} \underline{y}_j^2 = 1\}$$

together with the set $E = S^{4n-1} \cap (\mathbf{C}^{2n} \setminus CN_0)$.

Furthermore consider the fibration

$$p : E \longrightarrow S_1 \subset \mathbf{C},\ Q \in E \mapsto \frac{\sum_1^{2n} Q_j^2}{|\sum_1^{2n} Q_j^2|}.$$

and the set $F_0 = p^{-1}(1)$ given by

$$F_0 = \{Q = \underline{x} + i\underline{y} \in \mathbf{C}^{2n} | \underline{x} \cdot \underline{y} = 0, \sum_{j=1}^{2n} (x_j^2 - y_j^2) > 0,\ \sum_{j=1}^{2n} (x_j^2 + y_j^2) = 1\}.$$

So we have a chain of inclusions

$$S^{2n-1} \subset F_0 \subset E \subset \mathbf{C}^{2n} \setminus CN_0$$

and it is thus sufficient to prove that each of these inclusions induces an isomorphism of the corresponding $(2n-1)$-dimensional groups of homology.

(1) The set E is a deformational retract of $\mathbf{C}^{2n} \setminus CN_0$ (using, for example the projection along rays through the origin onto S^{4n-1}), whence the inclusion induces an isomorphism of the corresponding groups of homology.

(2) Using the homogeneity of $|\sum_1^{2m} Q_i^2|$, it follows from the Fibration Theorem in [57] that p is a locally trivial (smooth) fibration. Let

$$\ldots \longrightarrow H_n(F_0, \mathbf{Z}) \xrightarrow{w} H_n(F_0, \mathbf{Z}) \longrightarrow H_n(E_0, \mathbf{Z}) \longrightarrow H_{n-1}(F_0, \mathbf{Z}) \longrightarrow \ldots$$

be the Wang sequence of p which is exact (see [57], Lemma.8.4). Then Theorem 9.1, Lemma 8.1 and Theorem 6.3 in [57] tell us that $w = 0$ and that $H_{n-1}(F_0, \mathbf{Z}) = 0$. Hence $i_* : H_n(F_0, \mathbf{Z}) \longrightarrow H_n(E_0, \mathbf{Z})$ is an isomorphism.

(3) The sphere S^{2n-1} is a deformational retract of F_0. The deformation can be described explicitly: for $t \in [0,1], \alpha = \sqrt{\sum_1^{2m} x_i^2}$, $\beta = \sqrt{\sum_1^{2m} y_i^2}$, we define a map $\theta_t : F_0 \longrightarrow F_0$ by

$$\theta_t(Q) = [sx_1 + ity_1, \ldots, sx_{2n} + ity_{2n}],$$

where $s = s(t) = \sqrt{\frac{1-t^2\beta^2}{\alpha^2}}$.

It is easy to verify that θ_t is the deformation needed. ■

1.2 Cauchy integral formula

We are now in a position to state and to prove the Cauchy integral formula for solutions of the Dirac equation in $\mathbf{C}^{2n}$.

In previous chapters the contour of integration in the Cauchy Integral Formula was the boundary Σ of a domain Ω in R^m. An alternative formulation in the Euclidean situation is that the contour of integration is homologically trivial in the domain Ω, where Ω is the domain in which the corresponding function is monogenic. The idea of the proof is that the contour of integration is homological in $\Omega \setminus \{P\}$ to a small sphere around the point P, in which the monogenic function is evaluated. The situation is, however, quite different in the complexified case. The difficulty lies in the fact that if the contour of integration is homologically trivial in $\Omega \subset \mathbf{C}^{2n}$, it is no more true that Σ is homological in $\Omega \setminus \mathbf{C}N_P$ to a small sphere around P. To explain why, let us consider Ω consisting of two layers, one containing P and the other containing Σ. Then Σ is homologically trivial in Ω, but it is clearly not homological to a small sphere around P in $\Omega \setminus \mathbf{C}N_P$. The example can even be made more complicated when taking Ω connected.

Hence a restriction on Ω has to be imposed. The idea is that once Σ is near the cone $\mathbf{C}N_P$(during the deformation to a point) we should be able to follow rays on $\mathbf{C}N_P$ joining the point P. There is a simple condition which allows to do so.

Let us recall that a set A is said to be star-shaped with respect to a point P if for all $Q \in A$, the whole segment $\overline{PQ}$ belongs to A.

Theorem 1.2.1. (Elliptic integral formula) *Let f be a solution of the Dirac equation on a domain $\Omega \subset \mathbf{C}^{2n}$ and suppose that P is a point in Ω such that $\Omega \cap \mathbf{C}N_P$ is star-shaped with respect to P. Then for each $(2n-1)$-dimensional cycle Σ in $\Omega \setminus \mathbf{C}N_P$ which is homologically trivial in Ω*

$$f(P) \cdot \mathrm{Ind}_\Sigma P = \frac{1}{A_n} \int_\Sigma \frac{-(Q-P)}{|Q-P|^{2n}} DQ \cdot f(Q).$$

Proof. Without loss of generality we may take $P = 0$. Suppose first that there exists $k \in \mathbf{Z}$ such that $\Sigma \sim k \cdot S^{2n-1}$ in $\Omega \setminus \mathbf{C}N_P$. The form under the integral sign being a closed form on $\Omega \setminus \mathbf{C}N_P$, the theorem follows immediately from the Cauchy Integral Formula for a sphere in $\mathbf{R}^m$ (see I.0.2.5) and the fact that $\mathrm{Ind}_0 S^{2n-1} = 1$.

Now put $A = \mathbf{C}^{2n} \setminus \mathbf{C}N_P$; then we have $A \cap \Omega = \Omega \setminus \mathbf{C}N_P$ where by assumption $A \cap \Omega$ is star-shaped with respect to P and hence homologically trivial. The standard Mayer-Vietoris sequence (see [56]) gives us the exact sequence

$$0 \longrightarrow H_{2n-1}(A \cap \Omega, \mathbf{Z}) \xrightarrow{i_*} H_{2n-1}(A, \mathbf{Z}) \oplus H_{2n-1}(\Omega, \mathbf{Z}) \longrightarrow 0.$$

As $\Sigma \sim k \cdot S^{2n-1}$ in A (where $k = \mathrm{Ind}_\Sigma P$) and $\Sigma \sim 0 \sim k \cdot S^{2n-1}$ in Ω, the isomorphism i^* tells us that $\Sigma \sim k \cdot S^{2n-1}$ in $A \cap \Omega$. ∎

Note that it is not that difficult to verify that Ω is star-shaped and that convex domains are clearly star-shaped with respect to any of their points.

2 HYPERBOLIC INTEGRAL FORMULAE

The elliptic integral formula described in the previous section expresses the value of a solution in a point using values of it on a cycle of dimension $2n-1$ *outside* the complex null cone $\mathbb{C}N_P$. The Leray-Norguet residue, described in Appendix A, is the key to another type of integral formulae, called hyperbolic integral formulae. Their main feature is that the value of a solution in a point is computed using its values (and the values of its derivatives) on a cycle of dimension $2n-2$ *inside* the complex null cone $\mathbb{C}N_P$.

2.1 Hyperbolic integral formulae

The main tool needed to derive hyperbolic integral formulae from elliptic ones is the Leray-Norguet residue. Let N be a submanifold of a manifold M with $\dim M = m$ and $\dim N = n$ and let $p \geq m-n-1$. The Leray-Norguet residue is a map Res from $H^p(M \setminus N)$ to $H^{p-(m-n-1)}(N)$ and the Leray-Norguet cobord is the dual map from $H_{p-(m-n-1)}(N)$ to $H_p(M \setminus N)$. If $\omega \in \mathcal{E}^p(M \setminus N)$ is a closed form representing the corresponding class of cohomology, then the form $\operatorname{Res}\omega$ is represented by the integral of ω along fibers of the projection given by a chosen tubular neighbourhood of S in M. Different choices of tubular neighbourhoods can lead to different results of the integration along fibers, but all possible results differ only by an exact form, whence they represent the same class of cohomology. The Leray-Norguet cobord is defined simply as the dual map to Res, the duality being expressed by the formula

$$\int_{\delta C} \omega = \int_C \operatorname{Res}\omega,$$

where $\omega \in \mathcal{E}^{q+m-n-1}(M \setminus N)$ and C is a q-dimensional cycle in N. The Leray-Norguet theory is described in detail in Appendix A. By using the Leray Residue Theorem, Theorem 1.2.1. is translated into

Theorem 2.1.1. (Hyperbolic integral formula) *Let f be a solution of the (complex) Dirac equation on a domain $\Omega \subset \mathbb{C}^{2n}$ and let P be a point in Ω such that $\Omega \cap \mathbb{C}N_P$ is star-shaped with respect to P. Then for each $(2n-2)$-dimensional cycle C in $\mathbb{C}N_P \cap \Omega$ we have*

$$\operatorname{Ind}_{\delta C} P \cdot f(P) = \int_C \operatorname{Res}\omega, \tag{2.1}$$

where $\operatorname{Res}\omega$ denotes the Leray-Norguet residue of the form

$$\omega = \omega(P, Q; f) = \frac{1}{A_{2n-1}} \frac{-(Q-P)}{|Q-P|^{2n}} DQ \cdot f(Q)$$

appearing in Theorem 1.2.1.

Note that as well the elliptic and hyperbolic integral formulae as their translation by the Leray Residue Theorem may be summarized in one formula, namely

$$\operatorname{Ind}_{\delta C} P \cdot f(P) = \int_{\delta C} \omega = \int_C \operatorname{Res} \omega \tag{2.2}$$

and it is up to the reader to choose the one he prefers most.

We shall now use the Leray-Norguet theory in the special case where the manifold M is a domain in $\mathbf{C}^{2n}$ and S is a nonsingular complex hypersurface. It is shown in Appendix A that the choice of the representative for the residue can be fixed by a choice of a transversal vector field along the singularity surface and that the computation of the Leray-Norguet residue is reduced to the standard computation of the residue in complex analysis.

In our case, it is clear that the singularity of the form is determined by the denominator $|Q-P|^{2n}, Q \in \mathbf{C}N_P$. If we choose the vector field c along the singularity to be a null field. i.e. $(c,c)=0$, then, putting $b=Q-P$, the distance between the points $Q+\tau c(Q)$ and P is given by $|Q+\tau c(Q)-P|^2 = (b+\tau c, b+\tau c) = 2\tau(b,c)$ and, for the sake of simplicity, it is of course preferable to add the normalization condition $(b,c)=1/2$. To compute the Leray-Norguet residue by the integral along fibers, we shall compute the integral of a function depending on τ over a small circle around the origin, the latter one then being reduced to the computation of the corresponding residue. Notice that the normalization proposed yields an additional advantage, since the bivector $bc \in \mathbf{C}_{2n}$ is then an idempotent. Indeed, we have

$$bcbc = bc(2(b,c) - cb) = 2(b,c)bc - b^2c^2 = bc. \tag{2.3}$$

There are several alternatives how to describe a differential form. The standard one is to choose coordinates and to compute the coefficients of the form in the corresponding bases. Another one, which will be used below, is based on the fact that the differential form $\omega \in \mathcal{E}^p(M)$ at a point $Q \in M$ is an alternating map on $\underbrace{T_Q M \times \ldots \times T_Q M}_{p}$, whence it is sufficient to specify the numbers $\omega(t_1,\ldots,t_p)$ for any choice of vectors $t_i \in T_Q M$. If coordinates $(x_1,\ldots,x_m)$ are given on M, then the coefficients $\omega_I, I=\{i_1,\ldots,i_p\}$ of the form ω are given by $\omega(\partial_{x_{i_1}},\ldots,\partial_{x_{i_p}})$.

For a given vector field v on a subset of $\mathbf{C}N_P \setminus \{P\}$ and for a vector $t \in T_Q\mathbf{C}N_P$, we denote by $\nabla_t v$ the standard (flat) covariant derivative of v in the direction t. An explicit form of the form $\operatorname{Res}\omega$ is then given by the following

Theorem 2.1.2. *Let f be a solution of the Dirac equation on a domain Ω and let P be a point in Ω. Let $c(Q)$ be a smooth vector field on $U = \Omega \cap (\mathbf{C}N_P \setminus \{P\})$ such that $(c,c)=0$ and $(b,c)=1/2$, $b=Q-P$. Then for each $t_i \in T_Q\mathbf{C}N_P$, $i=1,\ldots,2n-2$,*

$$\operatorname{res}\omega[t_1,\ldots,t_{2n-2}] = \frac{-\mathrm{i}}{\pi^{n-1}}\cdot$$

$$\cdot\frac{\partial^{n-1}}{\partial\tau^{n-1}}\Big((b+\tau c)DQ[c,\nabla_{t_1}(b+\tau c),\ldots,\nabla_{t_{2n-2}}(b+\tau c)]f(b+\tau c)\Big)\big|_{\tau=0}$$

is a representative of the cohomology class $\operatorname{Res}\omega$.

Proof. Let us note first that for any vector $d \in T_Q\mathbf{C}N_P$, the relation $(d, b) = 0$ holds and so $c \notin T_Q\mathbf{C}N_P$.

Let $Q \in \mathbf{C}N_P$ be fixed. Then for any $\tau \in \mathbf{C}$ we have $|Q + \tau c(Q) - P|^2 = \tau$. The choice of the vector field c on $\mathbf{C}N_P \setminus \{P\}$ fixes a representative $\operatorname{res}\omega$ of the class $\operatorname{Res}\omega$. So for a given choice of vectors $t_1, \ldots, t_{2n-2} \in T_Q\mathbf{C}N_P$, we can rely upon the Theorem A.4.2 to compute $\operatorname{res}\omega(t_1, \ldots, t_{2n-2})$ using the ordinary residue at the origin of a holomorphic function of the variable $\tau \in \mathbf{C}$ (with values in the spinor space S) given in a neighbourhood of the point 0 by

$$\frac{-2\pi i}{A_{2n-1}} \frac{(b + \tau c)}{\tau^n} DQ[c, \nabla_{t_1}(b + \tau c), \ldots, \nabla_{t_{2n-2}}(b + \tau c)] f(b + \tau c).$$

Knowing that $A_{2n-1} = 2\pi^n/(n-1)!$ and using the standard formula for the residue at a pole of order n, we get the formula for $\operatorname{res}\omega$. ∎

Any choice of the null vector field c in Theorem 2.1.2 is possible, but if a contour of integration $C \subset \mathbf{C}N_P$ is given, some choices are more convenient. Suppose indeed that the cycle of integration C is represented by a submanifold $C \subset \mathbf{C}N_P$ without boundary. We shall show now that if c is chosen in such a way that for any point $Q \in C$, $c(Q)$ is orthogonal to the tangent space T_QC (see Fig. 2), then the formula for the form $\operatorname{res}\omega$ restricted to C can be substantially simplified.

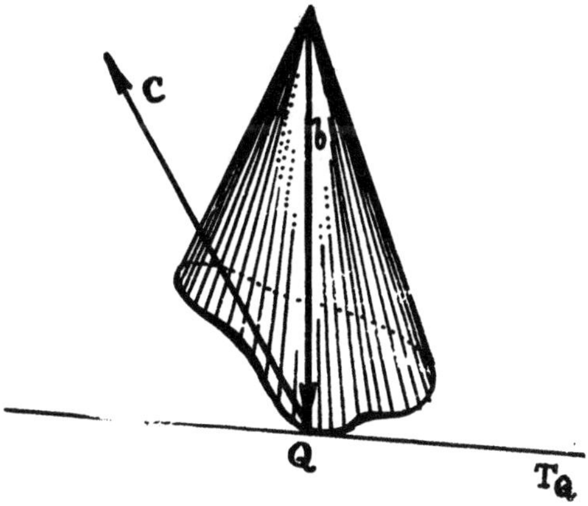

Figure 2. The vector c is orthogonal to T_QC.

To know the restriction $\operatorname{res}\omega|_C$, it is sufficient to know $\operatorname{res}\omega(t_1, \ldots, t_{2n-2})$ for vectors $t_j \in T_QC$ only. The following theorem then shows how the residue looks like in such cases.

Theorem 2.1.3. *Using the notation $dQ = dQ_1 \wedge \ldots \wedge dQ_{2n}$, let us define differential forms $F_j(\tau)$ of degree $2n-2$ on $\mathbf{C}N_P$ for τ near the origin in $\mathbf{C}$ by*

$$\begin{aligned} F(\tau)[t_1, \ldots, t_{2n-2}] &= dQ[b, c, \nabla_{t_1}(b+\tau c), \ldots, \nabla_{t_{2n-2}}(b+\tau c)] \\ &= \sum_{j=1}^{2n-2} F_j[t_1, \ldots, t_{2n-2}]\tau^j, \end{aligned}$$

$$t_i \in T_Q\mathbf{C}N_P,\ i = 1, \ldots, 2n-2.$$

Then for all vectors $t_i \in T_Q\mathbf{C}N_P$ orthogonal to the vector c

$$\begin{aligned} \operatorname{res}\omega[t_1, \ldots, t_{2n-2}] &= \frac{-2i}{\pi^{n-1}} bc\left(\frac{\partial^{n-1}}{\partial\tau^{n-1}}[fF]_{\tau=0}\right) \\ &= \frac{-2i}{\pi^{n-1}} bc\left(\sum_{j=0}^{n-1} \frac{(n-1)!}{j!}\frac{\partial^j f(0)}{\partial\tau^j} F_{n-1-j}\right). \end{aligned}$$

Proof. Let the vector fields $w_{ij} = w_{ij}(\tau, Q)$ for τ small and $Q \in \mathbf{C}N_P$ be given by

$$w_{ij} = (0, \ldots, 0, b_j + \tau c_j, 0, \ldots, 0),$$

where the only non zero term occurs at the i-th place.

Suppose now that all the vectors $t_i \in T_Q\mathbf{C}N_P$, $i = 1, \ldots, 2n-2$ are orthogonal to the vector c. Together with the relations $(b,b) = (c,c) = 0$, $(b,c) = 1/2$ and $\nabla_{t_i} b = t_i$, this implies that the vectors b and c are independent and orthogonal to all other vectors $\nabla_{t_1}(b+\tau c), \ldots, \nabla_{t_{2n-2}}(b+\tau c)$. We thus have that

$$\begin{aligned} &2bc\, dQ[b, c, \nabla_{t_1}(b+\tau c), \ldots, \nabla_{t_{2n-2}}(b+\tau c)] \\ &\quad = (b+\tau c) DQ[c, \nabla_{t_1}(b+\tau c), \ldots, \nabla_{t_{2n-2}}(b+\tau c)]. \end{aligned}$$

Indeed, if the vectors $\nabla_{t_1}(b+\tau c), \ldots, \nabla_{t_{2n-2}}(b+\tau c)$ are linearly dependent, then both sides are equal to zero. In the opposite case, the vectors

$$b, c, \nabla_{t_1}(b+\tau c), \ldots, \nabla_{t_{2n-2}}(b+\tau c)$$

form a basis in $\mathbf{C}^{2n}$ and the orthogonality relations imply that, for suitable numbers $a_2, \ldots, a_{2n-1}$,

$$w_{ij} = 2(w_{ij}, c)\, b + a_{2n-1}\, c + \sum_{i=1}^{2n-2} a_i \nabla_{t_1}(b+\tau c).$$

So we get

$$(b+\tau c)DQ[c,\nabla_{t_1}(b+\tau c),\ldots,\nabla_{t_{2n-2}}(b+\tau c)$$

$$=\sum_{i,j} e_je_i(-1)^{i+1}(b_j+\tau c_j)D\hat{Q}_i[c,\nabla_{t_1}(b+\tau c),\ldots,\nabla_{t_{2n-2}}(b+\tau c)]$$

$$=\sum_{i,j} e_je_i dQ[w_{ij},c,\nabla_{t_1}(b+\tau c),\ldots,\nabla_{t_{2n-2}}(b+\tau c)]$$

$$=2\sum_{i,j} e_je_i((b_j+\tau c_j)c_i)dQ[b,c,\nabla_{t_1}(b+\tau c),\ldots,\nabla_{t_{2n-2}}(b+\tau c)]$$

$$=2b\,c\,dQ[b,c,\nabla_{t_1}(b+\tau c),\ldots,\nabla_{t_{2n-2}}(b+\tau c)].\ \blacksquare$$

A variation of the computation made above will be needed later, namely

Theorem 2.1.4. *Let $\mathcal{M}=\{[P,Q]|P,Q\in \mathbf{C}^{2n};|P-Q|^2=0\}$. Then the form $\omega\wedge dP$ is closed on $\Omega\times\Omega\setminus\mathcal{M}$ and its Leray-Norguet residue is represented on $\mathcal{M}\setminus\Delta$, $\Delta=\{[P,Q]|P=Q\}$ by*

$$\operatorname{res}(\omega\wedge dP)=\operatorname{res}(\omega)\wedge dP.$$

Proof. Consider a vector field $\bar{c}(P,Q)=[0,c(Q)]$ on $\mathcal{M}$. The form $\operatorname{res}(\omega\wedge dP)$ is a $(4n-2)$-form on $\mathcal{M}$ and for proving the desired result, it is clearly sufficient to know its values on the vectors given by $[t'_1,\ldots,t'_{2n-2},s'_1,\ldots,s'_{2n}]$ with $t'_j=[0,t_j]$, $s'_j=[s_j,0]\in T_{[P,Q]}\mathcal{M}$. If $f:\mathcal{M}\times\mathbf{C}\to\Omega\times\Omega$ is the tubular neighbourhood of $\mathcal{M}$ induced by the vector field $\bar{c}$ (see Lemma A.4.1), then $f_*t'_j=[0,t_j+\tau\nabla_{t_j}c]$, $f_*s'_j=[s_j,0]$ and using Theorem A.4.2 we have

$$\begin{aligned}
&\operatorname{res}(\omega\wedge dP)[t'_1,\ldots,t'_{2n-2},s'_1,\ldots,s'_{2n}]\\
&= 2\pi\mathrm{i}\operatorname{res}_{\tau=0}(\omega\wedge dP)[\bar{c},t'_1,\ldots,t'_{2n-2},s'_1,\ldots,s'_{2n}])\\
&= 2\pi\mathrm{i}\operatorname{res}_{\tau=0}(\omega[\bar{c},t'_1,\ldots,t'_{2n-2}]dP[s'_1,\ldots,s'_{2n}])\\
&= (\operatorname{res}\omega)\wedge dP[t'_1,\ldots,t'_{2n-2},s'_1,\ldots,s'_{2n}]\ \blacksquare
\end{aligned}$$

Example 2.1.5. We shall compute now the form of the residue for a special cycle in $\mathbf{C}N_P$, given by the intersection with the Euclidean space $\mathbf{R}^{2n}$.

If $P\notin\mathbf{R}^{2n}$, then the complex null cone $\mathbf{C}N_P$ intersects the Euclidean space $\mathbf{R}^{2n}$ in a sphere S^{2n-2} of dimension $2n-2$. Indeed, if $P=X+iY$, $X,Y\in R^{2n}$, then $Q\in S^{2n-2}$ iff $|P-Q|^2=0$. But

$$|P-Q|^2=(X-Q+iY,X-Q+iY)=|X-Q|^2-|Y|^2+2i(X-Q,Y)$$

and so

$$S^{2n-2} = \{Q \in R^{2n} || Q - X|^2 = |Y|^2, (Q - X, Y) = 0\}.$$

Hence the sphere S^{2n-2} is always an intersection of a sphere of dimension $2n-1$ with center at X and radius $\sqrt{-|Y|^2}$ with a hyperplane (containing the center X) orthogonal to Y. So $|Y|$ is responsible for the size of the sphere and its position varies with the change of direction of Y (see Fig. 3).

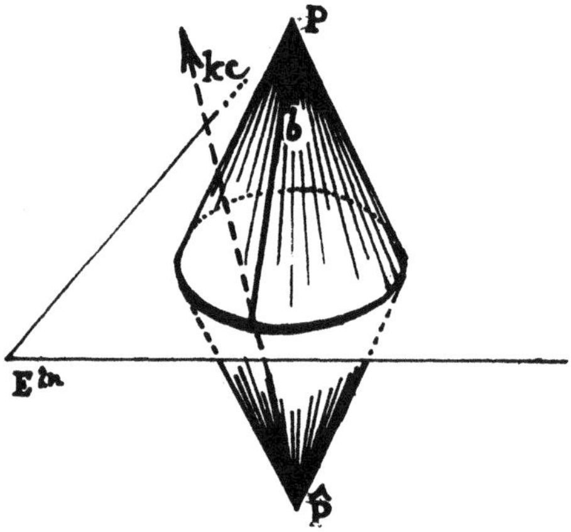

Figure 3.

Our choice for a vector field c in points $Q \in S^{2n-2}$ is fixed by the orthogonality condition $c \perp T_Q S^{2n-2}$ and the normalization condition $(b, c) = 1/2, b = Q - P$. The vector $\hat{b}$ again being an isotropic vector, the symmetry condition clearly shows that the direction c should be a multiple of $\hat{b}$ (see Fig. 4). If $kc = -\hat{b}, k \in \mathbf{R}$, then $\hat{P} - P = b - \hat{b} = b + kc$ and $-4|Y|^2 = (\hat{P} - P, \hat{P} - P) = (b + kc, b + kc) = k$. So we get

$$c = -(1/k)\,\hat{b} = 1/(4|Y|^2)\hat{b}.$$

Moreover, the vector $b + kc = \hat{P} - P$ being fixed, we have that $\nabla_t b = -k\nabla_t c$ for each $t \in T_Q S^{2n-2}$. By virtue of $\nabla_t b = t_i$, we obtain that

$$\begin{aligned} F(\tau)[t_1, \ldots, t_{2n-2}] &= dQ[b, c, \nabla_{t_1} b, \ldots, \nabla_{t_{2n-2}} b](1 - \frac{\tau}{k})^{n-2} \\ &= (1 - \frac{\tau}{k})^{n-2} F_0. \end{aligned}$$

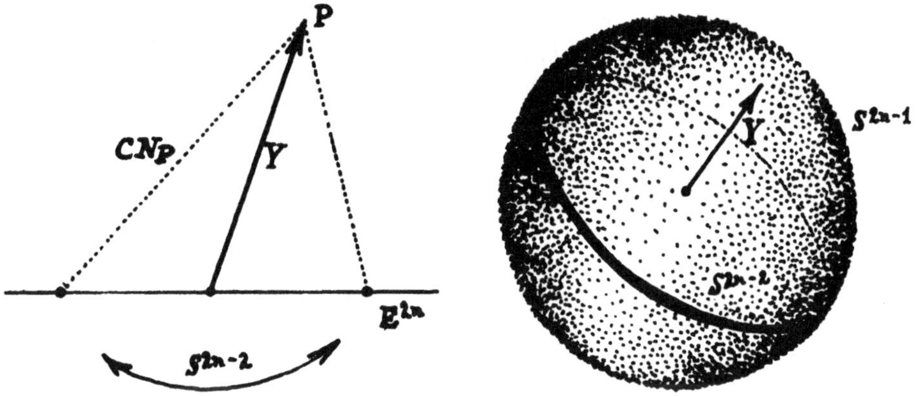

Figure 4.

We have for each $2n \times 2n$ matrix $A = [A^1, \dots, A^{2n}]$, where A^i are the columns of A, the standard relation $A^t A = G$, where G is the Gramm matrix associated with the columns of A. Taking into account the orthogonality relations, we thus obtain that

$$(\det[b, c, t_1, \dots, t_{2n-2}])^2 = (-1/4)\ \det(((t_i, t_j))_{i,j=1,\dots,2n-2}).$$

Hence (up to a plus or minus sign depending on a chosen orientation)

$$F_0 = \det[b, c, t_1, \dots, t_{2n-2}] = \pm\frac{\mathrm{i}}{2}\, dS,$$

where dS is the elementary surface element on the sphere S^{2n-2}. Note in passing that the sphere S^{2n-2} has its radius equal to $\sqrt{-|Y|^2}$, where $|Y|^2 = -k/4$. Hence $(-4/k)^{n-1} dS$ is the elementary surface element on the unit sphere. Moreover, as $b = -kc$, the derivatives $\partial^{n-1} f/\partial\tau^{n-1}$ can be written as $(-1/k)^j (\nabla_{\hat{b}})^j f$.

Summarizing, we get the following formula for the form $\operatorname{res}\omega$, restricted to S^{2n-2} :

$$\begin{aligned}\operatorname{res}\omega &= \frac{-2\mathrm{i}}{\pi^{n-1}} b\, c \left(\sum_{j=0}^{n-1} \left(\frac{-1}{k}\right)^j \left(\frac{-1}{k}\right)^{n-1-j} \binom{2n-2}{n-1-j} \frac{(n-1)!}{j!} (\nabla_{\hat{b}})^j f F_0\right) \\ &= \frac{-2\mathrm{i}}{\pi^{n-1}} b\, c \left(\sum_{j=0}^{n-1} \binom{2n-2}{n-1-j} \frac{(n-1)!}{j!} (\nabla_{\hat{b}})^j f\right) \left(\frac{-1}{k}\right)^{n-1} F_0\end{aligned}$$

$$= \pm\frac{1}{\pi^{n-1}}\left(\frac{1}{4}\right)^{n-1} bc\left(\sum_{j=0}^{n-1}\binom{2n-2}{n-1-j}\frac{(n-1)!}{j!}(\nabla_{\hat{b}})^j f\right) dS(1),$$

$dS(1)$ denoting the elementary surface element on the unit sphere.

The hyperbolic integral formula (Theorem 2.1.1) can now be used to express the value of a solution of the Dirac equation using the integral of the form res ω computed explicitly above. We have:

$$\mathrm{Ind}_{\delta S^{2n-2}} P \cdot f(P) = \pm\frac{1}{\pi^{n-1}}\left(\frac{1}{4}\right)^{n-1} \cdot$$

$$\cdot \int_{S^{2n-2}} bc\left(\sum_{j=0}^{n-1}\binom{2n-2}{n-1-j}\frac{(n-1)!}{j!}(\nabla_{\hat{b}})^j f\right) dS(1).$$

In its turn this gives rise to the question of computing the index $\mathrm{Ind}_{\delta S^{2n-2}}$. The simplest way to do so is to use the hyperbolic integral formula for the constant function $f \equiv 1$. Knowing that $bc = 1/2 + b \wedge c$, it is clear that the integral over the bivector part should vanish and that the scalar part leads to

$$\begin{aligned}\mathrm{Ind}_{\delta S^{2n-2}} P &= \pm\frac{1}{\pi^{n-1}}\left(\frac{1}{4}\right)^{n-1}\int_{S^{2n-2}} \frac{1}{2}\binom{2n-2}{n-1}(n-1)!\,dS(1)\\ &= \pm\frac{1}{2}\frac{1}{\pi^{n-1}}\frac{(2n-3)(2n-5)\ldots 1}{2^{n-1}}A_{2n-2} = \pm 1.\end{aligned}$$

So the index of the Leray cobord of the sphere S^{2n-2} with respect to P is ± 1, the sign depending on the orientation chosen.

2.2 Hyperbolic integral formula on Minkowski space

The description of solutions of the Dirac equation (or more generally of massless field equations) in twistor theory originated from integral formulae for these solutions established by R. Penrose in the case of (four-dimensional) Minkowski space (see [60]). These integral formulae , as well as their generalization to any even dimension (see [14]), can be deduced easily from the hyperbolic integral formula described above in the case of real analytic solutions. Note that it is also possible to use the hyperbolic integral formula to deduce integral formulae for solutions to the Dirac equation on ultrahyperbolic spaces of any signature (see [80]).

Consider the Minkowski space $M = M^{2n}$ embedded in $\mathbf{C}^{2n}$ and a (real analytic) solution of the Dirac equation on a domain in M. Let C be a submanifold of dimension $2n-2$ in the backward null cone N_0 with vertex at the origin $0 \in M$ such that every generator of N_0 is meeting C exactly in one point. Suppose for simplicity that the metric induced on C is nondegenerate (generators of the cone are not tangent to C). Then the orthogonal complement to $T_Q C$, $Q \in C$, contains exactly two null directions.

The first one is given by the generator $b = Q - P$ of the cone, the second one contains a unique null vector c satisfying the normalization condition $(b, c) = 1/2$. In such a way, we have defined a vector field c on C.

Let us also recall that the differential forms $F_j(\tau)$ of degree $2n - 2$ are given by the formulae (see Theorem 2.1.3)

$$\begin{aligned} F(\tau)[t_1, \ldots, t_{2n-2}] &= dQ[b, c, \nabla_{t_1}(b + \tau c), \ldots, \nabla_{t_{2n-2}}(b + \tau c)] \\ &= \sum_{j=1}^{2n-2} F_j[t_1, \ldots, t_{2n-2}]\tau^j, \end{aligned}$$

$$t_i \in T_Q \mathbf{C} N_P,\ i = 1, \ldots, 2n - 2.$$

After restriction to C, they are forms of the top degree on C. Using the relation $F_0[t_1, \ldots, t_{2n-2}] = dQ[b, c, t_1, \ldots, t_{2n-2}]$, the corresponding Gramm matrix and the orthogonality relations, we get (as above) that $F_0 = i/2 \cdot dS$, where dS is the elementary surface element induced on C by the bilinear form $(.,.)$. This implies that the form F_0 is nondegenerate at all points of C and we can express all other forms $F_j(\tau)$ in the form $G_j(\tau)F_0$, where $G_j(\tau)$ are functions on C. Taking into account the definition of the forms F_j, we can intuitively describe the functions G_j as the quotient of the elementary surface element dS_τ of C_τ at the point $Q + \tau c(Q)$ and dS of C at the point Q, where the shifted surface C_τ is the image of C under the map $Q \mapsto Q + \tau c(Q)$.

Theorem 2.2.1 *Let P be a point in the Minkowski space M^{2n} an let C be a submanifold in the backward null cone N_P with the properties described above. Furthermore define the function $G_j(\tau)$ on C by the relation $F_j(\tau) = G_j(\tau)F_0$. Then for every analytic solution f of the Dirac equation in a neighbourhood of the section of the cone N_P bounded by C and P we have*

$$f(P) = \frac{1}{\pi^{n-1}} \int_C b\, c\, \frac{\partial^k}{\partial \tau^k} [G \cdot f]_{\tau=0} dS.$$

Proof. It suffices to use Theorem 2.1.1, Theorem 2.1.3, the relation $F_0 = i/2 \cdot dS$ and the definition of the functions G_j. The manifold C can be deformed into the sphere S^{2m-2} in the null cone along the generators, whence $Ind_{\delta C} 0 = \pm 1$. To check the sign, it is sufficient to compute the integral for a constant function. We did it in Example 2.1.5. ∎

3 Isotropic flag manifolds

Isotropic flag manifolds are basic spaces when studying the Penrose transform for solutions to the Dirac equation.

3.1 Isotropic flag manifolds

Let $\mathbf{C}^{2n}$ be provided with a nondegenerate bilinear form $(.,.)$ determined by the matrix

$$\mathbf{E} = \begin{pmatrix} 0 & -I_n \\ -I_n & 0 \end{pmatrix}.$$

The corresponding quadratic form then reads

$$Q(Z) = -\sum_{j=1}^{n} Z_j Z_{n+j},\ Z \in \mathbf{C}^{2n}.$$

Definition 3.1.1. *The isotropic flag manifold $IG_{i_1,\dots,i_j;2n}$, $1 \le i_1 < \dots < i_j \le 2n$, is defined by*

$$IG_{i_1,\dots,i_j;2n} := \{[L_{i_1},\dots,L_{i_j}] | L_{i_1} \subset \dots \subset L_{i_j} \subset \mathbf{C}^{2n};\ Q|L_{i_j} = 0\},$$

where L_k stands for a subspace of $\mathbf{C}^{2n}$ of dimension k. An isotropic flag manifold of the form $IG_{i,2n}$ is called an isotropic Grassmannian. The dimension of the ambient space will be dropped when it is clearly understood from the context.

It is well known (see e.g. [41]) that the maximal dimension of an isotropic subspace in $\mathbf{C}^{2n}$ is n. Hence the bound $i_j \le 2n$ in the definition can be substituted as well by $i_j \le n$.

A basic property of isotropic flag manifolds is that they are homogeneous spaces of the group $\mathrm{SO}(2n;\mathbf{C})$, resp. $\mathrm{Spin}_+(2n;\mathbf{C})$. Let us describe the matrix group $\mathrm{SO}(2n;\mathbf{C})$ and its corresponding Lie algebra $\mathrm{so}(2n;\mathbf{C})$ explicitly by using block matrices. We have

$$\begin{aligned}\mathrm{SO}(2n;\mathbf{C}) &= \{g \in \mathbf{C}(2n) | g^t\mathbf{E}g = \mathbf{E}\} \\ &= \Big\{ g = \begin{pmatrix} A & B \\ C & D \end{pmatrix} \Big| A,B,C,D \in \mathbf{C}(n), \det g = 1, \\ &\qquad A^tC + C^tA = 0, B^tD + D^tB = 0, C^tB + A^tD = I_n \Big\},\end{aligned}$$

while

$$\mathrm{so}(2n;\mathbf{C}) =$$

$$= \Big\{ \begin{pmatrix} A & B \\ C & -A^t \end{pmatrix} \Big| A,B,C \in \mathbf{C}(n), C + C^t = 0, B^t + B = 0, \Big\}$$

Any isotropic flag manifold is a homogeneous space $SO(2n;\mathbf{C})/H$ for a suitable subgroup H.

In the sequel, we need three special cases of isotropic flag manifolds, namely the isotropic Grassmannians IG_1 and IG_n and the intermediate isotropic flag space $IG_{1,n}$. The isotropic Grassmannian IG_1 is nothing else but the complex nonsingular quadric in the projective space $\mathbf{P}(\mathbf{C}^{2n})$. It is a compactification of the complex Minkowski or Euclidean space. The space IG_n is the twistor space corresponding to IG_1 (see [62], vol.II); we met it already in I.4.7. It was shown there that the space IG_n of all maximal isotropic subspaces can be identified as well with the space of all almost complex structures on $\mathbf{R}^{2n}$ compatible with the standard Euclidean metric as with the space of all pure spinors. The space $IG_{1,n}$ is then the space of all elements from $IG_1 \times IG_n$ which are linked by the incidence relation $L_1 \subset L_n$. The spaces IG_n and $IG_{1,n}$ are not connected, they have two connected components. The action of $SO(2n;\mathbf{C})$ is transitive on these components and the whole spaces are, in fact, homogeneous spaces of the group $O(2n,\mathbf{C})$. In what follows, only one of the connected components will be systematically used and we shall denote it by the same symbol.

3.2 The isotropic Grassmannian IG_n

Let us now study the twistor space IG_n in more details. To describe it as a homogeneous space of the group $SO(2n;\mathbf{C})$, note that the isotropic group of the subspace W generated by the first n elements of the canonical basis is the subgroup H given by

$$\begin{aligned} H &= \left\{ \begin{pmatrix} A & B \\ C & D \end{pmatrix} \in SO(2n;\mathbf{C}) \middle| C = 0 \right\} \\ &= \left\{ \begin{pmatrix} A & B \\ 0 & (A^t)^{-1} \end{pmatrix} \middle| (A^{-1}B)^t + A^{-1}B = 0, A \in GL(n,\mathbf{C}) \right\}. \end{aligned}$$

So $IG_n = SO(2n;\mathbf{C})/H$ and in the sequel we shall consider the component of IG_n given by the orbit of W under the action of $SO(2n;\mathbf{C})$. We shall denote it by the same symbol.

The Lie algebra $\underline{h}$ of the group H is a subalgebra of $so(2n;\mathbf{C})$ given by

$$\underline{h} = \left\{ \begin{pmatrix} A & B \\ 0 & -A^t \end{pmatrix}, B^t = -B \right\}.$$

It is also possible to consider IG_n as a homogeneous space of $Spin_+(2n;\mathbf{C})$, which is a double covering of $SO(2n;\mathbf{C})$ (see I.4.9). We have

$$IG_n = Spin_+(2n;\mathbf{C})/\tilde{H},$$

where $\tilde{H}$ is the preimage of H in $Spin_+(2n;\mathbf{C})$. This way of considering IG_n is fundamental for our purposes and it will be systematically used below.

All isotropic flag manifolds share a very nice and important property, namely that they are not only homogeneous spaces of the complex group $G = \mathrm{SO}(2n;\mathbf{C})$, but that they are also homogeneous spaces of its compact real form $\widehat{\mathrm{SO}}(2n,\mathbf{R})$ (see Appendix B). Note that due to the choice of the quadratic form, the Euclidean space $\mathbf{R}^{2n} \subset \mathbf{C}^{2n}$ is the set of fixed points of the complex conjugation $\hat{}$ given by

$$\hat{}: P = [P', P''] \in \mathbf{C}^{2n} \mapsto \hat{P} = [\overline{P''}, \overline{P'}].$$

Indeed, if $P = \hat{P}$, then $(P,P) = (P,\hat{P}) = -\sum_{j=1}^{2n} |P_j|^2$. The compact real form $\widehat{\mathrm{SO}}(2n,\mathbf{R})$ is related to the group $\mathrm{SO}(2n,\mathbf{R})$ of real matrices in $\mathrm{SO}(2n;\mathbf{C})$by

$$\widehat{\mathrm{SO}}(2n,\mathbf{R}) = J^{-1}\,\mathrm{SO}(2n,\mathbf{R})J,$$

where $J^{-1} = J = \frac{1}{\sqrt{2i}}\begin{pmatrix} I_n & iI_n \\ iI_n & I_n \end{pmatrix}$. The twistor space IG_n is the homogeneous space $\widehat{\mathrm{SO}}(2n,\mathbf{R})/U(n)$.

Using the real Spin group $\widehat{\mathrm{Spin}}(2n;\mathbf{R})$, the isotropic Grassmannian is also a homogeneous space $\widehat{\mathrm{Spin}}(2n;\mathbf{R})/\widetilde{\mathrm{U}(n)}$, where $\widetilde{\mathrm{U}(n)}$ is the double covering of $\mathrm{U}(n)$ in $\widehat{\mathrm{Spin}}(2n;\mathbf{R})$. In order to deal with global objects on IG_n such as, for example, elements of cohomology groups with values in a line bundle, it is of fundamental importance to define suitable homogeneous coordinates on IG_n. The simplest and the most useful ones are given by the isotropic Stiefel manifolds.

The complex isotropic Stiefel manifold ISt_n is an example of a noncompact complex homogeneous space. It can be defined as the manifold of all ordered n-tuples of independent mutually orthogonal null vectors in $\mathbf{C}^{2n}$

$$ISt_n = \left\{ \mathcal{Z} \in \mathrm{C}(2n;n) | \operatorname{rank} \mathcal{Z} = n,\ \mathcal{Z}^t \mathbf{E} \mathcal{Z} = 0_n \right\}.$$

It is a homogeneous space of $\mathrm{SO}(2n;\mathbf{C})$ as well, namely $ISt_n = \mathrm{SO}(2n;\mathbf{C})/R$ with

$$R = \left\{ \begin{pmatrix} I & B \\ 0 & I \end{pmatrix} \middle| B^t + B = 0 \right\}.$$

There is a natural projection π from ISt_n onto IG_n, namely by mapping an element $\mathcal{Z} \in ISt_n$ to the span of the n columns of the matrix $\mathcal{Z}$. It is a principal fibre bundle with the group $GL(n,\mathbf{C})$ acting on fibers, mapping a basis for an isotropic subspace onto a similar basis.

Putting together all principal bundles over IG_n discussed above, we have the following picture :

$$\begin{array}{ccc}
\mathrm{Spin}_+(2n;\mathbf{C}) & \xrightarrow{\tilde{H}} & IG_n \\
\downarrow {\scriptstyle 2:1} & & \downarrow {\scriptstyle =} \\
\mathrm{SO}(2n,\mathbf{C}) & \xrightarrow{H} & IG_n \\
\downarrow & & \downarrow {\scriptstyle =} \\
ISt_n & \xrightarrow{\mathrm{GL}} & IG_n
\end{array}$$

All three spaces at the left hand side of the diagram can be used as homogeneous coordinates on IG_n. The simplest one is the Stiefel manifold ISt_n and no more is needed to describe the Penrose transform for the Laplace equation. On the other hand, in the case of the Dirac operator it will be absolutely necessary to work on the group $\mathrm{Spin}_+(2n;\mathbf{C})$ itself.

The twistor space IG_n enjoys a lot of nice properties, one of the most important ones being that it is a Kaehler manifold. To describe the Kaehler structure on IG_n, let us first define a (1,1)-form $\tilde{\phi}$ on the Stiefel manifold ISt_n by

$$\tilde{\phi} = -4i\partial\bar{\partial}\log\det(\overline{\mathcal{Z}}^t\mathcal{Z}), \mathcal{Z} \in ISt_n.$$

One may verify that $\tilde{\phi}$ is a pull-back of a closed $(1,1)$-form ϕ on IG_n, where ϕ is invariant with respect to the action of the group $\widetilde{\mathrm{SO}}(2n,\mathbf{R})$. The form ϕ is a Kaehler form on IG_n (for more details see [48],II,p.161). In fact the form is defined by the formula above on the Grassmannian $G_{n;2n}$ and we are considering its restriction to the complex submanifold IG_n. In such a way we are applying a general principle saying that a complex submanifold of a Kaehler manifold is again a Kaehler manifold with the Kaehler form given by the restriction (see [48],II,p.165).

3.3 Line bundles on IG_n

Considering the twistor space IG_n as a homogeneous space $\mathrm{Spin}_+(2n;\mathbf{C})/\tilde{H}$, then for every one-dimensional representation ρ of the isotropic group $\tilde{H}$ there is a line bundle L_ρ on IG_n associated to the representation ρ. The bundle L_ρ is defined as the quotient space obtained by defining the equivalence $[s,z] \sim [s\tilde{p}, \rho(\tilde{p}^{-1})z]$ on $\mathrm{Spin}_+(2n;\mathbf{C}) \times \mathbf{C}$. This is a special case of a homogeneous bundle on a homogeneous space (see Appendix B). If the representation ρ of $\tilde{H}$ is holomorphic, then L_ρ is a holomorphic line bundle.

There are some interesting one-dimensional representations on $\tilde{H}$. For example, considering the map $\tilde{\pi} : \tilde{H} \to GL(n,\mathbf{C})$ given by the composition of two projections $\tilde{H} \to H \to GL(n,\mathbf{C})$, where the last map is defined by

$$\begin{pmatrix} A & B \\ 0 & (A^t)^{-1} \end{pmatrix} \to A,$$

then clearly the map $\rho : \tilde{h} \mapsto \det A = \det(\tilde{\pi}(\tilde{h}))$ is a homomorphism of $\tilde{H}$ into the nonzero complex numbers.

We have even more. The group $\tilde{H}$ being a $2:1$ covering of H, we can construct a square root of the representation $\det(\tilde{\pi}(\tilde{h}))$. To this end, let us first recall that if $f_1, \ldots, f_n, \bar{f}_1, \ldots, \bar{f}_n$ is a canonical basis in $\mathbf{C}^{2n}$, then the idempotent $I = I_1 \ldots I_n$, $I_j = \bar{f}_j f_j$, was used for the realization of the spinor space S_{2n}^+ as $(\wedge^{\text{even}} W)I \subset \mathbf{C}_{2n}$ (see I.4.7.4).

Theorem 3.3.1. *For each $\tilde{h} \in \tilde{H}$ there exists a number $f(\tilde{h}) \in \mathbf{C}$ such that $\tilde{h}I = f(\tilde{h})I$. Moreover, $f(\tilde{h})^2 = \det \tilde{\pi}(\tilde{h})$, where $\tilde{\pi} : \tilde{H} \to \mathrm{GL}(n, \mathbf{C})$ is the projection defined above and the map $f : \tilde{H} \to \mathbf{C}^* = \mathbf{C} \setminus \{0\}$ is a one-dimensional holomorphic representation of $\tilde{H}$.*

Proof. Let us discuss first the simplest case from which all basic ideas in the proof will become transparent.

Let f be a null vector in $\mathbf{C}^{2n}$ normalized in such a way that $(f, \bar{f}) = 1/2$. Then $\bar{f}f$ is an idempotent (see section 2,(2.2)). Now consider the subalgebra of $\mathbf{C}_{2n}$ generated by f and $\bar{f}$. First note that for $u, v \in \mathbf{C}$,

$$(1 + u\bar{f}f)(1 + v\bar{f}f) = 1 \ \text{ iff } \ (1+u)(1+v) = 1.$$

So if $u \neq -1$, then the element $1 + u\bar{f}f$ belongs to the Clifford group. A simple calculation shows that the rotation corresponding to that element is given by the matrix $\pi(1 + u\bar{f}f) = \begin{pmatrix} 1+u & 0 \\ 0 & 1+v \end{pmatrix}$ whence an element in the Spin group corresponding to this matrix is obtained by normalizing $1 + u\bar{f}f$, where using the involution α (see I.3.3.1) the norm $\mathcal{N}(1 + u\bar{f}f)$ is given by

$$\mathcal{N}(1 + u\bar{f}f) = \alpha((1 + u\bar{f}f))(1 + u\bar{f}f) = (1 + uf\bar{f})(1 + u\bar{f}f) = 1 + u.$$

Consequently the elements of the Spin group corresponding to the matrix above are given by

$$\pm g = \pm \frac{1 + u\bar{f}f}{\sqrt{1+u}}$$

whence

$$gI = \pm \frac{1}{\sqrt{1+u}}(1 + u\bar{f}f)I = \pm \frac{1}{\sqrt{1+u}}(1+u)I = \pm\sqrt{1+u}\,I.$$

So we have proved the theorem for the simplest elements in $\tilde{H}$.

Let us consider now an element $\tilde{h} \in \tilde{H}$ such that the corresponding rotation is given by the matrix $\mathcal{A} = \begin{pmatrix} A & 0 \\ 0 & (A^t)^{-1} \end{pmatrix}$ where A is a diagonal matrix of the form

$$A = \mathrm{diag}(1 + a_1, \ldots, 1 + a_n),\ a_i \in \mathbf{C},\ a_i \neq -1.$$

Then, due to the fact that the elements $I_j = \bar{f}_j f_j$ mutually commute, we obtain by repeating the calculations made above that

$$\tilde{h}I = \pm\sqrt{\Pi_j(1 + a_j)}I = \pm\sqrt{\det A}I.$$

For an element $\tilde{h}$ in the preimage of $\mathcal{A}$ where A is diagonalizable, we can perform the same computations in terms of the basis with respect to which A has a diagonal form. By continuity, the theorem is then proved for any $\tilde{h}$ in the preimage of $\mathcal{A}$ where A is nonsingular.

On the other hand, an inverse image in the Spin group of $\beta = \begin{pmatrix} I_n & B \\ 0 & I_n, \end{pmatrix}$ with B antisymmetric is given by

$$\pm\tilde{\beta} = \exp\{\sum_{i,j=1}^{n} B_{ij} f_i f_j\}$$

(see Chap.1,Example 4.9.2). Then clearly $\pm\tilde{\beta}I = \pm I$, since the elements f_j are annihilation operators. By virtue of the decomposition

$$\begin{pmatrix} A & B \\ 0 & (A^t)^{-1} \end{pmatrix} = \begin{pmatrix} A & 0 \\ 0 & (A^t)^{-1} \end{pmatrix} \begin{pmatrix} I_n & A^{-1}B \\ 0 & I_n \end{pmatrix},$$

we have proved that $\tilde{h}I = \pm\sqrt{\det \tilde{\pi}(\tilde{h})}$ for any $\tilde{h} \in \tilde{H}$. Obviously, the function f defined in such a way is a holomorphic one-dimensional representation of $\tilde{H}$. ■

Note that the function f defined on $\tilde{H}$ is determined (up to a sign) by $\sqrt{\det \tilde{\pi}(\tilde{h})}$ and that the sign of the square root is well defined for points in $\tilde{H}$. This function will frequently appear in the sequel; we shall denote it simply by $\sqrt{\det}$.

Definition 3.3.2. *The function $\sqrt{\det}$ is defined on $\tilde{H}$ by the expression*

$$\tilde{h}I = \sqrt{\det}(\tilde{h})I. \tag{3.1}$$

By means of the function $\sqrt{\det}$, we are now able to define the important line bundle $\mathbf{L}$ on IG_n in the following way.

Definition 3.3.3. *The holomorphic line bundles $\mathbf{L}^k, k \in \mathbf{Z}$, are defined as the line bundles associated to the representation $(\sqrt{\det})^k$ of $\tilde{H}$.*

Notice that the bundle $\mathbf{L}$ may be characterized in several other ways (see e.g. [4] or [64]).

The space IG_n of maximal isotropic subspaces can be identified, as was explained in I.4.9, with the projective space of pure spinors $\mathbf{P}(S^{\text{pure}})$, $S^{\text{pure}} \subset S^+$. This identification can be alternatively described using Theorem 3.3.1. Indeed, Theorem 3.3.1 implies that the map from $\text{Spin}_+(2n; \mathbf{C})$ to S^+ given by the map $\mathcal{S} : g \mapsto gI$ induces a map from IG_n to $\mathbf{P}(S^+)$. The fact that the image of the induced map is the projective space of pure spinors and that the map is injective follows then from the properties of the map $\mathcal{S}$ listed in

Lemma 3.3.4 *The map* $\mathcal{S} : \text{Spin}_+(2n; \mathbf{C}) \to S^+$ *with* $\mathcal{S}(g) = gI$, *is a holomorphic map satisfying*

$$\mathcal{S}(g\tilde{h}) = \sqrt{\det}(\tilde{h})\mathcal{S}(g).$$

Moreover, if Z *is a vector in the isotropic subspace* $\pi(g) \in IG_n$, *then* $Z\mathcal{S}(g) = 0$ *in* $\mathbf{C}_{2n}$.

Proof. The first part follows immediately from Theorem 3.3.1. Furthermore, as the columns of the matrix in $\text{SO}(2n; \mathbf{C})$corresponding to g are given by

$$[gf_1g^{-1}, \ldots, gf_ng^{-1}, g\bar{f}_1g^{-1}, \ldots, g\bar{f}_ng^{-1}],$$

the isotropic subspace $\pi(g)$ is generated by gf_jg^{-1}. But $gf_jg^{-1}gI = gf_jI = 0$. ∎

The properties of the map $\mathcal{S}$ are quite important and will be substantially used below. Notice also that the values of the map $\mathcal{S}$ belong to the space of pure spinors; even more, the value at a point g is exactly a pure spinor annihilated by the vectors in the isotropic subspace corresponding to g.

Given a pure spinor $s \in S^{\text{pure}}$, we shall need an explicit formula for the projection from S^+ to the one-dimensional subspace generated by s in S^+. A simple description of it is given in

Lemma 3.3.5 *Let us consider an element* $[s] \in IG_n$ *represented by* $s \in \text{Spin}_+(2n; \mathbf{C})$. *Then the projection* $\Pi([s])$ *onto the line generated by the corresponding pure spinor defined by*

$$\Pi([s]) = \frac{1}{|s|^2} sI\bar{s}$$

does not depend on the choice of a representative s.

Proof. Let us note first that $|s|^2 I = I\bar{s}sI$. Consequently $\Pi([s])$ is really a projection and we know already that the image belongs necessarily to the line generated by the corresponding pure spinor. As for the independence on the choice of a representative, it is sufficient to apply Theorem 3.3.1. ∎

3.4 Invariant forms on isotropic Grassmannians

In studying the Penrose transform for solutions to the Dirac equation, some invariant forms on the isotropic Grassmannians IG_n will be quite useful, such as the invariant volume form on IG_n which will be denoted by κ_n.

The existence and uniqueness (up to normalization) of an invariant measure on a compact group (usually called Haar measure) is well known. The existence of invariant measures on homogeneous spaces of compact groups is more delicate. Nevertheless, if G is a compact Lie group and H its closed connected subgroup, then there exists (up to normalization) a unique left invariant volume form on the homogeneous space G/H (more details with full proofs can be found e.g. in [11],p.49).

The isotropic Grassmannian IG_n is, as we know, a homogeneous space of a compact group, namely $IG_n = \widetilde{\mathrm{Spin}}(2n;\mathbf{R})/\widetilde{\mathrm{U}(n)}$. So there is a unique left invariant volume form κ_n on IG_n, normalized by the condition

$$\int_{IG_n} \kappa_n = 1.$$

There is an alternative procedure how to construct the volume form, due to the fact that IG_n is a Kaehler manifold. Using the Kaehler form ϕ on IG_n given by

$$\tilde{\phi} = -4i\partial\bar{\partial}\log\det(\bar{\mathcal{Z}}^t \mathcal{Z}), \mathcal{Z} \in ISt_n,$$

we can construct the form $\underbrace{\phi \wedge \ldots \wedge \phi}_{n(n-1)/2}$, which is a form of top degree on IG_n and which is moreover invariant with respect to $\widetilde{\mathrm{SO}}(2n,\mathbf{R})$ whence it is equal (up to a normalization) to the volume form κ_n. This is not surprising since in general the top power of the Kaehler form is a volume form (see e.g. [91],p.259). So we have the explicit formula

$$\kappa_n = c_1 \cdot \underbrace{\phi \wedge \ldots \wedge \phi}_{n(n-1)/2},$$

where $c_1 = 1/(n!2^n) \cdot vol(IG_n)$ and $vol(IG_n)$ is the volume of IG_n with respect to the Riemannian metric corresponding to the Kaehler form.

In what follows we shall also make use of the form

$$\tau_n = \underbrace{\phi \wedge \ldots \wedge \phi}_{(n-1)(n-2)/2}. \tag{3.2}$$

The form τ_n is a volume form on any complex submanifold $N \subset IG_n$ of dimension $(n-1)(n-2)/2$, since the restriction of the form ϕ to N is again a Kaehler form. A typical example of such a submanifold N is the twistor space IG_{n-1} embedded in IG_n in a way described in section 4.1.

Let us now turn our attention to another problem concerning invariant forms. The form κ_n was characterized as an invariant form of the top degree, i.e. an element of

$\mathcal{E}^{(\binom{n}{2},\binom{n}{2})}(IG_n)$. The question is, if it is also possible to find an invariant form in the space $\mathcal{E}^{(\binom{n}{2},0)}(IG_n)$. Denoting the holomorphic cotangent space of IG_n by $T^{1,0*}$, we know that the top power $\Lambda^{\binom{n}{2}}(T^{1,0*})$ is a line bundle and that forms in $\mathcal{E}^{(\binom{n}{2},0)}(IG_n)$ are sections of it. In fact, there is no such invariant form, but we can find an invariant form of bidegree $(\binom{n}{2},0)$, if we assume such forms to have values in a suitable line bundle, as described in the following theorem.

Theorem 3.4.1. *There exists a unique (up to a nonzero multiple) holomorphic invariant form α_n of bidegree $(\binom{n}{2},0)$ on IG_n with values in the line bundle $\mathbf{L}^{2(n-1)}$.*

Proof. First note that the holomorphic cotangent bundle is a homogeneous bundle. More precisely, writing IG_n as the homogeneous space $\mathrm{SO}(2n,\mathbf{C})/H$ and denoting by $\mathrm{so}(2n;\mathbf{C})$, resp. $\underline{\mathrm{h}}$, the Lie algebras of $\mathrm{SO}(2n,\mathbf{C})$, resp. H, we may consider the adjoint representation Ad of H on the quotient space $\mathrm{so}(2n;\mathbf{C})/\underline{\mathrm{h}}$. It is known (see e.g. [33], p.209 or [4],p.26-29) that the holomorphic cotangent bundle is a homogeneous bundle given by the representation Ad on $(\mathrm{so}(2n;\mathbf{C})/\underline{\mathrm{h}})^*$.

Consequently, the top power $\Lambda^{\binom{n}{2}}(T^{1,0*})$ is a homogeneous bundle given by the representations $\Lambda^{\binom{n}{2}}Ad$ on the one-dimensional space $\Lambda^{\binom{n}{2}}((\mathrm{so}(2n;\mathbf{C})/\underline{\mathrm{h}})^*)$.

Using the explicit matrix description of $\mathrm{so}(2n;\mathbf{C})$ and $\underline{\mathrm{h}}$ discussed in the previous section, we get that the action of a matrix $\begin{pmatrix} A & B \\ 0 & A^{-t} \end{pmatrix} \in H$ on an element $m \in \mathrm{so}(2n;\mathbf{C})/\underline{\mathrm{h}}$, represented by an antisymmetric matrix, is given by $A^t m A$. This implies that the one-dimensional representation of H on $\Lambda^{\binom{n}{2}}((\mathrm{so}(2n;\mathbf{C})/\underline{\mathrm{h}})^*)$ is given by $(\det A)^{1-n}$. Indeed, only nontrivial one-dimensional representations of the group H are given by powers of the representation $\det A$ and the corresponding power can be computed using matrices of the form $A = \lambda \cdot I_n, \lambda \in \mathbf{C}$.

Knowing that the line bundle $\mathbf{L}^{2(n-1)}$ is a homogeneous bundle corresponding to the representation $(\det A)^{n-1}$ of H, we have obtained that the homogeneous bundle $\Lambda^{\binom{n}{2}}(T^{1,0*}) \otimes \mathbf{L}^{2(n-1)}$ is an associated bundle to the trivial representation. Moreover, elements in $\mathcal{E}^{(\binom{n}{2},0)}(IG_n, \mathbf{L}^{2(n-1)})$ are sections of the bundle $\Lambda^{\binom{n}{2}}(T^{1,0*}) \otimes \mathbf{L}^{2(n-1)}$. The form α_n is defined then as the form corresponding to a constant (nonzero) function on IG_n. ∎

There is an alternative, more explicit description of the form α_n. Let us define first an invariant form on a general Grassmannian manifold $G_{k,n}$ of k-dimensional subspaces in $\mathbf{C}^n$. We shall write it out in coordinates given by the Stiefel manifold $St_{k;n}$ of all $n \times k$ complex matrices of rank k.

Let $Z^i \in \mathbf{C}^n$, $i = 1, \dots, n$; then we put

$$\det[Z^1, \dots, Z^k, dZ^{k+1}, \dots, dZ^n] = \sum_{\pi} Z^1_{\pi_1} \cdot \ldots \cdot Z^k_{\pi_k} dZ^{k+1}_{\pi_{k+1}} \wedge \ldots \wedge Z^n_{\pi_n},$$

where the sum is taken over all permutations π of the set $\{1,\ldots,n\}$.

A special form $\omega_{k,n}$ on the Stiefel manifold $St_{k;n}$ is defined by

$$\omega_{k,n}(\mathcal{Z}) = \Lambda_{i=1}^{k}\det[Z^1,\ldots,Z^k,\underbrace{dZ^i,\ldots,dZ^i}_{l}],\ \mathcal{Z}\in St_{k;n}.$$

Under the transform

$$\mathcal{Z}\mapsto\mathcal{Z}\cdot g,\ d\mathcal{Z}\mapsto d\mathcal{Z}\cdot g+\mathcal{Z}\cdot dg,\ g\in \mathrm{GL}(k,\mathrm{C}),\mathcal{Z}\in St_{k;n}.$$

the form $\omega_{k,n}$ turns into $(\det g)^n\cdot\omega_{k,n}$.

The isotropic Stiefel manifold $ISt_n = ISt_{n;2n}$ is then defined by the set of equations $\varphi_{ij}(\mathcal{Z}) = (Z^i,Z^j) = 0, i,j, = 1,\ldots,n; i\geq j$ on the Stiefel manifold $St_{n;2n}$. As

$$d\varphi_{ij}\wedge\omega_n = 0;\ i,j, = 1,\ldots,n; i\geq j,$$

we may then define the form α_n by

$$\alpha_n = d\varphi_{11}\rfloor(d\varphi_{12}\rfloor\ldots(d\varphi_{nn}\rfloor\omega_n)\ldots)|_{ISt_n},$$

where the symbol $\rfloor$ denotes the division of differential forms.

In local coordinates $\mathcal{Z} = \begin{bmatrix} A \\ I_n \end{bmatrix}$, A being an antisymmetric matrix, it is easy to write the form $\omega_{k,n}$ explicitly and to perform the division. The result is (not quite surprisingly) the holomorphic form of top degree $\alpha_n = \Pi_{i<j}dA_{ij}$.

The form α_n transforms as $\alpha_n\mapsto(\det g)^{n-1}\alpha_n$ under the substitution

$$\mathcal{Z}\mapsto\mathcal{Z}\cdot g,\ d\mathcal{Z}\mapsto d\mathcal{Z}\cdot g+\mathcal{Z}\cdot dg,\ g\in \mathrm{GL}(n,\mathrm{C}),\mathcal{Z}\in ISt_n \tag{3.3}$$

whence it represents a nontrivial element in $\mathcal{E}^{(\binom{n}{2},0)}(IG_n,\mathrm{L}^{2(n-1)})$ (for more details see [15]).

The conjugated form $\overline{\alpha_n}$ belonging to $\mathcal{E}^{(0,l(\binom{n}{2}))}(IG_n,\overline{\mathrm{L}}^{2(n-1)})$ - where $\overline{\mathrm{L}}$ is an antiholomorphic line bundle given by the representation $\overline{\det A}^{2(n-1)}$ - can be constructed in a quite similar way. It is an antiholomorphic form of the top degree on IG_n.

If now $M_I(\mathcal{Z})$, $I\subset\{1,\ldots,2n\},|I| = n$, denotes the determinant of the corresponding $n\times n$ minor of the matrix $\mathcal{Z}$, we get still another description of the volume form κ_n on IG_n. Under the substitution $\mathcal{Z}\mapsto\mathcal{Z}\cdot g$, the determinants are transforming as $M_I\mapsto M_I\det g$, whence the form

$$\frac{\alpha_n(\mathcal{Z})\wedge\overline{\alpha_n(\mathcal{Z})}}{(\sum_I |M_I(\mathcal{Z})|^2)^{n-1}} \tag{3.4}$$

is invariant with respect to the substitution (3.3) and so it is a form on IG_n.

As the forms α_n and $\overline{\alpha}_n$ are invariant under $SO(2n, \mathbf{C})$ and the denominator is preserved under the action of the group $SO(2n, \mathbf{R})$, the form (3.4) coincides (after normalization) with the form κ_n.

Using the projection $\Pi(\alpha), \alpha \in IG_n$, defined in Lemma 3.3.5 we have

Theorem 3.4.2 *The map $T : S^+ \to S^+$ given by*

$$T = \int_{IG_n} \Pi(\alpha)\kappa_n(\alpha)$$

is a (nonzero) multiple of the identity map on the spinor space S^+.

Proof. Take any $g \in \widehat{\mathrm{Spin}}(2n; \mathbf{R})$. The form κ_n being invariant with respect to $\widehat{\mathrm{Spin}}(2n; \mathbf{R})$, we have that

$$T = \int_{IG_n} \Pi(g\alpha)\kappa_n(\alpha) = \int_{IG_n} g\Pi(\alpha)\frac{\bar{g}}{|g|^2}\kappa_n(\alpha) = gTg^{-1},$$

which implies that $T \in \mathbf{C}_{2n}^+$ commutes with all elements of $\widehat{\mathrm{Spin}}(2n; \mathbf{R})$ whence it is an intertwining map for the representation S^+ of $\widehat{\mathrm{Spin}}(2n; \mathbf{R})$. The representation S^+ being irreducible, the map T is a multiple of the identity.

To prove that T is nontrivial, consider nonhomogeneous coordinates on an open, dense subset of IG_n given by the set $\mathcal{A}$ of all $n \times n$ antisymmetric matrices. For any matrix $A \in \mathcal{A}$, a basis of the corresponding isotropic subspace is given by columns of the matrix $\mathcal{Z} = \begin{bmatrix} I_n \\ A \end{bmatrix}$. The form κ_n has top degree in these nonhomogeneous coordinates and it is of the form

$$\frac{\prod_{i<j}[dA_{ij} \wedge \overline{A}_{ij}]}{(\sum_J |M_J(\mathcal{Z})|^2)^{n-1}},$$

where $M_J, J \subset \{1, \ldots, 2m\}, |J| = n$, are the Plücker coordinates of $\mathcal{Z}$. As the element $s = \exp(\sum_{i<j} A_{ij}\bar{f}_i\bar{f}_j) \in \mathrm{Spin}_+(2n; \mathbf{C})$ is a preimage of $\mathcal{Z}$, we obtain that

$$ITI = \int_{\mathcal{A}} \frac{1}{|s|^2} \frac{\prod_{i<j}[dA_{ij} \wedge \overline{A}_{ij}]}{(\sum_J |M_J(\mathcal{Z})|^2)^{n-1}} \neq 0. \quad \blacksquare$$

3.5 Forms with coefficients in line bundles

The Penrose transform for the Dirac equation establishes a correspondence between elements of a certain cohomology group with coefficients in the sheaf of holomorphic sections of a holomorphic line bundle and holomorphic solutions to the Dirac equation. There are two basic approaches to these groups, one based on Čech cohomology with coefficients in sheaves, the other one based on Dolbeault cohomology groups represented by $\overline{\partial}$-closed differential forms with coefficients in a line bundle. Trying to keep the discussion on an elementary level, we want to avoid the use of sheaves and cohomology groups with coefficients in sheaves. Using the Dolbeault cohomology and suitable homogeneous coordinates, it is possible to describe these groups by working with ordinary (C-valued) forms satisfying an additional condition.

Suppose now that $\pi : G \to G/H$ is a homogeneous space and L_ρ is a bundle on G/H given by a representation ρ of the group H on a vector space V. Sections of L_ρ are in 1-1 correspondence with maps $f : G \to V$ satisfying the condition

$$f(gh) = \rho(h^{-1})f(g),\ g \in G,\ h \in H. \tag{3.5}$$

In our case, differential forms are, in fact, a special case of the described situation, since differential forms are sections of the outer powers of the cotangent bundle and the cotangent bundle is a homogeneous bundle corresponding to the adjoint representation on $\underline{g}/\underline{h}$, where $\underline{g}$, resp. $\underline{h}$, are the Lie algebras of G and H, respectively. Differential forms with values in a line bundle L_ρ fit well into this scheme, since from this point of view they are simply sections of the homogeneous bundle given by the tensor product of the cotangent bundle with L_ρ.

Hence it is not surprising that, as it was the case for sections of homogeneous bundles, it is possible to describe differential forms with values in L_ρ on subsets U in G/H by ordinary forms ω on $\pi_{-1}(U) \subset G$ transforming as $\omega \mapsto \rho(h_{-1})\omega$ under the substitution

$$g \to gh,\ dg \to dg \cdot h + gdh \tag{3.6}$$

(see Appendix B for more details)

Coming back to cohomology groups, let G and H be complex groups and let ρ be a holomorphic representation of H. Furthermore, let Ω be a domain in G/H and let $\mathcal{E}^{0,k}(\Omega, L_\rho)$ denote the space of all $(0,k)$-forms on Ω with values in L_ρ. Then we can define the Dolbeault complex by

$$\mathcal{E}^{0,0}(\Omega, L_\rho) \xrightarrow{\overline{\partial}_0} \dots \xrightarrow{\overline{\partial}_{k-1}} \mathcal{E}^{0,k}(\Omega, L_\rho) \xrightarrow{\overline{\partial}_k} \dots,$$

where $\overline{\partial}_k$ denotes the $\overline{\partial}$-operator acting on $(0,k)$-forms. The cohomology groups $H^q(\Omega, L_\rho)$ can then be defined by the homology of this Dolbeault complex, i.e.

$$H^q(\Omega, L_\rho) = \mathrm{Ker}\overline{\partial}_k / \mathrm{Im}\overline{\partial}_{k-1}.$$

Using the description of the differential forms with values in L_ρ by ordinary $\overline{\partial}$-forms on G satisfying the condition (3.5), it becomes easy to compute with cohomology classes in the same way as it is usually done for de Rham cohomology groups. We shall use this procedure in the next sections.

For more information we refer to [90].

4 TWISTOR CORRESPONDENCE

4.1 The basic twistor diagram

The Penrose transform is always based on a diagram of homogeneous spaces (see [3] and [4]). In our case (i.e. for the description of solutions of the Dirac equation in even dimensions), the diagram contains certain isotropic flag manifolds, namely isotropic Grassmannians IG_1, IG_{n+1} and an isotropic flag manifold $IG_{1,n+1}$ (see section 3). Together with the natural projections, they form the basic diagram

$$\begin{array}{ccc} & IG_{1,n+1} & \\ {}^{\mu}\swarrow & & \searrow^{\nu} \\ IG_{n+1} & & IG_1 \end{array}$$

This double fibration encodes principal information concerning the geometry of objects involved in the Penrose transform. The double fibration is used for the translation of objects from one bottom piece to the other using the middle part and the maps μ and ν. The compositions $\mu \circ (\nu)^{-1}$ and $\nu \circ (\mu)^{-1}$ are not 1-1 maps and their values are sets in the corresponding spaces. These sets play a major role in the geometry of the Penrose transform and it is worthwile to say more about them.

If we take a point α in the twistor space IG_{n+1} then the set $\mu \circ (\nu)^{-1}(\alpha)$ is the projective space of the corresponding isotropic subspace $L_{n+1}(\alpha)$ in $\mathbf{C}^{2n+2}$. These isotropic subspaces (as well as their projective versions) are usually called α-planes (there are, in fact two families of isotropic subspaces in the given quadric, the other family being usually called β-planes; for more details see e.g. [41]). So points in the twistor space are represented by α-planes in the quadric.

If we now take a point P in the quadric IG_1, then the set $\nu \circ (\mu)^{-1}(P)$ is a complex submanifold of IG_{n+1} which is isomorphic (as a complex manifold) to IG_n. Indeed, the set $\nu \circ (\mu)^{-1}(P)$ can be described as the set of all isotropic subspaces in IG_n containing the point P (more precisely containing the corresponding one-dimensional vector subspace in $\mathbf{C}^{2n}$). Using the real structure on $\mathbf{C}^{2n+2}$ given by

$$P = [P_1, P_2],\ P_i \in \mathbf{C}^{n+1} \mapsto \hat{P} = [\overline{P_2}, \overline{P_1}],$$

the space $\mathbf{C}^{2n}$ can be embedded into $\mathbf{C}^{2n+2}$ as the space of all vectors orthogonal both to P and $\hat{P}$. By restricting the bilinear form $(.,.)$ to $\mathbf{C}^{2n}$, the restriction of any isotropic subspace $\alpha \in IG_{n+1}$ containing the point P is then an isotropic subspace of dimension n in S^{2n}. The described map being one-to-one, we have identified the set $\nu \circ \mu^{-1}(P)$ with IG_n.

Our aim is to describe solutions of the (complexified) Dirac equation only on subsets of the (complexified) Minkowski space

$$\mathbf{C}M \subset IG_1, \mathbf{C}M = \{L = \text{span}\{Z\} | Z \in \mathbf{C}^{2n+2},\ Z_{n+1} \neq 0\},$$

the latter being an open dense subset of IG_1. So we shall consider nonhomogeneous coordinates (x, y) on $\mathbf{C}M$ by the identification

$$\mathbf{C}M = \{[x, 1, y, -x \cdot y]^t | x, y \in \mathbf{C}^n\},$$

where $x \cdot y = \sum_1^n x_j y_j$.

The basic twistor diagram will hence be restricted to the double fibration

$$\begin{array}{ccc} & \mathbf{F} & \\ {}^{\mu}\swarrow & & \searrow^{\nu} \\ \mathbf{T} & & \mathbf{C}M \end{array}$$

where the spaces involved are defined as

$$\mathbf{F} = \nu^{-1}(\mathbf{C}M) = \{[L_1, L_{n+1}] | L_1 \in \mathbf{C}M\},$$

$$\mathbf{T} = \mu(\mathbf{F}) = \{L_{n+1} | L_{n+1} \cap \mathbf{C}M \neq \emptyset\}.$$

4.2 Homogeneous coordinates.

To calculate with forms on the isotropic Grassmannian (resp. flag) manifolds discussed above, we shall use homogeneous coordinates. Suitable homogeneous coordinates on IG_{n+1} are given by the isotropic Stiefel manifolds ISt_{n+1}; the corresponding projection $\pi : ISt_{n+1} \mapsto IG_{n+1}$ is then a holomorphic map. The same Stiefel manifold ISt_{n+1} gives rise to homogeneous coordinates on $IG_{1,n+1}$ if we consider the first vector of a basis as the generator of the corresponding one-dimensional subspace. We shall use also the whole group $\mathrm{Spin}_+(2n+2, \mathbf{C})$ as homogeneous coordinates on IG_{n+1}.

Let us now consider the twistor diagram corresponding to the complex Minkowski space $\mathbf{C}M$. In this situation, it is very useful to restrict further bases allowed in ISt_{n+1}. Notice first that any basis of an element $[L_1, L_{n+1}] \in \mathbf{F}$ may be transformed into a basis of the form

$$\begin{bmatrix} x & x^1 & \dots & x^n \\ 1 & 0 & \dots & 0 \\ y & y^1 & \dots & y^n \\ -x \cdot y, & -x \cdot y^1 - x^1 \cdot y & \dots & -x \cdot y^n - x^n \cdot y \end{bmatrix}, \tag{4.1}$$

where $x^j, y^j \in \mathbf{C}^n$; $x^i \cdot y^j + x^j \cdot y^i = 0$; $i, j = 1, \dots, n$.

The vectors $Z^i = [x^i, y^i] \in \mathbf{C}^{2n}$ generate an n-dimensional isotropic subspace with respect to the form $(.,.)$ on $\mathbf{C}^{2n}$ whence the set $\mathbf{F}$ may be identified with the product $\mathbf{C}^{2n} \times IG_n$, an element $P \in \mathbf{C}^{2n}$ being given by $P = [x, y]$ and an isotropic subspace in $\mathbf{C}^{2n}$ being the span of $Z^i, i = 1, \dots, n$. The map ν is then the projection to the first factor and we shall use $[P, \mathcal{Z}] \in \tilde{\mathbf{F}} = \mathbf{C}^{2n} \times ISt_n$ as homogeneous coordinates for $\mathbf{F}$. The fibration $\pi : \tilde{\mathbf{F}} \mapsto \mathbf{F}$ is then a principal fibration with the group $\mathrm{GL}(n, \mathbf{C})$ acting on the second factor and the projection π is a holomorphic map.

As discussed above, points in IG_{n+1} correspond to α-planes in $\mathbf{C}^{2n+2}$ under the twistor correspondence. We want to find out now how α - planes look like after the restriction to $\mathbf{C}M$. To this end consider a point $L_{n+1} \in \mathbf{T}$. As before, we can find a basis in the form (4.1) for L_{n+1} and denoting by $[P, Z^1, \ldots, Z^n]$ the corresponding element in $\mathbf{C}^{2n} \times ISt_n$, we may describe the intersection of L_{n+1} with $\mathbf{C}M$ as the affine isotropic subspace $\{P' \in \mathbf{C}M | P' = P + \sum_{i=1}^n \lambda_i Z^i, \lambda_i \in \mathbf{C}\}$. If we define $\eta \in \mathbf{C}^n$ by $\eta^i = (P, Z^i), i = 1, \ldots, n$, then this affine isotropic subspace can be described by $\{P' \in \mathbf{C}M | \eta^i = (P', Z^i), i = 1, \ldots, n\}$. The space $\tilde{\mathbf{T}} = \mathbf{C}^n \times ISt_n$ with coordinates $[\eta, \mathcal{Z}], \mathcal{Z} = [Z^1, \ldots, Z^n]$ hence yields suitable homogeneous coordinates on $\mathbf{T}$. Note that the coordinates $[\eta, \mathcal{Z}]$ are exactly the nontrivial elements in the last n columns of the matrix in (4.1) and this since $\eta^i = (P, Z^i) = -(x \cdot y^i + y \cdot x^i)$.

So we have the principal fibre bundle $\pi : \tilde{\mathbf{T}} \mapsto \mathbf{T}$ with the group $GL(n, \mathbf{C})$, the action of the group $\mathrm{GL}(n, \mathbf{C})$ being $[\eta, \mathcal{Z}] \mapsto [\eta \cdot g, \mathcal{Z} \cdot g], g \in \mathrm{GL}(n, \mathbf{C})$. Moreover, the projection π is then a holomorphic map.

We would like to point out that if a point in $\tilde{\mathbf{T}}$ has coordinates $[\eta, \mathcal{Z}]$ and a point in $\mathbf{C}M$ has coordinates P, then the incidence relation (stating that the corresponding couple belongs to $\tilde{\mathbf{F}}$) is given by

$$\eta^i = (P, Z^i), i = 1, \ldots, n.$$

This relation describes completely the geometry of the twistor diagram. Indeed, if a point in $\mathbf{T}$ has coordinates $[\eta, \mathcal{Z}]$, then the corresponding α-plane in $\mathbf{C}^{2n}$ is given by $\{P' \in \mathbf{C}^{2n} | \eta^i = (P', Z^i), i = 1, \ldots, n.\}$ On the other hand, for a given point $P \in \mathbf{C}^{2n}$, the corresponding set $\mu \circ \nu^{-1}(P)$ in $\mathbf{T}$ is described in coordinates $[\eta, \mathcal{Z}] \in \mathbf{T}$ by the equations $\eta^i = (P, Z^i), i = 1, \ldots, n$. This shows at the same time that in $\mathbf{T}$ $\mu \circ \nu^{-1}(P) = IG_n$, the latter thus being holomorphically embedded into $\mathbf{T}$.

Working with holomorphic functions on $\tilde{\mathbf{F}}$, we shall need a characterization of those among them which are pull-back of functions on $\tilde{\mathbf{T}}$. These pull-backs are functions on $\tilde{\mathbf{F}}$ which are constant along α-planes, or alternatively, functions annihilated by the vector fields $\sum_{i=1}^{2n} Z_i^j \partial_{P_i}, \mathcal{Z} = [Z^1, \ldots, Z^n] \in ISt_n$. Indeed, holomorphic functions which are pull-backs of holomorphic functions on $\tilde{\mathbf{T}}$ can be written in the form $f = f((P, Z^i), \mathcal{Z})$, where $f = f(\eta^i, \mathcal{Z})$ is holomorphic on $\tilde{\mathbf{T}}$. So it is sufficient to notice that

$$\sum_{i=1}^{2n} \frac{\partial \eta^k}{\partial P_i} Z_i^j = (Z^k, Z^i) = 0$$

and to use the chain rule for differentiation.

Summarizing we have the following basic diagram

$$\begin{array}{ccc} & \tilde{\mathbf{F}} \sim [P,\mathcal{Z}] & \\ \tilde{\mu} : \eta^i = [P, Z^i] \swarrow & & \searrow \tilde{\nu} \\ \tilde{\mathbf{T}} \sim [\eta, \mathcal{Z}] & & \mathrm{CM} \sim [P] \end{array}$$

To deal effectively with forms having values in line bundles, as will be the case when discussing solutions of the Dirac equation, it will be necessary to consider as homogeneous coordinates even bigger spaces given by the fibrations

$$\tilde{\pi} : \tilde{\tilde{\mathbf{F}}} = \mathbf{C}^{2n} \times \mathrm{Spin}_+(2n, \mathbf{C}) \mapsto \mathbf{F} \; ; \; \tilde{\pi}' : \tilde{\tilde{\mathbf{T}}} = \mathbf{C}^n \times \mathrm{Spin}_+(2n, \mathbf{C}) \mapsto \mathbf{T}.$$

Coordinates in $\tilde{\tilde{\mathbf{F}}}$ will be usually denoted by $[P, g]$, those in $\tilde{\tilde{\mathbf{T}}}$ by $[\eta, g]$. The spaces of homogeneous coordinates just described are again linked by a diagram similar to the diagram above. Let us also recall that the embeddings $\mathrm{GL}(n, \mathbf{C}) \subset H \subset \mathrm{SO}(2n, \mathbf{C})$ are lifted to the embeddings $\widetilde{\mathrm{GL}} \subset \tilde{H} \subset \mathrm{Spin}_+(2n, \mathbf{C})$ and that we have the projections $\pi : H \mapsto \mathrm{GL}(n, \mathbf{C})$ and $\tilde{\pi} : \tilde{H} \mapsto \mathrm{GL}(n, \mathbf{C})$ (see section 3). The fibrations above are principal fiber bundles with respect to the group $\tilde{H}$ and the action of it in homogeneous coordinates is given for $\tilde{h} \in \tilde{H}$ by $[P, g] \mapsto [P, g\tilde{h}]$ and $[\eta, g] \mapsto [\eta\tilde{\pi}(g), g\tilde{h}]$.

If we want to describe solutions on an open subset $\Omega \subset CM$, we restrict ourselves to subsets $\Omega' = \nu^{-1}(\Omega) \subset \mathbf{F}$ and $\Omega'' = \mu(\Omega') \subset \mathbf{T}$. Elements of the corresponding cohomology groups are then represented by ordinary forms on the corresponding subsets of $\tilde{\tilde{\mathbf{T}}}$ and $\tilde{\tilde{\mathbf{F}}}$.

4.3 The twistor space for $\mathbf{R}^{2n}$

Let us recall that the Euclidean space $\mathbf{R}^{2n}$ is embedded into $\mathbf{C}^{2n}$ as the set of fixed points of the complex conjugation $\hat{}$ given by

$$\hat{} : P = [P', P''] \in \mathbf{C}^{2n} \mapsto \hat{P} = [\overline{P''}, \overline{P'}]$$

Corresponding to this real structure, there is a canonical fibration of the twistor space $\mathbf{T}$ over $\mathbf{R}^{2n}$. This property is so important, that we discuss it in more details.

As any α-plane intersects $\mathbf{R}^{2n}$ in just one point, it defines the fibration of $\mathbf{T}$ over $\mathbf{R}^{2n}$ by mapping an α-plane to the intersection. Fibers of this fibration are given by the sets $\mu \circ \nu^{-1}(X), X \in \mathbf{R}^{2n}$. In other words, the space $\mathbf{T}$ may be identified with the subset $\mathbf{R}^{2n} \times IG_n$ of $\mathbf{C}^{2n} \times IG_n$.

It should be noticed that this embedding is not holomorphic. To describe the complex structure on $\mathbf{T}$ in terms of coordinates $[X, \mathcal{Z}]$, we now determine the intersection point explicitly.

Consider an α-plane given by the coordinates $[\eta, \mathcal{Z}]$. Since the vectors $Z^1, \ldots, Z^n$ span an isotropic subspace W in $\mathbf{C}^{2n}$, we have that when denoting by $\hat{W}$ the span of the vectors $\hat{Z}^1, \ldots, \hat{Z}^n$, $W \cap \hat{W} = \{0\}$, Indeed, for any $X \in W \cap \hat{W}$ we have that $(X, \hat{X}) = 0$, whence $X = 0$. Consequently $\mathbf{C}^{2n} = W \oplus \hat{W}$.

Let us now define a dual basis $[\mathcal{Y}, \hat{\mathcal{Y}}]$ to the basis $[\mathcal{Z}, \hat{\mathcal{Z}}]$. As for all $i, j = 1, \ldots, n$ the relations

$$(Z^i, Y^j) = (\hat{Z}^i, \hat{Y}^j) = \delta_{ij}, (Z^i, \hat{Y}^j) = (\hat{Z}^i, Y^j) = 0$$

should hold and moreover W and $\hat{W}$ are maximal isotropic subspaces, it is clear that the span of $\{\hat{Y}^j\}$ is just W and the span of $\{Y^j\}$ is $\hat{W}$.

Any point $R \in \mathbf{C}^{2n}$ may be then written as

$$R = \sum_{i=1}^{n} [(R, Z^j) Y^j + (R, \hat{Z}^j) \hat{Y}^j].$$

Moreover the point X given by

$$X = \sum_{i=1}^{n} [(P, Z^j) Y^j + (\hat{P}, \hat{Z}^j) \hat{Y}^j]$$

clearly belongs to $\mathbf{R}^{2n}$ and the vector

$$P - X = \sum_{i=1}^{n} [(P - \hat{P}, \hat{Z}^j) \hat{Y}^j] \tag{4.2}$$

belongs to the span of $\{\hat{Y}^j\}$, i.e. to W. Hence X is the intersection with $\mathbf{R}^{2n}$ of the α-plane going through P in the direction $\mathcal{Z}$. So we have proved the following

Lemma 4.3.1 *If an α-plane is described by coordinates $[\eta, \mathcal{Z}]$, then its intersection with $\mathbf{R}^{2n}$ is given by the point*

$$X = \sum_{i=1}^{n} [\eta^j Y^j + \overline{\eta^j} \hat{Y}^j], \tag{4.3}$$

where $Y^j = Y^j(Z_l^k, \hat{Z}_l^k)$.

Note that the vectors Y^j depends on Z^j and $\hat{Z}^j$ in a very complicated way (in principle by expressions used for inverting a matrix).

So we have two sets of coordinates on $\tilde{\mathbf{T}}$, namely $[\eta, \mathcal{Z}]$ and $[X, \mathcal{Z}]$, and we know that $\eta^i = (X, Z^i)$ and that X may be calculated using $[\eta, \mathcal{Z}]$ by (4.3). Furthermore, the (0,1)-vector fields on $\tilde{\mathbf{T}}$ in the holomorphic coordinates $[\eta, \mathcal{Z}]$ are generated by $\frac{\partial}{\partial \overline{\eta}^i}$ and $\frac{\partial}{\partial \overline{Z}_i^j}$. We are now able to determine the form of $(0,1)$-vectors in terms of the nonholomorphic coordinates $[X, \mathcal{Z}]$.

Lemma 4.3.2 *In the coordinates $[X, \mathcal{Z}]$, (0,1)-vector fields are generated by $\frac{\partial}{\partial \overline{Z}_i^j}$ and by $\sum_{i=1}^{2n} Z_i^j \frac{\partial}{\partial x_i}$, $\mathcal{Z} = [Z^1, \ldots, Z^n] \in ISt_n$.*

Proof. As

$$\frac{\partial}{\partial \overline{\eta}^i} = \sum_i \frac{\partial x_i}{\partial \overline{\eta}^j} \frac{\partial}{\partial x_i} = \sum_i \hat{Y}_i^j \frac{\partial}{\partial x_i},$$

it suffices to note that $\operatorname{span}\{\hat{Y}^j\} = \operatorname{span}\{Z^j\} = W$. ■

5 The Penrose transform for the Dirac equation

5.1 Definition

Consider an open domain $\Omega \subset \mathbf{C}M$ and the corresponding domains $\Omega' = \nu^{-1}(\Omega) \subset \mathbf{F}$, and $\Omega'' = \mu(\Omega') \subset \mathbf{T}$.

In order to be able to represent cohomology classes on the twistor space by ordinary forms (satisfying a certain condition), we have to use homogeneous coordinates given by the fibrations

$$\tilde{\pi} : \mathbf{C}^{2n} \times \mathrm{Spin}_{+}(2n, \mathbf{C}) \mapsto \mathbf{C}^{2n} \times IG_n = \mathbf{F}$$

and

$$\tilde{\pi}' : \mathbf{C}^{n} \times \mathrm{Spin}_{+}(2n, \mathbf{C}) \mapsto \mathbf{C}^{n} \times IG_n = \mathbf{T}.$$

So the domain Ω' will be the base of a fibration

$$\widetilde{\Omega'} = \tilde{\pi}^{-1}(\Omega') \subset \mathbf{C}^{2n} \times \mathrm{Spin}_{+}(2n, \mathbf{C}) \mapsto \Omega'$$

and the domain Ω'' will be the base of a fibration

$$\widetilde{\Omega''} = \tilde{\pi}^{-1}(\Omega'') \subset \mathbf{C}^{n} \times \mathrm{Spin}_{+}(2n, \mathbf{C}) \mapsto \Omega''.$$

All these domains are connected by the twistor diagram

$$\begin{array}{ccc} & \widetilde{\Omega'} & \\ \tilde{\mu} \swarrow & & \searrow \tilde{\nu} \\ \widetilde{\Omega''} & & \Omega \end{array}$$

The holomorphic form of top degree $\alpha_n \wedge dP$, where $dP = dP_1 \wedge \ldots \wedge dP_{2n}$ on $\mathbf{C}^{2n} \times IG_n$, can be lifted to $\widetilde{\Omega'}$ and it will transform under the substitution $g \mapsto g\tilde{h}$, $g \in \mathrm{Spin}_{+}(2n, \mathbf{C})$, $\tilde{h} \in \widetilde{H}$ as

$$\alpha_n \wedge dP \mapsto (\det(\tilde{\pi}(\tilde{h}))^{n-1} \alpha_n \wedge dP.$$

Using the holomorphic map $\mathcal{S}(s)$ introduced in Lemma 3.3.4 we have

Theorem 5.1.1 *The map*

$$\begin{array}{rcl} \mathcal{E}^{(0,\binom{n}{2})}(\widetilde{\Omega''}) & \mapsto & \mathcal{E}^{n(n+1)}(\widetilde{\Omega'}) \\ \beta & \mapsto & \mathcal{S} \cdot \tilde{\mu}^{*}\beta \wedge \alpha_n \wedge dP \end{array} \qquad (5.1)$$

induces a well-defined map

$$H^{(0,\binom{n}{2})}(\Omega'', \mathbf{L}^{1-2n}) \mapsto H^{n(n+1)}_{DR}(\Omega', S^{+}),$$

where $H^{n(n+1)}_{DR}(\Omega', S^{+})$ *stands for the de Rham cohomology group with coefficients in* S^{+}.

Proof. Suppose that a cohomology class $\beta \in H^{(0,\binom{n}{2})}(\Omega'', \mathbf{L}^{1-2n})$ is represented by a $\bar{\partial}$-closed form β of degree $(0, \frac{n(n-1)}{2})$ on $\widetilde{\Omega}'$ transforming as $\beta \mapsto (\sqrt{\det(\tilde{h})})^{1-2n}\beta$ under the substitution

$$\eta \mapsto \eta\tilde{\pi}(\tilde{h}), g \mapsto g\tilde{h}, \ g \in \mathrm{Spin}_+(2n, \mathbf{C}), \ \tilde{h} \in \tilde{P}.$$

Then its pullback $\tilde{\mu}^*\beta$ is transformed in the same way.

Moreover, as the form α_n and the section S are transforming as

$$\alpha_n \mapsto (\sqrt{\det(\tilde{h})})^{2n-2}\alpha_n, \ S \mapsto (\sqrt{\det(\tilde{h})})S$$

under the same transformation, the product (5.1) is invariant and it represents a well-defined element in $H^{n(n+1)}_{DR}(\Omega', S)$.

Performing calculations on $\mathbf{C}^{2n} \times \mathrm{Spin}_+(2n, \mathbf{C})$, we get

$$\begin{aligned} d(S \cdot \tilde{\mu}^*\beta \wedge \alpha_n \wedge dP) &= \partial(S \cdot \tilde{\mu}^*\beta \wedge \alpha_n \wedge dP) + \bar{\partial}(S \cdot \tilde{\mu}^*\beta \wedge \alpha_n \wedge dP) \\ &= S\tilde{\mu}^*[\bar{\partial}(\beta)] \wedge \alpha_n \wedge dP, \end{aligned}$$

and this since the form $S \cdot \tilde{\mu}^*\beta \wedge \alpha_n \wedge dP$ is ∂-closed on $\mathbf{C}^{2n} \times IG_n$ (as it is of top degree in the space of holomorphic differentials) and S, α_n and dP are holomorphic.

If $\beta = \bar{\partial}\tau$, where τ enjoys the same transformation properties as β does, then the form $S \cdot \tilde{\mu}^*\tau \wedge \alpha_n \wedge dP$ represents a well-defined element in $H^{n(n+1)-1}_{DR}(\Omega', S)$ and

$$d(S \cdot \tilde{\mu}^*\tau \wedge \alpha_n \wedge dP) = S \cdot \tilde{\mu}^*[\bar{\partial}\tau] \wedge \alpha_n \wedge dP = S \cdot \tilde{\mu}^*\beta \wedge \alpha_n \wedge dP. \ \blacksquare$$

Theorem 5.1.1 now leads to the definition of the Penrose transform.

Definition 5.1.2 *If* $[\beta] \in H^{(0,\binom{n}{2})}(\Omega'', \mathbf{L}^{1-2n})$, *then the form*

$$S \cdot \tilde{\mu}^*\beta \wedge \alpha_n \wedge dP$$

represents an element in $H^{n(n+1)}_{DR}(\Omega', S)$ *and the direct image*

$$\nu_*(S \cdot \tilde{\mu}^*\beta \wedge \alpha_n \wedge dP)$$

is a well-defined closed holomorphic form of top degree on $\mathbf{C}M$ *with values in* S, *whence it can be written in the form* $f(P)dP$, $f(P) \in S$ *on* Ω. *Moreover, the map* ν_* *is the integration along fibers, i.e.*

$$f(P)dP = \int_{\nu^{-1}(P)} S \cdot \tilde{\mu}^*\beta \wedge \alpha_n \wedge dP.$$

The function f *defined in such a way will be called the Penrose transform of* β *and it will be denoted by* $\mathcal{P}(\beta)$.

Theorem 5.1.3 *The Penrose transform $f(P) = \mathcal{P}(\beta)(P)$ is a holomorphic function on the domain Ω with values in S^+ and it satisfies the (complexified) Dirac equation there.*

Proof. The form $\mathcal{S} \cdot \tilde{\mu}^*\beta \wedge \alpha_n \wedge dP$ is closed and so the same holds for its image under the map μ_*. This implies that f is holomorphic. If the class of cohomology is represented by another element, then the integral along fibers will not change, so that $f(P)$ is well-defined. To verify that $f(P)$ satisfies the Dirac equation, it clearly suffices to introduce coordinates and to use differentiation under the integral sign. By definition

$$f(P) = \int_{\nu^{-1}(P)} \mathcal{S}(g) \cdot \tilde{\mu}^*\beta(P, g) \wedge \alpha_n(g).$$

Now the form β on the twistor space $\tilde{\mathbf{T}}$ can be written as a linear combination of terms $\beta_{I,J}(\eta^i, Z_l^k) d\bar{\eta}_I \wedge d\bar{Z}_J$. Furthermore, the pull-back $\tilde{\mu}^*\beta$ restricted to the fiber is given by the substitution

$$\eta^i = (P, Z^i),\ d\bar{\eta}^i = (\hat{P}, d\hat{Z}^i).$$

Applying the Dirac operator $D = \sum_{j=1}^n [e_{j+n}\frac{\partial}{\partial P_j} + e_j \frac{\partial}{\partial P_{j+n}}]$ under the integral sign and observing that $D(P, Z^I) = \sum_j e_j Z_j^i$ whence

$$D\beta_{I,J}((P, Z^i), \mathcal{Z}) = \sum_i (\sum_j Z_j^i e_j) \mathcal{S}(\mathcal{Z}) \frac{\partial \beta_I, J}{\partial \eta_i}$$

we obtain that, as $(\sum_j Z_j^i e_j)\mathcal{S}(\mathcal{Z}) = 0$ (see Lemma 3.3.4), $Df(P) = 0$. ∎

5.2 The surjectivity

We wish to show now that the Penrose transform is a 1-1 map, on domains of a suitable type. To this end it will be convenient to choose Lie balls, they form a basis of neighbourhoods of any point.

Definition 5.2.1 *For $\epsilon > 0$ we define the Lie ball $L(\epsilon)$ by*

$$L \equiv L(\epsilon) = \{P \in \mathbf{C}^{2n} | CN_P \cap \mathbf{R}^{2n} \subset B(\epsilon)\},$$

where $B \equiv B(\epsilon) \subset \mathbf{R}^{2n}$ is the ball with center in the origin and radius ϵ.

A Lie ball $L \equiv L(X, \epsilon), X \in \mathbf{C}^{2n}$, is then defined as the Lie ball $L(\epsilon)$ shifted by X, i.e.

$$L(X, \epsilon) = \{P + X | P \in L(\epsilon)\}.$$

A Lie ball is the nicest example of a domain of holomorphy for solutions of the Dirac (or the Laplace) equation. If a ball $B(X,\epsilon) \subset \mathbf{R}^{2n}$ is given, then the corresponding Lie ball $L(X,\epsilon) \subset \mathbf{C}^{2n}$ may be characterized as the biggest domain in $\mathbf{C}^{2n}$ such that all solutions of the Dirac (resp. the Laplace) equation can be analytically extended from $B(X,\epsilon)$ to the whole of $L(X,\epsilon)$ (see [71], [2], [66], [49]). One part of the claim is, in fact, a direct consequence of the definition above and the hyperbolic integral formula described in section 5.2, since the value of the extended solution in a point $P \in L$ can be defined by means of the hyperbolic integral formula.

The set $L \in \mathbf{C}^{2n}$ then defines two other subsets $L' \subset \mathbf{F}$ and $L'' \subset \mathbf{T}$ by $L' = \nu^{-1}(L)$ and $L'' = \mu(L')$. The special form of the Lie ball L implies that the set L'' is the set of all isotropic subspaces (of dimension n) in all complex null cones $\mathbf{C}N_X$ with vertices $X \in B$. So in this case L'' can be identified with $B \times IG_n$.

The question whether or not the Penrose transform is injective is a topological question : it is sufficient to verify that $H^p(\mu^{-1}(x),\mathbf{C}) = 0, p = 0,\ldots,\binom{n}{2} - 1$ (see [12],[4]). This condition is clearly satisfied in the case when Ω is a Lie ball, the fibers then being convex sets. So we concentrate next to the surjectivity problem.

The simplest way to show that the Penrose transform is surjective on L'' is by generalizing the approach worked out by N. Woodhouse in dimension 4 (see [92]). In this paper, N. Woodhouse studies first the twistor spaces for Riemannian manifolds and the Penrose transform for this situation and he then extends the results to the complex situation. He also stresses a close interplay between the Penrose transform and the analytic continuation of solutions in complex domains, an idea which is also central in the discussion presented here.

Starting from a solution f of the Dirac equation on a Lie ball $L \subset \mathbf{C}^{2n}$, we can define a form β on the twistor space $L'' = B \times IG_n$ in the following way. As the solution $f = f(P), P \in L$, has values in the spinor space $S^+ = (\Lambda^{\mathrm{even}}W)I$, we shall henceforth write f as $f(P)I$. Next, if $g \in \mathrm{Spin}_+(2n,\mathbf{C})$, then the projection onto the line generated by the corresponding pure spinor s_g is given by $\Pi(g) = gI\overline{g}/|g|^2$ (see Lemma 3.3.5). Let us recall here that the projection Π on $\mathrm{Spin}_+(2n,\mathbf{C})$ is lifted from a projection define d on IG_n and that there is a (non-zero) constant c_1 such that

$$\int_{IG_n} \Pi(\mathcal{Z})\kappa_n(\mathcal{Z}) = 1/c_1 \cdot Id,$$

where Id is the identity operator on S^+. Defining the form β on $B \times \mathrm{Spin}_+(2n,\mathbf{C})$ by

$$\beta = I\frac{\overline{g}}{|g|^2}f(X)I\frac{1}{\sum_J |M_J(g)|^2)^{n-1}}\overline{\alpha_n}(g),$$

it is easily seen that it is of bidegree $(0,\binom{n}{2})$ and that, as expected, it is multiplied by $(\sqrt{\det(\tilde{p})})^{(1-2n)/2}$ under the substitution $g \mapsto g\tilde{h}$.

Noting that $S(g)I\overline{g}/|g|^2 = \Pi(g)$, it is clear that for all $X \in B$ we obtain by virtue of Theorem 3.4.1 that there exists a nonzero constant c_2 such that

$$\int_{\nu^{-1}(X)} S\tilde{\mu}^*\beta \wedge \alpha_n \wedge dP = \int_{\nu^{-1}(X)} \Pi(g) f(X) \kappa_n(g) = \frac{1}{c_2} f(X).$$

So for real points in the Lie ball L, the form $c_2\beta$ gives back the value $f(X)$.

If the form β is $\overline{\partial}$-closed on L'', then its image under the Penrose transform is a holomorphic function. As we have just seen, this image is equal to f in all real points of L whence by the principle of analytic continuation it is equal to f in L. So we have shown that f is in the image of the Penrose transform.

The main property to be proved in such approach, namely that the form β is $\overline{\partial}$-closed on L'', was proved in dimension 4 by Woodhouse using methods of differential geometry. Similar methods were not yet developed in higher dimensions. Instead we are going to use the properties of the hyperbolic integral formulae given by a Euclidean section (see Example 2.1.5). Let us discuss the main ideas of this approach in more details.

In Example 2.1.5, the value of a solution f in a point P was expressed by the values of the solution (and its derivatives) in the points of the intersection of the complex null cone $\mathbf{C}N_P$ with $\mathbf{R}^{2n}$ and as will be readily seen, there a strong connection between this integral formula and the Penrose transform. Let us first have a closer look at the four dimensional case. Each isotropic subspace in $\mathbf{C}N_P$ intersects $\mathbf{R}^{2n}$ in exactly one point and in dimension 4 such a subspace is uniquely characterised by this intersection. So, in fact, the integral formula in dimension 4 can be understood as the integral over all α-planes inside the null cone. If the form under the integral sign depended only on the twistor variables (i.e. only on suitable combinations of coordinates of P and coordinates for α-planes), by the integral formula we would thus find, in principle, the preimage of the solution under the Penrose transform. Although this requirement is not satisfied, the form under the integral sign is not far from being twistorial and its properties can be used to prove that the form β defined above is $\overline{\partial}$-closed.

The situation is similar in higher dimensions up to one important difference: an isotropic subspace of $\mathbf{C}N_P$ is not characterized by its intersection with $\mathbf{R}^{2n}$, since there are a lot of them going through the same point (as we saw in section 4.3). Nevertheless, we can define a map ϕ such that for each point $P \in L$ and each isotropic subspace $\mathcal{Z}$ in $\mathbf{C}N_P$, the image is given by the couple (P, Q), where Q is the corresponding intersection of $\mathcal{Z}$ with $\mathbf{R}^{2n}$. The pull back of the form under the integral sign in the hyperbolic integral formula by ϕ is then a form on the space $L \times IG_n$. It is clear that the degree of the form is wrong (the total degree being $2n-2$, while the dimension of IG_n is $\binom{n}{2}$). To improve the situation, we shall multiply the form by the form τ_n (see (3.2)), which is designed in such a way that the integral of it over the set of all α-planes going through the fixed point Q is equal to a constant.

In such a way, we get a form of a suitable degree on $L \times IG_n$, but it is still a spinor-valued form. To get an ordinary form and to come closer to the expected form $S\tilde{\mu}^*\beta \wedge \alpha_n \wedge dP$, we will have to apply the projection $\Pi(\mathcal{Z})$ to the values of the form. The result can then be written in the form $S\gamma \wedge \alpha_n \wedge dP$, where γ is a form of bidegree $\left(0, \binom{n}{2}\right)$. Let us now compute explicitly the result of the procedure suggested.

Denote by $\mathcal{M}$ the set

$$\mathcal{M} = \{[P,Q] \in L \times L | |P-Q|^2 = 0, P \neq Q\}$$

and define the map

$$\phi : (L \setminus B) \times ISt_n \mapsto \mathcal{M} \times ISt_n$$

in the following way. For any element$(P,[Z^1,\ldots,Z^n]) \in (L \setminus B) \times ISt_n$, the value $\phi(P,[Z^1,\ldots,Z^n])$ is determined by $(P,Q,[Z^1,\ldots,Z^n])$, where the point $Q \in \mathbf{R}^{2n}$ is characterized by the property that $P-Q$ belongs to the isotropic subspace generated by the vectors $Z^1,\ldots,Z^n$. So it is the intersection of the isotropic subspace given by $(P,[Z^1,\ldots,Z^n])$ with $\mathbf{R}^{2n}$.

Let us also recall that for the form

$$\omega = \frac{1}{A_{2n-1}} \frac{-(Q-P)}{|Q-P|^{2n}} DQ.f(Q)I$$

which appeared in Theorem 1.3, the Leray residue $res(\omega \wedge dP)$ was calculated in section 2 as a form on $\mathcal{M}$ and, for P fixed, as a form on the intersection S^{2n-2} of CN_P with $\mathbf{R}^{2n}$.

Using the form τ_n on ISt_n (see section 3), we want to compute now the form

$$\Pi(\mathcal{Z}) \circ \phi^*(\operatorname{Res}\omega) \wedge \tau_n, \; \mathcal{Z} \in IG_n,$$

where $\Pi(\mathcal{Z})$ is the projection onto the subspace of pure spinors corresponding to the isotropic subspace $\mathcal{Z}$. By means of the result computed in Example 2.1.5, we obtain that

$$\begin{aligned}\Pi(\mathcal{Z}) \circ \phi^*(\operatorname{Res}\omega) \wedge \tau_n \wedge dP &= \frac{1}{\pi^{n-1}} \left(\frac{1}{4}\right)^{n-1} \cdot \\ \cdot \Pi(\mathcal{Z}) I_b &\left(\sum_{j=0}^{n-1} \binom{2n-2}{n-1-j} \frac{(n-1)!}{j!} (\nabla_{\bar{b}})^j f(Q(P,\mathcal{Z}))I \right) \cdot \\ &\cdot \phi^*\{dS(1)\} \wedge \tau_n \wedge dP,\end{aligned}$$

where $I_b = I_b(P,\mathcal{Z}) = bc$ is an idempotent in the Clifford algebra (due to the fact that both b and c are null vectors and that they satisfy the normalization condition $(b,c) = 1/2$) and $dS(1)$ is the volume form on the unit sphere S^{2n-2}.

The multiplication by the idempotent I_b is a projection operator onto a subspace of the spinor space. The isotropic subspace $\mathcal{Z}$ contains b, which implies that $\Pi(\mathcal{Z})I_b(s) = \Pi(\mathcal{Z})s$, $s \in S^+$. The invariance properties with respect to $\mathrm{SO}(2n,\mathbf{R})$ show that the form $\phi^*\{dS(1)\} \wedge \tau_n$ is equal (up to a non-zero constant) to the invariant form κ_n on IG_n, i.e. there exists a nonzero number c_3 such that

$$\phi^*\{dS(1)\} \wedge \tau_n = c_3 \kappa_n.$$

Summarizing we have that for a suitable nonzero constant c_4

$$\Pi(\mathcal{Z})\circ\phi^*(\operatorname{Res}\omega)\wedge\tau_n\wedge dP = c_4\Pi(\mathcal{Z})\cdot$$
$$\cdot\left\{\sum_{j=0}^{n-1}\binom{2n-2}{n-1-j}\frac{(n-1)!}{j!}(\nabla_{\hat{b}})^j f(Q(P,\mathcal{Z}))I\right\}\kappa_n\wedge dP.$$

The form above is defined only on $(L\setminus B)\times IG_n$, but there is clearly a smooth extension of it to the whole domain $L\times IG_n$.

When a point P approaches $\mathbf{R}^{2n}$, the vectors $b = P - Q$ and $\hat{b}$ approach the zero vector. So all terms containing derivatives disappear in the limit and we can define it at points of $(X,\mathcal{Z})\in B\times IG_n$ by the formula

$$\Pi(\mathcal{Z})\circ\phi^*(\operatorname{Res}\omega)\wedge\tau_n(X,\mathcal{Z}) =$$
$$= c_4\binom{2n-2}{n-1}(n-1)!\Pi(\mathcal{Z})f(X)I\kappa_n\wedge dP.$$

Note that the restriction of the form above to $B\times IG_n$ coincides (up to a constant) with the form $\mathcal{S}\tilde{\mu}^*\beta\wedge\alpha_n\wedge dP$.

Consequently, we may use properties of the form

$$\Pi(\mathcal{Z})\circ\phi^*(\operatorname{Res}\omega)\wedge\tau_n(X,\mathcal{Z})$$

to show that the form β is $\overline{\partial}$-closed. Moreover, when working on $\mathbf{C}^{2n}\times\operatorname{Spin}_+(2n,\mathbf{C})$ we can write all parts of our formulas by using ordinary $\mathbf{C}$-valued forms. If we define the form γ on $L\times\operatorname{Spin}_+(2n,\mathbf{C})$ by

$$\gamma \;=\; I\frac{\overline{g}}{|g|^2}\left(\sum_{j=0}^{n-1}\binom{2n-2}{n-1-j}\frac{(n-1)!}{j!}(\nabla_{\hat{b}})^j f(Q(P,g))I\right)\cdot$$
$$\cdot\frac{\overline{\alpha_n(g)}}{(\sum_I |M_I(g)|^2)^{n-1}},$$

we thus have that

$$\Pi(\mathcal{Z})\circ\phi^*(\operatorname{Res}\omega)\wedge\tau_n\wedge dP = c_4\cdot\mathcal{S}\gamma\wedge\alpha_n\wedge dP.$$

As the left hand side is a closed form, γ is $\overline{\partial}$-closed. Unfortunately we may not derive from this that its restriction β is also $\overline{\partial}$-closed since, as we saw in section 3.3, the embedding of $\mathbf{T} = \mathbf{R}^{2n}\times IG_n$ i nto $\mathbf{F} = \mathbf{C}^{2n}\times IG_n$ is not holomorphic. On the other hand, the fact that γ is $\overline{\partial}$-closed implies that the coefficient at $\overline{\alpha}_n$ appearing in the expression for γ is holomorphic in P. Moreover, in points belonging to $B\times IG_n$, the whole sum reduces to just one summand which depends only on twistor variables. It is hence nothing else but the solution f restricted to B and then extended constantly along corresponding isotropic subspaces through points in B. The information just obtained is in fact already sufficient to prove that β is $\overline{\partial}$-closed on $B\times IG_n$ (as we shall see immediately), and this since the holomorphy of the coefficient implies that the complex derivatives at the points in $B \times IG_n$ are equal to the real derivatives. The requirement that β is $\overline{\partial}$-closed is thus translated into the requirement that the coefficient is constant along isotropic subspaces.

Theorem 5.2.2 *The Penrose transform $\mathcal{P}$ is a 1-1 map between $H^{(0,\binom{n}{2})}(L'', \mathbf{L}^{(1-2n)})$ and the space of all (holomorphic) solutions of the Dirac operator on the Lie ball L.*

Proof. To prove the surjectivity, let f be a solution of the Dirac operator on L. We have shown above that

$$\Pi(\mathcal{Z}) \circ \phi^*(\operatorname{Res}\omega) \wedge \tau_n \wedge dP = c_4 \Pi(\mathcal{Z}) \cdot \\ \cdot \left\{ \sum_{j=0}^{n-1} \begin{pmatrix} 2n-2 \\ n-1-j \end{pmatrix} \frac{(n-1)!}{j!} (\nabla_{\hat{b}})^j f(Q(P, \mathcal{Z}))I \right\} \kappa_n \wedge dP$$

and that the form can be smoothly extended to $B \times IG_n$ by

$$\Pi(\mathcal{Z}) \circ \phi^*(\operatorname{Res}\omega) \wedge \tau_n \wedge dP = \\ = c_4 (-1)^{n-1} \binom{2n-2}{n-1} \Pi(\mathcal{Z}) f(X) I \kappa_n \wedge dP, \ (X, \mathcal{Z}) \in B \times IG_n.$$

Considering the pullback of $\Pi(\mathcal{Z}) \circ \phi^*(\operatorname{Res}\omega) \wedge \tau_n \wedge dP$ to $L \times \operatorname{Spin}_+(2n, \mathbf{C})$ and defining the form γ by

$$\gamma = I \frac{\overline{g}}{|g|^2} \left(\sum_{j=0}^{n-1} \begin{pmatrix} 2n-2 \\ n-1-j \end{pmatrix} \frac{(n-1)!}{j!} (\nabla_{\hat{b}})^j f(Q(P, g))I \right) \cdot \\ \cdot \frac{\overline{\alpha_n}(g)}{(\sum_I |M_I(g)|^2)^{n-1}},$$

we have that

$$\Pi(\mathcal{Z}) \circ \phi^*(\operatorname{Res}\omega) \wedge \tau_n \wedge dP = c_4 \cdot \mathcal{S}\gamma \wedge \alpha_n \wedge dP.$$

The left hand side being closed, we get

$$\begin{aligned} 0 &= d(\mathcal{S}\gamma \wedge \alpha_n \wedge dP) \\ &= \overline{\partial}(\mathcal{S}\gamma \wedge \alpha_n \wedge dP) \\ &= \mathcal{S}\overline{\partial}(\gamma) \wedge \alpha_n \wedge dP, \end{aligned}$$

whence γ is $\overline{\partial}$-closed on $L \times \operatorname{Spin}_+(2n, \mathbf{C})$. The coordinates P, g being holomorphic coordinates, the coefficient at $\overline{\alpha}_n$ is holomorphic in P (which is a highly nontrivial information due to the complicated expression for the intersection point $Q(P, g)$).

Moreover, the form

$$\beta = I\frac{\overline{g}}{|g|^2}\binom{2n-2}{n-1}(n-1)!f(X)I\frac{1}{\sum_I |M_I(g)|^2)^{n-1}}\,\overline{\alpha_n}(g)$$

is the restriction of γ to $B \times \mathrm{Spin}_+(2n, \mathbf{C})$ and the known transformation properties of $\mathcal{S}$ and $\overline{\alpha}_n$ imply that as well the form β as the form γ are multiplied by $(\sqrt{\det(\tilde{p})})^{(1-2n)/2}$ under the substitution $g \mapsto g\tilde{h}$. Furthermore, both forms have bidegree $(0, \binom{n}{2})$ and so β is an element of $\mathcal{E}^{(0,\binom{n}{2})}(L'', \mathbf{L}^{1-2n})$. We now prove that $\overline{\partial}\beta = 0$ on L''. The form β has the top degree in $\overline{\mathcal{Z}}$ on L'', so it is sufficient to verify that for all points $\mathcal{Z} \in IG_n$ (represented by $[Z^1, \ldots, Z^n] \in ISt_n$) the vector field $\sum_j Z^i_j \partial_{x_j}$ annihilates the coefficient in the form β (details on the complex structure on $B \times \mathrm{Spin}_+(2n, \mathbf{C})$ were explained in section 3.3).

Now, the function $F(P, g) = f(Q(P, g))$ is holomorphic in the variable P at the points $X \in B$ since the form γ is $\overline{\partial}$-closed. Moreover, F depends on P, g only through the intersection Q of the corresponding isotropic plane with $\mathbf{R}^{2n}$, whence it is a pull-back of a function from L''. Consequently, the function F is annihilated by the vector fields $\sum_j Z^i_j \partial_{P_j}, i = 1, \ldots, n$, for each $\mathcal{Z} \in IG_n$. The holomorphy in P shows that F is annihilated by the vector fields $\sum_j Z^i_j \partial_{x_j}$ as well.

Having proved that β is $\overline{\partial}$-closed, it represents an element in $H^{(0,\binom{n}{2})}(L'', \mathbf{L}^{(1-2n)})$. Note that its pull-back $\tilde{\mu}^*\beta$ to $\widetilde{L'}$ does not coincide with γ. Moreover, its Penrose transform is given by

$$\mathcal{P}[\beta](P) = \int_{\nu^{-1}(P)} \mathcal{S} \cdot \tilde{\mu}^*\beta \wedge \alpha_n, \; P \in L,$$

and for real points $X \in B$ we get (up to a nonzero constant)

$$\begin{aligned}\mathcal{P}[\beta](X) &= \int_{\nu^{-1}(X)} P(\mathcal{Z})f(X)I\kappa_n(\mathcal{Z}) \\ &= f(X).\end{aligned}$$

Replacing β by a suitable multiple of β, when necessary, $\mathcal{P}[\beta](P)$ coincides with f on B and, by analyticity, on the whole L. ∎

Appendices

Appendix A

The Leray residue theory was originally introduced as a higher dimensional generalization of the standard residue theory in complex analysis (see [53]). The theory was formulated for closed differential forms having singularities on complex hypersurfaces in complex manifolds. A real version of the theory, treating differential forms with singularities on a real submanifold (of arbitrary codimension) of a real manifold was soon after worked out by Norguet and announced in [59]. For a review of recent developments in the study of the Leray residue theory, we refer to [26].

The Leray-Norguet residue theory is used at several places in this book. We present here a self-contained account of the theory along the lines of the original papers by Norguet (see [59]); it is essentially based on the notion of integrals along fibers which we present first. After recalling some properties of tubular neighbourhoods of submanifolds, we define the Leray-Norguet residue and cobord and state the Leray theorem. The case of a hypersurface in a complex manifold is discussed then and we prove a theorem showing how the residue can be computed in a special case needed in Chapter 5.

All manifolds in this appendix are supposed to be smooth and oriented. We are using singular homology theory for the homology groups under consideration and de Rham cohomology is always used for cohomology groups. In the standard treatment of singular homology, chains are defined by continuous maps, but on differentiable manifolds the corresponding homology groups coincide with those defined by differentiable maps. We shall use the differentiable version here; by Stokes' Theorem, de Rham cohomology groups are then dual to singular homology groups with real coefficients.

A full treatment of these questions can be found in [56].

A.1 Integration along fibers

Although integration of differential forms along fibers is very useful, it is not a standard part of courses on integration of differential forms. We define integration along fibers here in the simple case where the fibers are smooth and compact manifolds. A systematic description of integration along fibers can be found in [82]. Another approach, based on ideas going back to Andreotti, uses the language of currents;

details can be found in [69] or [24].

Let $f : M \to N$ be a fibration such that both M and N are oriented manifolds with $\dim M = m$, $\dim N = n$ and all fibers $f^{-1}(u), u \in N$, are diffeomorphic to a given compact manifold W of dimension $q = m-n$. Then there is a unique orientation induced on the fibers, i.e. we can choose an orientation on W such that there is a locally finite covering $\{U_\gamma\}$ of N for which

$$f^{-1}(U_\gamma) \simeq U_\gamma \times W,$$

where on the right hand side we consider the canonical orientation of the Cartesian product.

The operation f_* of integration along fibers will be defined for any form in $\mathcal{E}^p(M)$ with $p \geq q$; the result of it will be a form in $\mathcal{E}^{p-q}(N)$.

Let us consider first the simple case of the trivial fibration, i.e. let $M = N \times W$. The space $\mathcal{E}^p(M)$ of p-forms decomposes then as

$$\mathcal{E}^p(M) = \oplus_j \mathcal{E}^0(N \times W) \otimes \mathcal{E}^j(W) \otimes \mathcal{E}^{p-j}(N),$$

where the spaces $\mathcal{E}^j(W)$ and $\mathcal{E}^{p-j}(N)$ are embedded into $\mathcal{E}^p(M)$ using the pull-back by the corresponding projections. Any element in $\mathcal{E}^0(N \times W) \otimes \mathcal{E}^j(W) \otimes \mathcal{E}^{p-j}(N)$, j fixed, is a linear combination of forms

$$\omega = g\omega_W \wedge \omega_N,\ g \in \mathcal{E}^0(N \times W),\ \omega_W \in \mathcal{E}^j(W),\ \omega_N \in \mathcal{E}^{p-j}(N).$$

We want to define the integral along fibers for such forms as $(\int_W g\omega_W)\omega_N$. For $j = q$, it is completely meaningful; for other j we shall use the convention that if the degree of a form and the dimension of a manifold do not agree, then the integral of the form over the manifold is zero.

Definition A.1.1.

(i) Let N and W be manifolds of respective dimensions n and q and let $p \geq q$. Let $M = N \times W$ and let $f : M \to N$ be the projection. Then for any form $\omega = g\omega_W \wedge \omega_N$, $g \in \mathcal{E}^0(M)$ $\omega_W \in \mathcal{E}^j(W)$, $\omega_N \in \mathcal{E}^{p-j}(N)$, we define

$$f_*(\omega) = \Big(\int_W g\omega_W\Big)\omega_N.$$

The map f_* is extended by linearity to the whole space $\mathcal{E}^p(M)$.

(ii) Let $f : M \to N$ be a fibration such that both M and N are oriented manifolds with $\dim M = m$, $\dim N = n$ and all fibers $f^{-1}(u), u \in N$, are diffeomorphic to a given compact manifold W of dimension $q = m - n$. Let $\{U_\gamma\}$ be a locally finite covering of M for which

$$f^{-1}(U_\gamma) \simeq U_\gamma \times W$$

and let $\{\phi_\gamma\}$ be a partition of unity subordinated to $\{U_\gamma\}$. Then for any $\omega \in \mathcal{E}^p(M)$ we define

$$f_*\omega = \sum_\gamma f_*(\phi_\gamma \omega)$$

where the terms on the right hand side are defined using (i).

It is not difficult to show that the integral along fibers does not depend on the choices made in the definition. We shall use for it the following standard lemma, where $\mathcal{E}_c^q$ denotes the space of all smooth q-forms on M with compact support.

Lemma A.1.2. *Let M be a manifold of dimension m and let $\omega \in \mathcal{E}^p(M)$ be given. If $\int_M \omega \wedge \tau = 0$ for all $\tau \in \mathcal{E}_c^{m-p}(M)$, then $\omega = 0$.*

Proof. The simplest case, namely when $p = m$, is nothing else but a reformulation of a basic property in integration theory saying that if f is a function on $\mathbf{R}^m$ and if $\int fg = 0$ for all functions g, then $f = 0$. All other cases reduce easily to this simplest case. ■

Theorem A.1.3. *The map f_* defined above does not depend on the choices made.*

Proof. Suppose that $M = N \times W$ and $\omega = g\omega_W \wedge \omega_N$, $g \in \mathcal{E}^0(M)$, $\omega_W \in \mathcal{E}^q(W)$, $\omega_N \in \mathcal{E}^{p-q}(N)$. Then for every $\alpha \in \mathcal{E}_c^{n-(p-q)}(N)$, we have

$$\begin{aligned} \int_N f_*\omega \wedge \alpha &= \int_N (\int_W g\omega_W)\omega_N \wedge \alpha = \int_N \int_W (g\omega_W \wedge \omega_N \wedge f^*\alpha) \\ &= \int_M \omega \wedge f^*\alpha. \end{aligned}$$

In the general case, we can choose a covering giving a local trivialization of the bundle and a subordinate partition of unity. We then get that for any $\omega \in \mathcal{E}^p(M)$ and any $\alpha \in \mathcal{E}_c^{m-(p-q)}(N)$

$$\int_M f_*\omega \wedge \alpha = \int_M \omega \wedge f^*\alpha. \tag{A.1}$$

To prove the independence of f_* on the choices made, it is sufficient to use Lemma A.1.2 and the fact that the right hand side of (A.1) does not depend on any choice made. ■

The main properties of the integration along fibers are summarized in

Theorem A.1.4. *Let $f : M \to N$ be a fibration with $\dim M = m$, $\dim N = n$ and with fibers diffeomorphic to a given compact manifold W of dimension $q = m - n$. Let $\omega \in \mathcal{E}^p(M), p \geq q$. Then*

(i) $f_*\omega \wedge \omega'_N = f_*(\omega \wedge f^*\omega'_N)$ *for all* $\omega'_N \in \mathcal{E}^j(N)$;

(ii) *If* $\omega \in \mathcal{E}^m$, *then* $\int_M \omega = \int_N f_*\omega$;

(iii) $d(f_*\omega) = (-1)^q f_*(d\omega)$;

(iv) $(f \circ g)_* = f_* \circ g_*$.

Proof. Using linearity and a partition of unity, it is sufficient to prove the claims only in the case when $M = N \times W$ and $\omega = g\omega_W \wedge \omega_N$, $g \in \mathcal{E}^0(M)$, $\omega_W \in \mathcal{E}^q(W)$, $\omega_N \in \mathcal{E}^{p-q}(W)$.

(i) $f_*\omega \wedge \omega'_N = (\int_W g\omega_W)\omega_N \wedge \omega'_N = f_*(g\omega_W \wedge \omega_N \wedge \omega'_N)$

(ii) $\int_M \omega = \int_{N\times W} g\omega_W \wedge \omega_N = \int_N(\int_W g\omega_W)\omega_N = \int_N f_*\omega$.

For the proof of the other two properties, we rely upon the property (A.1). For any form α with compact support on N and having complementary degree we get:

(iii)

$$\begin{aligned}\int_N d(f_*\omega) \wedge \alpha &= (-1)^{p-q} \int_N f_*\omega \wedge d\alpha = (-1)^{p-q} \int_N \omega \wedge f^*(d\alpha) \\ &= (-1)^{p-q} \int_N \omega \wedge d(f^*\alpha) = (-1)^q \int_N f_*(d\omega) \wedge \alpha\end{aligned}$$

(iv)

$$\begin{aligned}\int_M (f \circ g)_*\omega \wedge \alpha &= \int_M \omega \wedge (f \circ g)^*\alpha = \int_M \omega \wedge g^*(f^*\alpha) \\ &= \int_M (f_* \circ g_*)\omega \wedge \alpha. \ \blacksquare\end{aligned}$$

The integration along fibers can be used also in the case of a fibration given by complex manifolds. In this case the sign $(-1)^q$ disappears in Theorem A.1.3(iii). Moreover, additional information is available due to the splitting of forms with respect to bidegrees, namely

Theorem A.1.5. *Let* $f : M \to N$ *be a holomorphic fibration where* M *and* N *are complex manifolds and the fibers are diffeomorphic to a given compact complex manifold* W *of (complex) dimension* q. *If* $\omega \in \mathcal{E}^{(r,s)}(M), r \geq q, s \geq q$, *then* $f_*\omega \in \mathcal{E}^{(r-q,s-q)}(N)$. *Moreover*

$$\partial f_*\omega = f_*(\partial\omega),\ \overline{\partial} f_*\omega = f_*(\overline{\partial}\omega).$$

and so holomorphic forms are mapped by f_* *onto holomorphic forms.*

The integration along fibers can be easily extended to forms with values in a fixed vector space V by a choice of a basis in V and by componentwise integration. The result is clearly independent of a choice of the basis.

A.2 Tubular neighbourhoods

The approach to the Leray-Norguet residue presented in the next section uses the notion of a tubular neighbourhood of a submanifold of a manifold. Properties of tubular neighbourhoods are described in detail in [37] (see also [81]). We shall use the properties proved in [37] in a slightly different situation, since we shall assume all vector bundles and manifolds to be oriented.

Suppose that M is a manifold and $S \subset M$ is a submanifold such that $\dim M = m$, $\dim S = n$. An oriented tubular neighbourhood of S in M is a pair (f, ξ), where $\xi = (\pi, E, S)$ is an oriented vector bundle over S and $f : E \mapsto M$ is an embedding such that

(i) f preserves the orientations;

(ii) $f|S = \mathbf{1}_S$ (S being identified with the zero section of E);

(iii) $f(E)$ is an open neighbourhood of S in M.

An oriented closed tubular neighbourhood of radius $\varepsilon > 0$ of S in M is the image $f(D_\varepsilon)$ determined by a tubular neighbourhood (f, ξ), where $\xi = (\pi, E, S, ||.||)$ is an oriented orthogonal vector bundle (i.e. there is a norm $||.||$ coming from a positive definite scalar product on fibers, depending smoothly on a point) and where $D_\varepsilon = \{x \in E| \, ||x|| \leq \varepsilon\}$ is the disk subbundle of ξ with radius ε.

The main theorem concerning the existence of tubular neighbourhoods reads as follows (see [37],p.116):

Theorem A.2.1. *Let S be an oriented submanifold of an oriented manifold M. Then there exists an oriented tubular neighbourhood of S in M.*

It is proved in [37] that it is always possible to choose an orthogonal structure on a fibre bundle; so closed tubular neighbourhoods exist as well. An important information concerning different tubular neighbourhoods is contained in the following theorem.

Theorem A.2.2. *Let S and M be oriented manifolds such that S is a submanifold of M. Let $(f_i, \xi_i, ||.||_i, \varepsilon_i), i = 0, 1$, be two oriented closed tubular neighbourhoods of S in M. Then these two tubular neighbourhoods are isotopic by an isotopy of oriented tubular neighbourhoods $F_t : E_0 \mapsto M, 0 \leq t \leq 1$, with $F_1(D_{\varepsilon_0}) = f_1(D_{\varepsilon_1})$.*

By the isotopy of oriented tubular neighbourhoods a homotopy $F : E_0 \times I \mapsto M$ is meant such that

(i) the related map $\hat{F} : E_0 \times I \mapsto M \times I, \hat{F}(x,t) = (F_t(x), t)$ *is an embedding;*

(ii) S *is left invariant by the isotopy;*

(iii) $F_0 = f_0$;

(iv) $F_1(E_0) = f_1(E_1)$;

(v) $f_1^{-1}F_1 : E_0 \mapsto E_1$ *is an isomorphism of oriented vector bundles.*

Notice that these two theorem are stated (and proved) in [37] without any assumption concerning orientations, but their proofs can be modified so that the orientations considered are conserved. Furthermore the fibers are always oriented in a way compatible with the orientations of S and M.

A.3 Leray-Norguet residues and cobord

We are now able to define the Leray-Norguet residue of a cohomology class and the cobord of a cycle. We shall see that the residue is defined, in principle, by the integral along fibers of the form representing the cohomology class and the notion of a cobord is defined then by using the duality between homology and cohomology.

Lemma A.3.1. *Let* S *be a submanifold of a manifold* M *with* $\dim M = m$, $\dim S = n$ *and let* $p \geq m - n - 1$. *Let* (f, ξ) *be an oriented closed tubular neighbourhood of* S *in* M *of radius* 1 *given by an oriented orthogonal vector bundle* $\xi = (\pi', E, S, ||.||)$ *and let* U *be the image* $f(D_1)$ *of the disk subbundle of radius* 1. *Then the map* $\pi = \pi' \circ f^{-1} : U \to S$ *induces the maps*

$$\pi_* : \mathcal{E}(\partial U, \Lambda^p) \mapsto \mathcal{E}(S, \Lambda^{p-(m-n-1)})$$

and

$$\pi_* : H^p(\partial U) \mapsto H^{p-(m-n-1)}(S).$$

Proof. The first map is given by the integration along fibers. The map π_* commutes with d, whence closed forms are mapped onto closed forms and exact forms onto exact forms. Hence the map π_* for cohomology groups is well-defined. ■

Theorem A.3.2. *Let* S *be a submanifold of* M *with* $\dim M = m, \dim S = n$. *Let* $U = f(D_1)$ *be an oriented closed tubular neighbourhood given by a choice of* $(f, \xi, ||\cdot||)$ *and let* $\pi : \partial U \mapsto S$ *be the induced projection and* i *denote the inclusion* $\partial U \subset M \setminus S$.

Then for all $p \geq m - n - 1$, *the map*

$$\operatorname{Res} = \pi_* \circ i^* : H^p(M \setminus S) \mapsto H^{p-(m-n-1)}(S)$$

is independent of all choices made.

Proof. Let $(f_i, \xi_i, ||.||_i), i = 0, 1$, be two possible choices. As a consequence of Theorem A.2.2 there exists an isomorphism $\Phi : \xi_0 \mapsto \xi_1$ between the two bundles which implies that it is possible to take $f_1' = f_1 \circ \Phi$ instead of f_1 and to define both tubular neighbourhoods using the same bundle ξ.

So let us suppose from the beginning that $\xi_0 = \xi_1 = \xi, \xi = \{\pi : E \mapsto S\}$. Putting $U_i = f(D_1)$, we have maps $\pi_i : \partial U_i \to S$ defining then the maps $(\pi_i)_*$. But, the maps f_i being isomorphisms of D_1 onto U_i,

$$(\pi_i)_*\omega = \pi_* \circ (f_i)^*\omega.$$

Theorem A.2.2 tells us that the maps f_i are homotopic with each other, whence the forms $(f_i)^*\omega$ are in the same class of cohomology. But then the same is also true for their integral along the fibers of the maps $\pi_* \circ (f_i)^*\omega$. ∎

Definition A.3.4. *Let $p \geq m - n - 1$. Then the map*

$$\text{Res } = \pi_* \circ i^* : H^p(M \setminus S) \mapsto H^{p-(m-n-1)}(S)$$

is called the Leray-Norguet residue.

The cohomology and homology groups being dual to each other, the Leray-Norguet cobord is defined to be the dual map to the Leray-Norguet residue.

Definition A.3.5. *Let $q \leq n$. Then the map*

$$\delta : H_q(S) \to H_{q+(m-n-1)}(M \setminus S)$$

dual to the map

$$\text{Res } : H^{q+(m-n-1)}(M \setminus S) \mapsto H^q(S)$$

is called the Leray-Norguet cobord.

In the original paper [56], the Leray residue and the Leray cobord were introduced independently of each other. Their connection was then given by the following theorem, which is just a restatement of the definition of the cobord in the approach presented here.

Theorem A.3.6. (Leray residue theorem) *Let $0 \leq p \leq n$. Then for each closed $(p + m - n - 1)$-form ω on $M \setminus S$ and for each p-dimensional cycle C on S, we have*

$$\int_{\delta C} \omega = \int_C \text{Res } \omega.$$

Notice that although δC as well as Res ω are, respectively, homology and cohomology classes, both integrals above are well-defined. The definition of the Leray-Norguet cobord map being somewhat abstract, it is useful, also for computations, to have a geometrical picture in mind.

So let us consider the simple situation in which the fiber bundle ξ, determining the tubular neighbourhood of S in M, is a trivial fibre bundle $S \times V$ and let $f : D_1 \mapsto U$ then be the corresponding closed tubular neighbourhood.

Furthermore, let a p-dimensional cycle C be given, for simplicity, by a smooth map $C : I_p \mapsto S$, where I_p is the p-dimensional interval and let us choose any map $b : I_{m-n-1} \mapsto S_1$ representing the unit sphere $S_1 \subset V$ as a cycle. Then $C \times b$ maps $I_{p+m-n-1}$ into $S \times S_1 \subset S \times V$. Applying the definition of the integration along fibers and using Theorem A.1.4(ii), we get for each form ω of degree $p + m - n - 1$ that

$$\int_C \text{Res } \omega = \int_{f \circ [C \times b]} \omega.$$

The Leray residue theorem then implies that

$$\int_{\delta C} \omega = \int_{f \circ [C \times b]} \omega$$

for all $\omega \in \mathcal{E}^{p+m-n-1}(M \setminus S)$, whence the map $f \circ [C \times b] : I_{p+m-n-1} \mapsto M \setminus S$ represents the Leray-Norguet cobord δC.

Geometrically speaking, the Leray-Norguet cobord map consists of replacing every point of a chain C by a $(m - n - 1)$-dimensional sphere having its center in the given point and going to a transversal direction (given by a choice of the tubular neighbourhood and its projection).

To compute the Leray-Norguet residue, a choice of a closed tubular neighbourhood reduces the computation of the Leray-Norguet residue to the computation of integrals along fibers, which can often be computed explicitely in special cases.

Finally notice that we shall use the residue theorem above in a slightly more general situation, namely in the case of vector-valued differential forms. To this end, let V be a fixed finite dimensional real vector space. The Leray-Norguet residue for V-valued forms may then be defined using analogous methods as above and it belongs to the cohomology space $H^p(\partial U, V)$ with coefficients in V. Note that by choosing a fixed basis for V, the Leray-Norguet residue can be defined and computed componentwise.

A.4 Leray residue

The Leray residue theory was first worked out for closed differential forms on a complex manifold M having singularities on a nonsingular complex hypersurface $S \subset M$; the Leray residue itself was defined using division of differential forms. From the point of view of the Leray-Norguet theory, it can be viewed as a special case where

S has real codimension 2 and the fibers appearing in the integration along fibers are circles.

There are some particular aspects in this complex situation. One of them is that in some special cases the integration along circles can be reduced to the computation of standard residues in complex analysis. We shall now discuss such a special case in more details; it will be needed in Chapter V.

Let $M = \Omega$ be a domain in $\mathbf{C}^m$ and let S be a nonsingular complex hypersurface in Ω. Furthermore consider a transversal vector field $c(s)$ on S, i.e. suppose that for each $s \in S$ the vector $c(s)$ does not belong to the tangent space $T_s S$. Then the induced closed tubular neighbourhood (f, ξ) of S in Ω is given by the following construction.

Take the trivial bundle $E = S \times \mathbf{C}$ and the map $f : E \to \Omega$ determined by $(s, \tau) \in S \times \mathbf{C} \mapsto f(s, \tau) = s + \tau c(s)$. The tangent space $T_s S$ is a complex hyperplane in Ω and $c(s) \notin T_s S$, whence $T_s S \oplus \mathbf{C} \cdot c(s) = \mathbf{C}^m$. The tangent map $T_s f$ at the point $s = (s, 0) \in S \times \{0\}$ is the identity map as well on $T_s S$ as on $\mathbf{C}$ and so the tangent map Tf has rank m at all points $s \in S$. Consequently, f is an immersion on a neighbourhood of $S = S \times \{0\}$. The restriction $f|_S$ being the identity, we can find a neighbourhood U of S in E such that $f|_U$ is an embedding (see e.g. [37],p.41). Now choosing any smooth family of maps $g = g_s : \{s\} \times \mathbf{C} \to \{s\} \times \mathbf{C}$ such that $g(E) \subset U$ and g is the identity on a neighbourhood of $S \subset E$, we have constructed a tubular neighbourhood $f \circ g : E \to \Omega$. So we have proved

Lemma A.4.1. *Let Ω be a domain in $\mathbf{C}^m$ and let S be a nonsingular complex hypersurface in Ω. Then each transversal vector field $c(s)$ on S induces a tubular neighbourhood $(f, \xi), \xi = (\pi, S \times \mathbf{C}, S)$ such that on a neighbourhood U of $S \times \{0\}$ we have that $f(s, \tau) = s + \tau c(s)$.*

Having chosen a tubular neighbourhood $(f, \xi), \xi = (\pi, S \times \mathbf{C}, S)$ corresponding to a vector field $c(s)$ on S, consider now a closed form $\omega \in \mathcal{E}^p(\Omega \setminus S)$. To compute a representative for the Leray-Norguet residue Res ω on S, we have to choose a closed tubular neighbourhood V in E, i.e. we have to choose an orthogonal structure on the fibers $\mathbf{C}$ smoothly depending on the points $s \in S$. To do so, it is sufficient to choose a smooth positive function $\epsilon(s), s \in S$, and to define a norm $||.||_s$ on $\{s\} \times \mathbf{C}$ by $||\tau||_s = (1/\epsilon(s))|\tau|$. Furthermore it is possible to choose the function $\epsilon(s)$ in such a way that $S \times D_1 \subset U, D_1 = \{\tau | \, ||\tau||_s \leq 1\}$ so that the disk bundle $S \times D_1$ is then mapped to Ω. Denoting by $\tilde{\pi}$ the projection $\tilde{\pi} : S \times \partial D_1 \to S$, the Leray-Norguet residue is then given by $\tilde{\pi}_*(f^*\omega)$. In this simple situation, the form $f^*\omega$ can be written on $S \times D_1$ as

$$f^*\omega = d\tau \wedge \alpha_1(\tau) + d\bar{\tau} \wedge \alpha_2(\tau) + \beta(\tau),$$

where $\alpha_1(\tau), \alpha_2(\tau) \in \mathcal{E}^{p-1}(S)$ and $\beta(\tau) \in \mathcal{E}^p(S)$. Only the first two terms are relevant in the computation of res ω; they are given for all $t_1, \ldots, t_{p-1} \in T_s S$ by

$$\alpha_1(\tau)[t_1,\ldots,t_{p-1}] = \omega[f_*\frac{\partial}{\partial\tau}, f_*t_1,\ldots,f_*t_{p-1}] = f^*\omega[\frac{\partial}{\partial\tau}, t_1,\ldots,t_{p-1}],$$

$$\alpha_2(\tau)[t_1,\ldots,t_{p-1}] = \omega[f_*\frac{\partial}{\partial\overline{\tau}}, f_*t_1,\ldots,f_*t_{p-1}].f^*\omega[\frac{\partial}{\partial\overline{\tau}}, t_1,\ldots,t_{p-1}].$$

Denoting the (flat) covariant derivative of a vector field c(s) on $S \subset \mathbf{C}^m$ in the direction $t \in T_sS$ by $\nabla_t c$, we have for a vector $t \in T_sS$ that

$$f_*(s,\tau)(t) = t + \tau\nabla_t c(s) = \nabla_t(s + \tau c(s))$$

and so Res ω is represented by the form

$$\int_{\partial D_1} [d\tau(\omega[f_*\frac{\partial}{\partial\tau}, \nabla_{t_1}(s+\tau c(s)),\ldots,\nabla_{t_{p-1}}(s+\tau c(s))])+$$

$$+ \quad d\overline{\tau}(\omega[f_*\frac{\partial}{\partial\overline{\tau}}, \nabla_{t_1}(s+\tau c(s)),\ldots,\nabla_{t_{p-1}}(s+\tau c(s))])].$$

If the form ω is holomorphic on $\Omega \setminus S$, the integral above is reduced to

$$\int_{\partial D_1} d\tau(\omega[c, \nabla_{t_1}(s+\tau c(s)),\ldots,\nabla_{t_{p-1}}(s+\tau c(s))])$$

and it can be computed by the residue calculus in standard complex analysis. It is also clear that the integral along fibers is then independent of the choice of the function $\epsilon(s)$. So we have proved

Theorem A.4.2. *Let Ω be a domain in $\mathbf{C}^m$ and let S be a nonsingular complex hypersurface in Ω. Furthermore suppose that the tubular neighbourhood (f,ξ), $\xi = (\pi, S \times \mathbf{C}, S)$ is induced by a transversal vector field $c(s)$ on S. Then for each $\omega \in \mathcal{E}^p(\Omega \setminus S)$ closed, the value of a representative* res ω *of* Res ω *on vectors $t_j \in T_sS$, $j = 1,\ldots,p-1$, is given by*

$$\operatorname{res}\omega[t_1,\ldots,t_{p-1}] = \int_{\partial D_1} [d\tau(\omega[f_*\frac{\partial}{\partial\tau}, \nabla_{t_1}(s+\tau c(s)),\ldots,\nabla_{t_{p-1}}(s+\tau c(s))])+$$

$$+ \quad d\overline{\tau}(\omega[f_*\frac{\partial}{\partial\overline{\tau}}, \nabla_{t_1}(s+\tau c(s)),\ldots,\nabla_{t_{p-1}}(s+\tau c(s))])].$$

Moreover, if the form ω is holomorphic on $\Omega \setminus S$, then

$$\operatorname{res}\omega[t_1,\ldots,t_{p-1}] = 2\pi i \operatorname{res}_{\tau=0}\left(\omega[c, \nabla_{t_1}(s+\tau c(s)),\ldots,\nabla_{t_{p-1}}(s+\tau c(s))]\right).$$

Appendix B

B.1 Homogeneous bundles

Homogeneous spaces are manifolds with a transitive action of a Lie group. The standard definition starts with a group of symmetry G and its closed subgroup H, the homogeneous space $M = G/H$ being their quotient. The group G acts then on the homogeneous space M from the left and $\pi : G \longrightarrow M$ is a principal fibre bundle on M with the group H.

Every representation τ of H on a vector space $\mathbf{E}$ defines a vector bundle on G/H as follows. Let the equivalence relation $\sim$ be defined on the product $G \times \mathbf{E}$ by $[g, v] \sim [gh, \tau(h^{-1})v], g \in G, v \in \mathbf{E}, h \in H$. The set $E = G \times_\tau \mathbf{E}$ of equivalence classes together with the natural projection to the first component is then a vector bundle on M with fiber $\mathbf{E}$. The construction described is a special case of a more general construction of vector bundles associated to a given principal fibre bundle with the group H and a given representation τ of H.

The bundle E constructed above is an example of a homogeneous vector bundle defined in

Definition B.1.1 *A vector bundle E over M is called a homogeneous vector bundle if G acts on E from the left and the action satisfies the following conditions :*

(i) For $g \in G$ and $x \in M$, $gE_x = E_{gx}$

(ii) The mapping from E_x onto E_{gx} is linear for $g \in G$ and $x \in M$.

The action of G on $G \times \mathbf{E}$ given by $\tilde{g}[g, v] = [\tilde{g}g, v]$, $\tilde{g}, g \in G, v \in \mathbf{E}$, induces the action of G on the vector bundle $E = G \times_\tau \mathbf{E}$, whence E is a homogeneous bundle. The vector space $\mathbf{E}$ can be identified with the fiber $E_0 = E_{eH}$ (where e denotes the identity element in G) by the map $v \in \mathbf{E} \mapsto [e, v] \in E_0$. Moreover, every homogeneous bundle on G/H can be described in such a way (up to an isomorphism).

On a homogeneous space, the tangent bundle, the cotangent bundle and all tensor bundles in general are homogeneous bundles.

Indeed, if $\underline{g}$, resp. $\underline{h}$, are the Lie algebras of the Lie groups G, resp. H, and if $(Ad_H, \underline{g}/\underline{h})$, is the adjoint representation of H on the quotient space $\underline{g}/\underline{h}$, then the tangent bundle $T(G/H)$ is associated to the representation $(Ad_H, \underline{g}/\underline{h})$. The cotangent bundle being associated to the dual representation Ad^*_H, all tensor bundles can be constructed accordingly.

To deal with sections of homogeneous bundles, it is often convenient to represent them as maps on the group G satisfying an additional condition, described in

Theorem B.1.2 *Let $\pi : E \to G/H$ be a homogeneous bundle given by a representation τ on $\mathbf{E}$. Furthermore, let U be a domain in G/H and let $\tilde{U} = \pi^{-1}(U)$ be its preimage. Then the space $\Gamma(U, E)$ of all smooth sections of E can be identified with the set $C_\tau(\tilde{U}, \mathbf{E})$ of all smooth maps f from $\tilde{U}$ to $\mathbf{E}$ satisfying the condition*

$$f(gh) = \tau(h^{-1})f(g), \; g \in G, h \in H.$$

For a given section $f \in \Gamma(U, E)$, the corresponding map $\tilde{f} \in C_\tau(\tilde{U}, \mathbf{E})$ is then given by $\tilde{f}(g) = g^{-1}f(gH) \in E_0$, $g \in \tilde{U}$. Conversely, given a map $\tilde{f} \in C_\tau(\tilde{U}, \mathbf{E})$, the map $gH \mapsto [g, \tilde{f}(g)]$ induces a well defined map $f \in \Gamma(U, E)$.

A section $f \in \Gamma(G/H, E)$ is called invariant if $f(gx) = gf(x), g \in G, x \in G/H$. More details on homogeneous spaces and homogenous bundles can be found e.g. in [48], [33] and [86].

The homogeneous spaces used in Chapter V are complex homogeneous spaces. If G is a complex Lie group and H is a closed complex subgroup, then the homogeneous space G/H is a complex manifold and the group G acts on it as a group of holomorphic diffeomorphisms (see [48]). If τ is now a holomorphic representation of the group H on a vector space $\mathbf{E}$, then the corresponding vector bundle E is a holomorphic vector bundle on G/H (see [38]).

There is a nice subclass of complex homogeneous spaces commonly called generalized flag manifolds. It is the class of all homogeneous spaces $M = G/P$ where G is a complex semisimple Lie group and P is a parabolic subgroup of G. This class can be characterized alternatively as being the class of all compact, simply connected, Kaehler complex homogeneous spaces (see [87]).

Besides having the structure of a complex homogeneous space, a generalized flag manifold M has also the structure of a compact real homogeneous space, i.e. if K is a maximal compact subgroup of G, then M can be identified with the real homogeneous space $K/K \cap H$ (see [4]).

B.2 Differential forms on homogeneous spaces

Consider now a homogeneous vector bundle $\pi : E \to G/H$ given by a representation τ on a vector space $\mathbf{E}$. To define differential forms on $M = G/H$ with values in E, first recall that $\mathbf{C}$-valued differential forms of order p can be interpreted as sections of the p-th outer power of the cotangent bundle $\Lambda^p(T^*M)$. E-valued p-forms are then defined as sections of the tensor product $\Lambda^p(T^*M) \otimes E$. The set of all smooth E-valued

p-forms on $U \subset M$ is denoted by $\mathcal{E}^p(U, E)$. A form is invariant if the corresponding section is invariant.

To deal with such forms, it is again useful to interpret them as differential forms on the group G. One possibility is simply to interpret them as sections of a homogeneous bundle and to use Theorem B.1.2. In such a way to any p-form on $U \subset M$ with values in E corresponds the differential form $\tilde{\omega}$ on $\tilde{U} = \pi^{-1}(U)$ with values in $\mathbf{E}$. The form $\tilde{\omega}$ may then be described by its values in the vectors $t_i \in T_gG$ as

$$\tilde{\omega}(g)[t_1, \ldots, t_p] = g^{-1}\omega(gH)[\pi_* t_1, \ldots, \pi_* t_p].$$

The problem is now to characterize those forms in $\mathcal{E}^p(\tilde{U}, \mathbf{E})$ which can be written as $\tilde{\omega}$ with $\omega \in \mathcal{E}^p(U, E)$. This can be done as follows.

Let us first give an intuitive description. Let $\pi : E_\tau \to G/H$ be a homogeneous bundle given by a representation τ of H on a vector space $\mathbf{E}$. Then there is a bijective correspondence between the space $\mathcal{E}^p(U, E), U \subset G/H,$ of all p-forms with values in E on U and the set $\mathcal{E}^p_\tau(\tilde{U}, \mathbf{E})$ of all $\mathbf{E}$-valued forms on $\tilde{U} = \pi^{-1}(U) \subset G$ transforming under the substitution

$$g \mapsto gh, \; dg \mapsto dg\, h + g\, dh$$

as

$$\tilde{\omega} \mapsto \tau(h^{-1})\tilde{\omega}.$$

To express this condition in a more precise language, denote by ϕ the multiplication map given by $\phi : G \times H \to G, \; \phi(g, h) = gh$ and let p be the projection of $G \times H$ onto the first factor. Then we have (see [15])

Theorem B.2.2 *Let U be a domain in G/H. Then there is a bijection between the set $\mathcal{E}^p(U, E)$ and the set $\mathcal{E}^p_\tau(\tilde{U}, \mathbf{E})$ of all p-forms on $\tilde{U}$ with values in $\mathbf{E}$ satisfying the condition*

$$\phi^*\tilde{\omega}(g, h) = \tau(h^{-1})p^*\tilde{\omega}(g, h), \; (g, h) \in \tilde{U} \times H.$$

Appendix C

1 Introduction

In recent years, an increasing number of scientists who are involved with mathematics have started to use special computer programs to assist them in manipulating large symbolic mathematical expressions. These programs, collectively known as computer algebra packages, are now available on a wide variety of computer hardware, including the omnipresent small, personal computers.

Today, a handful of these computer algebra packages represent the state of the art in symbolic computation: we shall mention MACSYMA, MAPLE, MATHEMATICA and, last but not least, REDUCE.

The purpose of this appendix is to describe three packages of REDUCE software, `3d_clif.red`, `4d_clif.red` and `clifford.red`, in which some relevant algorithms of Clifford algebra and analysis have been implemented, the first two of which containing programs for the special cases of $\mathbf{R}_{0,3}$ and $\mathbf{R}_{0,4}$ respectively. These are by no means complete packages, nor are they full implementation of the theory that was developed in the rest of this book.

Rather, they provide building blocks that may be useful for exploratory computer algebra (we have included a small real-life example of this), or for integration in a larger REDUCE application.

The source code furthermore illustrates some programming methods which, when studied along with the REDUCE manual and possibly some other publications concerning REDUCE (such as [8], [7] or [55]), may prove useful.

2 The Clifford algebra $\mathbf{R}_{0,3}$: `3d_clif.red`

Vectors

A general vector in three dimensional Euclidean space is represented by

$$\mathtt{v1} * \mathtt{e}(1) + \mathtt{v2} * \mathtt{e}(2) + \mathtt{v3} * \mathtt{e}(3),$$

the unit vectors `e(1)`, `e(2)` and `e(3)` forming an orthonormal right-handed basis.

Extracting the jth cartesian co-ordinate out of the vector v is done by the command

comp(j,v).

Next to addition and scalar multiplication, the fundamental operations on vectors are provided; the scalar product of the vectors v and w is given by

dot(v,w)

while their vector product is obtained by

cross(v,w).

Example:

```
v:=e(1)+2*e(2)+3*e(3);
w:=4*e(1)-5*e(2)-6*e(3);

dot(v,w);

-24

cross(v,w);

- 13*E(3) + 18*E(2) + 3*E(1)
```

(Conventionally, we have set input lines in a slanted face and output in a straight one.)

The length of a vector v is obtained by `lengte`(v) (notice that the operator `length` is a built-in prefix operator in REDUCE, so that another name is used here), the angle between the two vectors u and v is obtained by `angle`(u,v) and the decomposition of a vector u parallel and orthogonal to a vector v is obtained by the procedures `paral`(u,v) and `ortho`(u,v) respectively.

Example:

```
lengte v;

SQRT(14)

angle(v,w);

           - 24
ACOS(------------)
      7*SQRT(22)
```

```
vpar:=paral(v,w);

         144*E(3) + 120*E(2) - 96*E(1)
VPAR := ------------------------------
                     77

vorto:=ortho(v,w);

          87*E(3) + 34*E(2) + 173*E(1)
VORTO := ----------------------------
                     77
```

Clifford algebra

The operator e, representing the orthonormal basis, is declared to be non-commutative. The geometric product of vectors and Clifford numbers in general is then trivially obtained by means of the usual product infix operator '*'. In the example, the value of the variable c13 is an arbitrary Clifford number.

Example:

```
v*w;

3*(E(2)*E(3)) - 18*(E(1)*E(3)) - 13*(E(1)*E(2)) + 24

c13;

A3*E(3) + A23*(E(2)*E(3)) + A2*E(2) + A13*(E(1)*E(3))

 + A123*(E(1)*E(2)*E(3)) + A12*(E(1)*E(2)) + A1*E(1) + A0

y:= c13^2;

Y := 2*E(3)*(A0*A3 - A12*A123) + 2*E(2)*E(3)*(A0*A23 - A1*A123)

      + 2*E(2)*(A0*A2 + A13*A123) + 2*E(1)*E(3)*(A0*A13 + A2*A123)

      + 2*E(1)*E(2)*E(3)*(A0*A123 + A1*A23 - A2*A13 + A3*A12)

      + 2*E(1)*E(2)*(A0*A12 - A3*A123) + 2*E(1)*(A0*A1 - A23*A123)

          2     2     2     2      2      2      2      2
      + A0  - A1  - A2  - A3  - A12  - A13  - A23  + A123
```

The kth grade component of a Clifford number x is obtained by

grade(x,k).

Example:

```
grade(y,2);

2*E(2)*E(3)*(A0*A23 - A1*A123) + 2*E(1)*E(3)*(A0*A13 + A2*A123)

 + 2*E(1)*E(2)*(A0*A12 - A3*A123)
```

Next to the geometric product, some other useful products are also provided: the inner product, the outer product and the star product, i.e. the real part of the geometric product.The corresponding operators `inner`, `outer` and `star` are declared to be infix operators.

Example: A general vector, a general bivector and a general trivector are stored in the variables `vect3`, `biv3` and `triv3` respectively.

```
vect3;

V3*E(3) + V2*E(2) + V1*E(1)

biv3;

B23*(E(2)*E(3)) + B13*(E(1)*E(3)) + B12*(E(1)*E(2))

vect3 inner biv3;

 - E(3)*(V1*B13 + V2*B23) + E(2)*( - V1*B12 + V3*B23)

 + E(1)*(V2*B12 + V3*B13)

vect3 outer biv3;

E(1)*E(2)*E(3)*(V1*B23 - V2*B13 + V3*B12)

vect3 star biv3;

0
```

The relation between the outer product of two vectors and their cross product is easily verified:

```
x:=x1*e(1)+x2*e(2)+x3*e(3);
y:=y1*e(1)+y2*e(2)+y3*e(3);

(x outer y) + cross(x,y)*e(1)*e(2)*e(3)-x*y;


0
```

The three involutions of a Clifford number x are obtained as follows: `rev(x)` is the reversion of x, `mainvol(x)` is the main involution of x and `bar(x)` is the conjugate of x.

Example:

```
bar(vect3);


 - V3*E(3) - V2*E(2) - V1*E(1)

rev(biv3);


 - B23*(E(2)*E(3)) - B13*(E(1)*E(3)) - B12*(E(1)*E(2))

mainvol(triv3);


 - C123*(E(1)*E(2)*E(3))
```

Two norms are implemented: the Clifford norm, being the product of a Clifford number with its conjugate, and the euclidean norm, being the real part of the Clifford norm. The corresponding commands are `cnorm(x)` and `enorm(x)`.

Example:

```
cnorm(3+e(1)*e(2)*e(3));


6*(E(1)*E(2)*E(3)) + 10

enorm(3+e(1)*e(2)*e(3));


10
```

Groups

There is a series of predicates checking if a given Clifford number belongs to a certain subspace or group of the Clifford algebra; we enumerate them, their meaning being quite clear: `realp`, `vectorp`, `bivectorp`, `trivectorp`, `clifplusp`, `clifminp`, `gammap`, `gammaplusp`, `gammaminp`, `pinp`, `spinp`, `oddpinp`.

Given a Clifford number x, the command group3(x) retuns a message indicating to which group or subgroup x might belong.

Example:

*group3(3+e(1)*e(2)*e(3));*

```
this Clifford number is not in the Clifford group, however its
inverse might exist
```

Remark: it is indeed invertible since

$$(3 + e(1) * e(2) * e(3)) * (3 - e(1) * e(2) * e(3)) = 8.$$

*group3(e(1)+e(1)*e(2));*

```
this Clifford number is not in the Clifford group,
however it is invertible with inverse
                                     - E(1)*E(2) - E(1)
                                    ---------------------
                                              2
```

*group3(1 +e(1)*e(3));*

```
this Clifford number is in the group gammaplus(3) and

 E(1)*E(3) + 1
---------------  is in the  Spin(3)-group
    SQRT(2)
```

*group3(e(2)+e(1)*e(2)*e(3));*

```
this Clifford number is in the group gamma(3), more precisely in the
gammamin(3) subspace and

 E(2) + E(1)*E(2)*E(3)
-----------------------
        SQRT(2)

is in the Pin(3)-group, more precisely in the Oddpin(3) subspace
```

The geometry of Spin *and* Oddpin

A Pin element can be written as a product of two unit vectors (if it is Spin) or at most three unit vectors (if it is Oddpin). This decomposition, which is far from being unique, can be obtained by means of the procedure pin2prodv. The decomposition is written on the screen and the factors are stored in a list. If the given Clifford number is not in the Pin group, a warning message results.

Example:

pin2prodv(-1);

```
the Spin element -1 is the square of any unit vector
```

*pin2prodv((1+e(1)*e(3))/sqrt(2));*

```
 E(1)*E(3) + 1                   E(3) + E(1)
--------------- =  - E(3) * -------------
    SQRT(2)                       SQRT(2)


          E(3) + E(1)
{ - E(3),-------------}
             SQRT(2)
```

pin2prodv e(1);

```
this number is already a vector
```

*pin2prodv((e(2)+e(1)*e(2)*e(3))/sqrt(2));*

```
 E(2) + E(1)*E(2)*E(3)                  E(3) + E(1)
------------------------ =  - E(2) * ------------- * E(3)
        SQRT(2)                           SQRT(2)


{ - E(2),

  E(3) + E(1)
 -------------,
    SQRT(2)

 E(3)}
```

A Pin element represents an orthogonal transformation of three-dimensional Euclidean space; it is a rotation if it is Spin, whereas an Oddpin represents an anti-rotation, i.e. the composite of a rotation and the reflection w.r.t. the origin. Given a Clifford number x, its geometric meaning may be obtained by the command geom(x).

Example:

```
geom((1+e(1)*e(3))/sqrt(2));

 the action of this Spin element is the rotation about the axis
 - E(2)
 with rotation angle 2*theta, where theta is given by
                     1
cos(theta) = ---------
                  SQRT(2)
                             1
 and by sin(theta) = ---------
                          SQRT(2)

geom((e(2)+e(1)*e(2)*e(3))/sqrt(2));

 the action of this Oddpin element is the rotation about the axis
 - E(2)
 with rotation angle 2*theta where theta is given by cos(theta) =
   - 1
---------
 SQRT(2)
                             1
 and by sin(theta) = ---------
                          SQRT(2)
 followed by the reflection w.r.t. the origin
```

Conversely, if a rotation axis and angle are given, the corresponding Spin element is given by `rotator(`*axis*`,`*angle*`)`; similarly, given a plane, the corresponding Oddpin element is given by `reflector(`*plane*`)`. Here *axis* stands for a list of two vectors and *plane* for a list of three vectors, these vectors representing points in space.
Example:

```
rotator({0,e(3)},pi/2);

 SQRT(2)*(E(1)*E(2)) + SQRT(2)
--------------------------------
                2
```

which is verified by

```
geom ws;

the action of this Spin element is the rotation about the axis E(3)

 with rotation angle 2*theta, where theta is given by
```

```
               SQRT(2)
cos(theta) = ---------
                 2

                        1
 and by sin(theta) = ---------
                      SQRT(2)
```

```
reflector {e(1),e(2),e(3)};
```

```
 E(3) + E(2) + E(1)
--------------------
      SQRT(3)
```

which is verified by

```
geom ws;
```

```
the action of this vector is the reflection w.r.t.
the plane through the origin, perpendicular to that vector
```

Pin(3) and Spin(3) provide double coverings of O(3) and SO(3) respectively. The corresponding matrices of those orthogonal transformations are obtained by the command `clif32mat(`x`)`, x being a given Clifford number. Conversely, given an orthogonal matrix m, the corresponding Pin element is constructed by `mat2clif3(`m`)`.
Example:

```
clif32mat((1+e(1)*e(3))/sqrt(2));
```

```
[0  0  -1]
[        ]
[0  1  0 ]
[        ]
[1  0  0 ]
```

```
mat2clif3 ws;
```

```
  - E(1)*E(3) - 1
------------------
     SQRT(2)
```

```
clif32mat((e(2)+e(1)*e(2)*e(3))/sqrt(2));
```

```
[0    0   1]
[          ]
[0   -1   0]
[          ]
[-1   0   0]
```

mat2clif3 ws;

```
  - E(2) - E(1)*E(2)*E(3)
--------------------------
          SQRT(2)
```

Quaternions

The even subalgebra of $\mathbf{R}_{0,3}$ is isomorphic to the algebra of real quaternions, preserving the conjugation. This isomorphism is expressed by means of the procedures `clif3plus2h` and `h2clif3plus`. Conjugation in the algebra of quaternions is done by the procedure `hbar`, multiplication is done by the infix operator `hprod`, a natural power is taken by means of the prefix operator `hpow` and inversion is carried out by the procedure `hinv`.
Example:

cl3plus;

```
A23*(E(2)*E(3)) + A13*(E(1)*E(3)) + A12*(E(1)*E(2)) + A0
```

clif3plus2h(cl3plus);

```
I*A23 - J*A13 + K*A12 + A0
```

hbar ws;

```
 - I*A23 + J*A13 - K*A12 + A0
```

h2clif3plus ws;

```
- A23*(E(2)*E(3)) - A13*(E(1)*E(3)) - A12*(E(1)*E(2)) + A0
```

*a:=e(1)*e(2);*
*b:=e(2)*e(3);*
clif3plus2h(a) hprod clif3plus2h(b);

```
J
```

*clif3plus2h(a*b);*

```
J
```

```
c:=bar(cl3plus)/cnorm(cl3plus);       %this is the inverse of cl3plus
clif3plus2h(c);

  - I*A23 + J*A13 - K*A12 + A0
-------------------------------
     2      2      2      2
   A0  + A12  + A13  + A23

hinv(clif3plus2h cl3plus);

  - I*A23 + J*A13 - K*A12 + A0
-------------------------------
     2      2      2      2
   A0  + A12  + A13  + A23
```

The Clifford algebra $\mathbf{R}_{0,3}$ itself is isomorphic with the direct sum $\mathbf{H} \oplus \mathbf{H}$ For the addition and the multiplication in this ring the prefix operators `hohplus` and `hohprod` are available. The isomorphism is expressed by the procedures `clif32hoh` and `hoh2clif3`.

Example:

```
qq1:=clif32hoh(cl3);

QQ1 := { - I*A1 + I*A23 - J*A2 - J*A13 - K*A3 + K*A12 + A0 + A123,

        I*A1 + I*A23 + J*A2 - J*A13 + K*A3 + K*A12 + A0 - A123}

qq2:=clif32hoh(cl3plus);

QQ2 := {I*A23 - J*A13 + K*A12 + A0,

        I*A23 - J*A13 + K*A12 + A0}

qq3:=clif32hoh(cl3min);

QQ3 := { - I*A1 - J*A2 - K*A3 + A123,

        I*A1 + J*A2 + K*A3 - A123}

hoh2clif3(hohplus(qq2,qq3));
```

```
A3*E(3) + A23*(E(2)*E(3)) + A2*E(2) + A13*(E(1)*E(3))

 + A123*(E(1)*E(2)*E(3)) + A12*(E(1)*E(2)) + A1*E(1) + A0
```

Remark: this is, as was expected, the Clifford number c13.

hoh2clif3(hohprod(qq1,qq1));

```
2*E(3)*(A0*A3 - A12*A123) + 2*E(2)*E(3)*(A0*A23 - A1*A123)

 + 2*E(2)*(A0*A2 + A13*A123) + 2*E(1)*E(3)*(A0*A13 + A2*A123)

 + 2*E(1)*E(2)*E(3)*(A0*A123 + A1*A23 - A2*A13 + A3*A12)

                                                                    2
 + 2*E(1)*E(2)*(A0*A12 - A3*A123) + 2*E(1)*(A0*A1 - A23*A123) + A0

      2      2      2       2       2       2       2
 - A1   - A2   - A3   - A12   - A13   - A23   + A123
```

Remark: this is nothing else but c13^2.

3 The Clifford algebra $\mathbf{R}_{0,4}$: 4d_clif.red

Vectors

A general vector in four dimensional Euclidean space is represented by

$$v1 * e(1) + v2 * e(2) + v3 * e(3) + v4 * e(4),$$

the unit vectors e(1), e(2), e(3) and e(4) forming an orthonormal right-handed basis.

Extracting the jth cartesian co-ordinate out of the vector v is done by the command

comp(j,v).

Next to addition and scalar multiplication, the fundamental operations on vectors are provided; the scalar product of the vectors v and w is given by

dot(v,w)

while their vector product is obtained by

cross(v,w).

Example:

```
v:=e(1)+2*e(2)+3*e(3)-4*e(4);
w:=5*e(1)-6*e(2)-7*e(3)+8*e(4);

dot(v,w);

60

cross(v,w);

 - 16*E(3) - 28*(E(2)*E(3)*E(4)) + 22*E(2) - 8*(E(1)*E(3)*E(4))

 + 4*(E(1)*E(2)*E(4)) + 4*E(1)
```

Clifford algebra

The operator e, representing the orthonormal basis, is declared to be non-commutative. The geometric product of vectors and of general Clifford numbers is then trivially obtained by means of the usual product infix operator '*'.

The value of the variable cl4 is an arbitrary Clifford number.

Example:

```
v*w;

 - 4*(E(3)*E(4)) - 8*(E(2)*E(4)) + 4*(E(2)*E(3)) + 28*(E(1)*E(4))

 - 22*(E(1)*E(3)) - 16*(E(1)*E(2)) + 60

cl4;

A4*E(4) + A34*(E(3)*E(4)) + A3*E(3) + A24*(E(2)*E(4))

 + A234*(E(2)*E(3)*E(4)) + A23*(E(2)*E(3)) + A2*E(2)

 + A14*(E(1)*E(4)) + A134*(E(1)*E(3)*E(4)) + A13*(E(1)*E(3))

 + A124*(E(1)*E(2)*E(4)) + A1234*(E(1)*E(2)*E(3)*E(4))

 + A123*(E(1)*E(2)*E(3)) + A12*(E(1)*E(2)) + A1*E(1) + A0
```

The kth grade component of a Clifford number x is obtained by

$$\mathbf{grade}(x,k).$$

Example:

```
grade(cl4,2);

A34*(E(3)*E(4)) + A24*(E(2)*E(4)) + A23*(E(2)*E(3)) + A14*(E(1)*E(4))

 + A13*(E(1)*E(3)) + A12*(E(1)*E(2))
```

Next to the geometric product, some other useful products are also provided: the inner product, the outer product and the star product, i.e. the real part of the geometric product. The corresponding operators `inner`, `outer` and `star` are declared to be infix operators.

Example: A general vector, a general bivector, a general trivector and a general fourvector are stored as the values of the variables `vect4`, `biv4`, `triv4` and `fourv4` respectively:

```
vect4;

V4*E(4) + V3*E(3) + V2*E(2) + V1*E(1)

triv4;

C234*(E(2)*E(3)*E(4)) + C134*(E(1)*E(3)*E(4)) + C124*(E(1)*E(2)*E(4))

 + C123*(E(1)*E(2)*E(3))

vect4 inner triv4;

 - (E(3)*E(4))*(V1*C134 + V2*C234) + E(2)*E(4)*( - V1*C124 + V3*C234)

 - (E(2)*E(3))*(V1*C123 + V4*C234) + E(1)*E(4)*(V2*C124 + V3*C134)

 + E(1)*E(3)*(V2*C123 - V4*C134) - (E(1)*E(2))*(V3*C123 + V4*C124)

vect4 outer triv4;

E(1)*E(2)*E(3)*E(4)*(V1*C234 - V2*C134 + V3*C124 - V4*C123)

vect4 star triv4;

0
```

The three involutions of a Clifford number x are obtained as follows: `rev(`x`)` is the reversion of x, `mainvol(`x`)` is the main involution of x and `bar(`x`)` is the conjugate of x.

Example:

```
bar(vect4);

 - V4*E(4) - V3*E(3) - V2*E(2) - V1*E(1)

rev(biv4);

 - B34*(E(3)*E(4)) - B24*(E(2)*E(4)) - B23*(E(2)*E(3))

 - B14*(E(1)*E(4)) - B13*(E(1)*E(3)) - B12*(E(1)*E(2))

mainvol(triv4);

 - C234*(E(2)*E(3)*E(4)) - C134*(E(1)*E(3)*E(4))

 - C124*(E(1)*E(2)*E(4)) - C123*(E(1)*E(2)*E(3))
```

Two norms are implemented: the Clifford norm, being the product of a Clifford number with its conjugate and the euclidean norm, being the real part of the Clifford norm. The corresponding commands are cnorm(x) and enorm(x).
Example:

```
cnorm(e(1)*e(2)+e(3)*e(4));

 - 2*(E(1)*E(2)*E(3)*E(4)) + 2

enorm(e(1)*e(2)+e(3)*e(4));

2
```

Groups

There is a series of predicates checking if a given Clifford number belongs to a certain subspace or group of the Clifford algebra; we enumerate them, their meaning being quite clear: realp, realposp, vectorp, bivectorp, trivectorp, clifplusp, clifminp, gammap, gammaplusp, gammaminp, pinp, spinp, oddpinp.

Given a Clifford number x, the command group4(x) retuns a message indicating to which group or subgroup x might belong.
Example:

```
group4(e(1)*e(2)+e(3)*e(4));

this Clifford number is not in the Clifford group, however it
might be invertible
```

Remark: it is not invertible since

$$(e(1) * e(2) + e(3) * e(4)) * (e(1) * e(2) - e(3) * e(4)) = 0.$$

*group4(e(2)+e(2)*e(3)*e(4));*

```
this Clifford number is in the group gamma(4), more precisely in the
gammamin(4) subspace and

 E(2)*E(3)*E(4) + E(2)
-----------------------
        SQRT(2)

is in the Pin(4)-group, more precisely in the Oddpin(4) subspace
```

*group4(e(1)+e(4)+e(2)*e(3));*

```
this Clifford number is not in the Clifford group, however it
might be invertible
```

Remark: it is indeed invertible since

$$(e(1) + e(4) + e(2) * e(3)) * (-e(1) - e(4) + e(2) * e(3)) = 1.$$

*group4(e(1)*e(2)*e(3)*e(4));*

```
this Clifford number is in the Spin(4)-group
```

The vectorial representation of Spin(4) and Oddpin(4)

Pin(4) and Spin(4) provide a double covering of O(4) and SO(4) respectively. The corresponding matrices of those orthogonal transformations are obtained by the command `clif42mat(`x`)`, x being a given Clifford number.
Example:

*clif42mat(e(1)*e(2)*e(3)*e(4));*

```
[-1  0   0   0 ]
[              ]
[0   -1  0   0 ]
[              ]
[0   0   -1  0 ]
[              ]
[0   0   0   -1]
```

```
clif42mat(e(2)+e(2)*e(3)*e(4));

[2  0   0  0 ]
[            ]
[0  -2  0  0 ]
[            ]
[0  0   0  -2]
[            ]
[0  0   2  0 ]
```

Quaternions

The Clifford algebra $R_{0,4}$ is isomorphic with the real matrix algebra **H**(2), preserving conjugation. Matrices with quaternion entries are multiplied by means of the prefix operator h2prod; conjugation of such a matrix is done by means of h2bar. The isomorphism is expressed by the procedures clif42h2 and h22clif4.

Example:

```
clif42h2(cl3);       % cl3 is an arbitrary Clifford number in R(0,3)

[I*A23 - J*A13 + K*A12 + A0  I*A1 + J*A2 + K*A3 - A123 ]
[                                                      ]
[I*A1 + J*A2 + K*A3 - A123   I*A23 - J*A13 + K*A12 + A0]

mq1:=clif42h2(vect4);

       [            0                I*V1 + J*V2 + K*V3 + V4]
MQ1 := [                                                    ]
       [I*V1 + J*V2 + K*V3 - V4               0             ]

mq2:=clif42h2(triv4);

     [                 0                  I*C234 - J*C134 + K*C124 - C123]
MQ2:=[                                                                   ]
     [ - I*C234 + J*C134 - K*C124 - C123               0                 ]

mq12:=clif42h2(vect4*triv4);

MQ12 := MAT(( - I*V1*C123 - I*V2*C124 - I*V3*C134 - I*V4*C234

             + J*V1*C124 - J*V2*C123 - J*V3*C234 + J*V4*C134

             + K*V1*C134 + K*V2*C234 - K*V3*C123 - K*V4*C124
```

```
                + V1*C234 - V2*C134 + V3*C124 - V4*C123,0),

              (0, - I*V1*C123 + I*V2*C124 + I*V3*C134 - I*V4*C234

                  - J*V1*C124 - J*V2*C123 + J*V3*C234 + J*V4*C134

                  - K*V1*C134 - K*V2*C234 - K*V3*C123 - K*V4*C124

                  - V1*C234 + V2*C134 - V3*C124 + V4*C123))
```

mq:=h2prod(mq1,mq2);

```
MQ := MAT(( - I*V1*C123 - I*V2*C124 - I*V3*C134 - I*V4*C234

              + J*V1*C124 - J*V2*C123 - J*V3*C234 + J*V4*C134

              + K*V1*C134 + K*V2*C234 - K*V3*C123 - K*V4*C124 + V1*C234

              - V2*C134 + V3*C124 - V4*C123,0),

            (0, - I*V1*C123 + I*V2*C124 + I*V3*C134 - I*V4*C234

                - J*V1*C124 - J*V2*C123 + J*V3*C234 + J*V4*C134

                - K*V1*C134 - K*V2*C234 - K*V3*C123 - K*V4*C124

                - V1*C234 + V2*C134 - V3*C124 + V4*C123))
```

Remark: indeed the same matrix.

a:=1+e(1)+e(2)*e(3)+e(1)*e(3)*e(4)+e(1)*e(2)*e(3)*e(4);
ma:=clif42h2 a;

```
      [I + 2  I - J]
MA := [             ]
      [I + J    I  ]
```

mba:=clif42h2(bar a);

```
       [ - I + 2   - (I + J)]
MBA := [                    ]
       [ - I + J      - I   ]
```

h2bar(ma);

```
[ - I + 2   - (I + J)]
[                    ]
[ - I + J      - I   ]
```

4 The clifford.red PACKAGE

In this package, some general procedures are implemented for the Clifford algebras $\mathbf{R}_{0,n}$. The timings in the examples are taken from a Sun SparcStation 1+ running the PSL REDUCE 3.4 from ZIB (Berlin).

Examples

Let us consider a general element of a $\mathbf{R}_{0,2}$:

```
x:=a*e(0)+b*e(1)+c*e(2)+d*e(1)*e(2);

X := E(2)*C + E(1)*E(2)*D + E(1)*B + E(0)*A

Time: 17 ms
```

The package supports the usual operations on such symbolic expressions, in any dimension:

```
2*x;

2*(E(2)*C + E(1)*E(2)*D + E(1)*B + E(0)*A)

Time: 51 ms

x^2;

                                                     2          2
2*E(2)*A*C + 2*E(1)*E(2)*A*D + 2*E(1)*A*B + E(0)*A  - E(0)*B

         2        2
 - E(0)*C  - E(0)*D

Time: 255 ms
```

Note that we use the symbol E(0) as the multiplicative unity in Clifford algebras. The fundamental involutions are provided:

```
bar x;

 - E(2)*BAR(C) - E(1)*E(2)*BAR(D) - E(1)*BAR(B) + E(0)*BAR(A)

Time: 425 ms
```

In this result, BAR(C) and similar expressions are not yet simplified, because no properties of the variables C etc. are known yet.

If we do declare these variables to be real:

```
real a,b,c,d;

A,B,C,D declared real numbers.
Time: 34 ms
```

simplification will be carried out as expected:

```
bar x;

 - E(2)*C - E(1)*E(2)*D - E(1)*B + E(0)*A

Time: 374 ms
```

Our X being a general quaternion, we can verify the following:

```
x*bar(x);

       2          2          2          2
E(0)*A  + E(0)*B  + E(0)*C  + E(0)*D

Time: 493 ms  plus GC time: 306 ms
```

The package also provides transformation procedures from Clifford to matrix form and back.

The representation used is always of smallest complex dimension.

```
x2m(x,2);

Time: 289 ms

mm;

[A + D*I    - (B + C*I)]
[                      ]
[B - C*I     A - D*I   ]


Time: 34 ms
```

Let us invert this matrix:

```
mm:=1/mm;
```

```
       [       A - D*I              B + C*I       ]
       [-------------------  -------------------]
       [  2    2    2    2    2    2    2    2  ]
       [ A  + B  + C  + D    A  + B  + C  + D  ]
MM :=  [                                        ]
       [     - B + C*I              A + D*I       ]
       [-------------------  -------------------]
       [  2    2    2    2    2    2    2    2  ]
       [ A  + B  + C  + D    A  + B  + C  + D  ]
```

```
Time: 85 ms
```

Now, let us return to a Cliffordian expression:

m2x(2);

```
  - E(2)*C - E(1)*E(2)*D - E(1)*B + A
-----------------------------------------
            2    2    2    2
           A  + B  + C  + D
```

```
Time: 51 ms
```

... and verify (in REDUCE, ws always evaluates to the result obtained last):

*ws*x;*

```
E(0)
```

```
Time: 255 ms
```

Because of its usefulness, this combination of change to matrix form, inversion, and change back to Clifford form is provided in a single function, `inverse`:

inverse(x);

```
  - E(2)*C - E(1)*E(2)*D - E(1)*B + E(0)*A
----------------------------------------------
               2    2    2    2
              A  + B  + C  + D
```

```
Time: 476 ms
```

In fact, `inverse` will re-map the indices to the Clifford base elements when this allows the matrix form to be smaller. In the following example, there will not be a construction of a 2^{34} by 2^{34} matrix (which would take quite some time!), but the inverse of `a+b*e(1)` will be computed, and the `e(1)` in the result will be changed back to `e(67)`.

```
inverse(a+b*e(67));

  - E(67)*B + E(0)*A
----------------------
         2    2
        A  + B

Time: 153 ms
```

Now we shall introduce the symbol u to represent some as yet unspecified Clifford number.

```
clifford u;

U declared Clifford number.
Time: 34 ms
```

The procedure `inverse` is only meant to be used with expressions whose decomposition in terms of the conventional base of the embedding Clifford algebra is explicit; therefore, it yields silly results when applied to a variable such as u:

```
inverse(u);

 E(0)
------
 U()

Time: 51 ms
```

To represent an inverse abstractly, the operator `invclif` is provided instead:

```
invclif(u);

INVCLIF(U())

Time: 17 ms
```

In this and in the previous result, u has been replaced by `U()` which, in REDUCE is an essentially different expression. This has been done for technical reasons: a variable such as u can be made to stand for itself, i.e. be prevented from simplifying, but it will always be commutative in products. Clearly, this is not what we want for u, which is a general Clifford number. So we define an operator, which is also called u, but which is not commutative, and assign `U()` to u.

The 'inverse multiplicative property' of inversion is known too `invclif`:

```
invclif(u^2);

              2
INVCLIF(U())

Time: 34 ms
```

Some other important structures have been predefined:

```
spin ss;

SS declared element of the Spin group.

Time: 119 ms
```

The knowledge about ss will be used by other operators in the package:

```
invclif(ss);

SS()

Time: 34 ms
```

To illustrate these object types further, let us consider two variables declared real and positive (meaning non-zero, too):

```
realpos p1,p2;

P1,P2 declared positive real numbers.
Time: 85 ms

realposp(p1+p2);

T

Time: 0 ms
```

That was an easy test. In general, realposp's result must be interpreted as follows: true means that the expression is certainly real and positive. But false only means that the package cannot prove that the expression is real and positive; it may or may not be so.

```
realposp(p1*p2);

T

Time: 0 ms

realposp(p1-p2);

Time: 17 ms
```

(recall that, in algebraic mode, the value for false, NIL, is not printed out.

Two more examples:

```
realposp(1+p1^2+p2^2-2*p1*p2);

Time: 17 ms
```

of course the argument, $1+(p_1-p_2)^2$ will be real and positive as soon as p_1 and p_2 are real. But verifying this for general polynomials would be far too time-consuming.

```
realposp(1+p1^2+p2^2+2*p1*p2);

T

Time: 34 ms
```

Here, the expression was a sum of a real positive term, 1, and several real non-negative ones.

A similar example, involving a real non-negative variable:

```
realnneg nn1;

NN1 declared non-negative real number.
Time: 17 ms

realposp(p1+nn1);

T

Time: 17 ms
```

The package also provides some on-line help:

help general;

```
This is a REDUCE package for the definition of object types
and the symbolic manipulation of Clifford-valued expressions.

Written in 1991 by F. Brackx and D. Constales, at the
University of Ghent, Belgium.

Extra help on the following related subjects :
HELP RULES INVOLUTIONS INVCLIF PRODUCTS ACTION

Time: 17 ms
```

Let us follow one of the references:

help involutions;

```
The most common involutions on a Clifford algebra are implemented.

Extra help on the following related subjects :
BAR MAINVOL REV

Time: 17 ms
```

... and further still:

help bar;

```
BAR is the involution on the Clifford algebra defined by the following
rules :
BAR(1)=1, BAR(e(i))=-e(i) for i>0, BAR(e(0))=e(0),
BAR(xy)=BAR(y)*BAR(x) and real-linearity.

Time: 0 ms
```

Sometimes, it is useful to construct a generic Clifford number in an algebra of specified dimension, and which is of a given grade. This can be done using the function grade:

gclif(a,5,2);

```
A3,A5,A6,A9,A10,A12,A17,A18,A20,A24 declared real numbers.
E(4)*E(5)*A24 + E(3)*E(5)*A20 + E(3)*E(4)*A12 + E(2)*E(5)*A18

 + E(2)*E(4)*A10 + E(2)*E(3)*A6 + E(1)*E(5)*A17 + E(1)*E(4)*A9

 + E(1)*E(3)*A5 + E(1)*E(2)*A3

Time: 374 ms
```

Let us consider the following problem, which occured, for instance, in [19]: how can the inverse of an expression of the form

$$a_0 + x_1 y_1 + \ldots + x_k y_k$$

be computed, when a_0 is real and $x_i, y_i \in V_i$ where the V_i are pairwise orthogonal vector spaces?

First, we can assume that each pair x_i, y_i spans the plane $\langle e_{2i-1}, e_{2i} \rangle$. Splitting the products $x_i y_i$ in their real and bivector parts, the expression reduces to

$$a'_0 + a'_1 e_1 e_2 + \ldots + a'_k e_{2k-1} e_{2k}.$$

Now let us compute the inverse for $k = 2$:

```
y:=a*e(0)+b*e(1)*e(2)+c*e(3)*e(4);

Y := E(3)*E(4)*C + E(1)*E(2)*B + E(0)*A

Time: 51 ms

inverse(y);

                    2                    2                   3
( - E(3)*E(4)*A *C + E(3)*E(4)*B *C - E(3)*E(4)*C

                                                  2                   3
  + 2*E(1)*E(2)*E(3)*E(4)*A*B*C - E(1)*E(2)*A *B - E(1)*E(2)*B

                     2           3            2             2    4     2 2
  + E(1)*E(2)*B*C  + E(0)*A  + E(0)*A*B  + E(0)*A*C )/(A  + 2*A *B

          2 2    4      2 2    4
   + 2*A *C  + B  - 2*B *C  + C )

Time: 1054 ms
```

This answer is typical of much exploratory computer algebra: it will need some processing before any useful pattern can emerge. Matters get even more complicated when we add a term ($k = 3$):

*y:=a*e(0)+b*e(1)*e(2)+c*e(3)*e(4)+d*e(5)*e(6);*

```
Y := E(5)*E(6)*D + E(3)*E(4)*C + E(1)*E(2)*B + E(0)*A

Time: 85 ms
```

yinv:=inverse(y);

```
                          6                    4  2                        4  2
YINV := ( - E(5)*E(6)*A *D - E(5)*E(6)*A *B *D - E(5)*E(6)*A *C *D
```

... this result would fill six pages ...

```
                   4  4      2  6    8
            + 6*C *D  - 4*C *D  + D )

Time: 16728 ms  plus GC time: 2941 ms
```

We will check the inverse (and illustrate the speed of the package):

*y*yinv*

```
E(0)

Time: 4471 ms  plus GC time: 799 ms
```

It took less than five seconds.

Since we have computed the inverse of a polynomial expression, the denominator will give some information about existence of the inverse, and about the determinant of the matrix representation:

den yinv;

```
 8      6  2      6  2      6  2      4  4      4  2  2      4  2  2
A  + 4*A *B  + 4*A *C  + 4*A *D  + 6*A *B  + 4*A *B *C  + 4*A *B *D

     4  4      4  2  2      4  4      2  6      2  4  2      2  4  2
 + 6*A *C  + 4*A *C *D  + 6*A *D  + 4*A *B  - 4*A *B *C  - 4*A *B *D

     2  2  4       2  2  2  2      2  2  4      2  6      2  4  2
 - 4*A *B *C  + 40*A *B *C *D  - 4*A *B *D  + 4*A *C  - 4*A *C *D
```

```
      2  2  4       2  6     8       6  2       6  2       4  4
 - 4*A *C *D  + 4*A *D  + B  - 4*B *C  - 4*B *D  + 6*B *C

      4  2  2       4  4       2  6       2  4  2       2  2  4       2  6
 + 4*B *C *D  + 6*B *D  - 4*B *C  + 4*B *C *D  + 4*B *C *D  - 4*B *D

    8       6  2       4  4       2  6    8
 + C  - 4*C *D  + 6*C *D  - 4*C *D  + D
```

```
Time: 408 ms
```

This is the denominator. We shall factor it, as follows:

```
on factor;

Time: 0 ms

ws;

  2    2                    2            2
(A  + B  + 2*B*C + 2*B*D + C  + 2*C*D + D )

   2    2                    2            2
*(A  + B  + 2*B*C - 2*B*D + C  - 2*C*D + D )

   2    2                    2            2
*(A  + B  - 2*B*C + 2*B*D + C  - 2*C*D + D )

   2    2                    2            2
*(A  + B  - 2*B*C - 2*B*D + C  + 2*C*D + D )

Time: 2924 ms  plus GC time: 357 ms

off factor;

Time: 0 ms
```

There is definitely an easy pattern to be found here. Let us have a closer look at the matrix representation of the constituent terms:

```
x2m(e(1)*e(2),6);

Time: 408 ms
```

mm;

```
[I   0    0    0   0    0  0   0 ]
[                                ]
[0   - I  0    0   0    0  0   0 ]
[                                ]
[0   0    - I  0   0    0  0   0 ]
[                                ]
[0   0    0    I   0    0  0   0 ]
[                                ]
[0   0    0    0   - I  0  0   0 ]
[                                ]
[0   0    0    0   0    I  0   0 ]
[                                ]
[0   0    0    0   0    0  I   0 ]
[                                ]
[0   0    0    0   0    0  0  - I]
```

```
Time: 85 ms
```

Similarly, the other terms $e_{2k-1}e_{2k}$ can be seen to be represented by a diagonal matrix. Moving back to theoretical considerations, this fact can easily be proved for general k; our expression will therefore, as a whole, be represented by a diagonal matrix. To take the inverse, we must only decompose into primitive idempotents corresponding to the diagonal, invert each coefficient and recompose the result. (Further hints can be found e.g. in [19].)

The package allows one to decompose polynomials in harmonic or in monogenic parts (cf. [9]):

hd(x(1)^2,x,3,r2);

```
 1         1        2        2          2
---*R2 - ---*(X(3)  + X(2)  - 2*X(1) )
 3         3
```

meaning that, in three dimensions,

$$x_1^2 = \frac{1}{3}|x|^2 - \frac{1}{3}(x_3^2 + x_2^2 - 2x_1^2),$$

is a harmonic decomposition.

An application of this is the possibility to solve equations of the form $Df = g$ and $\Delta f = g$ where g is a polynomial. In the first case, the the unique solution f is produced for which $f = xh$ for some polynomial h; in the second case, the unique f which is a multiple of $|x|^2$.

id(x(1)^2,x,3);

```
 1            3              2                2
----*(E(3)*X(3)  + E(3)*X(3)*X(2)  - 4*E(3)*X(3)*X(1)
 35

                2               3                2
      + E(2)*X(3) *X(2) + E(2)*X(2)  - 4*E(2)*X(2)*X(1)

                 2                2                3
      - 4*E(1)*X(3) *X(1) - 4*E(1)*X(2) *X(1) - 9*E(1)*X(1) )
```

We can check this as follows:

dirac(ws,x,3);

```
    2
X(1) *E(0)
```

Finally, the package also provides limited support for Clifford basis symbols **e** with variable index:

e(k)*e(l);

```
     3    3              3    3
2*E(L) *E(K) *DELTA(L,K) - E(L) *E(K)
Time: 68 ms
```

(the cubes being introduced here to provide for both the Euclidean and anti-Euclidean cases).

e(l)*e(k);

```
E(L)*E(K)

Time: 17 ms
```

An overview of the functions in the `clifford.red` *package.*

The E *operator.*

This is a noncommutative operator used to represent the orthonormal basis of an anti-Euclidean Clifford algebra. Simplification of sums and products is automatic. You should make sure that quotients have real denominators. The computation of the inverse of Clifford expressions that are explicit in terms of the e_i can be carried out using the function `inverse`. For expressions involving variables declared to be of a Clifford type that is not necessarily real, the operator `invclif` should be used instead.

Types.

The following types can be declared:

`amin` odd Clifford numbers

`aplus` even Clifford numbers

`clifford` general Clifford numbers

`gamma` finite products of nonzero Clifford vectors

`gamma0` finite products of invertible Clifford numbers consisting of a real and a vector part

`gammamin` odd products of nonzero Clifford vectors

`gammaplus` even products of nonzero Clifford vectors

`oddpin` odd products of unit Clifford vectors

`pin` finite products of unit Clifford vectors

`pv` products of Clifford vectors

`pv0` finite products of Clifford numbers consisting of a real and a vector part

`pve` even products of Clifford vectors

`pvo` odd products of Clifford vectors

`real` real numbers

`realneg` negative real numbers

`realnneg` non-negative real numbers

`realnpos` non-positive real numbers

`realnzero` non-zero real numbers

`realpos` positive real numbers

`spin` even products of unit Clifford vectors

`spoin` even products of Clifford numbers consisting of a real and a vector part that have unit norm

`unitvec` unit Clifford vectors

`unitvec0` Clifford numbers consisting of a real and a vector part that have unit norm

`vec0nzero` non-zero Clifford numbers consisting of a real and a vector part

`vecnzero` non-zero Clifford vectors

`vector` Clifford vectors

`vector0` Clifford numbers consisting of a real and a vector part

`xxbar1` Clifford numbers for which $x\bar{x} = 1$

`xxbarnneg` Clifford numbers for which $x\bar{x}$ is non-negative real

`xxbarpos` Clifford numbers for which $x\bar{x}$ is positive real.

To each such type, there corresponds a declaration function (e.g. `amin`) and a predicate, (e.g. `aminp`) whose value is 'true' if the package can prove that the argument is of that type. If the predicate returns false, the argument still might be of that type, the package merely has not been able to prove it (because of practical limitations, or because of insufficient information).

Involutions.

The operators `bar`, `rev` and `mainvol` represent the corresponding involutions. When applied to variables whose type has been declared, simplification may occur.

Products.

The binary operators `ext`, `dot` and `star` correspond to the well-known special products in Clifford algebras.

Matrix representation.

The procedures `x2m` and `m2x` allow to translate a Clifford expression to matrix form, and back. `x2m` takes 2 arguments, the Clifford expression `x` and the dimension `n`; its result is produced in the global variable `mm`. `m2x` takes 1 argument, the dimension, and reads the matrix in the global variable `mm`.

Clifford action.

If we associate to e_{2k-1} the mapping $x \longrightarrow e_k x$ and to e_{2k} the mapping $x \longrightarrow i\hat{x}e_k$, a faithful algebra representation of the Clifford algebra as an action on itself is obtained. The operator `acn` represents this action: `acn(`x`,`y`)` being the action of x on y.

Generic Clifford numbers.

Given a root name, a dimension and a grade, `gclif` returns a generic Clifford number of the specified dimension and grade, with coefficients whose name is derived from the root name, and have been declared real.

Laplace and Dirac operators.

The operators `Lap` and `Dirac` take three arguments: an expression, the name of an operator storing the different co-ordinates, and a dimension. They return the result of applying the laplacian resp. the dirac operator to the expression.

Polynomial decomposition functions.

`HD` and `MD` take four arguments: an expression, the name of a co-ordinate operator, the dimension, and the name of a marker expression. They perform the polynomial decomposition of the expression in harmonic resp. monogenic polynomials; after decomposition, the $|x|^2$ resp. $\underline{x}$ powers are replaced by the marker expression. (Otherwise, simplification would reproduce the expression.)

Inversion of the Laplace and Dirac operators.

`IL` and `ID` take three arguments: an expression, a co-ordinate operator's name, and the dimension. They compute inverses to the Laplace resp. Dirac operators. In their result, the lowest-degree component w.r.t. $|x|^2$ resp. $\underline{x}$ will vanish.

Bibliography

[1] M. F. Atiyah, Classical groups and classical differential operators on manifolds (CIME, Corso tenuto a Varenna, 1975).

[2] V. Avanissian, Sur les fonctions harmoniques d'ordre quelconque et leur prolongement analytique dans C^n, in Lect. Notes in Math. 919 (Springer-Verlag, Berlin, 1981) 192-281.

[3] R. Baston, The algebraic construction of invariant differential operators (Ph. D. Thesis, Oxford University, 1985).

[4] R. J. Baston, M. G. Eastwood, *The Penrose Transform: its interaction with representation theory* (Oxford University Press, Oxford, 1989).

[5] I. M. Benn, Lorentzian applications of pure Spinors, in : Proceedings Workshop on Clifford Algebras and Their Applications in Mathematical Physics Eds J. S. R. Chisholm and A. Common (Reidel, Dordrecht, 1986) 257-264.

[6] F. Brackx, Rotaties en vlakspiegelingen in de driedimensionale euclidische ruimte, Wiskunde en Onderwijs 15, no 60 (1989) 417-441.

[7] F. Brackx, D. Constales, *Computer Algebra with LISP and REDUCE*, (Kluwer Academic Publishers, Dordrecht, 1991).

[8] F. Brackx, D. Constales, R. Delanghe, H. Serras, Clifford algebra with REDUCE, Rend. Circ. Mat. Palermo, ser II, No. 16 (1987) 11–19.

[9] F. Brackx, D. Constales, H. Serras, A. Ronveaux, On the harmonic and monogenic decomposition of polynomials, J. Symbolic Computation 8 (1989) 297–304.

[10] F. Brackx, R. Delanghe, F. Sommen, *Clifford Analysis* (Pitman, London, 1982).

[11] T. Bröcker, T. tom Dieck, *Representations of Compact Lie Groups* (Springer-Verlag, Berlin, 1985).

[12] N. Buchdahl, On the relative de Rham sequence, Proc. AMS 87 (1983) 363-366.

[13] P. Budinich, A. Trautman, *The Spinorial Chessboard* (Springer-Verlag, Berlin, 1988).

[14] J. Bureš, Integral formulae in complex Clifford analysis, in : Clifford algebras and Their Applications in Mathematical Physics (eds. J. S. R. Chisholm, A. Common), NATO ASI series C183 (Reidel, Dordrecht, 1985) 219-226.

[15] J. Bureš, V. Souček, The Penrose transform for orthogonal groups, to appear.

[16] M. Cahen, S. Gutt, L. Lemaire, P. Spindel, Killing spinors, Bull. Soc. Math. Belgique 38 (1986) 75-102.

[17] C. Chevalley, *The Algebraic Theory of Spinors* (Columbia Univ. Press, New York, 1954).

[18] J. Cnops, Orthogonal functions associated with the Dirac operator (Ph. D. Thesis, Univ. Ghent, 1988).

[19] D. Constales, The Bergman and Szegö kernels for separately monogenic functions, Zeitschrift für Analysis und Anwendungen 9 (2) (1990) 97–103.

[20] D. Constales, The relative position of L^2 domains in complex and in Clifford analysis (Ph. D. Thesis, Univ. Gent, 1989).

[21] R. Coqueraux, Spinors, reflections and Clifford algebras : A review (CPT - 85/PR. 1814).

[22] A. Crumeyrolle, *Orthogonal, Symplectic Clifford Algebras and Spinors* (Reidel, Dordrecht, 1990).

[23] M. L. Curtis, *Matrix Groups* (Springer-Verlag, Berlin, 1979).

[24] R. Delanghe, F. Sommen, V. Souček, Residues in Clifford analysis, to appear.

[25] R. Delanghe, V. Souček, On the structure of spinor-valued differential forms, to appear in Complex Variables.

[26] P. Dolbeault, Theory of residues in several variables in : Global Anal. and Appl. , (Int. Semin. Course, Trieste 1972, vol. 2, Vienna, 1974) 74-96.

[27] M. G. Eastwood, R. Penrose, R. O. Wells Jr. , Cohomology and massless fields, Comm. Math. Physics 78 (1981) 305-351.

[28] A. Erdélyi, W. Magnus, F. Oberhettinger, F. Tricomi, *Higher Transcendetal Functions* (Mc Graw-Hill, New York, 1953).

[29] G. Folland, *Introduction to Partial Differential Equations* (Princeton Univ. Press, Princeton, 1976).

[30] I. M. Gel'fand, G. E. Shilov, *Generalized Functions* (Ac. Press, New York, 1984).

[31] S. G. Gindikin, M. G. Henkin, The Penrose transform and complex integral geometry, Contemporary Problems in Math. , vol. 17 VINITI (1981) 57-111 [in Russian]

[32] H. Gluck, F. W. Warner, Great circle fibrations of the three-sphere, Duke Math. J. 50, no 1 (1983) 107-132.

[33] W. Greub, S. Halperin, R. Vanstone: *Connections, Curvature and Cohomology*, (Academic Press, New York, 1972).

[34] S. Helgason, *Groups and Geometric Analysis, Integral Geometry, Invariant Differential Operators and Spherical Functions* (Ac. Press, Orlando, 1984).

[35] D. Hestenes, *New Foundations for Classical Mechanics* (Reidel, Dordrecht, 1986).

[36] D. Hestenes, G. Sobczyk, *Clifford Algebra to Geometric Calculus* (Reidel, Dordrecht, 1984).

[37] M. W. Hirsch, *Differential topology*, (Springer-Verlag, Berlin, 1976).

[38] F. Hirzebruch, *Topological Methods in Algebraic Geometry*, (Springer-Verlag, Berlin, 1966).

[39] N. J. Hitchin, Linear fields on self-dual spaces, Proc. Royal Soc. London A 370, 173-191.

[40] H. Hochstadt, *The Functions of Mathematical Physics* (Wiley, New York, 1971).

[41] W. V. D. Hodge, D. Pedoe, *Methods of Algebraic Geometry*, vol. 2, (Cambridge Univ. Press, Cambridge, 1952).

[42] L. P. Hughston, T. R. Hurd, A $\mathbf{CP}^5$ calculus for space-time fields, Phys. Reports 100 (1983) 273-326.

[43] J. Humphreys, *Introduction to Lie Algebras and Representation Theory* (Springer-Verlag, Berlin, 1972).

[44] D. Husemoller, *Fibre Bundles* (McGraw-Hill, New York, 1966).

[45] F. John, *Plane Waves and Spherical Means* (Springer-Verlag, New York, 1955).

[46] E. Kaehler, Innere und ausserer Differentialkalkül, Abh. Deutsch. Akad. Wiss. Berlin (Math. -Phys.), no 4 (1960).

[47] E. Kähler, Der innere Differentialkalkül, Abh. Math. Sem. Univ. Hamburg 25 (1962) 192-205.

[48] S. Kobayashi, K. Nomizu, *Foundations of Differential Geometry* (Interscience, London, 1963).

[49] M. Kolář, Envelopes of holomorphy for solutions of the Laplace and Dirac equations, Comm. Math. Univ. Carolinae 32 (3) (1991) 1-16.

[50] G. Köthe, *Topological Vector Spaces I* (Springer-Verlag, Berlin, 1969).

[51] T. Y. Lam, *The Algebraic Theory of Quadratic Forms* (Benjamin, New York, 1973).

[52] H. Lawson, M. L. Michelsohn, *Spin Geometry* (Princeton Univ. Press, Princeton, 1989).

[53] J. Leray, Le calcul différentiel et intégral sur une variéte analytique complexe (Problème de Cauchy III), Bull. Soc. Math. France 87(1959) 81-180.

[54] P. Lounesto, On primitive idempotents of Clifford algebras, Report HTKK-MAT-A11 (1977).

[55] M. McCallum, F. Wright, *Algebraic Computing with REDUCE* (Oxford Science Publications, 1991).

[56] W. S. Massey, *Singular Homology Theory* (Springer-Verlag, Berlin, 1980).

[57] J. Milnor, *Singular points of complex hypersurfaces* (Princeton Univ. Press, Princeton, 1968).

[58] K. Nomizu, *Fundamentals of Linear Algebra* (McGraw-Hill, New York, 1966).

[59] F. Norguet, Sur la théorie des résidus, C. R. Acad. Sci. , sér. A, 248 (14) (1959) 2057-2059.

[60] R. Penrose, Null hypersurface initial data for classical fields of arbitrary spin and for general relativity, Gen. Relativity and Grav. 12 (1980) 225-264.

[61] R. Penrose, M. A. H. MacCallum, Twistor theory: an approach to the quantization of fields and space-time, Phys. Reports C (1972) 241-315.

[62] R. Penrose, W. Rindler, *Spinors and Space Time* (Cambridge Univ. Press, Cambridge, 1985).

[63] I. Porteous, *Topological Geometry* (Cambridge Univ. Press, Cambridge, 1981).

[64] A. Pressley, G. Segal, *Loop Groups* (Clarendon Press, Oxford, 1986).

[65] W. Rudin, *Function Theory in the Unit Ball of* $\mathbf{C}^n$ (Springer-Verlag, Berlin, 1980).

[66] J. Ryan, Cells of harmonicity and generalized Cauchy integral formula, Proc. London Math. Soc. 60 (3) (1990) 295-318.

[67] N. Salingaros, G. Wene, The Clifford algebra of differential forms, Acta Applicandae Mathematicae 4 (1985) 271-292.

[68] M. Sato, T. Miwa, M. Jimbo, Holonomic quantum fields I, Publ. RIMS, Kyoto Univ. 14 (1978) 223-267.

[69] L. Schwartz, *Théorie des distributions* (Hermann, Paris, 1966).

[70] J. Siciak, A characterization of analytic functions of n real variables, Studia Mathematica 35 (1970) 293-297.

[71] J. Siciak, Holomorphic continuation of harmonic functions, Ann. Polon. Math. 29 (1974) 67-73.

[72] F. Sommen, A Clifford Radon transform in Euclidean space, Proc. XV-DGM-Conference Clausthal (World Scientific, 1987) 377-389.

[73] F. Sommen, An extension of the Radon transform to Clifford analysis, Complex Variables 8 (1987) 243-266.

[74] F. Sommen, Clifford Radon transforms applied to massless fields, Simon Stevin 62 (1988) 293-319.

[75] F. Sommen, Plane wave decompositions of monogenic functions, Annal. Polon. Math. XLIX (1988) 101-114.

[76] F. Sommen, Power series expansions of monogenic functions, Complex Variables 11 (1989) 215-222.

[77] F. Sommen, Radon and X-ray transforms in Clifford analysis, Complex Variables 11 (1989) 49-70.

[78] F. Sommen, Special functions in Clifford analysis and axial symmetry, J. Math. Anal. Appl. 130 (1988) 110-133.

[79] F. Sommen, V. Souček, Monogenic differential forms, to appear in Applicable Analysis.

[80] V. Souček, Integral formulae for solutions of the Laplace and the Dirac equations on complex domains, to appear.

[81] M. Spivak, *A comprehensive introduction to differential geometry* (Publish or Perish, 1970).

[82] W. Stoll, Fiber integration and some of its applications, in : Lect. Notes in Math. 184 (Springer-Verlag, Berlin, 1971) 109-120.

[83] C. Thas, Projectiviteiten, affiniteiten en isometrieën, Simon Stevin 61 (Supplement December 1987).

[84] H. Triebel, *Höhere Analysis* (VEB, Berlin, 1972).

[85] N. Ja. Vilenkin, *Fonctions Spéciales et Théorie de la Représentation des Groupes* (Dunod, Paris, 1968).

[86] N. R. Wallach, *Harmonic Analysis on Homogeneous Spaces* (Marcel Dekker, 1973).

[87] H. C. Wang, Closed manifolds with complex homogeneous structure, Am. J. Math. 76 (1954) 1-32.

[88] R. S. Ward, R. O. Wells Jr. , *Twistor Geometry and Field Theory* (Cambridge Univ. Press, Cambridge, 1989).

[89] J. Weidmann, *Lineare Operatoren in Hilberträumen* (Teubner, Stuttgart, 1976).

[90] R. O. Wells Jr. , *Differential Analysis on Complex Manifolds* 2nd ed. (Springer-Verlag, Berlin, 1980).

[91] J. A. Wolf, *Spaces of constant curvature* (McGraw-Hill, New York, 1967).

[92] N. M. J. Woodhouse, Real methods in twistor theory, Class. Quantum Grav. 2 (1985) 257-291.

[93] Xu Zhenyuan, A function theory for the operator $D - \lambda$, Complex Variables 16 (1) (1991) 27-42.

[94] G. Zöll, Regular n-forms in Clifford analysis, their behaviour under change of variables and their residues, Complex Variables 11, (1989) 25-38.

Index

GPSR Compliance

The European Union's (EU) General Product Safety Regulation (GPSR) is a set of rules that requires consumer products to be safe and our obligations to ensure this.

If you have any concerns about our products, you can contact us on ProductSafety@springernature.com

In case Publisher is established outside the EU, the EU authorized representative is:

Springer Nature Customer Service Center GmbH
Europaplatz 3
69115 Heidelberg, Germany

Batch number: 10824484

Printed by Printforce, the Netherlands